Student Solutions Manual

Cheryl V. Roberts
Northern Virginia Community College

Intermediate Algebra
with Applications

SECOND EDITION

Linda L. Exley
Vincent K. Smith

DeKalb College
Clarkston, Georgia

PRENTICE HALL, Englewood Cliffs, New Jersey 07632

© 1994 by **PRENTICE-HALL, INC.**
A Paramount Communications Company
Englewood Cliffs, N.J. 07632

10 9 8 7 6 5 4

ISBN 0-13-475336-4

Printed in the United States of America

PREFACE

This manual contains solutions to every odd-numbered problem in <u>INTERMEDIATE ALGEBRA WITH APPLICATIONS</u> by Linda Exley and Vincent Smith. It also includes solutions to all problems in the Chapter Tests. The methods of solving have been modeled after those introduced in the examples.

I would like to thank Nancy Varner, my mother, for the many long hours spent at the computer typing this manual. Another thank-you goes to Ruth Lynch for proofreading and checking my work.

Cheryl V. Roberts
Northern Virginia Community College
Manassas Campus

TABLE OF CONTENTS

CHAPTER 0

Problem Set 0.1

1. True; a is an element of A.

3. True; every integer a can be written as
   $\frac{a}{1}$ which is a rational number.

5. False; $-\frac{1}{2} \notin \{\ldots -2, -1, 0, 1, 2, \ldots\}$

7. False; $0 \notin \{1, 2, 3, \ldots\}$

9. False; $A \cap \emptyset = \emptyset$

11. $A \cup B = \{-2, 0, 2, 4, 6, 8\}$
    (all elements which belong to A or B)

13. $\emptyset$, $\{-2\}$, $\{0\}$, $\{2\}$, $\{-2, 0\}$, $\{-2, 2\}$, $\{0, 2\}$,
    $\{-2, 0, 2\}$

15. $A = \{$red, white, blue$\}$

17. $A = \{a, b, c, d, e, f, g, h, i, j\}$

19. $A = \{x \mid x$ is an even number between 0 and 14$\}$

21. $A = \{x \mid x$ is a day of the week$\}$

23. $\frac{3}{4} = 3 \div 4 = 0.75\overline{0}$

25. $\frac{5}{8} = 5 \div 8 = 0.625\overline{0}$

27. $\frac{2}{3}$ is rational.

29. 2π is irrational since π is irrational.

31. $0.\overline{123}$ is a repeating decimal, thus it is rational.

33. False; $3 \in A$

35. True; the integers are a subset of the rational numbers.

37. True; ½ is a rational number and every rational number is also a real number.

39. True; 0 is a whole number and every whole number is also a rational number.

41. True; A and $\emptyset$ have no elements in common.

43. $A \cup B = \left\{\frac{1}{2}, 2, \frac{9}{4}, 5\right\}$
    (all elements which belong to A or B)

45. $\emptyset$, $\left\{\frac{9}{4}\right\}$, $\{5\}$, $\left\{\frac{9}{4}, 5\right\}$

47. $\{$Friday$\}$

49. $\{q, r, s, t, u, v, w, x, y, z\}$

51. $\{x \mid x$ is an integer between -3 and 3$\}$

53. $\{x \mid x$ is a suit in cards$\}$

55. 1.2 is rational since it is a terminating decimal.

57. $\sqrt{11}$ is irrational since 11 is not a perfect square.

59. 0.1111111 is rational since it is a terminating decimal.

61. $\frac{3}{5} = 3 \div 5 = 0.6\overline{0}$

63. $\frac{7}{8} = 7 \div 8 = 0.875\overline{0}$

65. $0.3 = \frac{3}{10}$

67. $0.655 = \frac{655}{1000} = \frac{131}{200}$

Problem Set 0.2

1. $-\sqrt{3}$
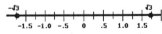

3. $\frac{1}{3}$
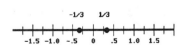

5. 0

1

7. -q

9. As $7 \cdot \frac{1}{7} = 1$, the reciprocal of 7 is $\frac{1}{7}$.

11. As $\frac{2}{3} \cdot \frac{3}{2} = 1$, the reciprocal of $\frac{2}{3}$ is $\frac{3}{2}$.

13. As $\frac{8}{5} \cdot \frac{5}{8} = 1$, the reciprocal of $\frac{8}{5}$ is $\frac{5}{8}$.

15. Reflexive Property of Equality

17. Associative Property for Addition

19. Identity for Addition

21. Identity for Multiplication

23. Multiplicative Inverse

25. Distributive Property

27. Principal of Substitution

29. 1) Commutative Property for Multiplication
    2) Distributive Property
    3) Addition of Real Numbers
    4) Commutative Property for Multiplication

31. Commutative Property for Addition

33. Symmetric Property of Equality

35. Distributive Property

37. Commutative Property for Addition

39. Additive Inverse

41. Multiplication by 0

43. Transitive Property

45. Associative Property for Addition

47. 2

49. $-\frac{2}{3}$

51. p

53. As $1 \cdot 1 = 1$, 1 is its own reciprocal.

55. As $\frac{1}{4} \cdot 4 = 1$, the reciprocal of $\frac{1}{4}$ is 4.

57. As $\frac{3}{7} \cdot \frac{7}{3} = 1$, the reciprocal of $\frac{3}{7}$ is $\frac{7}{3}$.

59. (1) Associative Property for Addition
    (2) Addition of Real Numbers

61.

Problem Set 0.3

1. The number 46 is 46 units from 0 on the number line, so $|46| = 46$.

3. $|7 - 11| = |-4| = 4$

5. The number 0 is 0 units from 0 on the number line, so $|0| = 0$.

7. $|54| - |16| = 54 - 16 = 38$

9. $-|13| = -13$

11. $-|32| - |-28| = -32 - 28 = -60$

13. $|a| = a$ since $a > 0$.

15. Won't simplify.

17. $|-a| = a$ since $a > 0$.

19. Won't simplify.

21.

$|-3 - 5| = |-8| = 8$ units

2

23.

$|-8 - (-6)| = |-2| = 2$ units

25.

$|-\sqrt{3} - (-\sqrt{5})| = |-\sqrt{3} + \sqrt{5}|$

$\phantom{|-\sqrt{3} - (-\sqrt{5})|} = -\sqrt{3} + \sqrt{5}$

$\phantom{|-\sqrt{3} - (-\sqrt{5})|} = (\sqrt{5} - \sqrt{3})$ units

(Note: $-\sqrt{3} > -\sqrt{5}$ thus $-\sqrt{3} - (-\sqrt{5}) > 0$.)

27. $|5y| = |5| \cdot |y| = 5|y|$

29. $-|-53k| = -|-53| \cdot |k| = -53|k|$

31. $\left|\dfrac{6}{z}\right| = \dfrac{|6|}{|z|} = \dfrac{6}{|z|}$

33. $|-16| = 16$

35. $|19 - 22| = |-3| = 3$

37. $|22| - |9| = 22 - 9 = 13$

39. $-|-47| = -47$

41. $|19| - |-22| = 19 - 22 = -3$

43. $-|-27| - |-81| = -27 - 81 = -108$

45. $|-21x| = |-21| \cdot |x| = 21|x|$

47. $-|81g| = -|81| \cdot |g| = -81|g|$

49. $\left|\dfrac{u}{9}\right| = \dfrac{|u|}{|9|} = \dfrac{|u|}{9}$

51. Won't simplify.

53. $|r| = r$ since $r > 0$.

55. $|s| = -s$ since $s < 0$.

57. $|-r| = r$ since $r > 0$.

59. $-|r| = -1 \cdot |r| = -1 \cdot r = -r$

61. $-|-r| = -1 \cdot |-r| = -1 \cdot r = -r$

63. False; $|-p| = p$

$ |-(-3)| = -3$

$ |3| = -3$

$ 3 = -3$

65. True; $|-p| = -p$

$ |-(-3)| = -(-3)$

$ |3| = 3$

$ 3 = 3$

67. True; $|q| = q$

$ |7| = 7$

$ 7 = 7$

69.

$|7 - (-1)| = |8| = 8$ units

71.

$|-5 - (-9)| = |4| = 4$ units

73.

$|-\sqrt{5} - (-\sqrt{7})| = |-\sqrt{5} + \sqrt{7}|$

$\phantom{|-\sqrt{5} - (-\sqrt{7})|} = -\sqrt{5} + \sqrt{7}$

$\phantom{|-\sqrt{5} - (-\sqrt{7})|} = (\sqrt{7} - \sqrt{5})$ units

(Note: $-\sqrt{5} > -\sqrt{7}$ thus $-\sqrt{5} - (-\sqrt{7}) > 0$.)

75. $|-8 + (-5)| = |-13| = 13$

77. $\dfrac{-8}{|-4|} = \dfrac{-8}{4} = -2$

3

79. $\left|-\sqrt{2}\right| + |\pi| = \sqrt{2} + \pi$

81. Since $3 - \pi < 0$, $|3 - \pi| = -(3 - \pi)$
$= -3 + \pi = \pi - 3.$

83. Since $-\pi - 3 < 0$, $|-\pi - 3| = -(-\pi - 3)$
$= \pi + 3.$

Problem Set 0.4

1. $6 - (-3) = 6 + (+3)$
$= 9$

3. $\dfrac{-16}{4} = -4$

5. $\dfrac{16}{0}$ is undefined

7. $|-12 + 4| \div 4 = |-8| \cdot \dfrac{1}{4}$

$= 8 \cdot \dfrac{1}{4}$

$= 2$

9. $\dfrac{1}{6} \div \left(-\dfrac{1}{3}\right)^2 = \dfrac{1}{6} \div \dfrac{1}{9}$

$= \dfrac{1}{6} \cdot \dfrac{9}{1}$

$= \dfrac{3}{2}$

11. $(2 - 3)^2 \cdot 5 = (-1)^2 \cdot 5$

$= 1 \cdot 5$

$= 5$

13. $15 \div (-3) \cdot \dfrac{-1}{5} = \dfrac{15}{1} \cdot \dfrac{1}{-3} \cdot \dfrac{-1}{5}$

$= \dfrac{-5}{1} \cdot \dfrac{-1}{5}$

$= 1$

15. $\dfrac{2 - 7 + 4}{3(2) - (-1)} = \dfrac{-5 + 4}{6 + 1} = \dfrac{-1}{7} = -\dfrac{1}{7}$

17. $5[6 + 2(-5 + 1)] = 5[6 + 2(-4)]$

$= 5[6 + (-8)]$

$= 5(-2)$

$= -10$

19. $(-6)^2 \cdot \left(-\dfrac{2}{3}\right) - (-7) = 36 \cdot \left(-\dfrac{2}{3}\right) + 7$

$= \left(\dfrac{36}{1} \cdot \dfrac{-2}{3}\right) + 7$

$= -24 + 7$

$= -17$

21. $(-13 + 4)^2 = (-9)^2 = 81$

23. $(-2)^2 - (-2)^3 = 4 - (-8)$

$= 4 + 8$

$= 12$

25. $7^2 - 8^2 = 49 - 64 = -15$

27. $-7 + (-7)^2 = -7 + 49 = 42$

29. $|-15 - (-3)| \div (-3) = |-15 + 3| \div (-3)$

$= |-12| \div (-3)$

$= 12 \div (-3)$

$= -4$

31. $\dfrac{\left(\dfrac{1}{8}\right)}{\left(\dfrac{-1}{4}\right)^3} = \dfrac{1}{8} \div \left(\dfrac{-1}{4}\right)^3$

$= \dfrac{1}{8} \div \dfrac{-1}{64}$

$= \dfrac{1}{8} \cdot \dfrac{64}{-1}$

$= -8$

33. $2^2 - 3 \cdot 5 = 4 - 3 \cdot 5$

$= 4 - 15$

$= -11$

35. $\dfrac{3^2 - 4^2}{(-2)(-1) - 1} = \dfrac{9 - 16}{2 - 1}$

$= \dfrac{-7}{1}$

$= -7$

37. $-2\left[6 - 3^2\left(\sqrt{5 + 4}\right)\right] = -2\left[6 - 9\left(\sqrt{9}\right)\right]$

$= -2[6 - 9(3)]$

$= -2[6 - 27]$

$= -2(-21)$

$= 42$

39. $-12 \div (-3)^3 \cdot \dfrac{-1}{3} = -12 \div (-27) \cdot \dfrac{-1}{3}$

$= \dfrac{-12}{-27} \cdot \dfrac{-1}{3}$

$= -\dfrac{4}{27}$

41. $(2 - 7)^2 - 7^2 = (-5)^2 - 7^2$

$= 25 - 49$

$= -24$

43. $\dfrac{3}{2} - \left(-\dfrac{1}{4} \div 4\right) = \dfrac{3}{2} - \left(-\dfrac{1}{4} \cdot \dfrac{1}{4}\right)$

$= \dfrac{3}{2} - \left(-\dfrac{1}{16}\right)$

$= \dfrac{24}{16} + \dfrac{1}{16}$

$= \dfrac{25}{16}$

45. $6\left(\dfrac{-2}{3} - 1\right) \div 3 = 6\left(-\dfrac{5}{3}\right) \div 3$

$= -10 \div 3$

$= -\dfrac{10}{3}$

47. $\dfrac{2\left[\dfrac{-1 - (-7)}{4 \cdot 2 - 10}\right] - 8 \cdot 3}{-(-5) - 7} = \dfrac{2\left(\dfrac{-1 + 7}{8 - 10}\right) - 24}{5 - 7}$

$= \dfrac{2\left(\dfrac{6}{-2}\right) - 24}{-2}$

$= \dfrac{2(-3) - 24}{-2}$

$= \dfrac{-6 - 24}{-2}$

$= \dfrac{-30}{-2}$

$= 15$

49. $\dfrac{-4\,|-9 - 2| + 8 \cdot 5}{2\left(\dfrac{-5 - 9}{5 + 2}\right) - \dfrac{6 - 2 \cdot 3}{0 - 2}} = \dfrac{-4\,|-11| + 40}{2\left(\dfrac{-14}{7}\right) - \dfrac{6 - 6}{-2}}$

$= \dfrac{-4(11) + 40}{2(-2) - \dfrac{0}{-2}}$

$= \dfrac{-44 + 40}{-4 - 0}$

$= \dfrac{-4}{-4}$

$= 1$

51. $\dfrac{3}{4}(-24 + 16) = \dfrac{3}{4}(-8)$

$= -6$

5

53. $\dfrac{\sqrt{-12 + 16}}{-2} = \dfrac{\sqrt{4}}{-2}$

$\qquad\qquad\quad = \dfrac{2}{-2}$

$\qquad\qquad\quad = -1$

55. $|-11| - [5 + (-7)] = 11 - (-2)$

$\qquad\qquad\qquad\quad = 11 + 2$

$\qquad\qquad\qquad\quad = 13$

57. $\dfrac{\dfrac{2}{3} - \dfrac{3}{7}}{\dfrac{1}{7} + \dfrac{1}{3}} = \dfrac{\dfrac{14}{21} - \dfrac{9}{21}}{\dfrac{3}{21} + \dfrac{7}{21}}$

$\qquad\qquad = \dfrac{\dfrac{5}{21}}{\dfrac{10}{21}}$

$\qquad\qquad = \dfrac{5}{21} \cdot \dfrac{21}{10}$

$\qquad\qquad = \dfrac{1}{2}$

Problem Set 0.5

1. $\quad \angle a = \angle d \qquad$ (Vertical angles)

$\qquad \angle a = 45°$

$\quad \angle a + \angle c = 180° \qquad$ (Supplementary angles)

$\quad 45° + \angle c = 180°$

$\qquad\quad \angle c = 135°$

$\qquad \angle b = \angle c \qquad$ (Vertical angles)

$\qquad \angle b = 135°$

$\qquad \angle q = \angle a \qquad$ (Corresponding angles)

$\qquad \angle q = 45°$

$\qquad \angle t = \angle q \qquad$ (Vertical angles)

$\qquad \angle t = 45°$

$\angle r = \angle b \qquad$ (Corresponding angles)

$\angle r = 135°$

$\angle s = \angle r \qquad$ (Vertical angles)

$\angle s = 135°$

$\angle i = \angle j = \angle k = \angle l = 90° \; (l_3 \perp l_4)$

$\angle b = \angle j + \angle g$

$135° = 90° + \angle g$

$45° = \angle g$

$\angle f = \angle g \qquad$ (Vertical angles)

$\angle f = 45°$

$\angle e + \angle f = 180° \qquad$ (Supplementary angles)

$\angle e + 45° = 180°$

$\angle e = 135°$

$\angle h = \angle e \qquad$ (Vertical angles)

$\angle h = 135°$

$\angle m = \angle e \qquad$ (Corresponding angles)

$\angle m = 135°$

$\angle p = \angle m \qquad$ (Vertical angles)

$\angle p = 135°$

$\angle n = \angle f \qquad$ (Corresponding angles)

$\angle n = 45°$

$\angle o = \angle n \qquad$ (Vertical angles)

$\angle o = 45°$

3. $\angle A + 30° + 30° = 180°$

$\qquad \angle A + 60° = 180°$

$\qquad\qquad \angle A = 120°$

5. $\angle A + 90° = 120°$

   $\angle A = 30°$

7. (a) Both triangles have a 90° angle and they share a common angle. If two angles of one triangle equal two angles of another triangle the triangles are similar.

   (b) Since l_1 and l_2 are parallel the two triangles have corresponding angles which are equal. Also, the two triangles share a common angle.

9. False; $\angle j + \angle k + \angle n = 180°$

11. **False;** $\angle g + \angle h = 180°$

    $\angle g + 105° = 180°$

    $\angle g = 75°$

13. True; $\angle f$ and $\angle h$ are supplementary. $\angle f$ and $\angle k$ are corresponding angles and thus equal. Thus $\angle h$ and $\angle k$ are also supplementary.

15. False; $\angle b = \angle c = 120°$

17. **False;** $\quad \angle d = 180° - \angle c$

    $\qquad\qquad = 180° - 120°$

    $\qquad\qquad = 60°$

    $\qquad \angle g = 180° - \angle h$

    $\qquad\qquad = 180° - 105°$

    $\qquad\qquad = 75°$

    $\qquad \angle j = 180° - (\angle d + \angle g)$

    $\qquad\qquad = 180° - (60° + 75°)$

    $\qquad\qquad = 180° - 135°$

    $\qquad\qquad = 45°$

    $\angle j + (\angle k + \angle n) = 180°$

    $45° + (\angle k + \angle n) = 180°$

    $\qquad \angle k + \angle n = 135° \neq 90°$

19. True; $\angle m + \angle n = \angle h$

    $\angle m + \angle n = 105°$

21. $\angle A + 45° + 85° = 180°$

    $\angle A + 130° = 180°$

    $\angle A = 50°$

23. $\angle A + 40° = 180°$

    $\angle A = 140°$

25. $115° = \angle A + (180° - 133°)$

    $115° = \angle A + 47°$

    $68° = \angle A$

27. $a^2 + b^2 = c^2$

    $6^2 + 8^2 = 10^2$

    $36 + 64 = 100$

    $100 = 100$

29. $c^2 = 3^2 + 4^2 \qquad c^2 = 6^2 + 8^2$

    $c^2 = 9 + 16 \qquad c^2 = 36 + 64$

    $c^2 = 25 \qquad\quad c^2 = 100$

    $c = 5 \qquad\qquad c = 10$

    $\dfrac{3}{6} = \dfrac{4}{8} = \dfrac{5}{10}$, corresponding sides are proportional, hence the triangles are similar.

31. True; each angle measures 60°.

33. True

35. False; a pentagon has 5 sides.

37. True; a triangle is a closed plane figure bounded by line segments.

39. True; the acute angles have a sum of $180° - 90° = 90°$.

41.

	number of vertices	number of triangles	sum of angles
Triangle	3	1	$1(180°) = 180°$
Quadrilateral	4	2	$2(180°) = 360°$
Pentagon	5	3	$3(180°) = 540°$
Hexagon	6	4	$4(180°) = 720°$
Polygon (n sides)	n	$n - 2$	$(n - 2)(180°)$

Problem Set 0.6

1. $P = a + b + c$

    $= 3 + 4 + 5$

    $= 12$
    12 in.

3. $P = 2a + 2b$

    $= 2(13) + 2(4)$

    $= 34$
    34 cm

5. $P = 2l + 2w$

    $= 2(20) + 2(17)$

    $= 74$
    74 yd

7. $P = 4s$

    $= 4(90)$

    $= 360$
    360 ft

9. $A = bh$

    $= (7)(4)$

    $= 28$
    28 m^2

11. $A = \frac{1}{2}bh$

    $= \frac{1}{2}(12)(10)$

    $= 60$
    60 cm^2

13. $V = s^3$

    $= 5^3$

    $= 125$
    125 cubic units

15. $V = lwh$

    $= (35)(25)(6)$

    $= 5250$
    5250 ft^3

17. $V = \frac{1}{3}\pi r^2 h$

    $= \frac{1}{3}\pi(1.1)^2(3.6)$

    $= 1.452\pi \approx 4.56$
    4.56 in^3

19. $SA = 4\pi r^2$

$\approx 4(3.14)(4000)^2$

$= 200{,}960{,}000$

$200{,}960{,}000 \text{ mi}^2$

21. $SA = 2wh + 2hl + 2wl$

$= 2\left(6\frac{1}{2}\right)\left(3\frac{1}{2}\right) + 2\left(3\frac{1}{2}\right)(14) + 2\left(6\frac{1}{2}\right)(14)$

$= 2\left(\frac{13}{2}\right)\left(\frac{7}{2}\right) + 2\left(\frac{7}{2}\right)(14) + 2\left(\frac{13}{2}\right)(14)$

$= 325.5$

325.5 in^2

23. $P = 3s$

$= 3(2.5)$

$= 7.5$

7.5 in

25. $P = 2a + 2b$

$= 2(23) + 2(8)$

$= 62$

62 ft

27. $P = 2l + 2w$

$= 2(20.3) + 2(17.23)$

$= 75.06$

75.06 yd

29. $C = 2\pi r$

$= 2\pi\left(\frac{8000}{2}\right)$

$= 8000\pi$

≈ 25120

25120 mi

31. $A = \frac{1}{2}\pi r^2$

$= \frac{1}{2}\pi\left(\frac{3}{2}\right)^2$

$= \frac{9}{8}\pi$

≈ 3.5325

3.5325 m^2

33. $V = s^3$

$= \left(2\frac{1}{3}\right)^3$

$= \left(\frac{7}{3}\right)^3$

$= \frac{343}{27}$

≈ 12.7

12.7 cubic units

35. $V = lwh$

$= \left(5\frac{1}{2}\right)\left(3\frac{1}{2}\right)\left(1\frac{1}{2}\right)$

$= \left(\frac{11}{2}\right)\left(\frac{7}{2}\right)\left(\frac{3}{2}\right)$

$= \frac{231}{8}$

$= 28\frac{7}{8}$

$28\frac{7}{8} \text{ in}^3$

37. $V = \frac{1}{3}\pi r^2 h$

$= \frac{1}{3}\pi(12)^2(5)$

$= 240\pi$

≈ 753.6

753.6 yd^3

39. $V = \dfrac{4}{3}\pi r^3$

$\quad = \dfrac{4}{3}\pi(4000)^3$

$\quad = 85333000000\pi$

$\quad \approx 2.68 \times 10^{11} \text{ mi}^3$

41. $6 \text{ in} = \dfrac{1}{2} \text{ ft}$

$\quad w = 2\dfrac{1}{2} - 2\left(\dfrac{1}{2}\right) = 1\dfrac{1}{2}$

$\quad l = 3 - 2\left(\dfrac{1}{2}\right) = 2$

$\quad h = \dfrac{1}{2}$

$\quad V = wlh$

$\qquad = \left(1\dfrac{1}{2}\right)(2)\left(\dfrac{1}{2}\right)$

$\qquad = 1\dfrac{1}{2} \text{ ft}^3$

43. $\quad r = \dfrac{1}{2}d$

$\qquad = \dfrac{1}{2}\left(2\dfrac{1}{2}\right) = \dfrac{5}{4}$

Surface Area of 1 can:

$SA = 2\pi r^2 + 2\pi rh$

$\quad = 2\pi\left(\dfrac{5}{4}\right)^2 + 2\pi\left(\dfrac{5}{4}\right)(4)$

$\quad = 13.125\pi \approx 41.2125$

Surface Area of 10000 cans:

$(10000)(41.2125) = 412125 \text{ in}^2$

45. $6 \text{ in} = \dfrac{1}{2} \text{ ft}$

$\quad w = l = 3 - 2\left(\dfrac{1}{2}\right) = 2$

$\quad h = \dfrac{1}{2}$

$\quad V = wlh$

$\qquad = (2)(2)\left(\dfrac{1}{2}\right)$

$\qquad = 2$

$\quad 2 \text{ ft}^3$

47.

$c^2 = 3^2 + 4^2$

$c^2 = 25$

$c = 5$ (width of rectangular side)

Area of 2 triangular ends: $2\left[\dfrac{1}{2}bh\right]$

$\quad = 2\left[\dfrac{1}{2}(6)(4)\right]$

$\quad = 24$

Area of 2 rectangular sides: $2[wl]$

$\quad = 2[(5)(15)]$

$\quad = 150$

Total Surface Area: $24 + 150$

$\quad = 174 \text{ ft}^2$

Chapter 0 Review Problems

1. $-(-7)[2 - (5 - 8)] = 7[2 - (-3)]$
$$= 7(5)$$
$$= 35$$

3. $\dfrac{-6(-7)}{(-2 - 19) - \sqrt{30 - 5}} = \dfrac{42}{-21 - \sqrt{25}}$
$$= \dfrac{42}{-21 - 5}$$
$$= \dfrac{42}{-26}$$
$$= -\dfrac{21}{13}$$

5. $\dfrac{(-3)(-4) - (-4)(-5)}{-5 - 2 + (-1)} = \dfrac{12 - 20}{-7 + (-1)}$
$$= \dfrac{-8}{-8}$$
$$= 1$$

7. $-|6 + (-11)| - 6^2 = -|-5| - 36$
$$= -5 - 36$$
$$= -41$$

9. $|(-4)(5)| - |6 - 19| = |-20| - |-13|$
$$= 20 - 13$$
$$= 7$$

11. $(6 - 8)^2 + \dfrac{9}{3} = (-2)^2 + \dfrac{9}{3}$
$$= 4 + 3$$
$$= 7$$

13. $(-9)^2 - [17 - (4 - 12)] = 81 - [17 - (-8)]$
$$= 81 - 25$$
$$= 56$$

15. Commutative Property for Addition

17. Distributive Property

19. Multiplicative Identity

21. Irrational; 11 is not a perfect square

23. Rational; 7.68 is a terminating decimal

25. Rational; $3.3\overline{7}$ is a repeating decimal

27. $A \cap B = \{c\}$; elements they have in common

29. $C = \{1, 3, 5\}$

31. $|-5| - |25| = 5 - 25$
$$= -20$$

33. $-|-25| = -1(25)$
$$= -25$$

35. $-|-4| - |-21| = -4 - 21$
$$= -25$$

37. $|b| = -b$, since $b < 0$.

39. $|-a| = a$; since $a > 0$.

41. $P = a + b + c + d + e$
$$= 6 + 8 + 8 + 10 + 8$$
$$= 40$$
40 in

A = Area of rectangle + Area of triangle
$$= wl + \dfrac{1}{2}bh$$
$$= (8)(10) + \dfrac{1}{2}(6)(8)$$
$$= 104$$
104 in^2

43. $C = 2\pi r$

   $= 2\pi (2.5)$

   $= 5\pi$

   ≈ 15.7 m

   $A = \pi r^2$

   $= \pi (2.5)^2$

   $= 6.25\pi$

   ≈ 19.625 m^2

45. $V = \frac{1}{3}\pi r^2 h$

   $= \frac{1}{3}\pi (3)^2(4)$

   $= 12\pi$

   ≈ 37.68 cubic units

   $SA = \pi r^2 + \pi r\sqrt{r^2 + h^2}$

   $= \pi (3)^2 + \pi (3)\sqrt{3^2 + 4^2}$

   $= 9\pi + 3\pi (5)$

   $= 24\pi$

   ≈ 75.36 square units

47. $V = \frac{4}{3}\pi r^3$

   $= \frac{4}{3}\pi (5)^3$

   $= \frac{500}{3}\pi$

   ≈ 523.3 cubic units

   $SA = 4\pi r^2$

   $= 4\pi (5)^2$

   $= 100\pi$

   ≈ 314 square units

49. Both triangles have three 60° angles and thus are similar.

Chapter 0 Test

1. B; $|A| = -A$ since $A < 0$.

2. D; the set of irrational numbers contains numbers that are not rational, integer, or natural.

3. B; $\frac{1}{6} = 1 \div 6 = 0.1\overline{6}$

4. C; $\emptyset$ is a subset of every set.

5. C

6. D; $|p - (-5)| = |p + 5|$

7. C; $-(2 - x) = -2 + x = x - 2$

8. C; if $p < 0$, then $-p > 0$ and $-p$ lies to the right of 0 on a number line.

9. B; $\frac{\frac{1}{3}}{\frac{3}{8}} = \frac{8}{3}$

10. $(-7)^2 - (-13) = 49 + 13$

   $= 62$

11. $\frac{-5}{9} \div (-15) = \frac{-5}{9} \cdot \frac{1}{-15}$

   $= \frac{1}{27}$

12. $-5^2 + 4 \cdot 6 = -25 + 24$

   $= -1$

13. $3\{2 - 3[5 + 2(8 - 9)]\}$

$= 3\{2 - 3[5 + 2(-1)]\}$

$= 3[2 - 3(5 - 2)]$

$= 3[2 - 3(3)]$

$= 3(2 - 9)$

$= 3(-7)$

$= -21$

14. $\dfrac{16 - 4^2}{8} = \dfrac{16 - 16}{8}$

$= \dfrac{0}{8}$

$= 0$

15. $-\dfrac{2}{3}\left(3 - \dfrac{9}{2}\right) = \left(-\dfrac{2}{3}\right)\left(\dfrac{3}{1}\right) - \left(-\dfrac{2}{3}\right)\left(\dfrac{9}{2}\right)$

$= -2 - (-3)$

$= -2 + 3$

$= 1$

16. $3\,|11 - 17| = 3\,|-6|$

$= 3(6)$

$= 18$

17. $6 + (-7) = -7 + 6$

18. $x(6 + y) = x(6) + xy$

$= 6x + xy$

19. $-5 + 5 = 0$

20. $2(x + t) = (x + t) \cdot 2$

21. $\angle b = 180° - \angle a$ (Supplementary angles)

$= 180° - 130°$

$= 50°$

$\angle f = \angle b$ (Vertical angles)

$= 50°$

$\angle g = \angle a$ (Vertical angles)

$= 130°$

$\angle b + \angle c + \angle d = 180°$

$50° + \angle c + 30° = 180°$

$\angle c = 100°$

$\angle e = 180° - \angle d$ (Supplementary angles)

$= 180° - 30°$

$= 150°$

22. $V = \pi r^2 h$

$= \pi (4)^2(12)$

$= 192\pi$

$\approx 602.88 \text{ ft}^3$

23. $P = 2w + 2l$

$= 2(22) + 2(28)$

$= 100$

100 m

24. $SA = 4\pi r^2$

$= 4\pi\left(3\dfrac{1}{2}\right)^2$

$= 49\pi$

$\approx 153.86 \text{ cm}^2$

25. $A = bh$

$= (28)(11)$

$= 308$

308 yd^2

CHAPTER 1

1. $2^4 = 2 \cdot 2 \cdot 2 \cdot 2 = 16$
base: 2

3. $-4^2 = -1 \cdot 4 \cdot 4 = -16$
base: 4

5. $-4^3 = -1 \cdot 4 \cdot 4 \cdot 4 = -64$
base: 4

7. $-3(-2)^2 = -3(-2 \cdot -2)$
$= -3(4)$
$= -12$
base: -2

9. $\left(\dfrac{2xy^2}{z^{11}}\right)^0 = 1$

11. $\dfrac{(-5x^0)^2}{(-5x^2)^0} = \dfrac{[(-5)(1)]^2}{1}$
$= (-5)^2$
$= 25$

13. $2^{-3} = \dfrac{1}{2^3}$
$= \dfrac{1}{8}$

15. $(-2)^{-4} = \dfrac{1}{(-2)^4}$
$= \dfrac{1}{16}$

17. $-2^{-3} = -1 \cdot \dfrac{1}{2^3}$
$= -\dfrac{1}{8}$

19. $2^3 \cdot 2^{-2} = 2^{3+(-2)} = 2^1 = 2$

21. $[(x+4)^2]^3 = (x+4)^{(2)(3)}$
$= (x+4)^6$

23. $\left(\dfrac{2x}{y^2}\right)^3 = \dfrac{(2x)^3}{(y^2)^3}$
$= \dfrac{2^3 x^3}{y^{(2)(3)}}$
$= \dfrac{8x^3}{y^6}$

25. $(-2t^2)^3 = (-2)^3(t^2)^3$
$= -8t^{(2)(3)}$
$= -8t^6$

27. $\dfrac{(c-a)^3}{(c-a)^5} = (c-a)^{3-5}$
$= (c-a)^{-2}$
$= \dfrac{1}{(c-a)^2}$

29. $(x^{-3}y^5)^3 = (x^{-3})^3(y^5)^3$
$= x^{-9}y^{15}$
$= \dfrac{1}{x^9} \cdot y^{15}$
$= \dfrac{y^{15}}{x^9}$

31. $(2 \cdot 3^2)^2 = (2)^2 \cdot (3^2)^2$
$= 4 \cdot 3^4$
$= 4 \cdot 81$
$= 324$

33. $\dfrac{2^{-3}}{3} = 2^{-3} \cdot \dfrac{1}{3}$

$\qquad = \dfrac{1}{2^3} \cdot \dfrac{1}{3}$

$\qquad = \dfrac{1}{8} \cdot \dfrac{1}{3}$

$\qquad = \dfrac{1}{24}$

35. $\dfrac{6^{-1}}{5^{-2}} = 6^{-1} \cdot \dfrac{1}{5^{-2}}$

$\qquad = \dfrac{1}{6} \cdot 5^2$

$\qquad = \dfrac{1}{6} \cdot 25$

$\qquad = \dfrac{25}{6}$

37. $\left(\dfrac{2}{3}\right)^{-2} = \left(\dfrac{3}{2}\right)^2$

$\qquad = \dfrac{9}{4}$

39. $\dfrac{2^{-1} + 2^{-2}}{2^{-1}} = \dfrac{\dfrac{1}{2} + \dfrac{1}{2^2}}{\dfrac{1}{2}}$

$\qquad = \dfrac{\dfrac{1}{2} + \dfrac{1}{4}}{\dfrac{1}{2}}$

$\qquad = \dfrac{\dfrac{3}{4}}{\dfrac{1}{2}}$

$\qquad = \dfrac{3}{4} \cdot \dfrac{2}{1} = \dfrac{3}{2}$

41. $0.000000021367 = 2.1367 \times 10^{-8}$

43. $-32000 = -3\,2000.$

$\qquad = -3.2 \times 10^4$

45. $1.609 \times 10^5 = 1.60900 \times 10^5$

$\qquad = 160{,}900$

47. $1.1 \times 10^{-9} = 0\,000000001.1 \times 10^{-9}$

$\qquad = 0.0000000011$

49. $(-3)^2 = (-3)(-3) = 9$

51. $(-3)^{-4} = \dfrac{1}{(-3)^4}$

$\qquad = \dfrac{1}{81}$

53. $\left(\dfrac{1}{2}\right)^{-3} = \left(\dfrac{2}{1}\right)^3$

$\qquad = 8$

55. $\left(-\dfrac{1}{10}\right)^{-1} = \left(-\dfrac{10}{1}\right)^1$

$\qquad = -10$

57. $-2^2(-3^3) = -1 \cdot 2^2(-1 \cdot 3^3)$

$\qquad = -1 \cdot 4(-1 \cdot 27)$

$\qquad = 108$

59. $\left(-\dfrac{1}{2} \cdot \dfrac{2}{3}\right)^3 = \left(-\dfrac{1}{3}\right)^3$

$\qquad = -\dfrac{1}{27}$

61. $3x^{-2} = 3 \cdot \dfrac{1}{x^2}$

$\qquad = \dfrac{3}{x^2}$

63. $\dfrac{-2x^{-2}}{y^{-3}} = -2 \cdot x^{-2} \cdot \dfrac{1}{y^{-3}}$

$= -2 \cdot \dfrac{1}{x^2} \cdot y^3$

$= -\dfrac{2y^3}{x^2}$

65. $\dfrac{(-5)^4}{5^5} = \dfrac{(-1)^4 5^4}{5^5}$

$= \dfrac{1}{5}$

67. $(x-2)^3(x-2)^5 = (x-2)^{3+5}$

$= (x-2)^8$

69. $4^{-1} + 4^{-2} = \dfrac{1}{4} + \dfrac{1}{4^2}$

$= \dfrac{1}{4} + \dfrac{1}{16}$

$= \dfrac{5}{16}$

71. $\dfrac{3^{-1} + 3^{-2}}{3^{-1}} = \dfrac{\dfrac{1}{3} + \dfrac{1}{3^2}}{\dfrac{1}{3}}$

$= \dfrac{\dfrac{1}{3} + \dfrac{1}{9}}{\dfrac{1}{3}}$

$= \dfrac{\dfrac{4}{9}}{\dfrac{1}{3}}$

$= \dfrac{4}{9} \cdot \dfrac{3}{1}$

$= \dfrac{4}{3}$

73. $\dfrac{1}{3^{-1}} + \left(\dfrac{1}{3}\right)^{-1} = 3 + \left(\dfrac{3}{1}\right)^1$

$= 6$

75. $(-4x)^3 = (-4)^3 x^3$

$= -64x^3$

77. $-(3x^{-2}y)^{-3} = -(3)^{-3} x^6 y^{-3}$

$= -\dfrac{x^6}{3^3 y^3}$

$= -\dfrac{x^6}{27y^3}$

79. $\left(\dfrac{3x}{y^3}\right)^2 = \dfrac{3^2 x^2}{(y^3)^2}$

$= \dfrac{9x^2}{y^6}$

81. $\left(\dfrac{x^{-1}}{3y}\right)^2 = \dfrac{(x^{-1})^2}{3^2 y^2}$

$= \dfrac{x^{-2}}{9y^2}$

$= \dfrac{1}{9x^2 y^2}$

83. $\dfrac{4^3}{4^6} = 4^{3-6}$

$= 4^{-3}$

$= \dfrac{1}{4^3}$

$= \dfrac{1}{64}$

99. 1 light year = 5.9 x 10^{12} miles
23,000 light years = (23000)(5.9 x 10^{12}) mil

$$= (2.3 \times 10^4)(5.9 \times 10^{12})$$

$$= 13.57 \times 10^{16} \text{ miles}$$

$$= 1.357 \times 10^{17} \text{ miles}$$

101. $(-5x^2z^4)^3 = (-5)^3(x^2)^3(z^4)^3$

$$= -125x^6z^{12}$$

103. $\left(\dfrac{12x^2y}{8uv^3}\right)^3 = \left(\dfrac{3x^2y}{2uv^3}\right)^3$

$$= \dfrac{3^3(x^2)^3y^3}{2^3u^3(v^3)^3}$$

$$= \dfrac{27x^6y^3}{8u^3v^9}$$

105. $A = s^2$

$$= (3x^3)^2$$

$$= 3^2(x^3)^2$$

$$= 9x^6$$

$9x^6 \text{ m}^2$

107. $V = s^3$

$$= (2t^2)^3$$

$$= 2^3(t^2)^3$$

$$= 8t^6$$

$8t^6 \text{ cm}^3$

109. $\dfrac{3^{3n}}{3^n} = 3^{3n-n}$

$$= 3^{2n}$$

111. $(5^2)^n = 5^{2n}$

113. $\dfrac{9^{n+3}}{9^{n+1}} = 9^{n+3-(n+1)}$

$$= 9^{n+3-n-1}$$

$$= 9^2$$

$$= 81$$

115. $x^2 \cdot x^n = x^{2+n}$

117. Demonstrate $(x^m)^n = x^{m \cdot n}$

$(x^{-3})^2 = \left(\dfrac{1}{x^3}\right)^2 = \dfrac{1^2}{(x^3)^2} = \dfrac{1}{x^6} = x^{-6} = x^{(-3)(2)}$

119. Demonstrate $\left(\dfrac{x}{y}\right)^n = \dfrac{x^n}{y^n}$

$\left(\dfrac{x}{y}\right)^{-1} = \dfrac{1}{\left(\dfrac{x}{y}\right)^1} = \dfrac{y}{x} = \dfrac{x^{-1}}{y^{-1}}$

Problem Set 1.2

1. $x + 6$
   degree: 1

3. $x^2 - 3x + 4$
   degree: 2

5. $-3x^3 + 4x^2 + \dfrac{1}{2}x - 7$
   degree: 3

7. $3(-2) + 7 = -6 + 7$

$$= 1$$

9. $2(-2)^3 - (-2) - 2 = 2(-8) + 2 - 2$

$$= -16 + 0$$

$$= -16$$

11. $(1)(-1) + 6 = -1 + 6$

    $\qquad\qquad = 5$

13. $2(1)^3(-1) - 5(1)^2(-1)^2 + (1)(-1) + (-1)^2$

    $= 2(-1) - 5(1) - 1 + 1$

    $= -2 - 5$

    $= -7$

15. $(4x - 5) + (3x - 9) = 4x + 3x - 5 - 9$

    $\qquad\qquad\qquad\qquad = 7x - 14$

17. $(x^2 + 4xy + y^2) + (4x^2 - 7xy - 4y^2)$

    $= x^2 + 4x^2 + 4xy - 7xy + y^2 - 4y^2$

    $= 5x^2 - 3xy - 3y^2$

19. $-(5x^6 - 3x^4 + 3) = -5x^6 - (-3x^4) - 3$

    $\qquad\qquad\qquad\qquad = -5x^6 + 3x^4 - 3$

21. $-(15x^4y - 12x^3y^2 + 7x^2y^2 - 12)$

    $= -15x^4y - (-12x^3y^2) - 7x^2y^2 - (-12)$

    $= -15x^4y + 12x^3y^2 - 7x^2y^2 + 12$

23. $(5x - 8) - (7x + 4)$

    $= 5x - 8 - 7x - 4$

    $= 5x - 7x - 8 - 4$

    $= -2x - 12$

25. $(2x^2 - 5xy + 3y^2) - (6x^2 + 3xy - 8y^2)$

    $= 2x^2 - 5xy + 3y^2 - 6x^2 - 3xy + 8y^2$

    $= 2x^2 - 6x^2 - 5xy - 3xy + 3y^2 + 8y^2$

    $= -4x^2 - 8xy + 11y^2$

27. $[(x^2 + 2x + 5) + (4x^2 - 8)] - (5x^2 - x)$

    $= (x^2 + 4x^2 + 2x + 5 - 8) - 5x^2 + x$

    $= 5x^2 + 2x - 3 - 5x^2 + x$

    $= 5x^2 - 5x^2 + 2x + x - 3$

    $= 3x - 3$

29. $[(5x^3y - x^2y^2 + 7xy + 5) - (3x^3y + 2x^2y^2 - 4xy)] + (2x^3y + 7)$

$= (5x^3y - x^2y^2 + 7xy + 5 - 3x^3y - 2x^2y^2 + 4xy) + 2x^3y + 7$

$= 2x^3y - 3x^2y^2 + 11xy + 5 + 2x^3y + 7$

$= 4x^3y - 3x^2y^2 + 11xy + 12$

31. $x - 4$; degree: 1; binomial

33. $x^2 - x + 1$; degree: 2; trinomial

35. $-2x^5 + 7x^4 + \dfrac{2}{5}x - 9$; degree: 5

37. $x^5 + y^5$; degree: 5; binomial

39. $x^7y^2 - x^4y^3 + y^7$; degree: 9; trinomial

41. $3(-3) + 7 = -9 + 7 = -2$

43. $8 - (-3) = 8 + 3 = 11$

45. $1 - (-3)^2 = 1 - 9 = -8$

47. $(-2)(3) + 6 = -6 + 6 = 0$

49. $(-2)^2 + 2(-2)(3) + 3^2 = 4 - 12 + 9 = 1$

51. $(-2)^2 - 3^2 = 4 - 9 = -5$

53. $(x - 5) + (23x + 9) = x + 23x - 5 + 9$

$\qquad\qquad = 24x + 4$

55. $(3y^2 + 4y - 8) - (3y^2 - 9)$

$= 3y^2 + 4y - 8 - 3y^2 + 9$

$= 3y^2 - 3y^2 + 4y - 8 + 9$

$= 4y + 1$

57. $(x^2 - 4xy - y^2) - (3x^2 + 7xy - 5y^2)$

$= x^2 - 4xy - y^2 - 3x^2 - 7xy + 5y^2$

$= x^2 - 3x^2 - 4xy - 7xy - y^2 + 5y^2$

$= -2x^2 - 11xy + 4y^2$

59. $(3x^3 - 4x^2 + 3x - 9) + (12x^3 + 6x^2 + x + 1)$

$= 3x^3 + 12x^3 - 4x^2 + 6x^2 + 3x + x - 9 + 1$

$= 15x^3 + 2x^2 + 4x - 8$

61. $(8x^5 - 4x^3 + 7x) - (3x^4 - 17x^3 + x^2 - 12)$

$= 8x^5 - 4x^3 + 7x - 3x^4 + 17x^3 - x^2 + 12$

$= 8x^5 - 3x^4 - 4x^3 + 17x^3 - x^2 + 7x + 12$

$= 8x^5 - 3x^4 + 13x^3 - x^2 + 7x + 12$

63. $(13x^5 - 4x^3 - 7x) + (11x^5 + 14x^4 - 3x)$

$= 13x^5 + 11x^5 + 14x^4 - 4x^3 - 7x - 3x$

$= 24x^5 + 14x^4 - 4x^3 - 10x$

65. $(9y + 8) + (10y + 6) - (y - 4)$

$= 9y + 8 + 10y + 6 - y + 4$

$= 18y + 18$

67. $(w^3 - 1) - (w^3 - 2w^2 + 2w - 2) - (w^2 + 4w - 4)$

$= w^3 - 1 - w^3 + 2w^2 - 2w + 2 - w^2 - 4w + 4$

$= w^2 - 6w + 5$

69. $(t^3 + 27) - [(t^3 - 27) - (27 - t^3)]$

$= t^3 + 27 - (t^3 - 27 - 27 + t^3)$

$= t^3 + 27 - (2t^3 - 54)$

$= t^3 + 27 - 2t^3 + 54$

$= -t^3 + 81$

71. $(6x^2 - 5x + 1) + (x + 7)$

$= 6x^2 - 5x + x + 1 + 7$

$= 6x^2 - 4x + 8$

73. $(3x^2 - 7x + 5) + (3x + 4)$

$= 3x^2 - 7x + 3x + 5 + 4$

$= 3x^2 - 4x + 9$

75. $[(x^2 - 1) + (x^2 - 2x - 1)] - (x^2 - x - 1)$

$= (2x^2 - 2x - 2) - x^2 + x + 1$

$= x^2 - x - 1$

77. $(y - x) - (x - y) = y - x - x + y$

$= 2y - 2x$

79. $(t^3 + 7t^2 - t - 5) - (-5t^3 - t^2 + t + 3)$

$= t^3 + 7t^2 - t - 5 + 5t^3 + t^2 - t - 3$

$= 6t^3 + 8t^2 - 2t - 8$

81. $P = a + b + c$

$= (s + 1) + (s^2 - 2s + 2) + (s^3 + 2s^2 + s - 9)$

$= s^3 + 3s^2 - 6$ units

83. $3t + (2t - 1) + (t^2 + 4t - 5)$

$= t^2 + 9t - 6$ yards

85. $64t - 16t^2$

When $t = 3$:

$64(3) - 16(3)^2 = 48$ feet

When $t = 4$: $64(4) - 16(4)^2 = 0$ feet

87. $180 - [(x + 30) + (3x - 80)]$

$= 180 - (4x - 50)$

$= 180 - 4x + 50$

$= 230 - 4x$ degrees

89. $4s - [16 + (2s - 7)]$

$= 4s - (2s + 9)$

$= 4s - 2s - 9$

$= 2s - 9$ feet

91. $90 - (x - 30) = 90 - x + 30$

$= 120 - x$ degrees

93. $x^n - 1 = (-1)^n - 1 = \begin{cases} -2 \text{ if } n \text{ is odd} \\ 0 \text{ if } n \text{ is even} \end{cases}$

95. $x^{2n+1} + 1 = (-1)^{2n+1} + 1$

$= -1 + 1$

$= 0$

($2n + 1$ is odd and -1 raised to an odd exponent equals -1.)

97. $(2x^{2n} - x^n + 1) + (5x^{2n} + 2x^n + 3)$

$= 7x^{2n} + x^n + 4$

99. $(5x^{2n} - 3x^n - 2) - (3x^{2n} - x^n - 4)$

$= 5x^{2n} - 3x^n - 2 - 3x^{2n} + x^n + 4$

$= 2x^{2n} - 2x^n + 2$

Problem Set 1.3

1. $(x + 2)(x + 1) = x^2 + x + 2x + 2$

$= x^2 + 3x + 2$

3. $\left[\dfrac{1}{2}x + 2\right]\left[\dfrac{1}{2}x + 1\right]$

$= \dfrac{1}{4}x^2 + \dfrac{1}{2}x + x + 2$

$= \dfrac{1}{4}x^2 + \dfrac{3}{2}x + 2$

5. $(x - 6)(x - 7) = x^2 - 7x - 6x + 42$

$= x^2 - 13x + 42$

7. $(x + 7)(x - 5) = x^2 - 5x + 7x - 35$
$$= x^2 + 2x - 35$$

9. $(4ab)(-3a^2b) = -12a^3b^2$

11. $(rs)s^3 = rs^4$

13. $-3x^2(1 - x)$
$$= -3x^2 \cdot 1 - (-3x^2) \cdot x$$
$$= -3x^2 + 3x^3$$

15. $(5x - 7)(3x - 8)$
$$= 15x^2 - 40x - 21x + 56$$
$$= 15x^2 - 61x + 56$$

17. $(x - y)(x^2 - 2xy^2 + y^3)$
$$= x \cdot x^2 - x \cdot 2xy^2 + x \cdot y^3$$
$$- y \cdot x^2 + y \cdot 2xy^2 - y \cdot y^3$$
$$= x^3 - 2x^2y^2 + xy^3 - x^2y + 2xy^3 - y^4$$
$$= x^3 - 2x^2y^2 - x^2y + 3xy^3 - y^4$$

19. $(x + 5)^2 = x^2 + 2(x)(5) + 5^2$
$$= x^2 + 10x + 25$$

21. $(3x - 4y)^2 = (3x)^2 - 2(3x)(4y) + (4y)^2$
$$= 9x^2 - 24xy + 16y^2$$

23. $(x - 2)(x + 2) = x^2 - 2^2$
$$= x^2 - 4$$

25. $(5x + 7)(5x - 7) = (5x)^2 - 7^2$
$$= 25x^2 - 49$$

27. $(x + 2)^3 = x^3 + 3x^2 \cdot 2 + 3x \cdot 2^2 + 2^3$
$$= x^3 + 6x^2 + 12x + 8$$

29. $(3a - b)^3$
$$= (3a)^3 - 3(3a)^2 \cdot b + 3(3a)b^2 - b^3$$
$$= 27a^3 - 27a^2b + 9ab^2 - b^3$$

31. $(x + 3)(x - 1) + 2x(x - 7)$
$$= x^2 - x + 3x - 3 + 2x^2 - 14x$$
$$= 3x^2 - 12x - 3$$

33. $5 - 2[3(x^2 - x) + (x + 1)(x + 3)]$
$$= 5 - 2(3x^2 - 3x + x^2 + 3x + x + 3)$$
$$= 5 - 2(4x^2 + x + 3)$$
$$= 5 - 8x^2 - 2x - 6$$
$$= -8x^2 - 2x - 1$$

35. $(2ab)(-4a^3b) = -8a^4b^2$

37. $(2rs)s^4 = 2rs^5$

39. $-2^3xy^3(x^4y) = -8x^5y^4$

41. $3b^5(-3^2abc)^3 = 3b^5(-9abc)^3$
$$= 3b^5(-729a^3b^3c^3)$$
$$= -2187a^3b^8c^3$$

43. $(-r^4t)^5 = (-1)^5r^{20}t^5$
$$= -r^{20}t^5$$

45. $(rst^5)^3(-3r^7)^2 = (r^3s^3t^{15})(9r^{14})$
$$= 9r^{17}s^3t^{15}$$

47. $(abc^5)(a^3bc)^7(-ab^2c)^4$
$$= (abc^5)(a^{21}b^7c^7)(a^4b^8c^4)$$
$$= a^{26}b^{16}c^{16}$$

49. $a^3bc(3a + 5ab^3c^6 - 4b)$

$= a^3bc \cdot 3a + a^3bc \cdot 5ab^3c^6 - a^3bc \cdot 4b$

$= 3a^4bc + 5a^4b^4c^7 - 4a^3b^2c$

51. $-\dfrac{3}{5}x^2(5x - 3y - 1)$

$= -\dfrac{3}{5}x^2 \cdot 5x + \dfrac{3}{5}x^2 \cdot 3y + \dfrac{3}{5}x^2 \cdot 1$

$= -3x^3 + \dfrac{9}{5}x^2y + \dfrac{3}{5}x^2$

53. $\dfrac{1}{3}x^4(3x^7 - 9x^6 + x^3 - 3)$

$= \dfrac{1}{3}x^4 \cdot 3x^7 - \dfrac{1}{3}x^4 \cdot 9x^6 + \dfrac{1}{3}x^4 \cdot x^3 - \dfrac{1}{3}x^4 \cdot 3$

$= x^{11} - 3x^{10} + \dfrac{1}{3}x^7 - x^4$

55. $(x + 1)(x + 9) = x^2 + 9x + x + 9$

$= x^2 + 10x + 9$

57. $\left(\dfrac{1}{2}x - 1\right)\left(\dfrac{1}{2}x + 1\right) = \left(\dfrac{1}{2}x\right)^2 - 1^2$

$= \dfrac{1}{4}x^2 - 1$

59. $(5 + x)(7 + x) = 35 + 5x + 7x + x^2$

$= 35 + 12x + x^2$

61. $(x^2 - 4)(x^2 + 6) = x^4 + 6x^2 - 4x^2 - 24$

$= x^4 + 2x^2 - 24$

63. $\left[x - \dfrac{1}{3}\right]\left[x + \dfrac{2}{3}\right] = x^2 + \dfrac{2}{3}x - \dfrac{1}{3}x - \dfrac{2}{9}$

$= x^2 + \dfrac{1}{3}x - \dfrac{2}{9}$

65. $(3 - x)(2 - x) = 6 - 3x - 2x + x^2$

$= 6 - 5x + x^2$

67. $(7 - x^2)(4 + x^2) = 28 + 7x^2 - 4x^2 - x^4$

$= 28 + 3x^2 - x^4$

69. $(4x + 5)(2x + 5) = 8x^2 + 20x + 10x + 25$

$= 8x^2 + 30x + 25$

71. $(11x + 5)(3x + 1) = 33x^2 + 11x + 15x + 5$

$= 33x^2 + 26x + 5$

73. $(4 - x)(5 - 4x) = 20 - 16x - 5x + 4x^2$

$= 20 - 21x + 4x^2$

75. $(5x^2 - 6)(x^2 - 7) = 5x^4 - 35x^2 - 6x^2 + 42$

$= 5x^4 - 41x^2 + 42$

77. $(c - d)(c^2 - 6cd + d^3)$

$= c \cdot c^2 - c \cdot 6cd + c \cdot d^3$

$\quad - d \cdot c^2 + d \cdot 6cd - d \cdot d^3$

$= c^3 - 6c^2d + cd^3 - c^2d + 6cd^2 - d^4$

$= c^3 - 7c^2d + cd^3 + 6cd^2 - d^4$

79. $(x + 3)(x - 5)(x + 7)$

$= (x^2 - 5x + 3x - 15)(x + 7)$

$= (x^2 - 2x - 15)(x + 7)$

$= x^2 \cdot x + x^2 \cdot 7 - 2x \cdot x - 2x \cdot 7 - 15 \cdot x - 15 \cdot 7$

$= x^3 + 7x^2 - 2x^2 - 14x - 15x - 105$

$= x^3 + 5x^2 - 29x - 105$

81. $(x - 3)(x + 3) = x^2 - 3^2$

$= x^2 - 9$

83. $(x + 6)^2 = x^2 + 2 \cdot x \cdot 6 + 6^2$

$= x^2 + 12x + 36$

85. $(9x - 8)^2 = (9x)^2 - 2(9x)(8) + 8^2$

$\qquad = 81x^2 - 144x + 64$

87. $(7x + 5)(7x - 5) = (7x)^2 - 5^2$

$\qquad = 49x^2 - 25$

89. $(5x - 3y)^2 = (5x)^2 - 2(5x)(3y) + (3y)^2$

$\qquad = 25x^2 - 30xy + 9y^2$

91. $(4x + 7y)^2 = (4x)^2 + 2(4x)(7y) + (7y)^2$

$\qquad = 16x^2 + 56xy + 49y^2$

93. $(x + 3)^3 = x^3 + 3 \cdot x^2 \cdot 3 + 3 \cdot x \cdot 3^2 + 3^3$

$\qquad = x^3 + 9x^2 + 27x + 27$

95. $(a - 2b)^3$

$\quad = a^3 - 3 \cdot a^2 \cdot 2b + 3 \cdot a \cdot (2b)^2 - (2b)^3$

$\quad = a^3 - 6a^2b + 12ab^2 - 8b^3$

97. $(x + 4)^2 - (x - 7)(2x - 3)$

$\quad = x^2 + 2 \cdot x \cdot 4 + 4^2 - (2x^2 - 3x - 14x + 21)$

$\quad = x^2 + 8x + 16 - 2x^2 + 17x - 21$

$\quad = -x^2 + 25x - 5$

99. $22 - [7 - (x + 3)^2 - (2x - 5)(3x - 2)]$

$\quad = 22 - [7 - (x^2 + 6x + 9) - (6x^2 - 19x + 10)]$

$\quad = 22 - (7 - x^2 - 6x - 9 - 6x^2 + 19x - 10)$

$\quad = 22 - (-7x^2 + 13x - 12)$

$\quad = 22 + 7x^2 - 13x + 12$

$\quad = 7x^2 - 13x + 34$

101. $N(N + 2) = N^2 + 2N$

103. $A = \pi r^2$

$\qquad = \pi(2x - 5)^2$

$\qquad = \pi(4x^2 - 20x + 25)$

$\qquad (4x^2 - 20x + 25)\pi \ \text{ft}^2$

105. $(N - 1)^3 = N^3 - 3N^2 \cdot 1 + 3N \cdot 1^2 - 1^3$

$\qquad = N^3 - 3N^2 + 3N - 1$

107. $A = \dfrac{1}{2}bh$

$\qquad = \dfrac{1}{2}(2W)(3W^2)$

$\qquad = 3W^3$

$\quad 3W^3 \ \text{yds}^2$

109. $x^n(x^2 - x + 1) = x^{n+2} - x^{n+1} + x^n$

111. $(x^n + 1)^2 = (x^n)^2 + 2 \cdot x^n \cdot 1 + 1^2$

$\qquad = x^{2n} + 2x^n + 1$

113. $(x^n + y^n)^2 = (x^n)^2 + 2x^n \cdot y^n + (y^n)^2$

$\qquad = x^{2n} + 2x^n y^n + y^{2n}$

115. $[(x - y) + 7][(x - y) - 7] = (x - y)^2 - 7^2$

$\qquad = (x - y)^2 - 49$

Problem Set 1.4

1. $(x + 1)^3 = x^3 + 3x^2 \cdot 1 + 3x \cdot 1^2 + 1^3$

$\qquad = x^3 + 3x^2 + 3x + 1$

3. $(x + 2)^4$

$\quad = x^4 + 4x^3 \cdot 2 + 6x^2 \cdot 2^2 \cdot 4x \cdot 2^3 + 2^4$

$\quad = x^4 + 8x^3 + 24x^2 + 32x + 16$

5. $(2x + 3y)^6$

$\quad = (2x)^6 + 6(2x)^5(3y) + 15(2x)^4(3y)^2$

$\qquad + 20(2x)^3(3y)^3 + 15(2x)^2(3y)^4$

$\qquad + 6(2x)(3y)^5 + (3y)^6$

5. (Con't.)

$$= 64x^6 + 576x^5y + 2160x^4y^2 + 4320x^3y^3$$
$$+ 4860x^2y^4 + 2916xy^5 + 729y^6$$

7. $(u - 2w)^6$

$$= u^6 - 6u^5(2w) + 15u^4(2w)^2 - 20u^3(2w)^3$$
$$+ 15u^2(2w)^4 - 6u(2w)^5 + (2w)^6$$
$$= u^6 - 12u^5w + 60u^4w^2 - 160u^3w^3$$
$$+ 240u^2w^4 - 192uw^5 + 64w^6$$

9. $(x - 2)^{10}$

$$= x^{10} - 10x^9 \cdot 2 + 45x^8 \cdot 2^2 - 120x^7 \cdot 2^3$$
$$+ 210x^6 \cdot 2^4 - 252x^5 \cdot 2^5 + 210x^4 \cdot 2^6$$
$$- 120x^3 \cdot 2^7 + 45x^2 \cdot 2^8 - 10x \cdot 2^9 + 2^{10}$$
$$= x^{10} - 20x^9 + 180x^8 - 960x^7 + 3360x^6$$
$$- 8064x^5 + 13440x^4 - 15360x^3 + 11520x^2$$
$$- 5120x + 1024$$

11. $(x - 1)^3 = x^3 - 3x^2 \cdot 1 + 3x \cdot 1^2 - 1^3$
$$= x^3 - 3x^2 + 3x - 1$$

13. $(y - 2)^4$

$$= y^4 - 4y^3 \cdot 2 + 6y^2 \cdot 2^2 - 4y \cdot 2^3 + 2^4$$
$$= y^4 - 8y^3 + 24y^2 - 32y + 16$$

15. $(b - 3)^5 = b^5 - 5b^4 \cdot 3 + 10b^3 \cdot 3^2$
$$- 10b^2 \cdot 3^3 + 5b \cdot 3^4 - 3^5$$
$$= b^5 - 15b^4 + 90b^3$$
$$- 270b^2 + 405b - 243$$

17. $(2x - 1)^4$

$$= (2x)^4 - 4(2x)^3 \cdot 1 + 6(2x)^2 \cdot 1^2$$
$$- 4(2x) \cdot 1^3 + 1^4$$
$$= 16x^4 - 32x^3 + 24x^2 - 8x + 1$$

19. $(a + 1)^5 = a^5 + 5a^4 \cdot 1 + 10a^3 \cdot 1^2 + 10a^2 \cdot 1^3 + 5a \cdot 1^4 + 1^5$

$\qquad = a^5 + 5a^4 + 10a^3 + 10a^2 + 5a + 1$

21. $(2x + 3)^4 = (2x)^4 + 4(2x)^3 \cdot 3 + 6(2x)^2 \cdot 3^2 + 4(2x) \cdot 3^3 + 3^4$

$\qquad = 16x^4 + 96x^3 + 216x^2 + 216x + 81$

23. $(3x - 2)^3 = (3x)^3 - 3(3x)^2 \cdot 2 + 3(3x) \cdot 2^2 - 2^3$

$\qquad = 27x^3 - 54x^2 + 36x - 8$

25. $(3x + 2)^5 = (3x)^5 + 5(3x)^4 \cdot 2 + 10(3x)^3 \cdot 2^2 + 10(3x)^2 \cdot 2^3 + 5(3x) \cdot 2^4 + 2^5$

$\qquad = 243x^5 + 810x^4 + 1080x^3 + 720x^2 + 240x + 32$

27. $(x - y)^6 = x^6 - 6x^5y + 15x^4y^2 - 20x^3y^3 + 15x^2y^4 - 6xy^5 + y^6$

29. $(x - 2)^8 = x^8 - 8x^7 \cdot 2 + 28x^6 \cdot 2^2 - 56x^5 \cdot 2^3 + 70x^4 \cdot 2^4 - 56x^3 \cdot 2^5 + 28x^2 \cdot 2^6 - 8x \cdot 2^7 + 2^8$

$\qquad = x^8 - 16x^7 + 112x^6 - 448x^5 + 1120x^4 - 1792x^3 + 1792x^2 - 1024x + 256$

31. $-165(3x)^8y^3 = -1082565x^8y^3$

33. $70(5z)^4(2)^4 = 700000z^4$

Problem Set 1.5

1. $64 = 2 \cdot 32$

$\qquad = 2 \cdot 2 \cdot 16$

$\qquad = 2 \cdot 2 \cdot 2 \cdot 8$

$\qquad = 2 \cdot 2 \cdot 2 \cdot 2 \cdot 4$

$\qquad = 2 \cdot 2 \cdot 2 \cdot 2 \cdot 2 \cdot 2$

$\qquad = 2^6$

3. $45 = 5 \cdot 9$

$\qquad = 5 \cdot 3 \cdot 3$

$\qquad = 3^2 \cdot 5$

5. $162 = 2 \cdot 81$

$\qquad = 2 \cdot 3 \cdot 27$

$\qquad = 2 \cdot 3 \cdot 3 \cdot 9$

$\qquad = 2 \cdot 3 \cdot 3 \cdot 3 \cdot 3$

$\qquad = 2 \cdot 3^4$

7. $243 = 3 \cdot 81$

$\qquad = 3 \cdot 3 \cdot 27$

$\qquad = 3 \cdot 3 \cdot 3 \cdot 9$

$\qquad = 3 \cdot 3 \cdot 3 \cdot 3 \cdot 3$

$\qquad = 3^5$

9. $216 = 2 \cdot 108$

$= 2 \cdot 2 \cdot 54$

$= 2 \cdot 2 \cdot 2 \cdot 27$

$= 2 \cdot 2 \cdot 2 \cdot 3 \cdot 9$

$= 2 \cdot 2 \cdot 2 \cdot 3 \cdot 3 \cdot 3$

$= 2^3 \cdot 3^3$

11. $24 = 2 \cdot 12$

$= 2 \cdot 3 \cdot 4$

$= 2 \cdot 3 \cdot 2 \cdot 2$

$= 2^3 \cdot 3$

$48 = 2 \cdot 24$

$= 2 \cdot 2 \cdot 12$

$= 2 \cdot 2 \cdot 3 \cdot 4$

$= 2 \cdot 2 \cdot 3 \cdot 2 \cdot 2$

$= 2^4 \cdot 3$

GCF: $2^3 \cdot 3 = 24$

13. $36 = 2 \cdot 18$

$= 2 \cdot 2 \cdot 9$

$= 2 \cdot 2 \cdot 3 \cdot 3$

$= 2^2 \cdot 3^2$

$54 = 2 \cdot 27$

$= 2 \cdot 3 \cdot 9$

$= 2 \cdot 3 \cdot 3 \cdot 3$

$= 2 \cdot 3^3$

GCF: $2 \cdot 3^2 = 18$

15. $x^3y^2; \ x^4y^6$

GCF: x^3y^2

17. $s^5t^5; \ st^2$

GCF: st^2

19. $96x^2y = 2^5 \cdot 3x^2y$

$80x = 2^4 \cdot 5x$

GCF: $2^4x = 16x$

21. $192x^2y^3 = 2^6 \cdot 3x^2y^3$

$48x^2y^4 = 2^4 \cdot 3x^2y^4$

$72x^2y^3 = 2^3 \cdot 3^2x^2y^3$

GCF: $2^3 \cdot 3x^2y^3 = 24x^2y^3$

23. $48a^2b^3c + 64a^3b^2c$

$= 2^4 \cdot 3a^2b^3c + 2^6a^3b^2c$

$= 2^4a^2b^2c(3b + 2^2a)$

$= 16a^2b^2c(3b + 4a)$

25. $20p^2q^7 - 28p^5q^5 + 36p^8q^3$

$= 2^2 \cdot 5p^2q^7 - 2^2 \cdot 7p^5q^5 + 2^2 \cdot 3^2p^8q^3$

$= 2^2p^2q^3(5q^4 - 7p^3q^2 + 3^2p^6)$

$= 4p^2q^3(5q^4 - 7p^3q^2 + 9p^6)$

27. $15t^3 - 15t^2 = 15t^2(t - 1)$

29. $15m^2n^3 - 9m^4n^4 + 6m^3n^5$

$= 3m^2n^3(5 - 3m^2n + 2mn^2)$

31. $-6x^2 + 3x = -3x(2x - 1)$

33. $-6uv + 3v = -3v(2u - 1)$

35. $m(p - q) + n(p - q) = (p - q)(m + n)$

37. $(x + y)^2 + 3(x + y)$

$= (x + y)[(x + y) + 3]$

$= (x + y)(x + y + 3)$

39. $5(1 - r)^4 + 3(1 - r)^3 - (1 - r)^2$

$\quad = (1 - r)^2[5(1 - r)^2 + 3(1 - r) - 1]$

41. $r(s - t) - u(t - s)$

$\quad = r(s - t) - u[-(s - t)]$

$\quad = r(s - t) + u(s - t)$

$\quad = (s - t)(r + u)$

43. $w(3u - v)^3 - 3u(3u - v)^2 + v(v - 3u)$

$\quad = w(3u - v)^3 - 3u(3u - v)^2 + v[-(3u - v)]$

$\quad = w(3u - v)^3 - 3u(3u - v)^2 - v(3u - v)$

$\quad = (3u - v)[w(3u - v)^2 - 3u(3u - v) - v]$

45. $7x + 7y + ax + ay$

$\quad = (7x + 7y) + (ax + ay)$

$\quad = 7(x + y) + a(x + y)$

$\quad = (x + y)(7 + a)$

47. $a^2 - ab + 2a - 2b$

$\quad = (a^2 - ab) + (2a - 2b)$

$\quad = a(a - b) + 2(a - b)$

$\quad = (a - b)(a + 2)$

49. $a^2 + ab + a + b$

$\quad = (a^2 + ab) + (a + b)$

$\quad = a(a + b) + 1(a + b)$

$\quad = (a + b)(a + 1)$

51. $5r + s^2 - 5s - rs$

$\quad = (5r - rs) + (s^2 - 5s)$

$\quad = r(5 - s) + s(s - 5)$

$\quad = r(5 - s) + s[-(5 - s)]$

$\quad = r(5 - s) - s(5 - s)$

$\quad = (5 - s)(r - s)$

53. $uv + wv - u - w$

$\quad = (uv + wv) + (-u - w)$

$\quad = v(u + w) - 1(u + w)$

$\quad = (u + w)(v - 1)$

55. $a^3 - a^2c + a^2b - abc$

$\quad = a(a^2 - ac + ab - bc)$

$\quad = a[(a^2 - ac) + (ab - bc)]$

$\quad = a[a(a - c) + b(a - c)]$

$\quad = a[(a - c)(a + b)]$

$\quad = a(a - c)(a + b)$

57. $xz^2 - z^2 + xz - z$

$\quad = z(xz - z + x - 1)$

$\quad = z[(xz - z) + (x - 1)]$

$\quad = z[z(x - 1) + 1(x - 1)]$

$\quad = z[(x - 1)(z + 1)]$

$\quad = z(x - 1)(z + 1)$

59. $15abx - 15aby - 25acx + 25acy$

$\quad = 5a[3bx - 3by - 5cx + 5cy]$

$\quad = 5a[(3bx - 3by) + (-5cx + 5cy)]$

$\quad = 5a[3b(x - y) - 5c(x - y)]$

$\quad = 5a[(x - y)(3b - 5c)]$

$\quad = 5a(x - y)(3b - 5c)$

61. $42a^4b^2c + 54a^2b^5c = 6a^2b^2c(7a^2 + 9b^3)$

63. $24r^4s^4 - 48r^3s^3 = 24r^3s^3(rs - 2)$

65. $27a^4b - 18ab^5 + 9ab = 9ab(3a^3 - 2b^4 + 1)$

67. $-16x^3 - 24x^6 = -8x^3(2 + 3x^3)$

69. $12t^4 - 12t^3 = 12t^3(t - 1)$

71. $2\sqrt{3}x - \sqrt{3}x^2 = \sqrt{3}x(2 - x)$

73. $2\sqrt{5}x + 4\sqrt{5}x^3 = 2\sqrt{5}x(1 + 2x^2)$

75. $c(d - 2) + b(d - 2) = (d - 2)(c + b)$

77. $\quad a(s - t) + u(t - s)$

$\quad = a(s - t) + u[-(s - t)]$

$\quad = a(s - t) - u(s - t)$

$\quad = (s - t)(a - u)$

79. $\quad (x + y)^3 - 5(x + y)^2$

$\quad = (x + y)^2[(x + y) - 5]$

$\quad = (x + y)^2(x + y - 5)$

81. $\quad 3(1 - r)^3 - 3(1 - r)^4 - 4(1 - r)^2$

$\quad = (1 - r)^2[3(1 - r) - 3(1 - r)^2 - 4]$

83. $\quad 2t(a - x) - t(x - a)$

$\quad = 2t(a - x) - t[-(a - x)]$

$\quad = 2t(a - x) + t(a - x)$

$\quad = (a - x)(2t + t)$

$\quad = (a - x)(3t)$

$\quad = 3t(a - x)$

85. $\quad c^2(5 + b)^4 - d^2(5 + b)^5 - 2(b + 5)^2$

$\quad = c^2(5 + b)^4 - d^2(5 + b)^5 - 2(5 + b)^2$

$\quad = (5 + b)^2[c^2(5 + b)^2 - d^2(5 + b)^3 - 2]$

87. $\quad 15(a - b)^4 + 45(a - b)^2 - 35(b - a)$

$\quad = 15(a - b)^4 + 45(a - b)^2 - 35[-(a - b)]$

$\quad = 15(a - b)^4 + 45(a - b)^2 + 35(a - b)$

$\quad = 5(a - b)[3(a - b)^3 + 9(a - b) + 7]$

89. $xy + 6y + x + 6 = (xy + 6y) + (x + 6)$

$\qquad\qquad\qquad\quad = y(x + 6) + 1(x + 6)$

$\qquad\qquad\qquad\quad = (x + 6)(y + 1)$

91. $\quad ar - br - as + bs$

$\quad = (ar - br) + (-as + bs)$

$\quad = r(a - b) - s(a - b)$

$\quad = (a - b)(r - s)$

93. $\quad a + b^2 - b - ab$

$\quad = (a - b) + (b^2 - ab)$

$\quad = 1(a - b) + b(b - a)$

$\quad = 1(a - b) + b[-(a - b)]$

$\quad = 1(a - b) - b(a - b)$

$\quad = (a - b)(1 - b)$

95. $\quad 4s + 12 - 3b - bs$

$\quad = (4s - bs) + (12 - 3b)$

$\quad = s(4 - b) + 3(4 - b)$

$\quad = (4 - b)(s + 3)$

97. $\quad t^5 + t^4 + s^4t + s^4$

$\quad = (t^5 + t^4) + (s^4t + s^4)$

$\quad = t^4(t + 1) + s^4(t + 1)$

$\quad = (t + 1)(t^4 + s^4)$

99. $\quad q^4 - q^2t + q^2 - t$

$\quad = (q^4 - q^2t) + (q^2 - t)$

$\quad = q^2(q^2 - t) + 1(q^2 - t)$

$\quad = (r^2 - t)(q^2 + 1)$

101. $ax^2 + b^2x^2 - ax - b^2x$

$= x(ax + b^2x - a - b^2)$

$= x[(ax + b^2x) + (-a - b^2)]$

$= x[x(a + b^2) - 1(a + b^2)]$

$= x[(a + b^2)(x - 1)]$

$= x(a + b^2)(x - 1)$

103. $16c^2r + 32c^2s - 8cr - 16cs$

$= 8c(2cr + 4cs - r - 2s)$

$= 8c[(2cr + 4cs) + (-r - 2s)]$

$= 8c[2c(r + 2s) - 1(r + 2s)]$

$= 8c[(r + 2s)(2c - 1)]$

$= 8c(r + 2s)(2c - 1)$

105. $w^3 - w^2t + w^3t - w^4$

$= w^2(w - t + wt - w^2)$

$= w^2[(w - t) + (wt - w^2)]$

$= w^2[1(w - t) + w(t - w)]$

$= w^2\{1(w - t) + w[-(w - t)]\}$

$= w^2[1(w - t) - w(w - t)]$

$= w^2[(w - t)(1 - w)]$

$= w^2(w - t)(1 - w)$

107. $x^{n+2} - x^{n+3} = x^{n+2} - x \cdot x^{n+2}$

$= x^{n+2}(1 - x)$

109. $x^{2mn} + x^{4mn} = x^{2mn} + (x^{2mn})^2$

$= x^{2mn}(1 + x^{2mn})$

Problem Set 1.6

1. $x^2 - 9 = x^2 - 3^2$

$= (x - 3)(x + 3)$

3. $9x^2 - 4 = (3x)^2 - 2^2$

$= (3x - 2)(3x + 2)$

5. $x^3 - 1 = x^3 - 1^3$

$= (x - 1)(x^2 + x + 1)$

7. $64 + t^3 = 4^3 + t^3$

$= (4 + t)(16 - 4t + t^2)$

9. $4x^2 - 64 = 4(x^2 - 16)$

$= 4(x^2 - 4^2)$

$= 4(x - 4)(x + 4)$

11. $3t^3 + 24s^3$

$= 3(t^3 + 8s^3)$

$= 3[t^3 + (2s)^3]$

$= 3(t + 2s)[t^2 - t(2s) + (2s)^2]$

$= 3(t + 2s)(t^2 - 2st + 4s^2)$

13. $x^2 + 3x + 2 = (x + 2)(x + 1)$

15. $-5x + x^2 + 6 = x^2 - 5x + 6$

$= (x - 2)(x - 3)$

17. $z^2 + z - 12 = (z + 4)(z - 3)$

19. $x^2 - 3x - 10 = (x - 5)(x + 2)$

21. $2x^2 + 3x + 1 = 2x^2 + 2x + x + 1$

$= 2x(x + 1) + 1(x + 1)$

$= (x + 1)(2x + 1)$

23. $8s^2 - 22st + 15t^2$

$= 8s^2 - 12st - 10st + 15t^2$

$= 4s(2s - 3t) - 5t(2s - 3t)$

$= (2s - 3t)(4s - 5t)$

25. $2x^2 - 7x + 6 = 2x^2 - 4x - 3x + 6$
$$= 2x(x - 2) - 3(x - 2)$$
$$= (x - 2)(2x - 3)$$

27. $6a^2 + 7a - 3 = 6a^2 + 9a - 2a - 3$
$$= 3a(2a + 3) - 1(2a + 3)$$
$$= (2a + 3)(3a - 1)$$

29. $4x^2 - 4x - 8 = 4(x^2 - x - 2)$
$$= 4(x - 2)(x + 1)$$

31. $-6y^3 + 6y^2 + 12y = -6y(y^2 - y - 2)$
$$= -6y(y - 2)(y + 1)$$

33. $x^2 - 81 = x^2 - 9^2$
$$= (x - 9)(x + 9)$$

35. $x^3 - 8 = x^3 - 2^3$
$$= (x - 2)(x^2 + 2x + 4)$$

37. $x^2 + 9x + 8 = (x + 8)(x + 1)$

39. $t^2 + t - 20 = (t + 5)(t - 4)$

41. $z^3 + 27 = z^3 + 3^3$
$$= (z + 3)(z^2 - 3z + 9)$$

43. $3x^2 - x - 10 = 3x^2 - 6x + 5x - 10$
$$= 3x(x - 2) + 5(x - 2)$$
$$= (x - 2)(3x + 5)$$

45. $16 - x^2 = 4^2 - x^2$
$$= (4 - x)(4 + x)$$

47. $27 - 8v^3$
$$= 3^3 - (2v)^3$$
$$= (3 - 2v)[3^2 + 3(2v) + (2v)^2]$$
$$= (3 - 2v)(9 + 6v + 4v^2)$$

49. $6y^2 + 25yz + 4z^2$
$$= 6y^2 + yz + 24yz + 4z^2$$
$$= y(6y + z) + 4z(6y + z)$$
$$= (6y + z)(y + 4z)$$

51. $5t^2 - 28t + 15 = 5t^2 - 3t - 25t + 15$
$$= t(5t - 3) - 5(5t - 3)$$
$$= (5t - 3)(t - 5)$$

53. $x^3 - 8y^3$
$$= x^3 - (2y)^3$$
$$= (x - 2y)[x^2 + x(2y) + (2y)^2]$$
$$= (x - 2y)(x^2 + 2xy + 4y^2)$$

55. $8x^3 - 2x = 2x(4x^2 - 1)$
$$= 2x[(2x)^2 - 1^2]$$
$$= 2x(2x - 1)(2x + 1)$$

57. $6y^3 + 6y^2 - 12y = 6y(y^2 + y - 2)$
$$= 6y(y + 2)(y - 1)$$

59. $2x^2 + 6xy + 4y^2 = 2(x^2 + 3xy + 2y^2)$
$$= 2(x + 2y)(x + y)$$

61. $81x^4 + 24x$
$$= 3x(27x^3 + 8)$$
$$= 3x[(3x)^3 + 2^3]$$
$$= 3x(3x + 2)[(3x)^2 - (3x)(2) + 2^2]$$
$$= 3x(3x + 2)(9x^2 - 6x + 4)$$

63. $7t^2 + 42t + 63 = 7(t^2 + 6t + 9)$

$\qquad\qquad\qquad = 7(t + 3)(t + 3)$

$\qquad\qquad\qquad = 7(t + 3)^2$

65. $x^4 - 2x^2 - 8 = (x^2 - 4)(x^2 + 2)$

$\qquad\qquad\qquad\ = (x - 2)(x + 2)(x^2 + 2)$

67. $z^4 - 3z^2 - 4 = (z^2 - 4)(z^2 + 1)$

$\qquad\qquad\qquad\ = (z - 2)(z + 2)(z^2 + 1)$

69. $x^4 - 1 = (x^2 - 1)(x^2 + 1)$

$\qquad\qquad = (x - 1)(x + 1)(x^2 + 1)$

71. $x^6 - 8x^3 + 15 = (x^3 - 5)(x^3 - 3)$

73. $4x^6 - 4x^3 - 8$

$\quad = 4(x^6 - x^3 - 2)$

$\quad = 4(x^3 - 2)(x^3 + 1)$

$\quad = 4(x^3 - 2)(x + 1)(x^2 - x + 1)$

75. $16x^2 - 16 = 16(x^2 - 1)$

$\qquad\qquad\ = 16(x - 1)(x + 1)$

77. $2ab + 2ax - by - xy$

$\quad = 2a(b + x) - y(b + x)$

$\quad = (b + x)(2a - y)$

79. Factored

81. $6s^2 + st - 15t^2$

$\quad = 6s^2 + 10st - 9st - 15t^2$

$\quad = 2s(3s + 5t) - 3t(3s + 5t)$

$\quad = (3s + 5t)(2s - 3t)$

83. Prime

85. $z^2 - k = (z + 4)(z - 4)$

$\qquad z^2 - k = z^2 - 16$

$\qquad\qquad k = 16$

87. $t^2 + kt - 7$

The factors of -7 are $1 \cdot (-7)$ and $(-1) \cdot 7$.

Hence $k = 1 + (-7) = -6$ or $k = -1 + 7 = 6$.

89. $-7 + 6 = -1$

$(s - 7)(s + 6) = s^2 - s - 42$

$k = -42$

91. Volume: $\pi x^2 h - \pi y^2 h$

$\qquad\qquad = \pi h(x^2 - y^2)$

$\qquad\qquad = \pi h(x - y)(x + y)$

93. $\pi y^2 - \pi x^2 = \pi(y^2 - x^2)$

$\qquad\qquad\qquad = \pi(y - x)(y + x)$

95. $(x + y)^2 - 4 = (x + y)^2 - 2^2$

$\qquad\qquad\qquad = [(x + y) - 2][(x + y) + 2]$

$\qquad\qquad\qquad = (x + y - 2)(x + y + 2)$

97. $125 - (s - t)^3$

$\quad = 5^3 - (s - t)^3$

$\quad = [5 - (s - t)][5^2 + 5(s - t) + (s - t)^2]$

$\quad = (5 - s + t)[25 + 5(s - t) + (s - t)^2]$

99. $1 - x^9$

$\quad = 1^3 - (x^3)^3$

$\quad = (1 - x^3)[1 + x^3 + (x^3)^2]$

$\quad = (1 - x)(1 + x + x^2)(1 + x^3 + x^6)$

101. $x^{2p} - 1 = (x^p)^2 - 1^2$

$\qquad\qquad\ = (x^p - 1)(x^p + 1)$

103. $x^{3p} + 1 = (x^p)^3 + 1^3$

$$= (x^p + 1)[(x^p)^2 - x^p(1) + 1^2]$$

$$= (x^p + 1)(x^{2p} - x^p + 1)$$

105. $x^6 - 1$

$$= (x^3)^2 - 1^2$$

$$= (x^3 - 1)(x^3 + 1)$$

$$= (x - 1)(x^2 + x + 1)(x + 1)(x^2 - x + 1)$$

$x^6 - 1$

$$= (x^2)^3 - 1^3$$

$$= (x^2 - 1)(x^4 + x^2 + 1)$$

$$= (x - 1)(x + 1)(x^4 + x^2 + 1)$$

107. $x^{2n} + 2x^n - 8 = (x^n)^2 + 2x^n - 8$

$$= (x^n + 4)(x^n - 2)$$

109. $x^{2n} + 2x^n y^m + y^{2m}$

$$= (x^n)^2 + 2x^n y^m + (y^m)^2$$

$$= (x^n + y^m)^2$$

111. (a.) $2^{32} - 1 = 4,295,000,000$

$3 \cdot 1,431,700,000 = 4,295,000,000$

$17 \cdot 252,650,000 = 4,295,000,000$

(b.) $2^{32} - 1$

$$= (2^{16})^2 - 1^2$$

$$= (2^{16} - 1)(2^{16} + 1)$$

$$= [(2^8)^2 - 1^2](2^{16} + 1)$$

$$= (2^8 - 1)(2^8 + 1)(2^{16} + 1)$$

$$= [(2^4)^2 - 1^2](2^8 + 1)(2^{16} + 1)$$

$$= (2^4 - 1)(2^4 + 1)(2^8 + 1)(2^{16} + 1)$$

$$= [(2^2)^2 - 1^2](2^4 + 1)(2^8 + 1)(2^{16} + 1)$$

$$= (2^2 - 1)(2^2 + 1)(2^4 + 1)(2^8 + 1)(2^{16} + 1)$$

$$= (2 - 1)(2 + 1)(2^2 + 1)(2^4 + 1)(2^8 + 1)(2^{16} + 1)$$

Two of the factors are:

$2 + 1 = 3$

$2^4 + 1 = 17$

Problem Set 1.7

1. $16x^2 - 16 = 16(x^2 - 1)$

$$= 16(x + 1)(x - 1)$$

3. $2w^3 + 250x^3 = 2(w^3 + 125x^3)$

$$= 2(w + 5x)(w^2 - 5wx + 25x^2)$$

5. $x^4 + x^3 - 2x^2 = x^2(x^2 + x - 2)$

$$= x^2(x + 2)(x - 1)$$

7. $2yz - 8y + 8 - 2z = 2(yz - 4y + 4 - z)$

$$= 2[(yz - 4y) - (z - 4)]$$

$$= 2[y(z - 4) - 1(z - 4)]$$

$$= 2(z - 4)(y - 1)$$

9. Prime

11. $16x^2 + 8x + 1 = (4x + 1)^2$

13. Factored

15. $18a^3 + 36b^3c^3 + 27b^2c^2$

$= 9(2a^3 + 4b^3c^3 + 3b^2c^2)$

17. $x^2 - 10x + 25 = (x - 5)^2$

19. $a^2z + a^2x - z - x = a^2(z + x) - 1(z + x)$

$$= (z + x)(a^2 - 1)$$

$$= (z + x)(a - 1)(a + 1)$$

21. $y^2 - yz - z + y = y(y - z) + 1(y - z)$

$$= (y - z)(y + 1)$$

23. $21x^2 + 24x - 36 = 3(7x^2 + 8x - 12)$

$\qquad = 3(7x - 6)(x + 2)$

25. $x^2 + 9x + 14 = (x + 7)(x + 2)$

27. $135r^3t^5 + 225r^2t^4 = 45r^2t^4(3rt + 5)$

29. $1 - 27a^3b^3 = (1 - 3ab)(1 + 3ab + 9a^2b^2)$

31. $x^2 - y^2 + x + y = (x + y)(x - y) + 1(x + y)$

$\qquad = (x + y)(x - y + 1)$

33. Factored

35. $32 - 2x^2 = 2(16 - x^2)$

$\qquad = 2(4 - x)(4 + x)$

37. $r^3(x - t) + (t - x) = r^3(x - t) - 1(x - t)$

$\qquad = (x - t)(r^3 - 1)$

$\qquad = (x - t)(r - 1)(r^2 + r + 1)$

39. $27r^3 + 64 = (3r + 4)(9r^2 - 12r + 16)$

41. $x(r + s) - y(s + r) = (r + s)(x - y)$

43. $x^8 - 16 = (x^4 + 4)(x^4 - 4)$

$\qquad = (x^4 + 4)(x - 2)(x + 2)$

45. $a(x - y) - b(y - x) = a(x - y) + b(x - y)$

$\qquad = (x - y)(a + b)$

47. Prime

49. $x^4 - x^3 + x^2 - x = x(x^3 - x^2 + x - 1)$

$\qquad = x[x^2(x - 1) + 1(x - 1)]$

$\qquad = x(x - 1)(x^2 + 1)$

51. $x^4 + x^2 - 20 = (x^2 + 5)(x^2 - 4)$

$\qquad = (x^2 + 5)(x - 2)(x + 2)$

53. $81 - 18x + x^2 = (9 - x)^2$

55. $a^9 + 8 = (a^3 + 2)(a^6 - 2a^3 + 4)$

57. $x^{4n} - 2x^{2n} + 1$

$\qquad = (x^{2n} - 1)(x^{2n} - 1)$

$\qquad = (x^n + 1)(x^n - 1)(x^n + 1)(x^n - 1)$

$\qquad = (x^n + 1)^2(x^n - 1)^2$

59. $x^{4n} - x^{2n} = x^{2n}(x^{2n} - 1)$

$\qquad = x^{2n}(x^n - 1)(x^n + 1)$

61. (a) $x^4 + 3x^2 + 4$

$\qquad = (x^4 + 3x^2 + 4) + (x^2 - x^2)$

$\qquad = (x^4 + 4x^2 + 4) - x^2$

$\qquad = (x^2 + 2)^2 - x^2$

$\qquad = [(x^2 + 2) - x][(x^2 + 2) + x]$

$\qquad = (x^2 - x + 2)(x^2 + x + 2)$

(b) $x^4 + 2x^2 + 9$

$\qquad = (x^4 + 2x^2 + 9) + (4x^2 - 4x^2)$

$\qquad = (x^4 + 6x^2 + 9) - 4x^2$

$\qquad = (x^2 + 3)^2 - 4x^2$

$\qquad = [(x^2 + 3)^2 - 2x][(x^2 + 3) + 2x]$

$\qquad = (x^2 - 2x + 3)(x^2 + 2x + 3)$

(c) $x^4 + 7x^2 + 16$

$\qquad = (x^4 + 7x^2 + 16) + (x^2 - x^2)$

$\qquad = (x^4 + 8x^2 + 16) - x^2$

$\qquad = (x^2 + 4)^2 - x^2$

$\qquad = [(x^2 + 4) - x][(x^2 + 4) + x]$

$\qquad = (x^2 - x + 4)(x^2 + x + 4)$

(d) $x^4 - 8x^2 + 4$

$\qquad = (x^4 - 8x^2 + 4) + (4x^2 - 4x^2)$

$\qquad = (x^4 - 4x^2 + 4) - 4x^2$

$\qquad = (x^2 - 2)^2 - 4x^2$

$\qquad = [(x^2 - 2) - 2x][(x^2 - 2) + 2x]$

$\qquad = (x^2 - 2x - 2)(x^2 + 2x - 2)$

(e) $x^4 - 11x^2 + 25$

$= (x^4 - 11x^2 + 25) + (x^2 - x^2)$

$= (x^4 - 10x^2 + 25) - x^2$

$= (x^2 - 5)^2 - x^2$

$= [(x^2 - 5) - x][(x^2 - 5) + x]$

$= (x^2 - x - 5)(x^2 + x - 5)$

Problem Set 1.8

1. $\dfrac{a^4 b^2}{b} = a^4 b$

3. $\dfrac{(x - y)^4 z^2}{(x - y)^2 z} = (x - y)^2 z$

5. $\dfrac{6x^2 + 12y}{6x} = \dfrac{6x^2}{6x} + \dfrac{12y}{6x}$

$\qquad\qquad = x + \dfrac{2y}{x}$

7.
$$
\begin{array}{r}
z - 6 \\
z + 2\overline{\smash{)}z^2 - 4z - 12} \\
\underline{z^2 + 2z} \\
-6z - 12 \\
\underline{-6z - 12} \\
0
\end{array}
$$

9.
$$
\begin{array}{r}
x - 3 \\
2x + 1\overline{\smash{)}2x^2 - 5x - 3} \\
\underline{2x^2 + x} \\
-6x - 3 \\
\underline{-6x - 3} \\
0
\end{array}
$$

11.
$$
\begin{array}{r}
x - 1 \\
x + 3\overline{\smash{)}x^2 + 2x - 4} \\
\underline{x^2 + 3x} \\
-x - 4 \\
\underline{-x - 3} \\
-1
\end{array}
$$

13.
$$
\begin{array}{r}
x^3 - x^2 - 4x - 7 \\
x^2 - 2x + 3\overline{\smash{)}x^5 - 3x^4 + x^3 - 2x^2 + 5x + 4} \\
\underline{x^5 - 2x^4 + 3x^3} \\
-x^4 - 2x^3 - 2x^2 + 5x + 4 \\
\underline{-x^4 + 2x^3 - 3x^2} \\
-4x^3 + x^2 + 5x + 4 \\
\underline{-4x^3 + 8x^2 - 12x} \\
-7x^2 + 17x + 4 \\
\underline{-7x^2 + 14x - 21} \\
3x + 25
\end{array}
$$

15.
$$
\begin{array}{r}
x^2 - xy + y^2 \\
x + 2y\overline{\smash{)}x^3 + x^2y - xy^2 + 2y^3} \\
\underline{x^3 + 2x^2y} \\
-x^2y - xy^2 + 2y^3 \\
\underline{-x^2y - 2xy^2} \\
xy^2 + 2y^3 \\
\underline{xy^2 + 2y^3} \\
0
\end{array}
$$

17.
$$
\begin{array}{r}
x^2 - x + 2 \\
x^2 + 0x - 1\overline{\smash{)}x^4 - x^3 + x^2 + 0x + 1} \\
\underline{x^4 + 0x^3 - x^2} \\
-x^3 + 2x^2 + 0x + 1 \\
\underline{-x^3 + 0x^2 + x} \\
2x^2 - x + 1 \\
\underline{2x^2 + 0x - 2} \\
-x + 3
\end{array}
$$

19. $\dfrac{27ab}{18b^2} = \dfrac{3a}{2b}$

21. $\dfrac{5x^2 + 10x - 15}{10y} = \dfrac{5x^2}{10y} + \dfrac{10x}{10y} - \dfrac{15}{10y}$

$\qquad\qquad\qquad = \dfrac{x^2}{2y} + \dfrac{x}{y} - \dfrac{3}{2y}$

23. $\dfrac{x^2 - xy + y^2}{xy} = \dfrac{x^2}{xy} - \dfrac{xy}{xy} + \dfrac{y^2}{xy}$

$\qquad\qquad\qquad = \dfrac{x}{y} - 1 + \dfrac{y}{x}$

25.

$$\begin{array}{r} 3x - 1 \\ 2x+3\overline{\smash{\big)}\,6x^2+7x-3} \\ \underline{6x^2+9x} \\ -2x-3 \\ \underline{-2x-3} \\ 0 \end{array}$$

27.

$$\begin{array}{r} y + 3 \\ y-1\overline{\smash{\big)}\,y^2+2y-3} \\ \underline{y^2-y} \\ 3y-3 \\ \underline{3y-3} \\ 0 \end{array}$$

29.

$$\begin{array}{r} x^2 - x - 1 \\ 2x+1\overline{\smash{\big)}\,2x^3-x^2-3x+3} \\ \underline{2x^3+x^2} \\ -2x^2-3x+3 \\ \underline{-2x^2-x} \\ -2x+3 \\ \underline{-2x-1} \\ 4 \end{array}$$

31.

$$\begin{array}{r} 2x^2 - x + 1 \\ 3x+2\overline{\smash{\big)}\,6x^3+x^2+x-5} \\ \underline{6x^3+4x^2} \\ -3x^2+x-5 \\ \underline{-3x^2-2x} \\ 3x-5 \\ \underline{3x+2} \\ -7 \end{array}$$

33.

$$\begin{array}{r} 2x + 1 \\ x-3\overline{\smash{\big)}\,2x^2-5x-3} \\ \underline{2x^2-6x} \\ x-3 \\ \underline{x-3} \\ 0 \end{array}$$

35.

$$\begin{array}{r} 3x + y \\ 2x+y\overline{\smash{\big)}\,6x^2+5xy+y^2} \\ \underline{6x^2+3xy} \\ 2xy+y^2 \\ \underline{2xy+y^2} \\ 0 \end{array}$$

37.

$$\begin{array}{r} x^2 - x + 1 \\ 3x^2+x-1\overline{\smash{\big)}\,3x^4-2x^3+x^2-2x+4} \\ \underline{3x^4+x^3-x^2} \\ -3x^3+2x^2-2x+4 \\ \underline{-3x^3-x^2+x} \\ 3x^2-3x+4 \\ \underline{3x^2+x-1} \\ -4x+5 \end{array}$$

39.

$$\begin{array}{r} 3x^2 - x + 1 \\ 7x+4\overline{\smash{\big)}\,21x^3+5x^2+3x+8} \\ \underline{21x^3+12x^2} \\ -7x^2+3x+8 \\ \underline{-7x^2-4x} \\ 7x+8 \\ \underline{7x+4} \\ 4 \end{array}$$

41.

$$\begin{array}{r} x^3 - 3x^2 + 9x - 28 \\ x+3\overline{\smash{\big)}\,x^4+0x^3+0x^2-x+2} \\ \underline{x^4+3x^3} \\ -3x^3+0x^2-x+2 \\ \underline{-3x^3-9x^2} \\ 9x^2-x+2 \\ \underline{9x^2+27x} \\ -28x+2 \\ \underline{-28x-84} \\ 86 \end{array}$$

43.

$$\begin{array}{r} x^3 + 2 \\ x^3+0x^2+0x-1\overline{\smash{\big)}\,x^6+0x^5+0x^4+x^3+0x^2+0x-3} \\ \underline{x^6+0x^5+0x^4-x^3} \\ 2x^3+0x^2+0x-3 \\ \underline{2x^3+0x^2+0x-2} \\ -1 \end{array}$$

45.

$$\begin{array}{r} 2x^3 - 4x + 1 \\ 2x^2+0x+1\overline{\smash{\big)}\,4x^5+0x^4-6x^3+2x^2+0x+1} \\ \underline{4x^5+0x^4+2x^3} \\ -8x^3+2x^2+0x+1 \\ \underline{-8x^3+0x^2-4x} \\ 2x^2+4x+1 \\ \underline{2x^2+0x+1} \\ 4x \end{array}$$

47.

$$\begin{array}{r} x^2 + 3x - 4 \\ x-2\overline{)x^3 + x^2 - 10x + 8} \end{array}$$

$$\underline{x^3 - 2x^2}$$

$$3x^2 - 10x + 8$$

$$\underline{3x^2 - 6x}$$

$$-4x + 8$$

$$\underline{-4x + 8}$$

$$0$$

49.

$$\begin{array}{r} 2t^2 - 3t + 5 \\ 3t-2\overline{)6t^3 - 13t^2 + 21t + 17} \end{array}$$

$$\underline{6t^3 - 4t^2}$$

$$-9t^2 + 21t + 17$$

$$\underline{-9t^2 + 6t}$$

$$15t + 17$$

$$\underline{15t - 10}$$

$$27$$

51.

$$\begin{array}{r} x^n - 3 \\ x^n+1\overline{)x^{2n} - 2x^n + 4} \end{array}$$

$$\underline{x^{2n} + x^n}$$

$$-3x^n + 4$$

$$\underline{-3x^n - 3}$$

$$7$$

53.

$$\begin{array}{r} x^2 - x + 4 \\ x+3\overline{)x^3 + 2x^2 + x + k} \end{array}$$

$$\underline{x^3 + 3x^2}$$

$$-x^2 + x + k$$

$$\underline{-x^2 - 3x}$$

$$4x + k$$

$$\underline{4x + 12}$$

If $k = 12$, the remainder will be 0.

Problem Set 1.9

1.

$$\begin{array}{r|rrr} 3 & 2 & -5 & -3 \\ & & 6 & 3 \\ \hline & 2 & 1 & 0 \end{array}$$

$$\frac{2x^2 - 5x - 3}{x - 3} = 2x + 1$$

3.

$$\begin{array}{r|rrrr} -1 & 1 & 3 & 3 & 1 \\ & & -1 & -2 & -1 \\ \hline & 1 & 2 & 1 & 0 \end{array}$$

$$\frac{x^3 + 3x^2 + 3x + 1}{x + 1} = x^2 + 2x + 1$$

5.

$$\begin{array}{r|rrrr} 4 & 1 & 0 & 0 & -64 \\ & & 4 & 16 & 64 \\ \hline & 1 & 4 & 16 & 0 \end{array}$$

The remainder is 0.

7.

$$\begin{array}{r|rrrrr} 1 & 1 & 0 & -2 & 0 & 1 \\ & & 1 & 1 & -1 & -1 \\ \hline & 1 & 1 & -1 & -1 & 0 \end{array}$$

$$\frac{x^4 - 2x^2 + 1}{x - 1} = x^3 + x^2 - x - 1$$

9.

$$\begin{array}{r|rrrr} -2 & 4 & 7 & 1 & 5 \\ & & -8 & 2 & -6 \\ \hline & 4 & -1 & 3 & -1 \end{array}$$

$$\frac{7x^2 + 4x^3 + x + 5}{x + 2} = 4x^2 - x + 3 - \frac{1}{x + 2}$$

11.

$$\begin{array}{r|rrr} 3 & 2 & -9 & 9 \\ & & 6 & -9 \\ \hline & 2 & -3 & 0 \end{array}$$

$$(2x^2 - 9x + 9) \div (x - 3) = 2x - 3$$

13.

$$\begin{array}{r|rrrrr} 2 & 1 & 0 & -2 & -5 & 3 \\ & & 2 & 4 & 4 & -2 \\ \hline & 1 & 2 & 2 & -1 & 1 \end{array}$$

$$\frac{x^4 - 2x^2 - 5x + 3}{x - 2} = x^3 + 2x^2 + 2x - 1 + \frac{1}{x - 2}$$

15.

$$\begin{array}{r|rrrr}
2 & 2 & -1 & -3 & -4 \\
  &   & 4 & 6 & 6 \\
\hline
  & 2 & 3 & 3 & 2
\end{array}$$

$$\frac{2x^3 - x^2 - 3x - 4}{x - 2} = 2x^2 + 3x + 3 + \frac{2}{x - 2}$$

17.

$$\begin{array}{r|rrrrr}
-1 & 1 & 0 & 0 & 0 & 1 \\
   &   & -1 & 1 & -1 & 1 \\
\hline
   & 1 & -1 & 1 & -1 & 2
\end{array}$$

$$\frac{(y^4 + 1)}{(y + 1)} = y^3 - y^2 + y - 1 + \frac{2}{y + 1}$$

19.

$$\begin{array}{r|rrrrr}
1 & 1 & 0 & 1 & 0 & -2 \\
  &   & 1 & 1 & 2 & 2 \\
\hline
  & 1 & 1 & 2 & 2 & 0
\end{array}$$

$$(x^4 + x^2 - 2) \div (x - 1) = x^3 + x^2 + 2x + 2$$

21.

$$\begin{array}{r|rrrrr}
1 & 1 & 1 & 1 & 1 & 1 \\
  &   & 1 & 2 & 3 & 4 \\
\hline
  & 1 & 2 & 3 & 4 & 5
\end{array}$$

$$\frac{x^4 + x^3 + x^2 + x + 1}{x - 1} = x^3 + 2x^2 + 3x + 4 + \frac{5}{x - 1}$$

23.

$$\begin{array}{r|rrrrr}
2 & 2 & -4 & 1 & -1 & -2 \\
  &   & 4 & 0 & 2 & 2 \\
\hline
  & 2 & 0 & 1 & 1 & 0
\end{array}$$

The remainder is 0.

Chapter 1 Review Problems

1. $1^3 - 1(-3)^2 + (-3)^3 = 1 - 9 - 27$
$$= -35$$

3. $(-5)^2 = (-5)(-5) = 25$

5. $(-2)^2(-3)^{-3} = \dfrac{(-2)^2}{(-3)^3}$
$$= \frac{4}{-27}$$
$$= -\frac{4}{27}$$

7. $(x^{-2}y^3)^4 = (x^{-2})^4(y^3)^4$
$$= x^{-8}y^{12}$$
$$= \frac{y^{12}}{x^8}$$

9. $(a + b)^4(a + b) = (a + b)^5$

11. $\left(\dfrac{-2x}{y}\right)^2 = \dfrac{(-2x)^2}{y^2}$
$$= \frac{4x^2}{y^2}$$

13. $\dfrac{-10^0}{10} = \dfrac{-1}{10} = -\dfrac{1}{10}$

15. $\dfrac{(-6)^4}{(-6)^5} = \dfrac{1}{(-6)^1}$
$$= -\frac{1}{6}$$

17. $\dfrac{-5^3}{-2^4} = \dfrac{-125}{-16}$
$$= \frac{125}{16}$$

19. $abc(a^2b^3c - ac^4)$

$= abc \cdot a^2b^3c - abc \cdot ac^4$

$= a^3b^4c^2 - a^2bc^5$

21. $(7p^3 - 4p^2 - 7) - (p^3 + 6p - 5)$

$= 7p^3 - 4p^2 - 7 - p^3 - 6p + 5$

$= 6p^3 - 4p^2 - 6p - 2$

23. $\dfrac{(xy^3z^2)}{(x^2yz)} = \dfrac{y^2z}{x}$

25. $(r^3s^2t)(r^3st^4) = r^6s^3t^5$

27. $(x^4 - x^3y + 3x^2y^2 - 5xy + y^5)$

$\quad - (2x^4 + 2x^3y - x^2y^2 + 7xy)$

$= x^4 - x^3y + 3x^2y^2 - 5xy + y^5$

$\quad -2x^4 - 2x^3y + x^2y^2 - 7xy$

$= -x^4 - 3x^3y + 4x^2y^2 - 12xy + y^5$

29.
$$
\begin{array}{r}
x^2 - 2x + 4 \\
x + 7\overline{\smash{)}x^3 + 5x^2 - 10x + 20} \\
\underline{x^3 + 7x^2} \\
-2x^2 - 10x + 20 \\
\underline{-2x^2 - 14x} \\
4x + 20 \\
\underline{4x + 28} \\
-8
\end{array}
$$

$$\dfrac{x^3 + 5x^2 - 10x + 20}{x + 7} = x^2 - 2x + 4 - \dfrac{8}{x+7}$$

31. $(t - 5)^3 = t^3 - 3t^2 \cdot 5 + 3t \cdot (5)^2 - 5^3$

$\qquad = t^3 - 15t^2 + 75t - 125$

33. $(7r - 2)(2r - 1) = 14r^2 - 7r - 4r + 2$

$\qquad\qquad\qquad = 14r^2 - 11r + 2$

35. $(c + 3d)(c - 3d) = c^2 - (3d)^2$

$\qquad\qquad\qquad = c^2 - 9d^2$

37. $(x - y) - (x + y) = x - y - x - y$

$\qquad\qquad\qquad\quad = -2y$

39. $xy^2(xy - 1) = xy^2 \cdot xy - xy^2 \cdot 1$

$\qquad\qquad\quad = x^2y^3 - xy^2$

41.
$$
\begin{array}{r}
4s^2 - 6s + 9 \\
2s + 3\overline{\smash{)}8s^3 + 0s^2 + 0s + 27} \\
\underline{8s^3 + 12s^2} \\
-12s^2 + 0s + 27 \\
\underline{-12s^2 - 18s} \\
18s + 27 \\
\underline{18s + 27} \\
0
\end{array}
$$

$(8s^3 + 27) \div (2s + 3) = 4s^2 - 6s + 9$

43. $(x + 3)^2 - (x - 2)^2$

$= x^2 + 6x + 9 - (x^2 - 4x + 4)$

$= x^2 + 6x + 9 - x^2 + 4x - 4$

$= 10x + 5$

45.
$$
\begin{array}{r}
x^2 + 3x \\
x^2 - x + 2\overline{\smash{)}x^4 + 2x^3 - x^2 + 0x + 6} \\
\underline{x^4 - x^3 + 2x^2} \\
3x^3 - 3x^2 + 0x + 6 \\
\underline{3x^3 - 3x^2 + 6x} \\
-6x + 6
\end{array}
$$

$$\dfrac{x^4 + 2x^3 - x^2 + 6}{x^2 - x + 2} = x^2 + 3x + \dfrac{-6x + 6}{x^2 - x + 2}$$

47. $r(r - t) + (2r - t)(2r + t)$

$= r^2 - rt + (2r)^2 - t^2$

$= r^2 - rt + 4r^2 - t^2$

$= 5r^2 - rt - t^2$

49. $6 - [x(x - 1) - 2(x + 3)]$

$= 6 - (x^2 - x - 2x - 6)$

$= 6 - (x^2 - 3x - 6)$

$= -x^2 + 3x + 12$

51. $x^2 - 7x - 18 = (x - 9)(x + 2)$

53. $64y^3 + 125$

$= (4y)^3 + 5^3$

$= (4y + 5)[(4y)^2 - (4y)(5) + 5^2]$

$= (4y + 5)(16y^2 - 20y + 25)$

55. Prime

57. $2x^2 + x - 1 = 2x^2 + 2x - x - 1$

$= 2x(x + 1) - 1(x + 1)$

$= (x + 1)(2x - 1)$

59. $4z^2(s - t) - 2z(t - s)$

$= 4z^2(s - t) - 2z[-(s - t)]$

$= 4z^2(s - t) + 2z(s - t)$

$= 2z(s - t)(2z + 1)$

61. Prime

63. $144 - x^2$

$= (12 - x)(12 + x)$

65. $30(u - v)^2 + 42(v - u)$

$= 30(u - v)^2 + 42[-(u - v)]$

$= 30(u - v)^2 - 42(u - v)$

$= 6(u - v)[5(u - v) - 7]$

$= 6(u - v)(5u - 5v - 7)$

67. $x^2 + 4$ is prime

69. $3y^2 - 11y + 8 = 3y^2 - 8y - 3y + 8$

$= y(3y - 8) + 1(3y - 8)$

$= (3y - 8)(y - 1)$

71. $a^6 - 125$

$= (a^2)^3 - 5^3$

$= (a^2 - 5)[(a^2)^2 + (a^2)(5) + 5^2]$

$= (a^2 - 5)(a^4 + 5a^2 + 25)$

73. $15 - x - 2x^2 = 15 - 6x + 5x - 2x^2$

$= 3(5 - 2x) + x(5 - 2x)$

$= (5 - 2x)(3 + x)$

75. $16x^2 + 8x + 1 = 16x^2 + 4x + 4x + 1$

$= 4x(4x + 1) + 1(4x + 1)$

$= (4x + 1)(4x + 1)$

$= (4x + 1)^2$

77. $27a^3 - 3a = 3a(9a^2 - 1)$

$= 3a[(3a)^2 - 1^2]$

$= 3a(3a - 1)(3a + 1)$

79. $x^2 + z^2 - 2xz - y^2$

$= (x^2 - 2xz + z^2) - y^2$

$= (x - z)^2 - y^2$

$= [(x - z) - y][(x - z) + y]$

$= (x - z - y)(x - z + y)$

81. height: x
    base: $2x$

$$A = \frac{1}{2}bh$$

$$= \frac{1}{2}(2x)(x)$$

$$= x^2$$

x^2 ft^2

83. $0.00098 = 9.98 \times 10^{-4}$

85. $123.456 = 1.23456 \times 10^2$

87. $1.001 \times 10^{11} = 1.00100000000 \times 10^{11}$

$$= 100,100,000,000$$

89. 2.006×10^{-7}

$$= 0.0000002.006 \times 10^{-7}$$

$$= 0.0000002006$$

91. 3 terms

93. 4 terms

95. $2x(x + 2) = 2x^2 + 4x$

97. $(x - 2)^2 = x^2 - 2(x)(2) + 2^2$

$$= x^2 - 4x + 4$$

99. $-3^2 = -1 \cdot 3 \cdot 3 = -9$

101. $(-3)^3 = -3 \cdot -3 \cdot -3 = -27$

103. $(-b)^3 = -b \cdot -b \cdot -b = -b^3$

105. $8x^3 + 125$

$$= (2x)^3 + 5^3$$

$$= (2x + 5)[(2x)^2 - (2x)(5) + 5^2]$$

$$= (2x + 5)(4x^2 - 10x + 25)$$

107. Factored

109. $a(1 - b) + (b - 1) = a(1 - b) - 1(1 - b)$

$$= (1 - b)(a - 1)$$

111. 1 term; $y + 7$ is a factor.

113. 3 terms; $y^2 + 14y + 49 = (y + 7)(y + 7)$
     $y + 7$ is a factor.

Chapter 1 Test

1. $\dfrac{-2x^{-1}}{y^{-2}} = \dfrac{-2y^2}{x}$

   A

2. $3(-1)^3 - 2(-1)^2(3)^2 = 3(-1) - 2(1)(9)$

   $$= -21$$

   D

3. $\left(\dfrac{-2a^5b}{b^4}\right)^2 = \left(\dfrac{-2a^5}{b^3}\right)^2 = \dfrac{(-2)^2(a^5)^2}{(b^3)^2} = \dfrac{4a^{10}}{b^6}$

   A

4. $\dfrac{(-3)^4}{(-3)^6} = \dfrac{1}{(-3)^2} = \dfrac{1}{9}$

   B

5. $-3^2ab^4(2ab)(-a^4) = 18a^6b^5$

   B

6. D

7. $\dfrac{6x^2 - 4x^3}{x^2} = \dfrac{6x^2}{x^2} - \dfrac{4x^3}{x^2}$

   $$= 6 - 4x$$

8. $(2x - 3)(5x + 7) = 10x^2 + 14x - 15x - 21$

   $$= 10x^2 - x - 21$$

9. $(5x^3y^2 - 4x^2y + 3xy)$

   $+ (-3x^3y^2 + x^2y - 7xy)$

   $= 2x^3y^2 - 3x^2y - 4xy$

10. $(4x - 7)^2 = (4x)^2 - 2(4x)(7) + 7^2$

    $= 16x^2 - 56x + 49$

11.
$$
\begin{array}{r}
6x^2 - 8x + 9 \\
x + 1 \overline{)6x^3 - 2x^2 + x - 2} \\
\underline{6x^3 + 6x^2} \\
-8x^2 + x - 2 \\
\underline{-8x^2 - 8x} \\
9x - 2 \\
\underline{9x + 9} \\
-11
\end{array}
$$

   $(6x^3 - 2x^2 + x - 2) \div (x + 1)$

   $= 6x^2 - 8x + 9 - \dfrac{11}{x + 1}$

12. $(5x^3 - x^2 + 4) - (x^3 - 2x - 7)$

    $= 5x^3 - x^2 + 4 - x^3 + 2x + 7$

    $= 4x^3 - x^2 + 2x + 11$

13. $7 - [2(x + 4) - (x + 2)(x - 2)]$

    $= 7 - [2x + 8 - (x^2 - 4)]$

    $= 7 - (2x + 8 - x^2 + 4)$

    $= 7 - (-x^2 + 2x + 12)$

    $= 7 + x^2 - 2x - 12$

    $= x^2 - 2x - 5$

14. $8a^3 - 27$

    $= (2a)^3 - 3^3$

    $= (2a - 3)[(2a)^2 + (2a)(3) + 3^2]$

    $= (2a - 3)(4a^2 + 6a + 9)$

15. $4x^2 + 18x - 10 = 2(2x^2 + 9x - 5)$

    $= 2(2x^2 + 10x - x - 5)$

    $= 2[2x(x + 5) - 1(x + 5)]$

    $= 2(x + 5)(2x - 1)$

16. $ax - bx + 2a - 2b = x(a - b) + 2(a - b)$

    $= (a - b)(x + 2)$

17. $x(p - q) - y(q - p)$

    $= x(p - q) - y[-(p - q)]$

    $= x(p - q) + y(p - q)$

    $= (p - q)(x + y)$

18. $a^4 - 2a^2 - 8 = (a^2 - 4)(a^2 + 2)$

    $= (a - 2)(a + 2)(a^2 + 2)$

19. $4y^2 - 25 = (2y)^2 - 5^2$

    $= (2y - 5)(2y + 5)$

20. $27a^2b^3 - 36a^3b^2 + 45a^4b^4$

    $= 9a^2b^2(3b - 4a + 5a^2b^2)$

21. $0.00001\,076 = 1.076 \times 10^{-5}$

22. $9\,009000 = 9.009 \times 10^6$

23. $5.22 \times 10^4 = 5.2200 \times 10^4$

    $= 52200$

24. $7.009 \times 10^{-4} = 0\,0007.009$

    $= 0.0007009$

CHAPTER 2

Problem Set 2.1

1. $x - 4 = 2$

   $x - 4 + 4 = 2 + 4$

   $x = 6$

   $\{6\}$

3. $2x = 10$

   $\dfrac{2x}{2} = \dfrac{10}{2}$

   $x = 5$

   $\{5\}$

5. $3x - 5 = 4$

   $3x - 5 + 5 = 4 + 5$

   $\dfrac{3x}{3} = \dfrac{9}{3}$

   $x = 3$

   $\{3\}$

7. $5 + 3x = 5x - 1$

   $3x - 5x = -1 - 5$

   $-2x = -6$

   $\dfrac{-2x}{-2} = \dfrac{-6}{-2}$

   $x = 3$

   $\{3\}$

9. $5x - 3 = 3x + 7$

   $5x - 3x = 7 + 3$

   $2x = 10$

   $\dfrac{2x}{2} = \dfrac{10}{2}$

   $x = 5$

   $\{5\}$

11. $6 + 3x = 3x + 6$

    $3x - 3x = 6 - 6$

    $0 = 0$

    $\Re$

13. $2(x + 1) = 3x - 4$

    $2x + 2 = 3x - 4$

    $2x - 3x = -4 - 2$

    $-x = -6$

    $\dfrac{-x}{-1} = \dfrac{-6}{-1}$

    $x = 6$

    $\{6\}$

15. $2(x - 6) - 3(2x - 2) = 0$

    $2x - 12 - 6x + 6 = 0$

    $-4x - 6 = 0$

    $-4x = 6$

    $\dfrac{-4x}{-4} = \dfrac{6}{-4}$

    $x = -\dfrac{3}{2}$

    $\left\{-\dfrac{3}{2}\right\}$

17. $-(1 - 2x) = 3 + 5x$

    $-1 + 2x = 3 + 5x$

    $2x - 5x = 3 + 1$

    $-3x = 4$

    $\dfrac{-3x}{-3} = \dfrac{4}{-3}$

    $x = -\dfrac{4}{3}$

    $\left\{-\dfrac{4}{3}\right\}$

19. $1 + 3x = 3(2 + x)$

$1 + 3x = 6 + 3x$

$3x - 3x = 6 - 1$

$0 = 5$

$\emptyset$

21. $2(x + 3) - 3[2(x - 3) + 4] = 0$

$2x + 6 - 3(2x - 6 + 4) = 0$

$2x + 6 - 3(2x - 2) = 0$

$2x + 6 - 6x + 6 = 0$

$-4x + 12 = 0$

$-4x = -12$

$x = 3$

23. $\dfrac{1}{2}x - 1 = \dfrac{1}{3}x$

$6\left(\dfrac{1}{2}x - 1\right) = 6\left(\dfrac{1}{3}x\right)$

$6\left(\dfrac{1}{2}x\right) - 6(1) = 2x$

$3x - 6 = 2x$

$-6 = -x$

$6 = x$

$\{6\}$

25. $\dfrac{8}{9}x - \dfrac{1}{3} = \dfrac{2}{3}x + 2$

$9\left(\dfrac{8}{9}x - \dfrac{1}{3}\right) = 9\left(\dfrac{2}{3}x + 2\right)$

$9\left(\dfrac{8}{9}x\right) - 9\left(\dfrac{1}{3}\right) = 9\left(\dfrac{2}{3}x\right) + 9(2)$

$8x - 3 = 6x + 18$

$2x = 21$

$x = \dfrac{21}{2}$

$\left\{\dfrac{21}{2}\right\}$

27. $\dfrac{x}{6} = \dfrac{5}{12}(x + 2)$

$12\left(\dfrac{x}{6}\right) = 12\left(\dfrac{5}{12}\right)(x + 2)$

$2x = 5(x + 2)$

$2x = 5x + 10$

$-3x = 10$

$x = -\dfrac{10}{3}$

$\left\{-\dfrac{10}{3}\right\}$

29. $\dfrac{x + 1}{2} = 2$

$2\left(\dfrac{x + 1}{2}\right) = 2(2)$

$x + 1 = 4$

$x = 3$

$\{3\}$

31. $\dfrac{1 - 3x}{2} - 3x = 5$

$2\left(\dfrac{1 - 3x}{2}\right) - 2(3x) = 2(5)$

$1 - 3x - 6x = 10$

$1 - 9x = 10$

$-9x = 9$

$x = -1$

$\{-1\}$

33. $\dfrac{7x - 3}{7} - \dfrac{5x + 7}{5} = 1$

$35\left(\dfrac{7x - 3}{7}\right) - 35\left(\dfrac{5x + 7}{5}\right) = 35(1)$

$5(7x - 3) - 7(5x + 7) = 35$

$$35x - 15 - 35x - 49 = 35$$
$$-64 = 35$$

$$\varnothing$$

35. $\quad x + 0.4 = 1.5$

$$x + \frac{4}{10} = \frac{15}{10}$$

$$10\left(x + \frac{4}{10}\right) = 10\left(\frac{15}{10}\right)$$

$$10x + 4 = 15$$

$$10x = 11$$

$$x = \frac{11}{10}$$

$$\{1.1\}$$

37. $\quad 1.3x - 6.1 = 2x + 5.8$

$$\frac{13}{10}x - \frac{61}{10} = 2x + \frac{58}{10}$$

$$10\left(\frac{13}{10}x - \frac{61}{10}\right) = 10\left(2x + \frac{58}{10}\right)$$

$$13x - 61 = 20x + 58$$

$$-119 = 7x$$

$$-17 = x$$

$$\{-17\}$$

39. $\quad 1.1x = 2.3(x + 2)$

$$\frac{11}{10}x = \frac{23}{10}(x + 2)$$

$$10\left(\frac{11}{10}x\right) = 10\left(\frac{23}{10}\right)(x + 2)$$

$$11x = 23(x + 2)$$

$$11x = 23x + 46$$

$$-12x = 46$$

$$x = -3.8\overline{3}$$

41. width: x

length: $x + 5$

$$2x + 2(x + 5) = 78$$
$$2x + 2x + 10 = 78$$
$$4x + 10 = 78$$
$$4x = 68$$
$$x = 17$$
$$x + 5 = 17 + 5 = 22$$

The dimensions are 17 m by 22 m.

43. 1st number: x

2nd number: $5x$

3rd number: $10 + x$

$$(x) + (5x) + (10 + x) = 24$$
$$7x + 10 = 24$$
$$7x = 14$$
$$x = 2$$
$$5x = 5(2) = 10$$
$$10 + x = 10 + 2 = 12$$

The numbers are 2, 10 and 12.

45. $x + 5 = 9$

$$x = 9 - 5$$
$$x = 4$$
$$\{4\}$$

47. $2x - 12 = 7$

$$2x = 7 + 12$$
$$2x = 19$$
$$x = \frac{19}{2}$$

$$\left\{\frac{19}{2}\right\}$$

49. $5 - 3x = 2$

$\qquad -3x = 2 - 5$

$\qquad -3x = -3$

$\qquad x = 1$

$\qquad \{1\}$

51. $5x + 8 = 11x$

$\qquad 8 = 11x - 5x$

$\qquad 8 = 6x$

$\qquad \dfrac{4}{3} = x$

$\qquad \left\{\dfrac{4}{3}\right\}$

53. $x + 7 = x + 3$

$\qquad x - x = 3 - 7$

$\qquad 0 = -4$

$\qquad \emptyset$

55. $2y + 3 = -y - 6$

$\qquad 2y + y = -6 - 3$

$\qquad 3y = -9$

$\qquad y = -3$

$\qquad \{-3\}$

57. $5 + 2x = 2x + 5$

$\qquad 2x - 2x = 5 - 5$

$\qquad 0 = 0$

$\qquad \Re$

59. $2(x + 1) = 1$

$\qquad 2x + 2 = 1$

$\qquad 2x = 1 - 2$

$\qquad 2x = -1$

$\qquad x = -\dfrac{1}{2}$

$\qquad \left\{-\dfrac{1}{2}\right\}$

61. $2(3 - x) = 5(x + 4)$

$\qquad 6 - 2x = 5x + 20$

$\qquad 6 - 20 = 5x + 2x$

$\qquad -14 = 7x$

$\qquad -2 = x$

$\qquad \{-2\}$

63. $(1 + s)3 = 7(-1 - s)$

$\qquad 3 + 3s = -7 - 7s$

$\qquad 3s + 7s = -7 - 3$

$\qquad 10s = -10$

$\qquad s = -1$

$\qquad \{-1\}$

65. $4(2x + 1) - (4 - x) = 18$

$\qquad 8x + 4 - 4 + x = 18$

$\qquad 9x = 18$

$\qquad x = 2$

$\qquad \{2\}$

67. $3(x + 1) - 3 = 2(x - 1) + (x + 2)$

$\qquad 3x + 3 - 3 = 2x - 2 + x + 2$

$\qquad 3x = 3x$

$\qquad 3x - 3x = 0$

$\qquad 0 = 0$

$\qquad \Re$

69. $2[1 - 3(x + 2)] = 0$

$\qquad 2(1 - 3x - 6) = 0$

$\qquad 2(-3x - 5) = 0$

$\qquad -6x - 10 = 0$

$\qquad -6x = 10$

47

$$x = -\frac{10}{6}$$

$$x = -\frac{5}{3}$$

$$\left\{-\frac{5}{3}\right\}$$

71. $-6\{x + 3[x + 3(2x - 5)]\} = 5 - 3(44x + 8)$

$$-6[x + 3(x + 6x - 15)] = 5 - 132x - 24$$

$$-6[x + 3(7x - 15)] = -132x - 19$$

$$-6(x + 21x - 45) = -132x - 19$$

$$-6(22x - 45) = -132x - 19$$

$$-132x + 270 = -132x - 19$$

$$-132x + 132x = -19 - 270$$

$$0 = -289$$

$$\emptyset$$

73. $\qquad \frac{1}{2}x - \frac{1}{4} = \frac{1}{3}x$

$$12\left(\frac{1}{2}x - \frac{1}{4}\right) = 12\left(\frac{1}{3}x\right)$$

$$12\left(\frac{1}{2}x\right) - 12\left(\frac{1}{4}\right) = 4x$$

$$6x - 3 = 4x$$

$$-3 = -2x$$

$$\frac{3}{2} = x$$

$$\left\{\frac{3}{2}\right\}$$

75. $\qquad \frac{1}{6} - \frac{1}{3}x = \frac{1}{9}(2x + 1)$

$$18\left(\frac{1}{6} - \frac{1}{3}x\right) = 18\left(\frac{1}{9}\right)(2x + 1)$$

$$18\left(\frac{1}{6}\right) - 18\left(\frac{1}{3}x\right) = 2(2x + 1)$$

$$3 - 6x = 4x + 2$$

$$-6x - 4x = 2 - 3$$

$$-10x = -1$$

$$x = \frac{1}{10}$$

$$\left\{\frac{1}{10}\right\}$$

77. $\qquad \frac{2x}{7} + \frac{x}{2} = 11$

$$14\left(\frac{2x}{7} + \frac{x}{2}\right) = 14(11)$$

$$14\left(\frac{2x}{7}\right) + 14\left(\frac{x}{2}\right) = 154$$

$$2(2x) + 7x = 154$$

$$4x + 7x = 154$$

$$11x = 154$$

$$x = 14$$

$$\{14\}$$

79. $\qquad \frac{3}{4}(x - 3) = \frac{5}{8}x$

$$8\left(\frac{3}{4}\right)(x - 3) = 8\left(\frac{5}{8}x\right)$$

$$6(x - 3) = 5x$$

$$6x - 18 = 5x$$

$$-18 = -x$$

$$18 = x$$

$$\{18\}$$

81. $\qquad \frac{x - 2}{3} = 4$

$$3\left(\frac{x - 2}{3}\right) = 3(4)$$

$$x - 2 = 12$$

$$x = 14$$

$$\{14\}$$

48

83.
$$\frac{x+3}{5} - \frac{2-x}{10} = 7$$

$$10\left(\frac{x+3}{5} - \frac{2-x}{10}\right) = 10(7)$$

$$10\left(\frac{x+3}{5}\right) - 10\left(\frac{2-x}{10}\right) = 70$$

$$2(x+3) - (2-x) = 70$$

$$2x + 6 - 2 + x = 70$$

$$3x + 4 = 70$$

$$3x = 66$$

$$x = 22$$

$$\{22\}$$

85.
$$\frac{3t-7}{7} = \frac{t+1}{2}$$

$$14\left(\frac{3t-7}{7}\right) = 14\left(\frac{t+1}{2}\right)$$

$$2(3t-7) = 7(t+1)$$

$$6t - 14 = 7t + 7$$

$$-21 = t$$

$$\{-21\}$$

87.
$$\frac{5x-10}{3} + \frac{x-8}{2} = -3$$

$$6\left(\frac{5x-10}{3}\right) + 6\left(\frac{x-8}{2}\right) = 6(-3)$$

$$2(5x-10) + 3(x-8) = -18$$

$$10x - 20 + 3x - 24 = -18$$

$$13x - 44 = -18$$

$$13x = 26$$

$$x = 2$$

$$\{2\}$$

89.
$$x + 2.3 = 1.1$$

$$x + \frac{23}{10} = \frac{11}{10}$$

$$10\left(x + \frac{23}{10}\right) = 10\left(\frac{11}{10}\right)$$

$$10x + 23 = 11$$

$$10x = -12$$

$$x = -1.2$$

$$\{-1.2\}$$

91. $3x + 6.03 = 3.42$

$$3x = -2.61$$

$$x = -0.87$$

$$\{-0.87\}$$

93.
$$1.4x - 0.5 = 1.1x + 0.5$$

$$\frac{14}{10}x - \frac{5}{10} = \frac{11}{10}x + \frac{5}{10}$$

$$10\left(\frac{14}{10}x - \frac{5}{10}\right) = 10\left(\frac{11}{10}x + \frac{5}{10}\right)$$

$$10\left(\frac{14}{10}x\right) - 10\left(\frac{5}{10}\right) = 10\left(\frac{11}{10}x\right) + 10\left(\frac{5}{10}\right)$$

$$14x - 5 = 11x + 5$$

$$3x = 10$$

$$x = 3.\overline{3}$$

$$\{3.\overline{3}\}$$

95. $1.5(x + 2) = 2.8 + x$

$$\frac{15}{10}(x + 2) = \frac{28}{10} + x$$

$$10\left(\frac{15}{10}\right)(x + 2) = 10\left(\frac{28}{10} + x\right)$$

$$15(x + 2) = 10\left(\frac{28}{10}\right) + 10x$$

$$15x + 30 = 28 + 10x$$

$$5x = -2$$

$$x = -0.4$$

$$\{-0.4\}$$

97. Number of Madonna shows: x

Number of Jackson shows: $1 + 2x$

$$1 + 2x = x + 38$$

$$x = 37$$

$$1 + 2x = 1 + 2(37) = 75$$

Janet Jackson performed 75 shows and Madonna performed 37 shows.

99. Number completed: x

Number attempted: $2x - 10$

$$2x - 10 = 36$$

$$2x = 46$$

$$x = 23$$

He completed 23 passes.

101. 1st angle = 2nd angle = 3rd angle = x

$$x + x + x = 180$$

$$3x = 180$$

$$x = 60$$

Each angle measures 60°.

103. 1st angle: x

2nd angle: $4x$

$$x + 4x = 180$$

$$5x = 180$$

$$x = 36$$

$$4x = 144$$

The angles measure 36° and 144°.

105. $2x - a = 4b$

$$2x = a + 4b$$

$$x = \frac{a + 4b}{2}$$

$$\left\{\frac{a + 4b}{2}\right\}$$

Problem Set 2.2

1. $C = 2\pi r$

$$\frac{1}{2\pi}(C) = \frac{1}{2\pi}(2\pi r)$$

$$\frac{C}{2\pi} = r$$

3. $x + a = 5$

$$x = 5 - a$$

5. $F = \frac{9}{5}C + 32$

$$F - 32 = \frac{9}{5}C$$

$$\frac{5}{9}(F - 32) = C$$

7. $5ax - 3 = 2ax$

$$-3 = -3ax$$

$$\frac{1}{a} = x$$

9. $a(x - b) = b(x + b)$

$ax - ab = bx + b^2$

$ax - bx = ab + b^2$

$x(a - b) = ab + b^2$

$x = \dfrac{ab + b^2}{a - b}$

11. $\dfrac{x}{2} = \dfrac{x - a}{3}$

$6\left(\dfrac{x}{2}\right) = 6\left(\dfrac{x - a}{3}\right)$

$3x = 2(x - a)$

$3x = 2x - 2a$

$x = -2a$

13. $A = lw$

$\dfrac{1}{w}(A) = \dfrac{1}{w}(lw)$

$\dfrac{A}{w} = l$

15. $V = lwh$

$\dfrac{1}{lh}(V) = \dfrac{1}{lh}(lwh)$

$\dfrac{V}{lh} = w$

17. $I = Prt$

$\dfrac{1}{Pr}(I) = \dfrac{1}{Pr}(Prt)$

$\dfrac{I}{Pr} = t$

19. $C = \pi d$

$\dfrac{1}{\pi}(C) = \dfrac{1}{\pi}(\pi d)$

$\dfrac{C}{\pi} = d$

21. $A = P + Prt$

$A = P(1 + rt)$

$\left(\dfrac{1}{1 + rt}\right)A = \left(\dfrac{1}{1 + rt}\right)P(1 + rt)$

$\dfrac{A}{1 + rt} = P$

23. $P = 2l + 2w$

$P - 2w = 2l$

$\dfrac{1}{2}(P - 2w) = \dfrac{1}{2}(2l)$

$\dfrac{P - 2w}{2} = l$

25. $S = 2\pi rh$

$\dfrac{1}{2\pi r}(S) = \dfrac{1}{2\pi r}(2\pi rh)$

$\dfrac{S}{2\pi r} = h$

27. $S = \dfrac{n}{2}[2a + (n - 1)d]$

$2S = n[2a + (n - 1)d]$

$2S = 2an + n(n - 1)d$

$2S = 2an + n^2d - nd$

$2S - n^2d + nd = 2an$

$\dfrac{1}{2n}(2S - n^2d + nd) = \dfrac{1}{2n}(2an)$

$\dfrac{2S - n^2d + nd}{2n} = a$

29. $A = wl$

       $46.7 = 6.6l$

       $7.0\overline{75} = l$

       $7.0\overline{75}$ ft

31. $r = \dfrac{D}{t}$

      $r = \dfrac{10}{55}$

      $r = \dfrac{2}{11}$

      $\dfrac{2}{11}$ km/min

33. $P = 2l + 2w$

    $110\dfrac{1}{4} = 2l + 2\left(27\dfrac{1}{2}\right)$

    $110\dfrac{1}{4} = 2l + 55$

      $55\dfrac{1}{4} = 2l$

      $\dfrac{221}{4} = 2l$

      $\dfrac{221}{8} = l$

      $27\dfrac{5}{8}$ ft

35. a) $F = 32°$

      $C = \dfrac{5}{9}(F - 32)$

        $= \dfrac{5}{9}(32 - 32)$

        $= \dfrac{5}{9}(0)$

        $= 0$

      $C = 0°$

b) $C = 5°$

      $F = \dfrac{9}{5}C + 32$

        $= \dfrac{9}{5}(5) + 32$

        $= 9 + 32$

        $= 41$

      $F = 41°$

c) $F = -31°$

      $C = \dfrac{5}{9}(F - 32)$

        $= \dfrac{5}{9}(-31 - 32)$

        $= \dfrac{5}{9}(-63)$

        $= -35$

      $C = -35°$

d) $C = 20°$

      $F = \dfrac{9}{5}C + 32$

        $= \dfrac{9}{5}(20) + 32$

        $= 36 + 32$

        $= 68$

      $F = 68°$

e) $C = -10°$

      $F = \dfrac{9}{5}C + 32$

        $= \dfrac{9}{5}(-10) + 32$

        $= -18 + 32$

        $= 14°$

      $F = 14°$

f) $F = 77°$

$$C = \frac{5}{9}(F - 32)$$

$$= \frac{5}{9}(77 - 32)$$

$$= \frac{5}{9}(45)$$

$$= 25$$

$$C = 25°$$

37. a)
$$S = 2wh + 2lh + 2wl$$
$$1152 = 2(12)h + 2(24)h + 2(12)(24)$$
$$1152 = 24h + 48h + 576$$
$$576 = 72h$$
$$8 = h$$
8 inches

b)
$$S = 2wh + 2lh + 2wl$$
$$1440 = 2(12)h + 2(24)h + 2(12)(24)$$
$$1440 = 24h + 48h + 576$$
$$864 = 72h$$
$$12 = h$$
12 inches

c)
$$S = 2wh + 2lh + 2wl$$
$$1512 = 2(12)h + 2(24)h + 2(12)(24)$$
$$1512 = 24h + 48h + 576$$
$$936 = 72h$$
$$13 = h$$
13 inches

39. $3x + 5 = a$
$$3x = a - 5$$
$$x = \frac{a - 5}{3}$$

41. $ax + ab = b$
$$ax = b - ab$$
$$x = \frac{b - ab}{a}$$

43. $\dfrac{x - a}{4} = \dfrac{3}{2}$
$$4\left(\frac{x - a}{4}\right) = 4\left(\frac{3}{2}\right)$$
$$x - a = 6$$
$$x = a + 6$$

45. $3(5 + ax) = 1$
$$15 + 3ax = 1$$
$$3ax = -14$$
$$x = -\frac{14}{3a}$$

47. $ax - 2 = bx$
$$ax - bx = 2$$
$$x(a - b) = 2$$
$$x = \frac{2}{a - b}$$

49. $2b - (x - a) = a(x - 2)$
$$2b - x + a = ax - 2a$$
$$2b + 3a = ax + x$$
$$2b + 3a = x(a + 1)$$
$$\frac{2b + 3a}{a + 1} = x$$

51. $bx + 4 - a^2 = ax$
$$4 - a^2 = ax - bx$$
$$4 - a^2 = x(a - b)$$
$$\frac{4 - a^2}{a - b} = x$$
$$\text{or} \quad \frac{a^2 - 4}{b - a} = x$$

53. $(x - b)(x + a) = x^2 + ab$

$$x^2 + ax - bx - ab = x^2 + ab$$

$$ax - bx = 2ab$$

$$x(a - b) = 2ab$$

$$x = \frac{2ab}{a - b}$$

55. $\frac{2}{3}bx + a + b = \frac{2}{5}x$

$$15\left(\frac{2}{3}bx + a + b\right) = 15\left(\frac{2}{5}x\right)$$

$$10bx + 15a + 15b = 6x$$

$$15a + 15b = 6x - 10bx$$

$$15a + 15b = x(6 - 10b)$$

$$\frac{15a + 15b}{6 - 10b} = x$$

57. $\frac{5}{6}(bx + 2a) = \frac{1}{6}ax + b^2$

$$6\left(\frac{5}{6}\right)(bx + 2a) = 6\left(\frac{1}{6}ax + b^2\right)$$

$$5bx + 10a = ax + 6b^2$$

$$10a - 6b^2 = ax - 5bx$$

$$10a - 6b^2 = x(a - 5b)$$

$$\frac{10a - 6b^2}{a - 5b} = x$$

59. $\frac{x + 2}{x - a} = \frac{x + a}{x - 1}$

$$(x - a)(x - 1)\left(\frac{x + 2}{x - a}\right) = (x - a)(x - 1)\left(\frac{x + a}{x - 1}\right)$$

$$(x - 1)(x + 2) = (x - a)(x + a)$$

$$x^2 + x - 2 = x^2 - a^2$$

$$x - 2 = -a^2$$

$$x = 2 - a^2; \quad x \neq a, \quad x \neq 1$$

Problem Set 2.3

1. $x - 3 \leq 4$

$$x \leq 7$$

$$\{x \mid x \leq 7\}$$

7

3. $x - 7 \geq 3$

$$x \geq 10$$

$$\{x \mid x \geq 10\}$$

10

5. $2x + 1 < -5$

$$2x < -6$$

$$x < -3$$

$$\{x \mid x < -3\}$$

−3

7. $-4x \geq 2$

$$x \leq -\frac{1}{2}$$

$$\left\{x \mid x \leq -\frac{1}{2}\right\}$$

9. $6(x - 2) \geq 2(x - 1)$

$$6x - 12 \geq 2x - 2$$

$$4x \geq 10$$

$$x \geq \frac{5}{2}$$

$$\left\{x \mid x \geq \frac{5}{2}\right\}$$

54

11. $1.1x - 0.5 > 0.7x$

$\qquad -0.5 > -0.4x$

$\qquad 1.25 < x$

$\qquad \{x \mid x > 1.25\}$

13. $\dfrac{2}{3}(2x - 1) < \dfrac{1}{6}x + 1$

$6\left(\dfrac{2}{3}\right)(2x - 1) < 6\left(\dfrac{1}{6}x + 1\right)$

$\qquad 4(2x - 1) < x + 6$

$\qquad 8x - 4 < x + 6$

$\qquad 7x < 10$

$\qquad x < \dfrac{10}{7}$

$\qquad \left\{x \mid x < \dfrac{10}{7}\right\}$

15. $x + 3 \le x + 4$

$\qquad 3 \le 4 \quad$ (Always true)

$\qquad \{x \mid x \text{ is a real number}\}$

17. $\{x \mid 1 < x < 2\}$

$\quad$ -1 $\qquad$ 2

19. $\{x \mid -1 < x \le 1\}$

$\quad$ -1 $\qquad$ 1

21. $\{x \mid x \le -7 \text{ or } x > -2\}$

$\quad$ -7 $\;$ -2

23. $2 < x + 3 \le 7$

$\qquad -1 < x \le 4$

$\quad$ -1 $\qquad$ 4

25. $1 < 2 - x < 5$

$\qquad -1 < -x < 3$

$\qquad 1 > x > -3$

$\qquad -3 < x < 1$

$\quad$ -3 $\qquad$ 1

27. $\{x \mid 2 < x < 3\}$

29. $\left\{x \mid -\dfrac{1}{2} < x \le 0\right\}$

31. $\{x \mid x < 5 \text{ or } x > 10\}$

33. $\{x \mid x > 5\}$

35. number: x

$\qquad 2x + 1 \le 7$

$\qquad 2x \le 6$

$\qquad x \le 3$

All numbers 3 or less.

37. width: w

length: $w + 4$

$\qquad 2w + 2(w + 4) \le 72$

$\qquad 2w + 2w + 8 \le 72$

$\qquad 4w + 8 \le 72$

$\qquad 4w \le 64$

$\qquad w \le 16$

The largest possible width is 16 ft.

39. $2 - x \geq 1$

$-x \geq -1$

$x \leq 1$

1

41. $4x < 2$

$x < \dfrac{2}{4}$

$x < \dfrac{1}{2}$

$\dfrac{1}{2}$

43. $\dfrac{1}{2}x > -5$

$2\left(\dfrac{1}{2}x\right) > 2(-5)$

$x > -10$

-10

45. $6x + 3 \leq 2x - 7$

$4x \leq -10$

$x \leq -\dfrac{10}{4}$

$x \leq -\dfrac{5}{2}$

$-\dfrac{5}{2}$

47. $\dfrac{1}{5}x - \dfrac{1}{3} \leq \dfrac{1}{5} + 2x$

$15\left(\dfrac{1}{5}x - \dfrac{1}{3}\right) \leq 15\left(\dfrac{1}{5} + 2x\right)$

$3x - 5 \leq 3 + 30x$

$-27x \leq 8$

$x \geq -\dfrac{8}{27}$

$-\dfrac{8}{27}$

49. $5(x + 1) < 2(1 - x)$

$5x + 5 < 2 - 2x$

$7x < -3$

$x < -\dfrac{3}{7}$

$-\dfrac{3}{7}$

51. $\dfrac{1}{3}(x + 2) > \dfrac{1}{4}(x - 2)$

$12\left(\dfrac{1}{3}\right)(x + 2) > 12\left(\dfrac{1}{4}\right)(x - 2)$

$4(x + 2) > 3(x - 2)$

$4x + 8 > 3x - 6$

$x > -14$

-14

53. $\dfrac{x}{7} - \dfrac{1}{3} < \dfrac{x}{3}$

$21\left(\dfrac{x}{7} - \dfrac{1}{3}\right) < 21\left(\dfrac{x}{3}\right)$

$3x - 7 < 7x$

$-7 < 4x$

$-\dfrac{7}{4} < x$

$-\dfrac{7}{4}$

55. $\dfrac{x - 3}{6} \le \dfrac{x}{6} - 1$

$6\left(\dfrac{x - 3}{6}\right) \le 6\left(\dfrac{x}{6} - 1\right)$

$x - 3 \le x - 6$

$-3 \le -6$

$\emptyset$

57. $\dfrac{1 - x}{3} > \dfrac{2}{3}(5 - x)$

$3\left(\dfrac{1 - x}{3}\right) > 3\left(\dfrac{2}{3}\right)(5 - x)$

$1 - x > 10 - 2x$

$x > 9$

9

59. $6(y + 3) > 2(y - 1)$

$6y + 18 > 2y - 2$

$4y > -20$

$y > -5$

-5

61. $5(x - 3) + 3 \le -2(2x - 4)$

$5x - 15 + 3 \le -4x + 8$

$5x - 12 \le -4x + 8$

$9x \le 20$

$x \le \dfrac{20}{9}$

$\dfrac{20}{9}$

63. $3[5 - (s + 2)] - 2s \ge 6s - 7$

$3(5 - s - 2) - 2s \ge 6s - 7$

$3(3 - s) - 2s \ge 6s - 7$

$9 - 3s - 2s \ge 6s - 7$

$9 - 5s \ge 6s - 7$

$16 \ge 11s$

$\dfrac{16}{11} \ge s$

$\dfrac{16}{11}$

65. $\dfrac{3}{4}(y + 1) < \dfrac{1}{8}y$

$8\left(\dfrac{3}{4}\right)(y + 1) < 8\left(\dfrac{1}{8}y\right)$

$6y + 6 < y$

$6 < -5y$

$-\dfrac{6}{5} > y$

$-\dfrac{6}{5}$

67.

$$\frac{x}{2} - \frac{1}{7} > \frac{1}{4}(2x - 9)$$

$$28\left(\frac{x}{2} - \frac{1}{7}\right) > 28\left(\frac{1}{4}\right)(2x - 9)$$

$$14x - 4 > 14x - 63$$

$$0 > -59$$

$$\longleftrightarrow$$
$$\Re$$

69.

$$\frac{2(x - 7)}{3} - \frac{3(x - 3)}{2} \geq 1 - x$$

$$6\left[\frac{2(x - 7)}{3} - \frac{3(x - 3)}{2}\right] \geq 6(1 - x)$$

$$4(x - 7) - 9(x - 3) \geq 6 - 6x$$

$$4x - 28 - 9x + 27 \geq 6 - 6x$$

$$-5x - 1 \geq 6 - 6x$$

$$x \geq 7$$

$$\xleftarrow{\qquad\;\;\mathsf{E}\quad\longrightarrow}$$
$$7$$

71.

$$0 < \frac{x - 7}{4}$$

$$4(0) < 4\left(\frac{x - 7}{4}\right)$$

$$0 < x - 7$$

$$7 < x$$

$$\xrightarrow{\quad\longleftarrow\quad\longrightarrow}$$
$$7$$

73.

$$(-5) < \frac{3(r - 1)}{2}$$

$$2(-5) < 2\left[\frac{3(r - 1)}{-2}\right]$$

$$-10 < -3(r - 1)$$

$$-10 < -3r + 3$$

$$-13 < -3r$$

$$\frac{13}{3} > r$$

$$\longleftrightarrow$$
$$\frac{13}{3}$$

75.

$$0.4x - 20.8 \leq 8.0 - 0.8x$$

$$10(0.4x - 20.8) \leq 10(8.0 - 0.8x)$$

$$4x - 208 \leq 80 - 8x$$

$$12x \leq 288$$

$$x \leq 24$$

$$\longleftarrow\!\!\mathsf{I}$$
$$24$$

77.

$$0.5 - 10.3x > 4.0x - 28.1$$

$$10(0.5 - 10.3x) > 10(4.0x - 28.1)$$

$$5 - 103x > 40x - 281$$

$$286 > 143x$$

$$2 > x$$

$$\longleftrightarrow$$
$$2$$

79. $2x + 5 > 2(x + 3)$

$$2x + 5 > 2x + 6$$

$$0 > 1$$

$$\emptyset$$

81. $3(1 - x) \leq 3(x + 1) - 6x$

$$3 - 3x \leq 3x + 3 - 6x$$

$$3 - 3x \leq -3x + 3$$

$$0 \leq 0$$

$$\longleftrightarrow$$
$$\Re$$

83. $\{x \mid 2 < x < 3\}$

    2   3

85. $\{x \mid x < 6 \text{ or } x > 11\}$

    6   11

87. $\{x \mid x \le -3 \text{ or } x > 3\}$

   -3   3

89. $\{x \mid 0.5 < x \le 1.5\}$

   0.5 1.5

91. Nonsense; a number cannot be both greater than -3 and less than -5.

93. Okay

95. Okay

97. $\{x \mid x < 0\}$

99. $\{x \mid 4 \le x \le 6\}$

101. $\{x \mid -11 < x < -8\}$

103. $\{x \mid x \le -3 \text{ or } x > 3\}$

105. $3 \le 2x + 1 \le 5$

    $2 \le 2x \le 4$

    $1 \le x \le 2$

    $\{x \mid 1 \le x \le 2\}$

107. $4 \le 1 - x \le 10$

    $3 \le -x \le 9$

   $-3 \ge x \ge -9$

    $\{x \mid -9 \le x \le -3\}$

109. $5 < \dfrac{1}{2}x - 2 \le 8$

    $7 < \dfrac{1}{2}x \le 10$

    $14 < x \le 20$

    $\{x \mid 14 < x \le 20\}$

111. $\quad \dfrac{1}{2} < 2x - 3 < 3$

   $2\left(\dfrac{1}{2}\right) < 2(2x - 3) < 2(3)$

      $1 < 4x - 6 < 6$

      $7 < 4x < 12$

      $\dfrac{7}{4} < x < 3$

      $\left\{x \mid \dfrac{7}{4} < x < 3\right\}$

113. Number: x

    $4x - 2 \le 10$

      $4x \le 12$

       $x \le 3$

The number can be at most 3.

115. 4th grade: x

   $\dfrac{85 + 65 + 73 + x}{4} \ge 80$

     $\dfrac{223 + x}{4} \ge 80$

   $4\left(\dfrac{223 + x}{4}\right) \ge 4(80)$

    $223 + x \ge 320$

       $x \ge 97$

She must score at least 97 on the fourth test.

117. time in hours: x

$$600x \geq 450$$

$$x \geq \frac{450}{600}$$

$$x \geq \frac{3}{4}$$

$\frac{3}{4}$ hour $= \frac{3}{4}(60) = 45$ minutes

It would take at least 45 minutes on the rowing machine.

119. number of games lost: x

number of games won: $2x$

$$x + 2x = 9$$

$$3x = 9$$

$$x = 3$$

$$2x = 6$$

number of additional games to win: y

$$6 + y > \frac{1}{2}(18)$$

$$6 + y > 9$$

$$y > 3$$

They must win at least 4 more games.

121. $x - 5 < b$

$$x < 5 + b$$

123. $3x - 2a > b$

$$3x > 2a + b$$

$$x > \frac{2a + b}{3}$$

125. $ax < 5;$ $a > 0$

$$x < \frac{5}{a}$$

127. $x \leq a$ or $x \geq b;$ $a < b$

$$\{x \mid x \leq a \text{ or } x \geq b\}$$

129. $1 < ax < 3;$ $a < 0$

$$\frac{1}{a} > x > \frac{3}{a}$$

131. $x < -1$ or $1 < x < 2$

Problem Set 2.4

1. $I = PRT$

$$300 = 1000R(3)$$

$$300 = 3000R$$

$$0.1 = R$$

The rate was 10%.

3. $A = P\left(1 + \dfrac{r}{n}\right)^{nt}$

$$= 10000\left(1 + \frac{0.065}{4}\right)^{4(2)}$$

$$= 11376.39$$

He will have $11,376.39.

5.   amount at 5.5%: x

amount at 6.5%: $25000 - x$

$$0.055x + 0.065(25000 - x) = 1470$$

$$1000[0.055x + 0.065(25000 - x)] = 1000(1470)$$

$$55x + 65(25000 - x) = 1470000$$

$$55x + 1625000 - 65x = 1470000$$

$$-10x = -155000$$

$$x = 15500$$

$$25000 - x = 9500$$

$15,500 was invested at 5.5% and $9,500 was invested at 6.5%.

7. amount at 5.25%: x

$$0.045(6500) + 0.0525x = 0.05(6500 + x)$$
$$292.5 + 0.0525x = 325 + 0.05x$$
$$0.0025x = 32.5$$
$$x = 13000$$

$13,000 should be invested at 5.25%.

9. original length: x

original area: x^2

new length: $(x + 4)$

new area: $(x + 4)^2$

$$(x + 4)^2 = x^2 + 40$$
$$x^2 + 8x + 16 = x^2 + 40$$
$$8x = 24$$
$$x = 3$$

The original length is 3 cm.

11. original width: x

original length: $2x$

original area: $2x^2$

new width: $x - 2$

new length: $2x - 2$

new area: $(x - 2)(2x - 2)$

$$(x - 2)(2x - 2) = 2x^2 - 38$$
$$2x^2 - 6x + 4 = 2x^2 - 38$$
$$-6x = -42$$
$$x = 7$$
$$2x = 14$$

The original dimensions are 7 cm by 14 cm.

13.

$$A = P\left(1 + \frac{r}{n}\right)^{nt}$$

$$2000 = 1000\left(1 + \frac{0.06}{12}\right)^{12t}$$

$$2000 = 1000(1.005)^{12t}$$

t	$1000(1.005)^{12t}$
1	1062
2	1127
5	1349
10	1819
11	1932
12	2051

It will take between 11 and 12 years.

15. 19; since each place increases by 4.

17. 243; since each place is tripled.

19. $2(4) - 1 = 7$

21.

Day	Number
1	$2(1000) = 2000$
2	$2 \cdot 2(1000) = 2^2 \cdot 1000 = 4000$
3	$2 \cdot 2^2 \cdot 1000 = 2^3 \cdot 1000 = 8000$
4	$2 \cdot 2^3 \cdot 1000 = 2^4 \cdot 1000 = 16000$
5	$2 \cdot 2^4 \cdot 1000 = 2^5 \cdot 1000 = 32000$
t	$2^t \cdot 1000$

23. (a) $A = P + PRT$

$$= 1000 + 1000(0.07)(25)$$
$$= 2750$$

$2,750

(b) $A = P\left(1 + \dfrac{r}{n}\right)^{nt}$

$= 1000\left(1 + \dfrac{0.07}{1}\right)^{1(25)}$

$= 5427.43$

$\$5{,}427.43$

(c) $A = P + \left(1 + \dfrac{r}{n}\right)^{nt}$

$= 1000\left(1 + \dfrac{0.07}{4}\right)^{4(25)}$

$= 5668.16$

$\$5{,}668.16$

25. (a) 8th: $I = PRT$

$= (10500)(0.05)(1)$

$= 525$

$A = P + I$

$= 10500 + 525$

$= 11025$

9th: $I = PRT$

$= (11025)(0.05)(1)$

$= 551.25$

$A = P + I$

$= 11025 + 551.25$

$= 11576.25$

10th: $I = PRT$

$= (11576.25)(0.05)(1)$

$= 578.81$

$A = P + I$

$= 11576.25 + 578.81$

$= 12155.06$

11th: $I = PRT$

$= (12155.06)(0.05)(1)$

$= 607.75$

$A = P + I$

$= 12155.06 + 607.75$

$= 12762.81$

(b) $A = P\left(1 + \dfrac{r}{n}\right)^{nt}$

$= 10000\left(1 + \dfrac{0.05}{1}\right)^{1(5)}$

$= 12762.81$

$\$12{,}762.81$

(c) $A = P\left(1 + \dfrac{r}{n}\right)^{nt}$

$= 10000\left(1 + \dfrac{0.05}{1}\right)^{1(10)}$

$= 16288.95$

$\$16{,}288.95$

27. amount at 5.5%: x

amount at 6.5%: $12000 - x$

$0.055x + 0.065(12000 - x) = 700$

$0.055x + 780 - 0.065x = 700$

$-0.01x = -80$

$x = 8000$

$12000 - x = 4000$

He invested $8,000 at 5.5% and $4,000 at o.5%.

29. amount at 10%: x

amount at 8% : $100000 - x$

$0.10x + 0.08(100000 - x) = 9400$

$0.10x + 8000 - 0.08x = 9400$

$$0.02x = 1400$$
$$x = 70000$$
$$100000 - 70000 = 30000$$

$70,000 was invested at 10% and $30,000 was invested at 8%.

31. amount in AAAA: x

amount in AA: $100000 - x$

$$0.0425x + 0.0525(100000 - x) = 4500$$
$$0.0425x + 5250 - 0.0525x = 4500$$
$$-0.01x = -750$$
$$x = 75000$$
$$100000 - 75000 = 25000$$

$75,000 should be in AAAA and $25,000 in AA.

33. 1st deal: x

2nd deal: $240000 - x$

$$0.20x - 0.05(240000 - x) = 0.1375(240000)$$
$$0.20x - 12000 + 0.05x = 33000$$
$$0.25x = 45000$$
$$x = 180000$$
$$240000 - 180000 = 60000$$

He invested $180,000 in the 1st deal that made a profit and $60,000 in the 2nd deal that had a loss.

35. original length of side: x
original area: x^2
new length: $x + 2$
new area: $(x + 2)^2$

$$(x + 2)^2 = x^2 + 48$$
$$x^2 + 4x + 4 = x^2 + 48$$
$$4x = 44$$
$$x = 11$$

The side is 11 in.

37. number of tickets given away: x

number of Bonnie's tickets: $\frac{1}{2}x$

number of Jackie's tickets: $\frac{1}{5}x$

$$\frac{1}{2}x = 3 + \frac{1}{5}x$$
$$10\left(\frac{1}{2}x\right) = 10\left(3 + \frac{1}{5}x\right)$$
$$5x = 30 + 2x$$
$$3x = 30$$
$$x = 10$$

He gave away 10 tickets.

39. number: x

$$\frac{1}{2}x + \frac{2}{3}x = 3 + x$$
$$6\left(\frac{1}{2}x + \frac{2}{3}x\right) = 6(3 + x)$$
$$3x + 4x = 18 + 6x$$
$$7x = 18 + 6x$$
$$x = 18$$

The number is 18.

41. number: x

$$\frac{1}{3}x - \frac{1}{5}x = 1 + \frac{1}{10}x$$
$$30\left(\frac{1}{3}x - \frac{1}{5}x\right) = 30\left(1 + \frac{1}{10}x\right)$$
$$10x - 6x = 30 + 3x$$
$$4x = 30 + 3x$$
$$x = 30$$

The number is 30.

43. Adding odd numbers:

$$1 + 3 + 5 + 7 + 9 = 25$$
$$1 + 3 + 5 + 7 + 9 + 11 = 36$$

45.

t	$50000 \cdot 3^{-0.05t}$
1	47328
10	28868
12	25864
13	24482
29	10166
30	9623

Between 12 and 13 years; between 29 and 30 years.

47. (a) 1992:

$$5.4 + (0.015)(5.4) = 5.4(1 + 0.015)$$
$$= 5.4(1.015) \approx 5.48 \text{ billion}$$

1993: $5.4(1.015) + (0.015)[5.4(1.015)] = (5.4)(1.015)(1 + 0.015)$
$$= (5.4)(1.015)(1.015)$$
$$= 5.4(1.015)^2 \approx 5.56 \text{ billion}$$

1994: $5.4(1.015)^2 + (0.015)[5.4(1.015)^2] = 5.4(1.015)^2(1 + 0.015)$
$$= 5.4(1.015)^2(1.015)$$
$$= 5.4(1.015)^3 \approx 5.65 \text{ billion}$$

1995: $5.4(1.015)^3 + (0.015)[5.4(1.015)^3] = 5.4(1.015)^3(1 + 0.015)$
$$= 5.4(1.015)^3(1.015)$$
$$= 5.4(1.015)^4 \approx 5.73 \text{ billion}$$

(b) $5.4(1.015)^t$

49. (a) 1: Population: $100000 + 0.05(100000) = 100000(1 + 0.05) = 100000(1.05) = 105,000$

Food-Prod: $200000 + 50000 = 250000$

2: Population: $100000(1.05) + 0.05[100000(1.05)]$

$$= 100000(1.05)[1 + 0.05]$$

$$= 100000(1.05)(1.05)$$

$$= 100000(1.05)^2 = 110,250$$

Food-Prod: $250000 + 50000 = 300,000$

3: Population: $100000(1.05)^3 = 115,762$

Food-Prod: $300000 + 50000 = 350,000$

4: Population: $100000(1.05)^4 = 121,550$

Food-Prod: $350000 + 50000 = 400,000$

5: Population: $100000(1.05)^5 = 127,628$

Food-Prod: $400000 + 50000 = 450,000$

6: Population: $100000(1.05)^6 = 134,009$

Food-Prod: $450000 + 50000 = 500,000$

t: Population: $100000(1.05)^t$

Food-Prod: $200000 + 50000t$

(b)

t	$100000(1.05)^t$	$200000 + 50000t$
75	3,883,268	3,950,000
76	4,077,432	4,000,000

It will happen between 75 and 76 years after 1992 which is between the years 2067 and 2068.

1.

$$\boxed{x \text{ gal}} + \boxed{30 \text{ gal}} = \boxed{(x + 30) \text{ gal}}$$

   60%   10%   40%

$$x \cdot \frac{60}{100} + 30 \cdot \frac{10}{100} = (x + 30) \cdot \frac{40}{100}$$

$$60x + 300 = 40(x + 30)$$

$$60x + 300 = 40x + 1200$$

$$20x = 900$$

$$x = 45$$

45 gal

3.

$$\boxed{10 \text{ gal}} + \boxed{x \text{ gal}} = \boxed{(10 + x) \text{ gal}}$$

   30%   100%   40%

$$10 \cdot \frac{30}{100} + x \cdot \frac{100}{100} = (10 + x) \cdot \frac{40}{100}$$

$$300 + 100x = 40(10 + x)$$

$$300 + 100x = 400 + 40x$$

$$60x = 100$$

$$x = \frac{5}{3}$$

$\frac{5}{3}$ **gal**

5. pipe A rate: $\frac{1}{2}$

   pipe B rate: $\frac{1}{4}$

   time together: x

$$\frac{1}{2}x + \frac{1}{4}x = 1$$

$$8\left(\frac{1}{2}x + \frac{1}{4}x\right) = 8(1)$$

$$4x + 2x = 8$$

$$6x = 8$$

$$x = \frac{4}{3}$$

$\frac{4}{3}$ hr.

7. Rowland's rate: $\frac{1}{21}$

   Don's rate: $\frac{1}{28}$

   time together: x

$$\frac{1}{21}x + \frac{1}{28}x = 1$$

$$84\left(\frac{1}{21}x + \frac{1}{28}x\right) = 84(1)$$

$$4x + 3x = 84$$

$$7x = 84$$

$$x = 12$$

12 min.

9.

	R	T	D = RT
Frank	2	t	$2t$
Chuck	3	t	$3t$

$$2t + 3t = 20$$

$$5t = 20$$

$$t = 4$$

$$2t = 8$$

$$3t = 12$$

They will meet in 4 hrs. Frank has
walked 8 mi and Chuck has walked 12 mi.

11.

	R	T	D
Lois	6 2/3	t	20/3 t
Joan	6	t	6t

$$\frac{20}{3}t + 6t = 38$$

$$20t + 18t = 114$$

$$38t = 114$$

$$t = 3$$

3 hrs.

13.

$$\boxed{30 \text{ oz}} \quad + \quad \boxed{x \text{ oz}} \quad = \quad \boxed{(30 + x) \text{ oz}}$$

$$15\% \qquad\qquad 50\% \qquad\qquad 20\%$$

$$30 \cdot \frac{15}{100} + x \cdot \frac{50}{100} = (30 + x) \cdot \frac{20}{100}$$

$$450 + 50x = 20(30 + x)$$

$$450 + 50x = 600 + 20x$$

$$30x = 150$$

$$x = 5$$

5oz.

15.

$$\boxed{10 \text{ mL}} \quad + \quad \boxed{x \text{ mL}} \quad = \quad \boxed{(10 + x) \text{ mL}}$$

$$6\% \qquad\qquad 0\% \qquad\qquad 3\%$$

$$10 \cdot \frac{6}{100} + x \cdot \frac{0}{100} = (10 + x) \cdot \frac{3}{100}$$

$$60 + 0 = 3(10 + x)$$

$$60 = 30 + 3x$$

$$30 = 3x$$

$$10 = x$$

10 mL

17.

$$\boxed{532 \text{ kg}} \quad + \quad \boxed{x \text{ kg}} \quad = \quad \boxed{(532 + x) \text{ kg}}$$

$$20\% \qquad\qquad 100\% \qquad\qquad 24\%$$

$$532 \cdot \frac{20}{100} + x \cdot \frac{100}{100} = (532 + x) \cdot \frac{24}{100}$$

$$10640 + 100x = 24(532 + x)$$

$$10640 + 100x = 12768 + 24x$$

$$76x = 2128$$

$$x = 28$$

28 kg

19. rate of pipe A: $\frac{1}{5}$

rate of pipe B: $\frac{1}{6}$

time together: x

$$\frac{1}{5}x + \frac{1}{6}x = 1$$

$$30\left(\frac{1}{5}x + \frac{1}{6}x\right) = 30(1)$$

$$6x + 5x = 30$$

$$11x = 30$$

$$x = \frac{30}{11}$$

$\frac{30}{11}$ hr

21. Jose's rate: $\dfrac{1}{3}$

tire's rate: $\dfrac{1}{5}$

time to fill faulty tire: x

$$\frac{1}{3}x - \frac{1}{5}x = 1$$

$$15\left(\frac{1}{3}x - \frac{1}{5}x\right) = 15(1)$$

$$5x - 3x = 15$$

$$2x = 15$$

$$x = \frac{15}{2} = 7\frac{1}{2}$$

$7\frac{1}{2}$ min

23. cold water's rate: $\dfrac{1}{3}$

hot water's rate: $\dfrac{1}{4}$

time together: x

$$\frac{1}{3}x + \frac{1}{4}x = \frac{1}{2}$$

$$12\left(\frac{1}{3}x + \frac{1}{4}x\right) = 12\left(\frac{1}{2}\right)$$

$$4x + 3x = 6$$

$$7x = 6$$

$$x = \frac{6}{7}$$

$\dfrac{6}{7}$ min

25. left valve's rate: $\dfrac{1}{5}$

right valve's rate: $\dfrac{1}{10}$

drain's rate: $\dfrac{1}{2\frac{1}{2}} = \dfrac{2}{5}$

time for all three together: x

$$-\frac{1}{5}x - \frac{1}{10}x + \frac{2}{5}x = 1$$

$$10\left(-\frac{1}{5}x - \frac{1}{10}x + \frac{2}{5}x\right) = 10(1)$$

$$-2x - x + 4x = 10$$

$$x = 10$$

10 hrs

27. Joanne's rate: $\dfrac{1}{2}\left(\dfrac{1}{42}\right) = \dfrac{1}{84}$

John's rate: $\dfrac{1}{2}\left(\dfrac{1}{30}\right) = \dfrac{1}{60}$

time together: x

$$\frac{1}{60}x + \frac{1}{84}x = 1$$

$$420\left(\frac{1}{60}x + \frac{1}{84}x\right) = 420(1)$$

$$7x + 5x = 420$$

$$12x = 420$$

$$x = 35$$

35 min

29.

	R	T	D = RT
from home	50	t	$50t$
to home	60	$5.5 - t$	$60(5.5 - t)$

$$50t = 60(5.5 - t)$$

$$50t = 330 - 60t$$

$$110t = 330$$

$$t = 3$$

$$50(3) = 150$$

150 mi

31.

	T	R	D = RT
Express	4	$2r$	$8r$
Slow Freight	2	r	$2r$

$8r = 210 + 2r$

$6r = 210$

$r = 35$

$2r = 70$

70 mi

33.

	R	T	D = RT
John	1/6	t	$1/6t$
Jim	1/8	$t + 3$	$1/8(t + 3)$

$\dfrac{1}{6}t = \dfrac{1}{8}(t + 3)$

$4t = 3t + 9$

$t = 9$

9 min

35.

	R	T	D = RT
going	5	t	$5t$
returning	4	$4 - t$	$4(4 - t)$

$5t = 4(4 - t)$

$5t = 16 - 4t$

$9t = 16$

$t = \dfrac{16}{9}$

$5\left(\dfrac{16}{9}\right) = \dfrac{80}{9} = 8\dfrac{8}{9}$

$8\dfrac{8}{9}$ mi

If he doubled his rental time he could go twice as far.

37. 50 min $= \dfrac{5}{6}$ hr

Rate for 10 km in 1 hr: $10\ \dfrac{\text{km}}{\text{hr}}$

$RT = D$

$R\left(\dfrac{5}{6}\right) = 10$

$R = 12$

His new rate would be 12 km/hr which is $12 - 10 = 2$ km/hr faster than his current rate.

39.

	R	T	D = RT
Ray	6	t	$6t$
Mike	12	$t - 1$	$12(t - 1)$

$6t = 12(t - 1)$

$6t = 12t - 12$

$-6t = -12$

$t = 2$

$6t = 6(2) = 12$

2 hrs; 12 mi

41. Speed of current: x

	R	D	T = D/R
upstream	$4 - x$	20	$20/(4 - x)$
downstream	$4 + x$	20	$20/(4 + x)$

$\dfrac{20}{4 - x} = 2\left(\dfrac{20}{4 + x}\right)$

$(4 - x)(4 + x)\left(\dfrac{20}{4 - x}\right) = (4 - x)(4 + x)(2)\left(\dfrac{20}{4 + x}\right)$

$20(4 + x) = 40(4 - x)$

$80 + 20x = 160 - 40x$

$60x = 80$

$x = \dfrac{4}{3}$

$\dfrac{4}{3}$ km/hr

1. $|x| = 3$

   $x = 3$ or $x = -3$

   $\{3, \ -3\}$

3. $|2x| = 5$

   $2x = 5$ or $2x = -5$

   $x = \dfrac{5}{2}$ or $x = -\dfrac{5}{2}$

   $\left\{\dfrac{5}{2}, \ -\dfrac{5}{2}\right\}$

5. $|1 - 2x| = 4$

   $1 - 2x = 4$ or $1 - 2x = -4$

   $-2x = 3$ or $-2x = -5$

   $x = -\dfrac{3}{2}$ or $x = \dfrac{5}{2}$

   $\left\{-\dfrac{3}{2}, \ \dfrac{5}{2}\right\}$

7. $|x| = -2$

   $-2 < 0$

   $\emptyset$

9. $|x| = 0$

   $x = 0$

   $\{0\}$

11. $|2x - 5| = 0$

   $2x - 5 = 0$

   $2x = 5$

   $x = \dfrac{5}{2}$

   $\left\{\dfrac{5}{2}\right\}$

13. $|2x - 3| - 5 = 0$

   $|2x - 3| = 5$

   $2x - 3 = 5$ or $2x - 3 = -5$

   $2x = 8$ or $2x = -2$

   $x = 4$ or $x = -1$

   $\{4, \ -1\}$

15. $|5 - 2x| + 2 = 5$

   $|5 - 2x| = 3$

   $5 - 2x = 3$ or $5 - 2x = -3$

   $-2x = -2$ or $-2x = -8$

   $x = 1$ or $x = 4$

   $\{1, \ 4\}$

17. $|x + 2| = |x|$

   $x + 2 = x$ or $x + 2 = -x$

   $2 = 0$ or $2 = -2x$

   $-1 = x$

   $\{-1\}$

19. $|2x - 1| = |x + 3|$

   $2x - 1 = x + 3$ or $2x - 1 = -(x + 3)$

   $x = 4$ or $2x - 1 = -x - 3$

   $3x = -2$

   $x = -\dfrac{2}{3}$

   $\left\{4, \ -\dfrac{2}{3}\right\}$

21. $x = 2000 - 500$ or $x = 2000 + 500$

   $x = 1500$ or $x = 2500$

   $|x - 2000| = 500$

70

23. $|x| = 6$

$\quad x = 6 \quad$ or $\quad x = -6$

$\quad\quad \{6, \; -6\}$

25. $|2x| = 0$

$\quad\quad 2x = 0$

$\quad\quad x = 0$

$\quad\quad \{0\}$

27. $|x - 4| = 2$

$\quad x - 4 = 2 \quad$ or $\quad x - 4 = -2$

$\quad\quad x = 6 \quad$ or $\quad\quad x = 2$

$\quad\quad\quad \{6, \; 2\}$

29. $|x - 6| = -7$

$\quad -7 < 0$

$\quad\quad \emptyset$

31. $|x + 13| = 7$

$\quad x + 13 = 7 \quad$ or $\quad x + 13 = -7$

$\quad\quad x = -6 \quad$ or $\quad\quad x = -20$

$\quad\quad\quad \{-6, \; -20\}$

33. $|4x + 5| - 4 = 4$

$\quad\quad |4x + 5| = 8$

$\quad 4x + 5 = 8 \quad$ or $\quad 4x + 5 = -8$

$\quad\quad 4x = 3 \quad$ or $\quad\quad 4x = -13$

$\quad\quad x = \dfrac{3}{4} \quad$ or $\quad\quad x = -\dfrac{13}{4}$

$\quad\quad\quad \left\{\dfrac{3}{4}, \; -\dfrac{13}{4}\right\}$

35. $|10x - 3| = 0$

$\quad\quad 10x - 3 = 0$

$\quad\quad 10x = 3$

$\quad\quad x = \dfrac{3}{10}$

$\quad\quad \left\{\dfrac{3}{10}\right\}$

37. $|5x + 11| = 9$

$\quad 5x + 11 = 9 \quad$ or $\quad 5x + 11 = -9$

$\quad\quad 5x = -2 \quad$ or $\quad\quad 5x = -20$

$\quad\quad x = -\dfrac{2}{5} \quad$ or $\quad\quad x = -4$

$\quad\quad \left\{-\dfrac{2}{5}, \; -4\right\}$

39. $6 - |2.5x - 4| = 0$

$\quad\quad\quad 6 = |2.5x - 4|$

$\quad 6 = 2.5x - 4 \quad$ or $\quad -6 = 2.5x - 4$

$\quad 10 = 2.5x \quad$ or $\quad -2 = 2.5x$

$\quad 4 = x \quad$ or $\quad -0.8 = x$

$\quad\quad \{4, \; -0.8\}$

41. $|2.3 - x| = 4.5$

$\quad 2.3 - x = 4.5 \quad$ or $\quad 2.3 - x = -4.5$

$\quad -x = 2.2 \quad$ or $\quad -x = -6.8$

$\quad x = -2.2 \quad$ or $\quad x = 6.8$

$\quad\quad \{-2.2, \; 6.8\}$

43. $|2s| = |s + 3|$

$\quad 2s = s + 3 \quad$ or $\quad 2s = -(s + 3)$

$\quad s = 3 \quad$ or $\quad 2s = -s - 3$

$\quad\quad\quad 3s = -3$

$\quad\quad\quad s = -1$

$\quad\quad \{3, \; -1\}$

45. $|x + 7| = |x - 4|$

$x + 7 = x - 4$ or $x + 7 = -(x - 4)$

$7 = -4$ or $x + 7 = -x + 4$

$2x = -3$

$x = -\dfrac{3}{2}$

$\left\{-\dfrac{3}{2}\right\}$

47. $|3x + 4| = |x - 2|$

$3x + 4 = x - 2$ or $3x + 4 = -(x - 2)$

$2x = -6$ or $3x + 4 = -x + 2$

$x = -3$ or $4x = -2$

$x = -\dfrac{1}{2}$

$\left\{-3, \ -\dfrac{1}{2}\right\}$

49. $|x + 1| = |2x + 1|$

$x + 1 = 2x + 1$ or $x + 1 = -(2x + 1)$

$0 = x$ or $x + 1 = -2x - 1$

$3x = -2$

$x = -\dfrac{2}{3}$

$\left\{0, \ -\dfrac{2}{3}\right\}$

51. $|2x - 5| = |4x + 3|$

$2x - 5 = 4x + 3$ or $2x - 5 = -(4x + 3)$

$-8 = 2x$ or $2x - 5 = -4x - 3$

$-4 = x$ or $6x = 2$

$x = \dfrac{1}{3}$

$\left\{\dfrac{1}{3}, \ -4\right\}$

53. $|2x + 9| = |7 - x|$

$2x + 9 = 7 - x$ or $2x + 9 = -(7 - x)$

$3x = -2$ or $2x + 9 = -7 + x$

$x = -\dfrac{2}{3}$ or $x = -16$

$\left\{-\dfrac{2}{3}, \ -16\right\}$

55. $|9x - 5| = |x + 3|$

$9x - 5 = x + 3$ or $9x - 5 = -(x + 3)$

$8x = 8$ or $9x - 5 = -x - 3$

$x = 1$ or $10x = 2$

$x = \dfrac{1}{5}$

$\left\{1, \ \dfrac{1}{5}\right\}$

57. $|x + 1| = \left|\dfrac{1}{2}x\right|$

$x + 1 = \dfrac{1}{2}x$ or $x + 1 = -\dfrac{1}{2}x$

$1 = -\dfrac{1}{2}x$ or $1 = -\dfrac{3}{2}x$

$-2 = x$ or $-\dfrac{2}{3} = x$

$\left\{-2, \ -\dfrac{2}{3}\right\}$

59. $|x| = 3$

$x = 3$ or $x = -3$

The numbers are 3 and -3.

61. $|4x - 3| = 1$

$4x - 3 = 1$ or $4x - 3 = -1$

$4x = 4$ or $4x = 2$

$x = 1$ or $x = \dfrac{1}{2}$

The numbers are 1 and $\dfrac{1}{2}$.

63. $|x - 40| = 0.025(40) = 1$

$x - 40 = 1$ or $x - 40 = -1$

$x = 41$ or $x = 39$

The lowest temperature is 39° F and the highest temperature *is* 41° *F.*

65. $|x + 5| = 3$

$x + 5 = 3$ or $x + 5 = -3$

$x = -2$ or $x = -8$

Yes.

67. $|x - a| = 1$

$x - a = 1$ or $x - a = -1$

$x = a + 1$ or $x = a - 1$

$\{a + 1, \ a - 1\}$

69. $|x + a| = |x|$

$x + a = x$ or $x + a = -x$

$a = 0$ or $a = -2x$

$-\dfrac{a}{2} = x$

$\left\{-\dfrac{a}{2}\right\}$

71. $|2x + 1| = x$

$2x + 1 = x$ or $2x + 1 = -x$

$1 = -x$ or $1 = -3x$

$-1 = x$ or $-\dfrac{1}{3} = x$

Check: $x = -1$

LS: $|2(-1) + 1| = |-1| = 1$

RS: -1

Does not check.

Check: $x = -\dfrac{1}{3}$

LS: $\left|2\left(-\dfrac{1}{3}\right) + 1\right| = \left|\dfrac{1}{3}\right| = \dfrac{1}{3}$

RS: $-\dfrac{1}{3}$

Does not check.

$\emptyset$

73. $|x - 2| = x + 3$

$x - 2 = x + 3$ or $x - 2 = -(x + 3)$

$-2 = 3$ or $x - 2 = -x - 3$

$2x = -1$

$x = -\dfrac{1}{2}$

Check: $x = -\dfrac{1}{2}$

LS: $\left|-\dfrac{1}{2} - 2\right| = \left|-\dfrac{5}{2}\right| = \dfrac{5}{2}$

RS: $-\dfrac{1}{2} + 3 = \dfrac{5}{2}$

Checks.

$\left\{-\dfrac{1}{2}\right\}$

Problem Set 2.7

1. $|x - 1| < 5$

Boundary numbers:

$|x - 1| = 5$

$x - 1 = 5$ or $x - 1 = -5$

$x = 6$ or $x = -4$

$$\underline{\quad A \quad | \quad B \quad | \quad C \quad}$$
$${-4}6$$

A: $x = -5$; $\quad |-5 - 1| < 5$

$$6 < 5 \quad \text{F}$$

B: $x = 0$; $\quad |0 - 1| < 5$

$$1 < 5 \quad \text{T}$$

C: $x = 7$; $\quad |7 - 1| < 5$

$$6 < 5 \quad \text{F}$$

$$\{x \mid -4 < x < 6\}$$

3. $|x + 2| \geq 8$

   Boundary numbers:

   $|x + 2| = 8$

   $x + 2 = 8 \quad \text{or} \quad x + 2 = -8$

   $ x = 6 \quad \text{or} \quad x = -10$

$$\underline{\quad A \quad | \quad B \quad | \quad C \quad}$$
$${-10}6$$

A: $x = -11$; $\quad |-11 + 2| \geq 8$

$$9 \geq 8 \quad \text{T}$$

B: $x = 0$; $\quad |0 + 2| \geq 8$

$$2 \geq 8 \quad \text{F}$$

C: $x = 7$; $\quad |7 + 2| \geq 8$

$$9 \geq 8 \quad \text{T}$$

$$\{x \mid x \leq -10 \text{ or } x \geq 6\}$$

5. $|5x - 7| - 2 < 0$

   Boundary numbers:

   $|5x - 7| - 2 = 0$

   $|5x - 7| = 2$

   $5x - 7 = 2 \quad \text{or} \quad 5x - 7 = -2$

   $5x = 9 \quad \text{or} \quad 5x = 5$

   $x = \dfrac{9}{5} \quad \text{or} \quad x = 1$

$$\underline{\quad A \quad | \quad B \quad | \quad C \quad}$$
$$1\dfrac{9}{5}$$

A: $x = 0$; $\quad |5(0) - 7| - 2 < 0$

$$5 < 0 \quad \text{F}$$

B: $x = 1\dfrac{1}{2}$; $\quad \left|5\left(\dfrac{3}{2}\right) - 7\right| - 2 < 0$

$$-\dfrac{3}{2} < 0 \quad \text{T}$$

C: $x = 2$; $\quad |5(2) - 7| - 2 < 0$

$$1 < 0 \quad \text{F}$$

$$\left\{x \mid 1 < x < \dfrac{9}{5}\right\}$$

7. $|x + 1| > |2x + 3|$

   Boundary numbers:

   $|x + 1| = |2x + 3|$

   $x + 1 = 2x + 3 \quad \text{or} \quad x + 1 = -(2x + 3)$

   $-2 = x \quad \text{or} \quad x + 1 = -2x - 3$

   $$3x = -4$$

   $$x = -\dfrac{4}{3}$$

$$\underline{\quad A \quad | \quad B \quad | \quad C \quad}$$
$${-2}-\dfrac{4}{3}$$

A: $x = -3$; $\quad |-3 + 1| > |2(-3) + 3|$

$$2 > 3 \quad \text{F}$$

B: $x = -\dfrac{5}{3}$; $\quad \left|-\dfrac{5}{3} + 1\right| > \left|2\left(-\dfrac{5}{3}\right) + 3\right|$

$$\dfrac{2}{3} > \dfrac{1}{3} \quad \text{T}$$

C: $x = 0$; $\quad |0 + 1| > |2(0) + 3|$

$$1 > 3 \quad \text{F}$$

$$\left(-2, \ -\dfrac{4}{3}\right)$$

9. $|2x + 3| < |5x - 3|$

Boundary numbers:

$|2x + 3| = |5x - 3|$

$2x + 3 = 5x - 3$ or $2x + 3 = -(5x - 3)$

$\quad\quad 6 = 3x$ or $2x + 3 = -5x + 3$

$\quad\quad 2 = x$ or $7x = 0$

$\quad\quad\quad\quad\quad\quad\quad\quad\quad x = 0$

$$\underset{0}{\overset{A}{\rule{2cm}{0.4pt}}}|\underset{}{\overset{B}{\rule{1.5cm}{0.4pt}}}\underset{2}{|}\overset{C}{\rule{1.5cm}{0.4pt}}$$

A: $x = -1$; $|2(-1) + 3| < |5(-1) - 3|$

$\quad\quad\quad\quad\quad\quad\quad\quad 1 < 8 \quad\quad\quad\quad$ T

B: $x = 1$; $|2(1) + 3| < |5(1) - 3|$

$\quad\quad\quad\quad\quad\quad\quad\quad 5 < 2 \quad\quad\quad\quad$ F

C: $x = 3$; $|2(3) + 3| < |5(3) - 3|$

$\quad\quad\quad\quad\quad\quad\quad\quad 9 < 12 \quad\quad\quad$ T

$$(-\infty, 0) \cup (2, \infty)$$

11. $|x - 7| \geq 0$

Boundary numbers:

$|x - 7| = 0$

$\quad x - 7 = 0$

$\quad\quad\quad x = 7$

$$\underset{7}{\overset{A}{\rule{2cm}{0.4pt}}}|\overset{B}{\rule{1.5cm}{0.4pt}}$$

A: $x = 0$; $|0 - 7| \geq 0$

$\quad\quad\quad\quad\quad\quad\quad\quad 7 \geq 0 \quad$ T

B: $x = 8$; $|8 - 7| \geq 0$

$\quad\quad\quad\quad\quad\quad\quad\quad 1 \geq 0 \quad$ T

$\{x \mid x \text{ is a real number}\}$

13. $|x - 5| < -7$

Boundary numbers:

$|x - 5| = -7$

No solution, no boundary numbers.

A: $x = 0$; $|0 - 5| < -7$

$\quad\quad\quad\quad\quad\quad\quad\quad 5 < -7 \quad\quad$ F

$\quad\quad\quad\quad\quad\quad \emptyset$

15. $|x - 7| \leq 5$

$-5 \leq x - 7 \leq 5$

$2 \leq x \leq 12$

17. $\left|2x + \dfrac{1}{2}\right| \geq 7$

$2x + \dfrac{1}{2} \leq -7$ or $2x + \dfrac{1}{2} \geq 7$

$\quad\quad 2x \leq -\dfrac{15}{2}$ or $2x \geq \dfrac{13}{2}$

$\quad\quad\quad x \leq -\dfrac{15}{4}$ or $x \geq \dfrac{13}{4}$

19. $|x - (-4)| \leq 2$

$|x + 4| \leq 2$

$-2 \leq x + 4 \leq 2$

$-6 \leq x \leq -2$

All numbers between -6 and -2, inclusive.

21. $\left|x - \dfrac{1}{2}\right| < \dfrac{1}{3}$

$-\dfrac{1}{3} < x - \dfrac{1}{2} < \dfrac{1}{3}$

$\dfrac{1}{6} < x < \dfrac{5}{6}$

All numbers between $\dfrac{1}{6}$ and $\dfrac{5}{6}$.

23. $|x| > 7$

$x < -7$ or $x > 7$

All numbers greater than 7 or less than -7.

25. $|x| \le 3$

$-3 \le x \le 3$

All numbers between -3 and 3, inclusive.

27. $|x| \ge 8$

$x \le -8$ or $x \ge 8$

$\{x \mid x \le -8 \text{ or } x \ge 8\}$

29. $|x - 3| < 9$

$-9 < x - 3 < 9$

$-6 < x < 12$

$\{x \mid -6 < x < 12\}$

31. $|x + 5| \le 3$

$-3 \le x + 5 \le 3$

$-8 \le x \le -2$

$\{x \mid -8 \le x \le -2\}$

33. $|2x + 5| > 7$

$2x + 5 < -7$ or $2x + 5 > 7$

$2x < -12$ or $2x > 2$

$x < -6$ or $x > 1$

$\{x \mid x < -6 \text{ or } x > 1\}$

35. $|2x - 9| - 1 \le 0$

$|2x - 9| \le 1$

$-1 \le 2x - 9 \le 1$

$8 \le 2x \le 10$

$4 \le x \le 5$

$\{x \mid 4 \le x \le 5\}$

37. $|5x - 8| > 9$

$5x - 8 < -9$ or $5x - 8 > 9$

$5x < -1$ or $5x > 17$

$x < -\dfrac{1}{5}$ or $x > \dfrac{17}{5}$

$\left\{x \mid x < -\dfrac{1}{5} \text{ or } x > \dfrac{17}{5}\right\}$

39. $|10y + 3| - 1 \ge 0$

$|10y + 3| \ge 1$

$10y + 3 \le -1$ or $10y + 3 \ge 1$

$10y \le -4$ or $10y \ge -2$

$y \le -\dfrac{2}{5}$ or $y \ge -\dfrac{1}{5}$

$\left\{y \mid y \le -\dfrac{2}{5} \text{ or } y \ge -\dfrac{1}{5}\right\}$

41. $3(|x - 1| + 3) < 15$

$|x - 1| + 3 < 5$

$|x - 1| < 2$

$-2 < x - 1 < 2$

$-1 < x < 3$

$\{x \mid -1 < x < 3\}$

43. $10 \le 3(4 - |x + 3|)$

$\dfrac{10}{3} \le 4 - |x + 3|$

$|x + 3| \le \dfrac{2}{3}$

$-\dfrac{2}{3} \le x + 3 \le \dfrac{2}{3}$

$-\dfrac{11}{3} \le x \le -\dfrac{7}{3}$

$\left\{x \mid -\dfrac{11}{3} \le x \le -\dfrac{7}{3}\right\}$

45. $|t - 2.5| < 6.3$

$-6.3 < t - 2.5 < 6.3$

$-3.8 < t < 8.8$

$(-3.8, 8.8)$

47. $|x - 2.73| > 6.72$

$x - 2.73 < -6.72$ or $x - 2.73 > 6.72$

$x < -3.99$ or $x > 9.45$

$(-\infty, -3.99) \cup (9.45, \infty)$

49. $\left| \dfrac{2x - 1}{3} \right| \le 4$

$-4 \le \dfrac{2x - 1}{3} \le 4$

$-12 \le 2x - 1 \le 12$

$-11 \le 2x \le 13$

$-\dfrac{11}{2} \le x \le \dfrac{13}{2}$

$\left[-\dfrac{11}{2}, \dfrac{13}{2} \right]$

51. $|2 - x| + 3 \ge 4$

$|2 - x| \ge 1$

$2 - x \le -1$ or $2 - x \ge 1$

$-x \le -3$ or $-x \ge -1$

$x \ge 3$ or $x \le 1$

$(-\infty, 1] \cup [3, \infty)$

53. $|x + 1| \ge 0$

$x + 1 \le -0$ or $x + 1 \ge 0$

$x \le -1$ or $x \ge -1$

$\Re$

55. $|z - 3| < -5$

Boundary numbers:

$|z - 3| = -5$

No solution, no boundary numbers.

A: $z = 0$; $|0 - 3| < -5$

$3 < -5$ F

$\emptyset$

57. $|2x + 3| \le -1$

Boundary numbers:

$|2x + 3| = -1$

No solution, no boundary numbers.

A: $x = 0$; $|2(0) + 3| \le -1$

$3 \le -1$ F

$\emptyset$

59. $|3x| \ge |x - 4|$

Boundary numbers:

$|3x| = |x - 4|$

$3x = x - 4$ or $3x = -(x - 4)$

$2x = -4$ or $3x = -x + 4$

$x = -2$ or $4x = 4$

$x = 1$

$\underset{-2}{\underline{A|B|C}}$
${-2}{1}$

A: $x = -3$; $|3(-3)| \ge |-3 - 4|$

$9 \ge 7$ T

B: $x = 0$; $|3(0)| \ge |0 - 4|$

$0 \ge 4$ F

C: $x = 2$; $|3(2)| \ge |2 - 4|$

$6 \ge 2$ T

$(-\infty, -2] \cup [1, \infty)$

61. $|4x + 3| \le |3x + 18|$

Boundary numbers:

$|4x + 3| = |3x + 18|$

$4x + 3 = 3x + 18$ or $4x + 3 = -(3x + 18)$

$x = 15$ or $4x + 3 = -3x - 18$

$7x = -21$

$x = -3$

$$\underset{-3}{\underline{\quad A \quad}}\Big|\underset{15}{\underline{\quad B \quad}}\Big|\underline{\quad C \quad}$$

A: $x = -4$; $\;|4(-4) + 3| \le |3(-4) + 18|$

$\qquad\qquad\qquad\qquad 13 \le 6 \qquad\qquad$ F

B: $x = 0$; $\quad |4(0) + 3| \le |3(0) + 18|$

$\qquad\qquad\qquad\qquad 3 \le 18 \qquad\qquad$ T

C: $x = 16$; $\;|4(16) + 3| \le |3(16) + 18|$

$\qquad\qquad\qquad\qquad 67 \le 66 \qquad\qquad$ F

$$[-3, 15]$$

63. $|x + 1| \le 3$

$\quad -3 \le x + 1 \le 3$

$\quad -4 \le x \le 2$

65. $|x| > 4.3$

$\quad x < -4.3 \quad$ or $\quad x > 4.3$

67. $|x + 3| < 2$

$\quad -2 < x + 3 < 2$

$\quad -5 < x < -1$

69. $\left|2x - \dfrac{1}{3}\right| < 2$

$\quad -2 < 2x - \dfrac{1}{3} < 2$

$\quad -6 < 6x - 1 < 6$

$\quad -5 < 6x < 7$

$\quad -\dfrac{5}{6} < x < \dfrac{7}{6}$

71. $|x| < 4$

$\quad -4 < x < 4$

All numbers between -4 and 4.

73. $|x| \le 8$

$\quad -8 \le x \le 8$

All numbers between -8 and 8, inclusive.

75. $|x - 1| < 2$

$\quad -2 < x - 1 < 2$

$\quad -1 < x < 3$

All numbers between -1 and 3.

77. $|x - 270| \le 4$

$\quad -4 \le x - 270 \le 4$

$\quad 266 \le x \le 274$

Between $266°$ and $274°$, inclusive.

79. $|x - 22| \le 5$

$\quad -5 \le x - 22 \le 5$

$\quad 17 \le x \le 27$

Between 17% and 27%, inclusive.

81. $|x - a| \le 4$

$\quad -4 \le x - a \le 4$

$\quad a - 4 \le x \le a + 4$

$\quad \{x \,|\, a - 4 \le x \le a + 4\}$

83. $|x - a| < d$; $\;d > 0$

$\quad -d < x - a < d$

$\quad a - d < x < a + d$

$\quad \{x \,|\, a - d < x < a + d\}$

85. $|x - a| < a$

$\quad -a < x - a < a$

$\quad 0 < x < 2a$

$\quad \{x \,|\, 0 < x < 2a\}$

87. $|x - a| \le 0$

$\quad 0 \le x - a \le 0$

$\quad a \le x \le a$

$\qquad \{a\}$

89. $|x + a| < a$

$-a < x + a < a$

$-2a < x < 0$

$\{x \mid -2a < x < 0\}$

91. $|x - a| \le s$

$-s \le x - a \le s$

$a - s \le x \le a + s$

$\{x \mid a - s \le x \le a + s\}$

Chapter 2 Review Problems

1. $6(x - 5) - (2 - x) = 2x + 1$

$6x - 30 - 2 + x = 2x + 1$

$7x - 32 = 2x + 1$

$5x = 33$

$x = \dfrac{33}{5}$

$\left\{\dfrac{33}{5}\right\}$

3. $|2z - 3| = 4$

$2z - 3 = 4$ or $2z - 3 = -4$

$2z = 7$ or $2z = -1$

$z = \dfrac{7}{2}$ or $z = -\dfrac{1}{2}$

$\left\{\dfrac{7}{2}, -\dfrac{1}{2}\right\}$

5. $1 < 2x < 4$

$\dfrac{1}{2} < x < 2$

$\left(\dfrac{1}{2}, 2\right)$

7. $|x + 16| > 8$

$x + 16 < -8$ or $x + 16 > 8$

$x < -24$ or $x > -8$

-24   -8

$(-\infty, -24) \cup (8, \infty)$

9. $2(x - 7) + 5(3 - x) \le 0$

$2x - 14 + 15 - 5x \le 0$

$-3x + 1 \le 0$

$1 \le 3x$

$\dfrac{1}{3} \le x$

$\left[\dfrac{1}{3}, \infty\right)$

11. $\left|\dfrac{3 + y}{2}\right| = 5$

$\dfrac{3 + y}{2} = 5$ or $\dfrac{3 + y}{2} = -5$

$3 + y = 10$ or $3 + y = -10$

$y = 7$ or $y = -13$

$\{7, -13\}$

13. $\dfrac{3}{2}(x + 2) = \dfrac{1}{2}(x - 2)$

$3(x + 2) = x - 2$

$3x + 6 = x - 2$

$2x = -8$

$x = -4$

$\{-4\}$

15. $\dfrac{1}{2}x - \dfrac{3}{4} < \dfrac{1}{4}x + 1$

$2x - 3 < x + 4$

$x < 7$

$(-\infty, 7)$

79

17. $\dfrac{1}{x} - \dfrac{1}{3x} = \dfrac{1}{2}$

$6x\left(\dfrac{1}{x} - \dfrac{1}{3x}\right) = 6x\left(\dfrac{1}{2}\right)$

$6 - 2 = 3x$

$4 = 3x$

$\dfrac{4}{3} = x$

$\left\{\dfrac{4}{3}\right\}$

19. $|6x - 7| \geq 0$

$6x - 7 \leq 0$ or $6x - 7 \geq 0$

$6x \leq 7$ or $6x \geq 7$

$x \leq \dfrac{7}{6}$ or $x \geq \dfrac{7}{6}$

$\Re$

$\longleftrightarrow$

21. $\dfrac{5}{2}(1 - x) = \dfrac{1}{3}(2x + 1)$

$15(1 - x) + 2(2x + 1)$

$15 - 15x = 4x + 2$

$13 = 19x$

$\dfrac{13}{19} = x$

$\left\{\dfrac{13}{19}\right\}$

23. $2x + 1 < 5$ and $x + 2 > 5$

$2x < 4$ and $x > 3$

$x < 2$ and $x > 3$

$\emptyset$

25. $\dfrac{1}{2}x - 2 \leq \dfrac{1}{4}(x - 3)$

$2x - 8 \leq x - 3$

$x \leq 5$

$\longleftarrow\quad\rbrack$

5

$(-\infty, 5]$

27. $\left|\dfrac{3x + 1}{2}\right| \leq 2$

$-2 \leq \dfrac{3x + 1}{2} \leq 2$

$-4 \leq 3x + 1 \leq 4$

$-5 \leq 3x \leq 3$

$\dfrac{-5}{3} \leq x \leq 1$

$\lbrack\quad\rbrack$

$\dfrac{-5}{3}\qquad 1$

$\left[-\dfrac{5}{3}, 1\right]$

29. $6x + 5 = 2x - 3$

$4x = -8$

$x = -2$

$\{-2\}$

31. $|3x + 5| = 6$

$3x + 5 = 6$ or $3x + 5 = -6$

$3x = 1$ or $3x = -11$

$x = \dfrac{1}{3}$ or $x = -\dfrac{11}{3}$

$\left\{\dfrac{1}{3}, -\dfrac{11}{3}\right\}$

33. $2(x + 1) + 3 < 2x - 1$

$2x + 2 + 3 < 2x - 1$

$2x + 5 < 2x - 1$

$5 < -1$

$\emptyset$

35. $\left| \dfrac{1 - x}{4} \right| = \dfrac{1}{2}$

$\dfrac{1 - x}{4} = \dfrac{1}{2}$ or $\dfrac{1 - x}{4} = -\dfrac{1}{2}$

$1 - x = 2$ or $1 - x = -2$

$-x = 1$ or $-x = -3$

$x = -1$ or $x = 3$

$\{-1,\ 3\}$

37. $\left| 3x - 5 \right| = 0$

$3x - 5 = 0$

$3x = 5$

$x = \dfrac{5}{3}$

$\left\{ \dfrac{5}{3} \right\}$

39. $3\left(x - \dfrac{2}{3} \right) + 4 > 1 + 3x$

$3x - 2 + 4 > 1 + 3x$

$3x + 2 > 1 + 3x$

$2 > 1$

$\Re$

$\longleftrightarrow$

41. $\left| 2x - 7 \right| = -3$

$-3 < 0$

$\emptyset$

43. length: x

width: $\dfrac{2}{3}x$

$2x + 2\left(\dfrac{2}{3}x \right) = 60$

$2x + \dfrac{4}{3}x = 60$

$6x + 4x = 180$

$10x = 180$

$x = 18$

$\dfrac{2}{3}x = 12$

12 in by 18 in.

45. 1st integer: x

2nd integer: $x + 2$

$x + x + 2 = 50$

$2x = 48$

$x = 24$

$x + 2 = 26$

The numbers are 24 and 26.

47. 1st integer: x

2nd integer: $x + 2$

3rd integer: $x + 4$

$x + 2(x + 2) + 3(x + 4) = 94$

$x + 2x + 4 + 3x + 12 = 94$

$6x + 16 = 94$

$6x = 78$

$$x = 13$$
$$x + 2 = 15$$
$$x + 4 = 17$$

The numbers are 13, 15 and 17.

49.

15 gal	+	x gal	=	$(15 + x)$ gal
8%		0%		5%

$$15 \cdot \frac{8}{100} + x \cdot \frac{0}{100} = (15 + x) \cdot \frac{5}{100}$$
$$120 + 0 = 5(15 + x)$$
$$120 = 75 + 5x$$
$$45 = 5x$$
$$9 = x$$

9 gal

51. width: w

length: $2w - 2$

$$2w + 2(2w - 2) = 86$$
$$2w + 4w - 4 = 86$$
$$6w = 90$$
$$w = 15$$
$$2w - 2 = 28$$

15 m by 28 m.

53. number: x

$$2x - 5 \le 17$$
$$2x \le 22$$
$$x \le 11$$

All numbers less than or equal to 11.

55. side length: x

$$4x \le 100$$
$$x \le 25$$

25 ft or less.

57. fifth test score: x

$$\frac{75 + 86 + 88 + 62 + x}{5} \ge 80$$
$$\frac{311 + x}{5} \ge 80$$
$$311 + x \ge 400$$
$$x \ge 89$$

She must score at least 89.

59. Factored

61. $(x + y) - 4(x + y)^2$

$= (x + y)[1 - 4(x + y)]$

$= (x + y)(1 - 4x - 4y)$

63. Factored

65. One term;

$x + 2$ is a factor.

67. One term;

$x + 2$ is a factor.

69. 3 terms;

$x^2 + 4x + 4 = (x + 2)^2$

$x + 2$ is a factor.

71. $(-8)^2 = 64$

73. $(a + 2b)^3$

$= a^3 + 3a^2(2b) + 3a(2b)^2 + (2b)^3$

$= a^3 + 6a^2b + 12ab^2 + 8b^3$

75. $x^3(x^2y - xy)$

$= x^3 \cdot x^2y - x^3 \cdot xy$

$= x^5y - x^4y$

77. Inequality;

$$2x - 5 < x - 3(4 - x)$$

$$2x - 5 < x - 12 + 3x$$

$$2x - 5 < 4x - 12$$

$$7 < 2x$$

$$\frac{7}{2} < x$$

$$\left\{ x \,\middle|\, x > \frac{7}{2} \right\}$$

79. Inequality;

$$|3x + 2| \leq 6$$

$$-6 \leq 3x + 2 \leq 6$$

$$-8 \leq 3x \leq 4$$

$$\frac{-8}{3} \leq x \leq \frac{4}{3}$$

$$\left\{ x \,\middle|\, -\frac{8}{3} \leq x \leq \frac{4}{3} \right\}$$

Chapter 2 Test

1. $3(2x - 1) - 4x = 3(5 - x)$

$$6x - 3 - 4x = 15 - 3x$$

$$2x - 3 = 15 - 3x$$

$$5x = 18$$

$$x = \frac{18}{5}$$

C

2. B

3. $b(1 - 6x) = 2ax$

$$b - 6bx = 2ax$$

$$b = 2ax + 6bx$$

$$b = x(2a + 6b)$$

$$\frac{b}{2a + 6b} = x$$

B

4. $\dfrac{x - 3}{6} \leq \dfrac{x}{2} - 1$

$$x - 3 \leq 3x - 6$$

$$3 \leq 2x$$

$$\frac{3}{2} \leq x$$

C

5. 1st: x

2nd: $x + 2$

3rd: $x + 4$

$$x + 2(x + 2) + 4(x + 4)$$

C

6. A

7. $|x + 3| < 0$

Boundary numbers:

$$|x + 3| = 0$$

$$x + 3 = 0$$

$$x = -3$$

$$\underline{\quad\quad A \quad\underset{-3}{\big|}\quad B \quad\quad}$$

A: $x = -4$; $|-4 + 3| < 0$

$$1 < 0 \quad \text{F}$$

B: $x = 0$; $|0 + 3| < 0$

$$3 < 0 \quad \text{F}$$

$$\emptyset$$

C

8. $|2x + 3| = 7$

$$2x + 3 = 7 \quad \text{or} \quad 2x + 3 = -7$$

$$2x = 4 \quad \text{or} \quad 2x = -10$$

$$x = 2 \quad \text{or} \quad x = -5$$

$$\{2, \; -5\}$$

9. $1 - 2x + 2(x - 7) < 3x$

$1 - 2x + 2x - 14 < 3x$

$-13 < 3x$

$-\dfrac{13}{3} < x$

$\left\{ x \mid x > -\dfrac{13}{3} \right\}$

10. $\dfrac{2}{3}(x - 1) = \dfrac{1}{6} + x$

$4(x - 1) = 1 + 6x$

$4x - 4 = 1 + 6x$

$-5 = 2x$

$-\dfrac{5}{2} = x$

$\left\{ -\dfrac{5}{2} \right\}$

11. $|1 - 2x| = -3$

$-3 < 0$

$\emptyset$

12. $|3x + 1| > 10$

$3x + 1 < -10$ or $3x + 1 > 10$

$3x < -11$ or $3x > 9$

$x < -\dfrac{11}{3}$ or $x > 3$

$\left\{ x \mid x < -\dfrac{11}{3} \text{ or } x > 3 \right\}$

13.

	T	R	D = RT
Honda	5	r	$5r$
bus	9	$r + 20$	$9(r + 20)$

$5r + 9(r + 20) = 600$

$5r + 9r + 180 = 600$

$14r = 420$

$r = 30$

30 mph

14. width: w

length: $w + 5$

$2w + 2(w + 5) = 134$

$2w + 2w + 10 = 134$

$4w = 124$

$w = 31$

$w + 5 = 36$

31 in by 36 in.

15.

$$\boxed{7 \text{ qts}} + \boxed{x \text{ qts}} = \boxed{(7 + x) \text{ qts}}$$

5%   20%   13%

$7 \cdot \dfrac{5}{100} + x \cdot \dfrac{20}{100} = (7 + x) \cdot \dfrac{13}{100}$

$35 + 20x + 13(7 + x)$

$35 + 20x = 91 + 13x$

$7x = 56$

$x = 8$

8 qts

CHAPTER 3

<u>Problem Set 3.1</u>

1. $x - 3 = 0$

   $x = 3$

3. $x - 5 = 0$

   $x = 5$

5. $6 = 0$

   $\emptyset$
   None

7. $x^2 + 1 = 0$

   $x^2 = -1$

   $\emptyset$
   None

9. $x - 2 = 0$

   $x = 2$

11. $(x + 3)(x - 1) = 0$

    $x + 3 = 0 \qquad x - 1 = 0$

    $x = -3 \qquad\qquad x = 1$

13. $x(x - 2) = 0$

    $x = 0 \qquad x - 2 = 0$

    $\qquad\qquad\qquad x = 2$

15. $x^2 - 4 = 0$

    $x^2 = 4$

    $x = 2 \qquad x = -2$

17. $\dfrac{-3}{x} = \dfrac{3}{-x} = -\dfrac{3}{x}$

19. $-\dfrac{7}{x} = \dfrac{-7}{x} = \dfrac{7}{-x}$

21. $\dfrac{2x}{-(x + 5)} = \dfrac{-2x}{x + 5} = -\dfrac{2x}{x + 5}$

23. $\dfrac{x - 1}{-x^2} = \dfrac{-(x - 1)}{x^2} = -\dfrac{x - 1}{x^2}$

25. $\dfrac{x^7}{x^2} = \dfrac{x^2 \cdot x^5}{x^2} = x^5$

27. $\dfrac{-2^3 xy}{2^2 x^2 y^2} = \dfrac{-2 \cdot 2^2 xy}{2^2 xy \cdot x \cdot y}$

    $= \dfrac{-2}{xy}$

    $= -\dfrac{2}{xy}$

29. $\dfrac{5x^2(x + y)^2}{10x(x + y)^3} = \dfrac{5x \cdot x \cdot (x + y)^2}{2 \cdot 5x \cdot (x + y)(x + y)^2}$

    $= \dfrac{x}{2(x + y)}$

31. $\dfrac{12x^2 - 12x}{18x^2} = \dfrac{12x(x - 1)}{18x^2}$

    $= \dfrac{2 \cdot 6x(x - 1)}{3 \cdot 6x \cdot x}$

    $= \dfrac{2(x - 1)}{3x}$

33. $\dfrac{x + 1}{x^3 + 1} = \dfrac{x + 1}{(x + 1)(x^2 - x + 1)}$

    $= \dfrac{1}{x^2 - x + 1}$

35. $\dfrac{8x^3 - 8}{2x^2 - 6x + 4} = \dfrac{8(x^3 - 1)}{2(x^2 - 3x + 2)}$

    $= \dfrac{4 \cdot 2(x - 1)(x^2 + x + 1)}{2(x - 1)(x - 2)}$

    $= \dfrac{4(x^2 + x + 1)}{x - 2}$

37. $\dfrac{s - 1}{1 - s} = \dfrac{-(1 - s)}{1 - s}$

    $= -1$

39. $\dfrac{(2x-1)^2}{(1-2x)(x+2)} = \dfrac{(2x-1)^2}{-(2x-1)(x+2)}$

$\qquad = -\dfrac{2x-1}{x+2}$

$\qquad$ or $\dfrac{1-2x}{x+2}$

41. $\dfrac{x^2-3x+2}{3-2x-x^2} = \dfrac{(x-2)(x-1)}{-(x^2+2x-3)}$

$\qquad = \dfrac{(x-2)(x-1)}{-(x+3)(x-1)}$

$\qquad = -\dfrac{x-2}{x+3}$

$\qquad$ or $\dfrac{2-x}{x+3}$

43. $\dfrac{5}{x} = \dfrac{5}{x} \cdot \dfrac{10}{10}$

$\qquad = \dfrac{50}{10x}$

45. $\dfrac{r^2}{6} = \dfrac{r^2}{6} \cdot \dfrac{m+n}{m+n}$

$\qquad = \dfrac{r^2(m+n)}{6m+6n}$

47. $x - x^2 = -x(-1+x)$

$\qquad = -x(x-1)$

$\dfrac{x+1}{x-1} = \dfrac{x+1}{x-1} \cdot \dfrac{-x}{-x}$

$\qquad = \dfrac{-x(x+1)}{x-x^2}$

49. $x^2 - 9 = (x-3)(x+3)$

$\qquad = -(3-x)(x+3)$

$\dfrac{x}{3-x} = \dfrac{x}{3-x} \cdot \dfrac{-(x+3)}{-(x+3)}$

$\qquad = \dfrac{-x(x+3)}{(x^2-9)}$

51. $x = 0$

53. $x^2 = 0$

$\quad x = 0$

55. $x - 7 = 0$

$\quad x = 7$

57. $(x-5)(x-2) = 0$

$\quad x - 5 = 0 \qquad x - 2 = 0$

$\qquad x = 5 \qquad\quad x = 2$

59. $x^2 - 1 = 0$

$\quad x^2 = 1$

$\quad x = 1 \qquad x = -1$

61. $\dfrac{x^8}{x^3} = \dfrac{x^5 \cdot x^3}{x^3}$

$\qquad = x^5$

63. $\dfrac{48t^3}{16t} = \dfrac{3 \cdot 16t \cdot t^2}{16t}$

$\qquad = 3t^2$

65. $\dfrac{-2x^2y}{-8xy} = \dfrac{-2x \cdot xy}{-2 \cdot 4xy}$

$\qquad = \dfrac{x}{4}$

67. $\dfrac{-3^3 xy}{3^2 x^3 y^3} = \dfrac{-3 \cdot 3^2 xy}{3^2 xy \cdot x^2 \cdot y^2}$

$\qquad = \dfrac{-3}{x^2 y^2}$

$\qquad = -\dfrac{3}{x^2 y^2}$

69. $\dfrac{7^2 5^3 x^2 y}{7^3 5^2 xy^3} = \dfrac{7^2 \cdot 5^2 \cdot 5x \cdot xy}{7^2 \cdot 7 \cdot 5^2 xy \cdot y^2}$

$\qquad = \dfrac{5x}{7y^2}$

71. $\dfrac{-n^5(n-m)}{n^4(n-m)^4} = \dfrac{-n \cdot n^4(n-m)}{n^4(n-m)(n-m)^3}$

$\qquad = \dfrac{-n}{(n-m)^3}$

$\qquad = -\dfrac{n}{(n-m)^3}$

73. $\dfrac{(x+2)(5+x)}{(x-1)(x+5)^2} = \dfrac{(x+2)(x+5)}{(x-1)(x+5)(x+5)}$

$\qquad = \dfrac{(x+2)}{(x-1)(x+5)}$

75. $\dfrac{(a-3)(a+3)}{(3-a)(1+a)} = \dfrac{(a-3)(a+3)}{-(a-3)(1+a)}$

$\qquad = \dfrac{a+3}{-(1+a)}$

$\qquad = -\dfrac{a+3}{1+a}$

77. $\dfrac{4a^2 - 8a}{8 - 4a} = \dfrac{4a(a-2)}{-4(-2+a)}$

$\qquad = \dfrac{4a(a-2)}{-4(a-2)}$

$\qquad = \dfrac{a}{-1}$

$\qquad = -a$

79. $\dfrac{5x}{25x - 15x^2} = \dfrac{5x}{5x(5-3x)}$

$\qquad = \dfrac{1}{5-3x}$

81. $\dfrac{3y + 12y^4}{15y^3 - 25y} = \dfrac{3y(1+4y^3)}{5y(3y^2 - 5)}$

$\qquad = \dfrac{3(1+4y^3)}{5(3y^2 - 5)}$

83. $\dfrac{x+3}{x^2 + 5x + 6} = \dfrac{x+3}{(x+3)(x+2)}$

$\qquad = \dfrac{1}{x+2}$

85. $\dfrac{x-1}{x^3 - 1} = \dfrac{x-1}{(x-1)(x^2 + x + 1)}$

$\qquad = \dfrac{1}{x^2 + x + 1}$

87. $\dfrac{x^2 - 8x + 16}{x - 4} = \dfrac{(x-4)^2}{(x-4)}$

$\qquad = x - 4$

89. $\dfrac{x^2 - x - 12}{x^2 - 3x - 4} = \dfrac{(x-4)(x+3)}{(x-4)(x+1)}$

$\qquad = \dfrac{x+3}{x+1}$

91. $\dfrac{p^3 + 27}{p^2 + 5p + 6} = \dfrac{(p+3)(p^2 - 3p + 9)}{(p+3)(p+2)}$

$\qquad = \dfrac{p^2 - 3p + 9}{p+2}$

93. $\dfrac{x^2 + 3x - 10}{8 - 2x - x^2} = \dfrac{x^2 + 3x - 10}{-(x^2 + 2x - 8)}$

$\qquad = \dfrac{(x+5)(x-2)}{-(x+4)(x-2)}$

$\qquad = -\dfrac{x+5}{x+4}$

95. $\dfrac{s^2 - 1}{1 - s^3} = \dfrac{(s - 1)(s + 1)}{-(s^3 - 1)}$

$\phantom{95. \dfrac{s^2 - 1}{1 - s^3}} = \dfrac{(s - 1)(s + 1)}{-(s - 1)(s^2 + s + 1)}$

$\phantom{95. \dfrac{s^2 - 1}{1 - s^3}} = -\dfrac{s + 1}{s^2 + s + 1}$

97. $\dfrac{3x^4 - 3x}{6x^3 - 6x} = \dfrac{3x(x^3 - 1)}{6x(x^2 - 1)}$

$\phantom{97. \dfrac{3x^4 - 3x}{6x^3 - 6x}} = \dfrac{3x(x - 1)(x^2 + x + 1)}{6x(x - 1)(x + 1)}$

$\phantom{97. \dfrac{3x^4 - 3x}{6x^3 - 6x}} = \dfrac{x^2 + x + 1}{2(x + 1)}$

99. $\dfrac{x^4 - x^2 - 2}{x^4 + x^2 - 6} = \dfrac{(x^2 - 2)(x^2 + 1)}{(x^2 + 3)(x^2 - 2)}$

$\phantom{99. \dfrac{x^4 - x^2 - 2}{x^4 + x^2 - 6}} = \dfrac{x^2 + 1}{x^2 + 3}$

101. $\dfrac{9 - 4x^2}{2x^2 - x - 3} = \dfrac{(3 - 2x)(3 + 2x)}{(2x - 3)(x + 1)}$

$\phantom{101. \dfrac{9 - 4x^2}{2x^2 - x - 3}} = \dfrac{-(2x - 3)(3 + 2x)}{(2x - 3)(x + 1)}$

$\phantom{101. \dfrac{9 - 4x^2}{2x^2 - x - 3}} = -\dfrac{3 + 2x}{x + 1}$

103. $\dfrac{am + 3m - 2a - 6}{cm - 2c + 6m - 12}$

$ = \dfrac{m(a + 3) - 2(a + 3)}{c(m - 2) + 6(m - 2)}$

$ = \dfrac{(a + 3)(m - 2)}{(m - 2)(c + 6)}$

$ = \dfrac{a + 3}{c + 6}$

105. $\dfrac{ps - pt - qs + qt}{pt + qt - ps - qs} = \dfrac{p(s - t) - q(s - t)}{t(p + q) - s(p + q)}$

$\phantom{105. \dfrac{ps - pt - qs + qt}{pt + qt - ps - qs}} = \dfrac{(s - t)(p - q)}{(p + q)(t - s)}$

$\phantom{105. \dfrac{ps - pt - qs + qt}{pt + qt - ps - qs}} = \dfrac{-(t - s)(p - q)}{(p + q)(t - s)}$

$\phantom{105. \dfrac{ps - pt - qs + qt}{pt + qt - ps - qs}} = -\dfrac{p - q}{p + q}$

$\text{or } \dfrac{q - p}{p + q}$

107. $\dfrac{x^3 - xy + x^2y - y^2}{x^3 - x^2y - xy + y^2}$

$ = \dfrac{x(x^2 - y) + y(x^2 - y)}{x^2(x - y) - y(x - y)}$

$ = \dfrac{(x^2 - y)(x + y)}{(x - y)(x^2 - y)}$

$ = \dfrac{x + y}{x - y}$

109. $\dfrac{5}{x} = \dfrac{5}{x} \cdot \dfrac{12x}{12x}$

$ = \dfrac{60x}{12x^2}$

111. $\dfrac{15b}{2c^2} = \dfrac{15b}{2c^2} \cdot \dfrac{5b}{5b}$

$ = \dfrac{75b^2}{10bc^2}$

113. $\dfrac{p}{p - 5} = \dfrac{p}{p - 5} \cdot \dfrac{2}{2}$

$ = \dfrac{2p}{2p - 10}$

115. $x^2 - 1 = (x - 1)(x + 1)$

$$\frac{-3x}{x - 1} = \frac{-3x}{x - 1} \cdot \frac{x + 1}{x + 1}$$

$$= \frac{-3x(x + 1)}{x^2 - 1}$$

$$= \frac{-3x^2 - 3x}{x^2 - 1}$$

117. $z^2 + 6z + 9 = (z + 3)^2$

$$\frac{2z}{z + 3} = \frac{2z}{z + 3} \cdot \frac{z + 3}{z + 3}$$

$$= \frac{2z(z + 3)}{z^2 + 6z + 9}$$

$$= \frac{2z^2 + 6z}{z^2 + 6z + 9}$$

119. $\dfrac{3}{x - y} = \dfrac{3}{x - y} \cdot \dfrac{-1}{-1}$

$$= \frac{-3}{y - x}$$

121. $x^2 - 4 = (x - 2)(x + 2)$

$$= -(2 - x)(x + 2)$$

$$\frac{x}{2 - x} = \frac{x}{2 - x} \cdot \frac{-(x + 2)}{-(x + 2)}$$

$$= \frac{-x(x + 2)}{x^2 - 4}$$

123. $x^3 + y^3 = (x + y)(x^2 - xy + y^2)$

$$\frac{2x}{x + y} = \frac{2x}{x + y} \cdot \frac{x^2 - xy + y^2}{x^2 - xy + y^2}$$

$$= \frac{2x(x^2 - xy + y^2)}{x^3 + y^3}$$

125. $x^3 - y^3 = (x - y)(x^2 + xy + y^2)$

$$\frac{2x}{x - y} = \frac{2x}{x - y} \cdot \frac{x^2 + xy + y^2}{x^2 + xy + y^2}$$

$$= \frac{2x(x^2 + xy + y^2)}{x^3 - y^3}$$

127. $a - 5 = \dfrac{a - 5}{1} \cdot \dfrac{3}{3}$

$$= \frac{3(a - 5)}{3}$$

$$= \frac{3a - 15}{3}$$

129. $x - 7 = \dfrac{x - 7}{1} \cdot \dfrac{x}{x}$

$$= \frac{x(x - 7)}{x}$$

131. $\dfrac{\text{distance}}{\text{time}}: \quad \dfrac{160x}{x + 12}$

133. $\dfrac{2N}{8 - N}$

135. $\quad C = \dfrac{88x}{500 - 5x}$

$x = 10: \quad C = \dfrac{88(10)}{500 - 5(10)} = 1.96$ million dollars

$x = 30: \quad C = \dfrac{88(30)}{500 - 5(30)} = 7.54$ million dollars

$x = 50: \quad C = \dfrac{88(50)}{500 - 5(50)} = 17.6$ million dollars

$x = 75: \quad C = \dfrac{88(75)}{500 - 5(75)} = 52.8$ million dollars

$x = 95: \quad C = \dfrac{88(95)}{500 - 5(95)} = 334.4$ million dollars

$x = 100: \quad C = \dfrac{88(100)}{500 - 5(100)} = \dfrac{8800}{0}$ undefined

137. $\dfrac{(x+2)^{n+1}}{(x+2)^n} = \dfrac{(x+2)^n(x+2)}{(x+2)^n}$

$= x + 2$

139. $x^2 + 4x + 3 = 0$

$(x + 3)(x + 1) = 0$

$x + 3 = 0 \qquad x + 1 = 0$

$x = -3 \qquad x = -1$

141. $x^4 - 16 = 0$

$(x^2 + 4)(x^2 - 4) = 0$

$x^2 - 4 = 0$

$x^2 = 4$

$x = 2 \qquad x = -2$

Problem Set 3.2

1. $\dfrac{xy^2}{z} \cdot \dfrac{z^2}{x^4y} = \dfrac{xy^2z^2}{x^4yz}$

$= \dfrac{yz}{x^3}$

3. $\dfrac{17a + 17b}{3} \cdot \dfrac{-39}{51a + 51b}$

$= \dfrac{17(a + b)(-39)}{3(51)(a + b)}$

$= -\dfrac{13}{3}$

5. $\dfrac{-4}{a^4 - 5a^2 + 6} \cdot (a^4 - 4)$

$= \dfrac{-4(a^2 - 2)(a^2 + 2)}{(a^2 - 3)(a^2 - 2)}$

$= -\dfrac{4(a^2 + 2)}{a^2 - 3}$

7. $\dfrac{2x^2 + 2x - 12}{x^2 - x - 12} \cdot \dfrac{x^2 - 3x - 4}{4x^2 - 4x - 8}$

$= \dfrac{2(x^2 + x - 6)}{(x - 4)(x + 3)} \cdot \dfrac{(x - 4)(x + 1)}{4(x^2 - x - 2)}$

$= \dfrac{2(x + 3)(x - 2)(x - 4)(x + 1)}{(x - 4)(x + 3)(4)(x - 2)(x + 1)}$

$= \dfrac{1}{2}$

9. $\dfrac{s - t}{t^2} \cdot \dfrac{t}{t - s} = \dfrac{-(t - s) \cdot t}{t^2(t - s)}$

$= -\dfrac{1}{t}$

11. $(x^2 - x) \cdot \dfrac{y^2}{x^2 - x^3}$

$= \dfrac{x(x - 1)}{1} \cdot \dfrac{y^2}{x^2(1 - x)}$

$= \dfrac{-x(1 - x)y^2}{x^2(1 - x)}$

$= -\dfrac{y^2}{x}$

13. $\dfrac{abc^2}{xyz^2} \div \dfrac{ab^2c}{xy^4}$

$= \dfrac{abc^2}{xyz^2} \cdot \dfrac{xy^4}{ab^2c}$

$= \dfrac{(abc^2)(xy^4)}{(xyz^2)(ab^2c)}$

$= \dfrac{cy^3}{bz^2}$

15. $\dfrac{(x+a)^2}{4a^2} \div (x+a) = \dfrac{(x+a)^2}{4a^2} \cdot \dfrac{1}{x+a}$

$\qquad\qquad\qquad = \dfrac{(x+a)^2}{4a^2(x+a)}$

$\qquad\qquad\qquad = \dfrac{x+a}{4a^2}$

17. $\dfrac{x^2+3x-4}{x^2+4x+4} \div \dfrac{x^2+2x-3}{x^2+3x+2}$

$\quad = \dfrac{x^2+3x-4}{x^2+4x+4} \cdot \dfrac{x^2+3x+2}{x^2+2x-3}$

$\quad = \dfrac{(x+4)(x-1)}{(x+2)(x+2)} \cdot \dfrac{(x+2)(x+1)}{(x+3)(x-1)}$

$\quad = \dfrac{(x+4)(x-1)(x+2)(x+1)}{(x+2)(x+2)(x+3)(x-1)}$

$\quad = \dfrac{(x+4)(x+1)}{(x+2)(x+3)}$

19. $\dfrac{b^3+8c^3}{x^4-1} \div \dfrac{ab+2ac}{x^2+1}$

$\quad = \dfrac{b^3+8c^3}{x^4-1} \cdot \dfrac{x^2+1}{ab+2ac}$

$\quad = \dfrac{(b+2c)(b^2-2bc+4c^2)}{(x+1)(x-1)(x^2+1)} \cdot \dfrac{x^2+1}{a(b+2c)}$

$\quad = \dfrac{(b+2c)(b^2-2bc+4c^2)(x^2+1)}{(x+1)(x-1)(x^2+1)a(b+2c)}$

$\quad = \dfrac{b^2-2bc+4c^2}{a(x+1)(x-1)}$

21. $(x-y) \div \dfrac{y-x}{y} = \dfrac{x-y}{1} \cdot \dfrac{y}{y-x}$

$\qquad\qquad\qquad = \dfrac{-(y-x)(y)}{y-x}$

$\qquad\qquad\qquad = -y$

23. $\dfrac{xy^3}{z} \cdot \dfrac{z^4}{x^2y} = \dfrac{xy^3z^4}{x^2yz}$

$\qquad\qquad\quad = \dfrac{y^2z^3}{x}$

25. $\dfrac{-3^3m}{n} \cdot \dfrac{n}{(-3)^2m^4} = \dfrac{-3^3mn}{3^2m^4n}$

$\qquad\qquad\qquad = -\dfrac{3}{m^3}$

27. $\dfrac{z}{x^4y} \cdot \left(\dfrac{-xy}{z^2}\right)^3 = \dfrac{z}{x^4y} \cdot \dfrac{-x^3y^3}{z^6}$

$\qquad\qquad\qquad = \dfrac{-x^3y^3z}{x^4yz^6}$

$\qquad\qquad\qquad = -\dfrac{y^2}{xz^5}$

29. $\dfrac{-2^3m}{n} \cdot \dfrac{n}{(-2)^2m^3} = \dfrac{-2^3mn}{2^2m^3n}$

$\qquad\qquad\qquad = -\dfrac{2}{m^2}$

31. $\dfrac{z}{x^2y} \cdot \left(\dfrac{-xy}{z}\right)^3 = \dfrac{z}{x^2y} \cdot \dfrac{-x^3y^3}{z^3}$

$\qquad\qquad\qquad = \dfrac{-x^3y^3z}{x^2yz^3}$

$\qquad\qquad\qquad = -\dfrac{xy^2}{z^2}$

33. $\dfrac{(ab^4)^3}{c(dx)^2} \cdot \dfrac{(c^3d)^2}{(-a)^3b^3}$

$\quad = \dfrac{a^3b^{12}}{cd^2x^2} \cdot \dfrac{c^6d^2}{-a^3b^3}$

$\quad = \dfrac{a^3b^{12}c^6d^2}{-a^3b^3cd^2x^2}$

$\quad = -\dfrac{b^9c^5}{x^2}$

91

35. $\dfrac{x-1}{x} \cdot \dfrac{x^2}{x-1} = \dfrac{(x-1)x^2}{x(x-1)}$

$\qquad\qquad\qquad\quad = x$

37. $\dfrac{13a + 13b}{5} \cdot \dfrac{-55}{39a + 39b}$

$\quad = \dfrac{13(a+b)}{5} \cdot \dfrac{-55}{39(a+b)}$

$\quad = \dfrac{13(-55)(a+b)}{5(39)(a+b)}$

$\quad = -\dfrac{11}{3}$

39. $\dfrac{a^4 + a^2}{b^8} \cdot \dfrac{b^2}{a^3 + a}$

$\quad = \dfrac{a^2(a^2+1)}{b^8} = \dfrac{b^2}{a(a^2+1)}$

$\quad = \dfrac{a^2 b^2(a^2+1)}{ab^8(a^2+1)}$

$\quad = \dfrac{a}{b^6}$

41. $\dfrac{(z+2)^2}{z^2-1} \cdot \dfrac{z-1}{z+2}$

$\quad = \dfrac{(z+2)^2}{(z+1)(z-1)} \cdot \dfrac{z-1}{z+2}$

$\quad = \dfrac{(z+2)^2(z-1)}{(z+1)(z-1)(z+2)}$

$\quad = \dfrac{z+2}{z+1}$

43. $\dfrac{c^2 - 16}{2c - 8} \cdot \dfrac{-8}{c^2 + 3c - 4}$

$\quad = \dfrac{(c+4)(c-4)}{2(c-4)} \cdot \dfrac{-8}{(c+4)(c-1)}$

$\quad = \dfrac{-8(c+4)(c-4)}{2(c-4)(c+4)(c-1)}$

$\qquad = -\dfrac{4}{c-1}$

$\quad \text{or } \dfrac{4}{1-c}$

45. $(x^3 - 125) \cdot \dfrac{x^3}{5-x}$

$\quad = \dfrac{(x-5)(x^2 + 5x + 25)}{1} \cdot \dfrac{x^3}{-(x-5)}$

$\quad = -\dfrac{x^3(x-5)(x^2 + 5x + 25)}{x-5}$

$\quad = -x^3(x^2 + 5x + 25)$

47. $\dfrac{2x^2 + 6x + 4}{x^2 - 4x + 3} \cdot \dfrac{x^2 - x - 6}{4x^2 - 4x - 8}$

$\quad = \dfrac{2(x^2 + 3x + 2)}{(x-3)(x-1)} \cdot \dfrac{(x-3)(x+2)}{4(x^2 - x - 2)}$

$\quad = \dfrac{2(x+2)(x+1)}{(x-3)(x-1)} \cdot \dfrac{(x-3)(x+2)}{4(x-2)(x+1)}$

$\quad = \dfrac{2(x+2)(x+1)(x-3)(x+2)}{4(x-3)(x-1)(x-2)(x+1)}$

$\quad = \dfrac{(x+2)^2}{2(x-1)(x-2)}$

49. $\dfrac{ax - az + bx - bz}{ax - 2bx + az - 2bz} \cdot \dfrac{ax + az + bx + bz}{ax - az - bx + bz}$

$\quad = \dfrac{a(x-z) + b(x-z)}{x(a-2b) + z(a-2b)} \cdot \dfrac{a(x+z) + b(x+z)}{a(x-z) - b(x-z)}$

$\quad = \dfrac{(x-z)(a+b)}{(a-2b)(x+z)} \cdot \dfrac{(x+z)(a+b)}{(x-z)(a-b)}$

$\quad = \dfrac{(x-z)(a+b)(x+z)(a+b)}{(a-2b)(x+z)(x-z)(a-b)}$

$\quad = \dfrac{(a+b)^2}{(a-2b)(a-b)}$

51. $\dfrac{6x^2 + 7x - 3}{10x^2 - x - 2} \cdot \dfrac{5x^2 - 3x - 2}{4x^2 + 4x - 3} \cdot \dfrac{6x^2 - 5x + 1}{3x^2 - 4x + 1}$

$= \dfrac{(3x - 1)(2x + 3)}{(5x + 2)(2x - 1)} \cdot \dfrac{(5x + 2)(x - 1)}{(2x + 3)(2x - 1)} \cdot \dfrac{(3x - 1)(2x - 1)}{(3x - 1)(x - 1)}$

$= \dfrac{(3x - 1)(2x + 3)(5x + 2)(x - 1)(3x - 1)(2x - 1)}{(5x + 2)(2x - 1)(2x + 3)(2x - 1)(3x - 1)(x - 1)}$

$= \dfrac{3x - 1}{2x - 1}$

53. $\dfrac{abc^3}{xyz^4} \div \dfrac{ab^2c}{xy^3}$

$= \dfrac{abc^3}{xyz^4} \cdot \dfrac{xy^3}{ab^2c}$

$= \dfrac{abc^3xy^3}{ab^2cxyz^4}$

$= \dfrac{c^2y^2}{bz^4}$

55. $\dfrac{m^5(n^3p)^2}{rst^3} \div \dfrac{(mnp)^3}{(r^3s^2t^2}$

$= \dfrac{m^5n^6p^2}{rst^3} \div \dfrac{m^3n^3p^3}{r^3s^2t^2}$

$= \dfrac{m^5n^6p^2}{rst^3} \cdot \dfrac{r^3s^2t^2}{m^3n^3p^3}$

$= \dfrac{m^5n^6p^2r^3s^2t^2}{m^3n^3p^3rst^3}$

$= \dfrac{m^2n^3r^2s}{pt}$

57. $\dfrac{-5^2x}{ab} \div \dfrac{5^3x^2y}{a^3}$

$= \dfrac{-5^2x}{ab} \cdot \dfrac{a^3}{5^3x^2y}$

$= \dfrac{-5^2xa^3}{5^3x^2yab}$

$= -\dfrac{a^2}{5xyb}$

59. $\dfrac{(2x)^2y}{15} \div \dfrac{(2x)^3}{75}$

$= \dfrac{2^2x^2y}{15} \div \dfrac{2^3x^3}{75}$

$= \dfrac{2^2x^2y}{15} \cdot \dfrac{75}{2^3x^3}$

$= \dfrac{2^2(75)x^2y}{2^3(15)x^3}$

$= \dfrac{5y}{2x}$

61. $\dfrac{(-ab)^3}{r^2s} \div \dfrac{a^2b}{(rs^2)^2}$

$= \dfrac{-a^3b^3}{r^2s} \div \dfrac{a^2b}{r^2s^4}$

$= \dfrac{-a^3b^3}{r^2s} \cdot \dfrac{r^2s^4}{a^2b}$

$= \dfrac{-a^3b^3r^2s^4}{a^2br^2s}$

$= -ab^2s^3$

63. $\dfrac{(-ab)^3}{r^3s} \div \dfrac{a^5b}{(rs^3)^4}$

$= \dfrac{-a^3b^3}{r^3s} \div \dfrac{a^5b}{r^4s^{12}}$

$= \dfrac{-a^3b^3}{r^3s} \cdot \dfrac{r^4s^{12}}{a^5b}$

$= \dfrac{-a^3b^3r^4s^{12}}{a^5br^3s}$

$= -\dfrac{b^2rs^{11}}{a^2}$

65. $\dfrac{p}{p-q} \div \dfrac{q}{q-p} = \dfrac{p}{p-q} \div \dfrac{q}{-(p-q)}$

$= \dfrac{p}{p-q} \cdot \dfrac{-(p-q)}{q}$

$= \dfrac{-p(p-q)}{q(p-q)}$

$= -\dfrac{p}{q}$

67. $\dfrac{ax-bx}{ay+by} \div \dfrac{a^2-ab}{2a+2b}$

$= \dfrac{ax-bx}{ay+by} \cdot \dfrac{2a+2b}{a^2-ab}$

$= \dfrac{x(a-b)}{y(a+b)} \cdot \dfrac{2(a+b)}{a(a-b)}$

$= \dfrac{2x(a-b)(a+b)}{ay(a-b)(a+b)}$

$= \dfrac{2x}{ay}$

69. $\dfrac{p^2+pt}{p-p^2} \div \dfrac{pq+qt}{pq-q}$

$= \dfrac{p^2+pt}{p-p^2} \cdot \dfrac{pq-q}{pq+qt}$

$= \dfrac{p(p+t)}{p(1-p)} \cdot \dfrac{q(p-1)}{q(p+t)}$

$= \dfrac{p(p+t)}{-p(p-1)} \cdot \dfrac{q(p-1)}{q(p+t)}$

$= \dfrac{pq(p+t)(p-1)}{-pq(p-1)(p+t)}$

$= \dfrac{1}{-1}$

$= -1$

71. $\dfrac{16x^2-1}{x^2+8x+16} \div \dfrac{64x^3-1}{x+4}$

$= \dfrac{16x^2-1}{x^2+8x+16} \cdot \dfrac{x+4}{64x^3-1}$

$= \dfrac{(4x-1)(4x+1)}{(x+4)(x+4)} \cdot \dfrac{x+4}{(4x-1)(16x^2+4x+1)}$

$= \dfrac{(4x-1)(4x+1)(x+4)}{(x+4)(x+4)(4x-1)(16x^2+4x+1)}$

$= \dfrac{4x+1}{(x+4)(16x^2+4x+1)}$

73. $\dfrac{x^2-4x-12}{x^2-4x-5} \div \dfrac{x^2-3x-18}{x^2-7x+10}$

$= \dfrac{x^2-4x-12}{x^2-4x-5} \cdot \dfrac{x^2-7x+10}{x^2-3x-18}$

$= \dfrac{(x-6)(x+2)}{(x-5)(x+1)} \cdot \dfrac{(x-5)(x-2)}{(x-6)(x+3)}$

$= \dfrac{(x-6)(x+2)(x-5)(x-2)}{(x-5)(x+1)(x-6)(x+3)}$

$= \dfrac{(x+2)(x-2)}{(x+1)(x+3)}$

75. $\dfrac{x^2-3x-10}{2x^2+5x+3} \div \dfrac{5+4x-x^2}{2x^2+7x+6}$

$= \dfrac{x^2-3x-10}{2x^2+5x+3} \cdot \dfrac{2x^2+7x+6}{5+4x-x^2}$

$= \dfrac{(x-5)(x+2)}{(2x+3)(x+1)} \cdot \dfrac{(2x+3)(x+2)}{(5-x)(1+x)}$

$= \dfrac{(x-5)(x+2)}{(2x+3)(x+1)} \cdot \dfrac{(2x+3)(x+2)}{-(x-5)(1+x)}$

$= \dfrac{(x-5)(x+2)(2x+3)(x+2)}{-(2x+3)(x+1)(x-5)(1+x)}$

$= -\dfrac{(x+2)^2}{(x+1)^2}$

77. $\dfrac{x^2 + 10x + 21}{x^2 - 4x + 3} \div \dfrac{x^3 + 7x^2}{x^2 - 2x + 1}$

$= \dfrac{x^2 + 10x + 21}{x^2 - 4x + 3} \cdot \dfrac{x^2 - 2x + 1}{x^3 + 7x^2}$

$= \dfrac{(x + 7)(x + 3)}{(x - 3)(x - 1)} \cdot \dfrac{(x - 1)(x - 1)}{x^2(x + 7)}$

$= \dfrac{(x + 7)(x + 3)(x - 1)(x - 1)}{x^2(x - 3)(x - 1)(x + 7)}$

$= \dfrac{(x + 3)(x - 1)}{x^2(x - 3)}$

79. $\dfrac{2x^2 + 5x - 3}{2x^2 + 5x + 3} \div \dfrac{2x^3 - x^2}{2x^2 + x - 3}$

$= \dfrac{2x^2 + 5x - 3}{2x^2 + 5x + 3} \cdot \dfrac{2x^2 + x - 3}{2x^3 - x^2}$

$= \dfrac{(2x - 1)(x + 3)}{(2x + 3)(x + 1)} \cdot \dfrac{(2x + 3)(x - 1)}{x^2(2x - 1)}$

$= \dfrac{(2x - 1)(x + 3)(2x + 3)(x - 1)}{x^2(2x + 3)(x + 1)(2x - 1)}$

$= \dfrac{(x + 3)(x - 1)}{x^2(x + 1)}$

81. $\dfrac{b^3 - 2b^2 + 3b - 6}{b^2 - b - 6} \div \dfrac{b^4 - 9}{b^2 - 9}$

$= \dfrac{b^3 - 2b^2 + 3b - 6}{b^2 - b - 6} \cdot \dfrac{b^2 - 9}{b^4 - 9}$

$= \dfrac{b^2(b - 2) + 3(b - 2)}{(b - 3)(b + 2)} \cdot \dfrac{(b + 3)(b - 3)}{(b^2 - 3)(b^2 + 3)}$

$= \dfrac{(b - 2)(b^2 + 3)}{(b - 3)(b + 2)} \cdot \dfrac{(b + 3)(b - 3)}{(b^2 - 3)(b^2 + 3)}$

$= \dfrac{(b - 2)(b^2 + 3)(b + 3)(b - 3)}{(b - 3)(b + 2)(b^2 - 3)(b^2 + 3)}$

$= \dfrac{(b - 2)(b + 3)}{(b + 2)(b^2 - 3)}$

83. $\dfrac{ac + bc + 2ad + 2bd}{a^2 - d^2} \div \dfrac{ac - bc + 2ad - 2bd}{a^2 + 2ab - ad - 2bd}$

$= \dfrac{ac + bc + 2ad + 2bd}{a^2 - d^2} \cdot \dfrac{a^2 + 2ab - ad - 2bd}{ac - bc + 2ad - 2bd}$

$= \dfrac{c(a + b) + 2d(a + b)}{(a - d)(a + d)} \cdot \dfrac{a(a + 2b) - d(a + 2b)}{c(a - b) + 2d(a - b)}$

$= \dfrac{(a + b)(c + 2d)}{(a - d)(a + d)} \cdot \dfrac{(a + 2b)(a - d)}{(a - b)(c + 2d)}$

$= \dfrac{(a + b)(c + 2d)(a + 2b)(a - d)}{(a - d)(a + d)(a - b)(c + 2d)}$

$= \dfrac{(a + b)(a + 2b)}{(a + d)(a - b)}$

85. $\dfrac{2x^2 + 5x - 3}{3x^2 + 2x - 5} \cdot \dfrac{3x^2 - x - 10}{x^2 + 2x - 3} \cdot \dfrac{x^2 - 2x + 1}{2x^2 + x - 1}$

$= \dfrac{(2x - 1)(x + 3)}{(3x + 5)(x - 1)} \cdot \dfrac{(3x + 5)(x - 2)}{(x + 3)(x - 1)} \cdot \dfrac{(x - 1)(x - 1)}{(2x - 1)(x + 1)}$

$= \dfrac{(2x - 1)(x + 3)(3x + 5)(x - 2)(x - 1)^2}{(3x + 5)(x - 1)(x + 3)(x - 1)(2x - 1)(x + 1)}$

$= \dfrac{x - 2}{x + 1}$

87. $\dfrac{p^2 + pt - 2t^2}{2p^2 - 5pt - 3t^2} \cdot \dfrac{p^2 - 2pt - 3t^2}{p^2 - 3pt + 2t^2} \cdot \dfrac{4p^2 + 4pt + t^2}{2p^2 + 3pt - 2t^2}$

$= \dfrac{(p + 2t)(p - t)}{(2p + t)(p - 3t)} \cdot \dfrac{(p - 3t)(p + t)}{(p - 2t)(p - t)} \cdot \dfrac{(2p + t)(2p + t)}{(2p - t)(p + 2t)}$

$= \dfrac{(p + 2t)(p - t)(p - 3t)(p + t)(2p + t)^2}{(2p + t)(p - 3t)(p - 2t)(p - t)(2p - t)(p + 2t)}$

$= \dfrac{(p + t)(2p + t)}{(p - 2t)(2p - t)}$

89.

$$\frac{x^2 - y^2}{2x^2 + 3xy - 2y^2} \cdot \frac{x^3 + 2x^2y}{x^2 + 2xy + y^2} \div \frac{x^2 + xy - 2y^2}{x^3 + x^2y + x + y}$$

$$= \frac{x^2 - y^2}{2x^2 + 3xy - 2y^2} \cdot \frac{x^3 + 2x^2y}{x^2 + 2xy + y^2} \cdot \frac{x^3 + x^2y + x + y}{x^2 + xy - 2y^2}$$

$$= \frac{(x - y)(x + y)}{(2x - y)(x + 2y)} \cdot \frac{x^2(x + 2y)}{(x + y)^2} \cdot \frac{(x + y)(x^2 + 1)}{(x + 2y)(x - y)}$$

$$= \frac{(x - y)(x + y)(x^2)(x + 2y)(x + y)(x^2 + 1)}{(2x - y)(x + 2y)(x + y)^2(x + 2y)(x - y)}$$

$$= \frac{x^2(x^2 + 1)}{(2x - y)(x + 2y)}$$

91.

$$\frac{x^{2n} - x^n - 6}{x^{2p} + 2x^p - 3} \cdot \frac{x^{2p} - 1}{x^{2n} + 4x^n + 4}$$

$$= \frac{(x^n - 3)(x^n + 2)}{(x^p + 3)(x^p - 1)} \cdot \frac{(x^p + 1)(x^p - 1)}{(x^n + 2)(x^n + 2)}$$

$$= \frac{(x^n - 3)(x^n + 2)(x^p + 1)(x^p - 1)}{(x^p + 3)(x^p - 1)(x^n + 2)(x^n + 2)}$$

$$= \frac{(x^n - 3)(x^p + 1)}{(x^p + 3)(x^n + 2)}$$

Problem Set 3.3

1.

$$\frac{7}{x^2} - \frac{x + 2}{x^2}$$

$$= \frac{7 - (x + 2)}{x^2}$$

$$= \frac{7 - x - 2}{x^2}$$

$$= \frac{5 - x}{x^2}$$

3.

$$\frac{x + y}{xy} + \frac{x - y}{xy}$$

$$= \frac{x + y + x - y}{xy}$$

$$= \frac{2x}{xy}$$

$$= \frac{2}{y}$$

5.

$$\frac{x}{x + y} + \frac{x + 2y}{x + y}$$

$$= \frac{x + x + 2y}{x + y}$$

$$= \frac{2x + 2y}{x + y}$$

$$= \frac{2(x + y)}{x + y}$$

$$= 2$$

7.

$$\frac{3}{m - 2} - \frac{2 - m}{m - 2}$$

$$= \frac{3 - (2 - m)}{m - 2}$$

$$= \frac{3 - 2 + m}{m - 2}$$

$$= \frac{1 + m}{m - 2}$$

9.

$$\frac{3y}{x - 5} - \frac{2y}{5 - x}$$

$$= \frac{3y}{x - 5} - \frac{2y}{-(x - 5)}$$

$$= \frac{3y}{x - 5} + \frac{2y}{x - 5}$$

$$= \frac{3y + 2y}{x - 5}$$

$$= \frac{5y}{x - 5}$$

11. $\dfrac{p}{pq^2} + \dfrac{q}{p^2q}$

$= \dfrac{p(p)}{pq^2(p)} + \dfrac{q(q)}{p^2q(q)}$

$= \dfrac{p^2}{q^2p^2} + \dfrac{q^2}{p^2q^2}$

$= \dfrac{p^2 + q^2}{p^2q^2}$

13. $\dfrac{1}{r + 2} + \dfrac{2}{r - 3}$

$= \dfrac{1(r - 3)}{(r + 2)(r - 3)} + \dfrac{2(r + 2)}{(r - 3)(r + 2)}$

$= \dfrac{r - 3 + 2(r + 2)}{(r + 2)(r - 3)}$

$= \dfrac{r - 3 + 2r + 4}{(r + 2)(r - 3)}$

$= \dfrac{3r + 1}{(r + 2)(r - 3)}$

15. $\dfrac{2}{5a + 10} + \dfrac{7}{3a + 6}$

$= \dfrac{2}{5(a + 2)} + \dfrac{7}{3(a + 2)}$

$= \dfrac{2(3)}{5(a + 2)(3)} + \dfrac{7(5)}{3(a + 2)(5)}$

$= \dfrac{6}{15(a + 2)} + \dfrac{35}{15(a + 2)}$

$= \dfrac{6 + 35}{15(a + 2)}$

$= \dfrac{41}{15(a + 2)}$

17. $\dfrac{x + 5}{x^2 - 2x - 15} - \dfrac{x}{x^2 - 6x + 5}$

$= \dfrac{x + 5}{(x - 5)(x + 3)} - \dfrac{x}{(x - 5)(x - 1)}$

$= \dfrac{(x + 5)(x - 1)}{(x - 5)(x + 3)(x - 1)} - \dfrac{x(x + 3)}{(x - 5)(x - 1)(x + 3)}$

$= \dfrac{x^2 + 4x - 5}{(x - 5)(x + 3)(x - 1)} - \dfrac{x^2 + 3x}{(x - 5)(x + 3)(x - 1)}$

$= \dfrac{x^2 + 4x - 5 - (x^2 + 3x)}{(x - 5)(x + 3)(x - 1)}$

$= \dfrac{x^2 + 4x - 5 - x^2 - 3x}{(x - 5)(x + 3)(x - 1)}$

$= \dfrac{x - 5}{(x - 5)(x + 3)(x - 1)}$

$= \dfrac{1}{(x + 3)(x - 1)}$

19. $x + \dfrac{2}{x}$

$= \dfrac{x(x)}{1(x)} + \dfrac{2}{x}$

$= \dfrac{x^2}{x} + \dfrac{2}{x}$

$= \dfrac{x^2 + 2}{x}$

21. $\dfrac{t}{t - 1} - 1$

$= \dfrac{t}{t - 1} - \dfrac{1(t - 1)}{t - 1}$

$= \dfrac{t}{t - 1} - \dfrac{t - 1}{t - 1}$

$= \dfrac{t - (t - 1)}{t - 1}$

$= \dfrac{t - t + 1}{t - 1}$

$= \dfrac{1}{t - 1}$

23. $\dfrac{10}{4 - 2a} - \dfrac{12}{3a - 6}$

$= \dfrac{10}{2(2 - a)} - \dfrac{12}{3(a - 2)}$

$= \dfrac{5}{2 - a} - \dfrac{4}{a - 2}$

$= \dfrac{5}{-(a - 2)} - \dfrac{4}{a - 2}$

$= \dfrac{-5}{a - 2} - \dfrac{4}{a - 2}$

$= \dfrac{-5 - 4}{a - 2}$

$= -\dfrac{9}{a - 2}$

$= \dfrac{9}{2 - a}$

25. $\dfrac{x}{x^2 - 6x + 8} - \dfrac{2}{2 + x - x^2}$

$= \dfrac{x}{x^2 - 6x + 8} - \dfrac{2}{-(x^2 - x - 2)}$

$= \dfrac{x}{x^2 - 6x + 8} + \dfrac{2}{x^2 - x - 2}$

$= \dfrac{x}{(x - 2)(x - 4)} + \dfrac{2}{(x - 2)(x + 1)}$

$= \dfrac{x(x + 1)}{(x - 2)(x - 4)(x + 1)} + \dfrac{2(x - 4)}{(x - 2)(x + 1)(x - 4)}$

$= \dfrac{x^2 + x}{(x - 2)(x - 4)(x + 1)} + \dfrac{2x - 8}{(x - 2)(x + 1)(x - 4)}$

$= \dfrac{x^2 + x + 2x - 8}{(x - 2)(x - 4)(x + 1)}$

$= \dfrac{x^2 + 3x - 8}{(x - 2)(x - 4)(x + 1)}$

27. $\dfrac{8}{y^3} - \dfrac{x + 2}{y^3}$

$= \dfrac{8 - (x + 2)}{y^3}$

$= \dfrac{8 - x - 2}{y^3}$

$= \dfrac{6 - x}{y^3}$

29. $\dfrac{a}{b^3c} + \dfrac{b}{bc^2}$

$= \dfrac{a(c)}{b^3c(c)} + \dfrac{b(b^2)}{bc^2(b^2)}$

$= \dfrac{ac}{b^3c^2} + \dfrac{b^3}{b^3c^2}$

$= \dfrac{ac + b^3}{b^3c^2}$

31. $\dfrac{2}{x^2y} - \dfrac{4}{xyz} + \dfrac{1}{x}$

$= \dfrac{2(z)}{x^2y(z)} - \dfrac{4(x)}{xyz(x)} + \dfrac{1(xyz)}{x(xyz)}$

$= \dfrac{2z}{x^2yz} - \dfrac{4x}{x^2yz} + \dfrac{xyz}{x^2yz}$

$= \dfrac{2z - 4x + xyz}{x^2yz}$

33. $\dfrac{2}{y - 1} + \dfrac{1}{y}$

$= \dfrac{2(y)}{(y - 1)(y)} + \dfrac{1(y - 1)}{y(y - 1)}$

$= \dfrac{2y}{(y - 1)y} + \dfrac{(y - 1)}{y(y - 1)}$

$= \dfrac{2y + y - 1}{y(y - 1)}$

$= \dfrac{3y - 1}{y(y - 1)}$

35. $\dfrac{3}{s+1} - \dfrac{1}{s-1}$

$= \dfrac{3(s-1)}{(s+1)(s-1)} - \dfrac{1(s+1)}{(s-1)(s+1)}$

$= \dfrac{3s-3}{(s+1)(s-1)} - \dfrac{s+1}{(s-1)(s+1)}$

$= \dfrac{3s-3-(s+1)}{(s+1)(s-1)}$

$= \dfrac{3s-3-s-1}{(s+1)(s-1)}$

$= \dfrac{2s-4}{(s+1)(s-1)}$

37. $\dfrac{q}{q-4} + \dfrac{q}{q+6}$

$= \dfrac{q(q+6)}{(q-4)(q+6)} + \dfrac{q(q-4)}{(q+6)(q-4)}$

$= \dfrac{q^2+6q}{(q-4)(q+6)} + \dfrac{q^2-4q}{(q+6)(q-4)}$

$= \dfrac{q^2+6q+q^2-4q}{(q-4)(q+6)}$

$= \dfrac{2q^2+2q}{(q-4)(q+6)}$

39. $\dfrac{a}{x-1} + \dfrac{4a}{1-x}$

$= \dfrac{a}{x-1} + \dfrac{4a}{-(x-1)}$

$= \dfrac{a}{x-1} - \dfrac{4a}{x-1}$

$= \dfrac{a-4a}{x-1}$

$= \dfrac{-3a}{x-1}$

$= -\dfrac{3a}{x-1}$

41. $\dfrac{2}{3r+6} + \dfrac{5}{4r+8}$

$= \dfrac{2}{3(r+2)} + \dfrac{5}{4(r+2)}$

$= \dfrac{2(4)}{3(r+2)(4)} + \dfrac{5(3)}{4(r+2)(3)}$

$= \dfrac{8}{12(r+2)} + \dfrac{15}{12(r+2)}$

$= \dfrac{8+15}{12(r+2)}$

$= \dfrac{23}{12(r+2)}$

43. $\dfrac{3}{5m-5n} + \dfrac{4}{2n-2m}$

$= \dfrac{3}{5(m-n)} + \dfrac{4}{-2(m-n)}$

$= \dfrac{3}{5(m-n)} - \dfrac{2}{m-n}$

$= \dfrac{3}{5(m-n)} - \dfrac{2(5)}{(m-n)(5)}$

$= \dfrac{3-10}{5(m-n)}$

$= \dfrac{-7}{5(m-n)}$

$= -\dfrac{7}{5(m-n)}$

45. $\dfrac{2t}{t - 1} - \dfrac{t^2 - 4t + 3}{t^2 - 2t + 1}$

$= \dfrac{2t}{t - 1} - \dfrac{(t - 3)(t - 1)}{(t - 1)^2}$

$= \dfrac{2t}{t - 1} - \dfrac{t - 3}{t - 1}$

$= \dfrac{2t - (t - 3)}{t - 1}$

$= \dfrac{2t - t + 3}{t - 1}$

$= \dfrac{t + 3}{t - 1}$

47. $\dfrac{d}{d + 3} + \dfrac{5d + 9}{d^2 + 4d + 3}$

$= \dfrac{d}{d + 3} + \dfrac{5d + 9}{(d + 3)(d + 1)}$

$= \dfrac{d(d + 1)}{(d + 3)(d + 1)} + \dfrac{5d + 9}{(d + 3)(d + 1)}$

$= \dfrac{d^2 + d}{(d + 3)(d + 1)} + \dfrac{5d + 9}{(d + 3)(d + 1)}$

$= \dfrac{d^2 + d + 5d + 9}{(d + 3)(d + 1)}$

$= \dfrac{d^2 + 6d + 9}{(d + 3)(d + 1)}$

$= \dfrac{(d + 3)^2}{(d + 3)(d + 1)}$

$= \dfrac{d + 3}{d + 1}$

49. $\dfrac{x^2 - x}{x^2 - 9} + \dfrac{1}{3 - x}$

$= \dfrac{x^2 - x}{(x + 3)(x - 3)} + \dfrac{1}{-(x - 3)}$

$= \dfrac{x^2 - x}{(x + 3)(x - 3)} - \dfrac{1}{(x - 3)}$

$= \dfrac{x^2 - x}{(x + 3)(x - 3)} - \dfrac{1(x + 3)}{(x - 3)(x + 3)}$

$= \dfrac{x^2 - x - (x + 3)}{(x + 3)(x - 3)}$

$= \dfrac{x^2 - x - x - 3}{(x + 3)(x - 3)}$

$= \dfrac{x^2 - 2x - 3}{(x + 3)(x - 3)}$

$= \dfrac{(x - 3)(x + 1)}{(x + 3)(x - 3)}$

$= \dfrac{x + 1}{x + 3}$

51. $\dfrac{1}{x^2 - 7x + 12} + \dfrac{x}{x^2 - 12x + 32}$

$= \dfrac{1}{(x - 4)(x - 3)} + \dfrac{x}{(x - 4)(x - 8)}$

$= \dfrac{1(x - 8)}{(x - 4)(x - 3)(x - 8)} + \dfrac{x(x - 3)}{(x - 4)(x - 8)(x - 3)}$

$= \dfrac{x - 8}{(x - 4)(x - 3)(x - 8)} + \dfrac{x^2 - 3x}{(x - 4)(x - 8)(x - 3)}$

$= \dfrac{x - 8 + x^2 - 3x}{(x - 4)(x - 3)(x - 8)}$

$= \dfrac{x^2 - 2x - 8}{(x - 4)(x - 3)(x - 8)}$

$= \dfrac{(x - 4)(x + 2)}{(x - 4)(x - 3)(x - 8)}$

$= \dfrac{x + 2}{(x - 3)(x - 8)}$

53. $\dfrac{z+2}{2z^2-21z+10} + \dfrac{1}{2z^2-7z+3}$

$= \dfrac{z+2}{(2z-1)(z-10)} + \dfrac{1}{(2z-1)(z-3)}$

$= \dfrac{(z+2)(z-3)}{(2z-1)(z-10)(z-3)} + \dfrac{1(z-10)}{(2z-1)(z-3)(z-10)}$

$= \dfrac{z^2-z-6}{(2z-1)(z-10)(z-3)} + \dfrac{z-10}{(2z-1)(z-3)(z-10)}$

$= \dfrac{z^2-z-6+z-10}{(2z-1)(z-10)(z-3)}$

$= \dfrac{z^2-16}{(2z-1)(z-10)(z-3)}$

55. $\dfrac{-3y+9}{y^3+27} + \dfrac{4}{y^2-9}$

$= \dfrac{-3y+9}{(y+3)(y^2-3y+9)} + \dfrac{4}{(y+3)(y-3)}$

$= \dfrac{(-3y+9)(y-3)}{(y+3)(y^2-3y+9)(y-3)} + \dfrac{4(y^2-3y+9)}{(y+3)(y-3)(y^2-3y+9)}$

$= \dfrac{-3y^2+18y-27}{(y+3)(y^2-3y+9)(y-3)} + \dfrac{4y^2-12y+36}{(y+3)(y-3)(y^2-3y+9)}$

$= \dfrac{-3y^2+18y-27+4y^2-12y+36}{(y+3)(y-3)(y^2-3y+9)}$

$= \dfrac{y^2+6y+9}{(y+3)(y-3)(y^2-3y+9)}$

$= \dfrac{(y+3)^2}{(y+3)(y-3)(y^2-3y+9)}$

$= \dfrac{y+3}{(y-3)(y^2-3y+9)}$

57. $\dfrac{3z + 17}{2z^2 + z - 3} + \dfrac{3z + 7}{2z^2 + 7z + 6}$

$= \dfrac{3z + 17}{(2z + 3)(z - 1)} + \dfrac{3z + 7}{(2z + 3)(z + 2)}$

$= \dfrac{(3z + 17)(z + 2)}{(2z + 3)(z - 1)(z + 2)} + \dfrac{(3z + 7)(z - 1)}{(2z + 3)(z + 2)(z - 1)}$

$= \dfrac{3z^2 + 23z + 34}{(2z + 3)(z - 1)(z + 2)} + \dfrac{3z^2 + 4z - 7}{(2z + 3)(z + 2)(z - 1)}$

$= \dfrac{3z^2 + 23z + 34 + 3z^2 + 4z - 7}{(2z + 3)(z - 1)(z + 2)}$

$= \dfrac{6z^2 + 27z + 27}{(2z + 3)(z - 1)(z + 2)}$

$= \dfrac{3(2z^2 + 9z + 9)}{(2z + 3)(z - 1)(z + 2)}$

$= \dfrac{3(2z + 3)(z + 3)}{(2z + 3)(z - 1)(z + 2)}$

$= \dfrac{3(z + 3)}{(z - 1)(z + 2)}$

59. $\dfrac{y}{y^2 - 10y + 25} - \dfrac{y + 1}{y^2 - 25}$

$= \dfrac{y}{(y - 5)^2} - \dfrac{y + 1}{(y - 5)(y + 5)}$

$= \dfrac{y(y + 5)}{(y - 5)^2(y + 5)} - \dfrac{(y + 1)(y - 5)}{(y - 5)(y + 5)(y - 5)}$

$= \dfrac{y^2 + 5y}{(y - 5)^2(y + 5)} - \dfrac{y^2 - 4y - 5}{(y - 5)^2(y + 5)}$

$= \dfrac{y^2 + 5y - (y^2 - 4y - 5)}{(y - 5)^2(y + 5)}$

$= \dfrac{y^2 + 5y - y^2 + 4y + 5}{(y - 5)^2(y + 5)}$

$= \dfrac{9y + 5}{(y - 5)^2(y + 5)}$

61. $\dfrac{s}{s^2 - 5s - 24} - \dfrac{s - 1}{s^2 - 10s + 16}$

$= \dfrac{s}{(s - 8)(s + 3)} - \dfrac{s - 1}{(s - 8)(s - 2)}$

$= \dfrac{s(s - 2)}{(s - 8)(s + 3)(s - 2)} - \dfrac{(s - 1)(s + 3)}{(s - 8)(s - 2)(s + 3)}$

$= \dfrac{s^2 - 2s}{(s - 8)(s + 3)(s - 2)} - \dfrac{s^2 + 2s - 3}{(s - 8)(s - 2)(s + 3)}$

$= \dfrac{s^2 - 2s - (s^2 + 2s - 3)}{(s - 8)(s + 3)(s - 2)}$

$= \dfrac{s^2 - 2s - s^2 - 2s + 3}{(s - 8)(s + 3)(s - 2)}$

$= \dfrac{-4s + 3}{(s - 8)(s + 3)(s - 2)}$

63. $\dfrac{u^2}{u - 1} - u$

$= \dfrac{u^2}{u - 1} - \dfrac{u(u - 1)}{u - 1}$

$= \dfrac{u^2}{u - 1} - \dfrac{u^2 - u}{u - 1}$

$= \dfrac{u^2 - (u^2 - u)}{u - 1}$

$= \dfrac{u^2 - u^2 + u}{u - 1}$

$= \dfrac{u}{u - 1}$

65. $\dfrac{a}{a + b} - 2$

$= \dfrac{a}{a + b} - \dfrac{2(a + b)}{a + b}$

$= \dfrac{a}{a + b} - \dfrac{2a + 2b}{a + b}$

$= \dfrac{a - (2a + 2b)}{a + b}$

$= \dfrac{a - 2a - 2b}{a + b}$

$= \dfrac{-a - 2b}{a + b}$

67. $\dfrac{2}{x} - \dfrac{3}{x + 2} + \dfrac{4}{x^2 + 2x}$

$= \dfrac{2}{x} - \dfrac{3}{x + 2} + \dfrac{4}{x(x + 2)}$

$= \dfrac{2(x + 2)}{x(x + 2)} - \dfrac{3x}{(x + 2)x} + \dfrac{4}{x(x + 2)}$

$= \dfrac{2x + 4}{x(x + 2)} - \dfrac{3x}{(x + 2)x} + \dfrac{4}{x(x + 2)}$

$= \dfrac{2x + 4 - 3x + 4}{x(x + 2)}$

$= \dfrac{-x + 8}{x(x + 2)}$

69.

$$\frac{1}{x + 4} - \frac{1}{x + 3} - \frac{1}{x + 2}$$

$$= \frac{1(x + 3)(x + 2)}{(x + 4)(x + 3)(x + 2)} - \frac{1(x + 4)(x + 2)}{(x + 3)(x + 4)(x + 2)} - \frac{1(x + 3)(x + 4)}{(x + 2)(x + 3)(x + 4)}$$

$$= \frac{x^2 + 5x + 6}{(x + 4)(x + 3)(x + 2)} - \frac{x^2 + 6x + 8}{(x + 3)(x + 4)(x + 2)} - \frac{x^2 + 7x + 12}{(x + 2)(x + 3)(x + 4)}$$

$$= \frac{x^2 + 5x + 6 - (x^2 + 6x + 8) - (x^2 + 7x + 12)}{(x + 4)(x + 3)(x + 2)}$$

$$= \frac{x^2 + 5x + 6 - x^2 - 6x - 8 - x^2 - 7x - 12}{(x + 4)(x + 3)(x + 2)}$$

$$= \frac{-x^2 - 8x - 14}{(x + 4)(x + 3)(x + 2)} = -\frac{x^2 + 8x + 14}{(x + 4)(x + 3)(x + 2)}$$

71.

$$\frac{qt}{q^2 - qt} + \frac{1}{q - t} + 1$$

$$= \frac{qt}{q(q - t)} + \frac{1}{q - t} + 1$$

$$= \frac{t}{q - t} + \frac{1}{q - t} + \frac{1(q - t)}{q - t}$$

$$= \frac{t + 1 + q - t}{q - t}$$

$$= \frac{q + 1}{q - t}$$

73.

$$\frac{3}{z + 1} - \frac{1}{z - 1} + \frac{1}{z + 2}$$

$$= \frac{3(z - 1)(z + 2)}{(z + 1)(z - 1)(z + 2)} - \frac{1(z + 1)(z + 2)}{(z - 1)(z + 1)(z + 2)} + \frac{1(z + 1)(z - 1)}{(z + 2)(z + 1)(z - 1)}$$

$$= \frac{3z^2 + 3z - 6}{(z + 1)(z - 1)(z + 2)} - \frac{z^2 + 3z + 2}{(z - 1)(z + 1)(z + 2)} + \frac{z^2 - 1}{(z + 2)(z + 1)(z - 1)}$$

$$= \frac{3z^2 + 3z - 6 - (z^2 + 3z + 2) + z^2 - 1}{(z + 1)(z - 1)(z + 2)}$$

$$= \frac{3z^2 + 3z - 6 - z^2 - 3z - 2 + z^2 - 1}{(z + 1)(z - 1)(z + 2)}$$

$$= \frac{3z^2 - 9}{(z + 1)(z - 1)(z + 2)}$$

75. $\dfrac{7}{x^2 + 5x + 6} - \dfrac{49}{x^2 - x - 12} + \dfrac{x}{x^2 - 2x - 8}$

$= \dfrac{7}{(x + 2)(x + 3)} - \dfrac{49}{(x - 4)(x + 3)} + \dfrac{x}{(x - 4)(x + 2)}$

$= \dfrac{7(x - 4)}{(x + 2)(x + 3)(x - 4)} - \dfrac{49(x + 2)}{(x - 4)(x + 3)(x + 2)} + \dfrac{x(x + 3)}{(x - 4)(x + 2)(x + 3)}$

$= \dfrac{7x - 28}{(x + 2)(x + 3)(x - 4)} - \dfrac{49x + 98}{(x - 4)(x + 3)(x + 2)} + \dfrac{x^2 + 3x}{(x - 4)(x + 2)(x + 3)}$

$= \dfrac{7x - 28 - (49x + 98) + x^2 + 3x}{(x + 2)(x + 3)(x - 4)}$

$= \dfrac{7x - 28 - 49x - 98 + x^2 + 3x}{(x + 2)(x + 3)(x - 4)}$

$= \dfrac{x^2 - 39x - 126}{(x + 2)(x + 3)(x - 4)}$

$= \dfrac{(x - 42)(x + 3)}{(x + 2)(x + 3)(x - 4)}$

$= \dfrac{x - 42}{(x + 2)(x - 4)}$

77. $\dfrac{1}{f} = \dfrac{1}{\text{object distance}} + \dfrac{1}{\text{image distance}}$

$= \dfrac{1}{x} + \dfrac{1}{x + 5}$

$= \dfrac{1(x + 5)}{x(x + 5)} + \dfrac{1(x)}{(x + 5)(x)}$

$= \dfrac{x + 5 + x}{x(x + 5)}$

$= \dfrac{2x + 5}{x(x + 5)} \quad \text{cm}$

79. $\dfrac{1}{R_t} = \dfrac{1}{R_1} + \dfrac{1}{R_2}$

$= \dfrac{1}{x + 1} + \dfrac{1}{2x - 3}$

$= \dfrac{1(2x - 3)}{(x + 1)(2x - 3)} + \dfrac{1(x + 1)}{(2x - 3)(x + 1)}$

$$= \frac{2x - 3 + x + 1}{(x + 1)(2x - 3)}$$

$$= \frac{3x - 2}{(x + 1)(2x - 3)} \quad \text{ohms}$$

81. $(x + 2)\left(\dfrac{3}{x + 2} + \dfrac{1}{x - 1}\right)$

$$= \frac{x + 2}{1} \cdot \frac{3}{x + 2} + \frac{x + 2}{1} \cdot \frac{1}{x - 1}$$

$$= 3 + \frac{x + 2}{x - 1}$$

$$= \frac{3(x - 1)}{x - 1} + \frac{x + 2}{x - 1}$$

$$= \frac{3x - 3}{x - 1} + \frac{x + 2}{x - 1}$$

$$= \frac{3x - 3 + x + 2}{x - 1}$$

$$= \frac{4x - 1}{x - 1}$$

83. $\left(\dfrac{1}{x - 3} - \dfrac{1}{x + 3}\right)\left(\dfrac{1}{x + 3} + \dfrac{1}{x - 3}\right)$

$$= \left[\frac{1(x + 3)}{(x - 3)(x + 3)} - \frac{1(x - 3)}{(x + 3)(x - 3)}\right]\left[\frac{1(x - 3)}{(x + 3)(x - 3)} + \frac{1(x + 3)}{(x - 3)(x + 3)}\right]$$

$$= \left[\frac{x + 3 - (x - 3)}{(x + 3)(x - 3)}\right]\left[\frac{x - 3 + x + 3}{(x + 3)(x - 3)}\right]$$

$$= \left[\frac{x + 3 - x + 3}{(x + 3)(x - 3)}\right]\left[\frac{2x}{(x + 3)(x - 3)}\right]$$

$$= \left[\frac{6}{(x + 3)(x - 3)}\right]\left[\frac{2x}{(x + 3)(x - 3)}\right]$$

$$= \frac{12x}{(x + 3)^2(x - 3)^2}$$

Problem Set 3.4

1. $\dfrac{1}{x} + 2 = 3$

$x\left(\dfrac{1}{x} + 2\right) = x(3)$

$x\left(\dfrac{1}{x}\right) + 2x = 3x$

$1 + 2x = 3x$

$1 = x$

$\{1\}$

3. $\dfrac{3}{x} - 1 = \dfrac{1}{2}$

$2x\left(\dfrac{3}{x} - 1\right) = 2x\left(\dfrac{1}{2}\right)$

$2x\left(\dfrac{3}{x}\right) - 2x(1) = x$

$6 - 2x = x$

$6 = 3x$

$2 = x$

$\{2\}$

5. $\dfrac{1}{2x} + \dfrac{1}{x} = \dfrac{1}{2}$

$2x\left(\dfrac{1}{2x} + \dfrac{1}{x}\right) = 2x\left(\dfrac{1}{2}\right)$

$2x\left(\dfrac{1}{2x}\right) + 2x\left(\dfrac{1}{x}\right) = x$

$1 + 2 = x$

$3 = x$

$\{3\}$

7. $\dfrac{2}{3t} + \dfrac{1}{4} = \dfrac{3}{4t}$

$12t\left(\dfrac{2}{3t} + \dfrac{1}{4}\right) = 12t\left(\dfrac{3}{4t}\right)$

$12t\left(\dfrac{2}{3t}\right) + 12t\left(\dfrac{1}{4}\right) = 9$

$8 + 3t = 9$

$3t = 1$

$t = \dfrac{1}{3}$

$\left\{\dfrac{1}{3}\right\}$

9. $\dfrac{3}{x - 1} + 2 = \dfrac{5}{x - 1}$

$(x - 1)\left(\dfrac{3}{x - 1} + 2\right) = (x - 1)\left(\dfrac{5}{x - 1}\right)$

$(x - 1)\left(\dfrac{3}{x - 1}\right) + (x - 1)(2) = 5$

$3 + 2x - 2 = 5$

$2x + 1 = 5$

$2x = 4$

$x = 2$

$\{2\}$

11. $\dfrac{5}{2x + 3} = \dfrac{1}{2x + 3} + 1$

$(2x + 3)\left(\dfrac{5}{2x + 3}\right) = (2x + 3)\left(\dfrac{1}{2x + 3} + 1\right)$

$5 = (2x + 3)\left(\dfrac{1}{2x + 3}\right) + (2x + 3)(1)$

$5 = 1 + 2x + 3$

$5 = 2x + 4$

$1 = 2x$

$\dfrac{1}{2} = x$

$\left\{\dfrac{1}{2}\right\}$

107

13. $$\frac{1.4}{x} + \frac{3.2}{2x} = -1.2$$

$$2x\left(\frac{1.4}{x} + \frac{3.2}{2x}\right) = 2x(-1.2)$$

$$2x\left(\frac{1.4}{x}\right) + 2x\left(\frac{3.2}{2x}\right) = -2.4x$$

$$2.8 + 3.2 = -2.4x$$

$$6 = -2.4x$$

$$-2.5 = x$$

$$\{-2.5\}$$

15. $$\frac{1}{x + 2} + \frac{1}{x} = \frac{12}{x^2 + 2x}$$

$$\frac{1}{x + 2} + \frac{1}{x} = \frac{12}{x(x + 2)}$$

$$x(x + 2)\left(\frac{1}{x + 2} + \frac{1}{x}\right) = x(x + 2)\left[\frac{12}{x(x + 2)}\right]$$

$$x(x + 2)\left(\frac{1}{x + 2}\right) + x(x + 2)\left(\frac{1}{x}\right) = 12$$

$$x + x + 2 = 12$$

$$2x + 2 = 12$$

$$2x = 10$$

$$x = 5$$

$$\{5\}$$

17. $$\frac{2}{x - 3} = \frac{4}{x + 4}$$

$$(x - 3)(x + 4)\left(\frac{2}{x - 3}\right) = (x - 3)(x + 4)\left(\frac{4}{x + 4}\right)$$

$$2(x + 4) = 4(x - 3)$$

$$2x + 8 = 4x - 12$$

$$20 = 2x$$

$$10 = x$$

$$\{10\}$$

19.

$$\frac{6}{x-3} - \frac{3}{x+2} = \frac{12}{x^2 - x - 6}$$

$$\frac{6}{x-3} - \frac{3}{x+2} = \frac{12}{(x-3)(x+2)}$$

$$(x-3)(x+2)\left(\frac{6}{x-3} - \frac{3}{x+12}\right) = (x-3)(x+2)\left[\frac{12}{(x-3)(x+2)}\right]$$

$$(x-3)(x+2)\left(\frac{6}{x-3}\right) - (x-3)(x+2)\left(\frac{3}{x+2}\right) = 12$$

$$6(x+2) - 3(x-3) = 12$$

$$6x + 12 - 3x + 9 = 12$$

$$3x + 21 = 12$$

$$3x = -9$$

$$x = -3$$

$$\{-3\}$$

21.

$$\frac{-4}{5(x+2)} = \frac{3}{x+2}$$

$$5(x+2)\left[\frac{-4}{5(x+2)}\right] = 5(x+2)\left(\frac{3}{(x+2)}\right)$$

$$-4 = 15$$

$$\emptyset$$

23.

$$\frac{4}{1-x} + \frac{1}{x} = \frac{5}{x^2 - x}$$

$$\frac{4}{-(x-1)} + \frac{1}{x} = \frac{5}{x(x-1)}$$

$$x(x-1)\left(\frac{-4}{x-1} + \frac{1}{x}\right) = x(x-1)\left[\frac{5}{x(x-1)}\right]$$

$$x(x-1)\left(\frac{-4}{x-1}\right) + x(x-1)\left(\frac{1}{x}\right) = 5$$

$$-4x + x - 1 = 5$$

$$-3x - 1 = 5$$

$$-3x = 6$$

$$x = -2$$

$$\{-2\}$$

25.

$$\frac{3}{2 + x} + \frac{2}{2 - x} = \frac{2x}{x^2 - 4}$$

$$\frac{3}{2 + x} + \frac{2}{-(x - 2)} = \frac{2x}{(x + 2)(x - 2)}$$

$$(x + 2)(x - 2)\left(\frac{3}{2 + x} - \frac{2}{x - 2}\right) = (x + 2)(x - 2)\left[\frac{2x}{(x + 2)(x - 2)}\right]$$

$$(x + 2)(x - 2)\left(\frac{3}{2 + x}\right) - (x + 2)(x - 2)\left(\frac{2}{x - 2}\right) = 2x$$

$$3(x - 2) - 2(x + 2) = 2x$$

$$3x - 6 - 2x - 4 = 2x$$

$$x - 10 = 2x$$

$$-10 = x$$

$$\{-10\}$$

27. $\dfrac{x - 4}{1 - x} \leq 0$

Free boundary number: $1 - x = 0$

$$1 = x$$

Other boundary numbers:

$$\frac{x - 4}{1 - x} = 0$$

$$(1 - x)\left(\frac{x - 4}{1 - x}\right) = (1 - x)(0)$$

$$x - 4 = 0$$

$$x = 4$$

$$\underline{\quad A \quad}_{1}\,\underline{\quad B \quad}_{4}\,\underline{\quad C \quad}$$

A: $x = 0$; $\dfrac{0 - 4}{1 - 0} \leq 0$

$$-4 \leq 0 \quad T$$

B: $x = 2$; $\dfrac{2 - 4}{1 - 2} \leq 0$

$$2 \leq 0 \quad F$$

110

C: $x = 5$; $\dfrac{5 - 4}{1 - 5} \le 0$

$-\dfrac{1}{4} \le 0$ T

$(-\infty, 1) \cup [4, \infty)$

29. $\dfrac{3 - x}{x + 6} \ge 2$

Free boundary number: $x + 6 = 0$

$x = -6$

Other boundary numbers:

$\dfrac{3 - x}{x + 6} = 2$

$(x + 6)\left(\dfrac{3 - x}{x + 6}\right) = (x + 6)(2)$

$3 - x = 2x + 12$

$-3x = 9$

$x = -3$

$$\underset{\qquad -6 \qquad\quad -3 \qquad}{\overset{A \qquad\quad B \qquad\quad C}{\rule{4cm}{0.4pt}}}$$

A: $x = -7$; $\dfrac{3 - (-7)}{-7 + 6} \ge 2$

$-10 \ge 2$ F

B: $x = -4$; $\dfrac{3 - (-4)}{-4 + 6} \ge 2$

$\dfrac{7}{2} \ge 2$ T

C: $x = 0$; $\dfrac{3 - 0}{0 + 6} \ge 2$

$\dfrac{1}{2} \ge 2$ F

$[-6, -3]$

31. $\dfrac{1}{x - 1} < \dfrac{2}{x + 7}$

Free boundary numbers:

$x - 1 = 0$ $x + 7 = 0$
$x = 1$ $x = -7$

Other boundary numbers:

$\dfrac{1}{x - 1} = \dfrac{2}{x + 7}$

$(x + 7) = 2(x - 1)$

$x + 7 = 2x - 2$

$9 = x$

$$\underset{\qquad -7 \quad\ 1 \quad\ 9 \qquad}{\overset{A \quad\ B \quad\ C \quad\ D}{\rule{4cm}{0.4pt}}}$$

A: $x = -8$; $\dfrac{1}{-8 - 1} < \dfrac{2}{-8 + 7}$

$-\dfrac{1}{9} < -2$ F

B: $x = 0$; $\dfrac{1}{0 - 1} < \dfrac{2}{0 + 7}$

$-1 < \dfrac{2}{7}$ T

C: $x = 2$; $\dfrac{1}{2 - 1} < \dfrac{2}{2 + 7}$

$1 < \dfrac{2}{9}$ F

D: $x = 10$; $\dfrac{1}{10 - 1} < \dfrac{2}{10 + 7}$

$\dfrac{1}{9} < \dfrac{2}{17}$ T

$(-7, 1) \cup (9, \infty)$

33.

	R	D	T = D/R
Kenneth	r	49	$49/r$
John Phillip	$r + 2$	56	$56/(r+2)$

Kenneth's time = John's time

$$\frac{49}{r} = \frac{56}{r + 2}$$

$$r(r + 2)\left(\frac{49}{r}\right) = r(r + 2)\left(\frac{56}{r + 2}\right)$$

$$49(r + 2) = 56r$$

$$49r + 98 = 56r$$

$$98 = 7r$$

$$14 = r$$

$$r + 2 = 14 + 2 = 16$$

John Phillip averages 16 mph.

35. $\dfrac{\text{boys}}{\text{girls}}$: $\dfrac{4}{5} = \dfrac{32}{x}$

$$5x\left(\frac{4}{5}\right) = 5x\left(\frac{32}{x}\right)$$

$$4x = 160$$

$$x = 40$$

There are 40 girls in the class.

37. $\dfrac{\text{inches}}{\text{centimeters}}$: $\dfrac{1}{2.54} = \dfrac{12}{x}$

$$2.54x\left(\frac{1}{2.54}\right) = 2.54x\left(\frac{12}{x}\right)$$

$$x = 30.48$$

There are 30.48 cm in 1 foot.

39. 1st number: x

2nd number: $2x$

$$\frac{1}{x} + \frac{1}{2x} > \frac{1}{2}$$

Free boundary numbers:

$$x = 0$$

Other boundary numbers:

$$\frac{1}{x} + \frac{1}{2x} = \frac{1}{2}$$

$$2x\left(\frac{1}{x} + \frac{1}{2x}\right) = 2x\left(\frac{1}{2}\right)$$

$$2x\left(\frac{1}{x}\right) + 2x\left(\frac{1}{2x}\right) = x$$

$$2 + 1 = x$$

$$3 = x$$

$$\underset{0}{\underline{\quad A \quad|\quad}} \underset{3}{\underline{B \quad|\quad}} \underline{C \quad}$$

A: $x = -1$;

$$\frac{1}{-1} + \frac{1}{2(-1)} > \frac{1}{2}$$

$$-\frac{3}{2} > \frac{1}{2} \quad F$$

B: $x = 1$;

$$\frac{1}{1} + \frac{1}{2(1)} > \frac{1}{2}$$

$$\frac{3}{2} > \frac{1}{2} \quad T$$

C: $x = 4$;

$$\frac{1}{4} + \frac{1}{2(4)} > \frac{1}{2}$$

$$\frac{3}{8} > \frac{1}{2} \quad F$$

The number must be between 0 and 3.

41.
$$\frac{1}{x} + \frac{1}{3} = \frac{1}{4}$$

$$12x\left(\frac{1}{x} + \frac{1}{3}\right) = 12x\left(\frac{1}{4}\right)$$

$$12x\left(\frac{1}{x}\right) + 12x\left(\frac{1}{3}\right) = 3x$$

$$12 + 4x = 3x$$

$$12 = -x$$

$$-12 = x$$

$$\{-12\}$$

43. $\dfrac{2}{y} - \dfrac{1}{5} = \dfrac{2}{y} \cdot \dfrac{5}{5} - \dfrac{1}{5} \cdot \dfrac{y}{y}$

$$= \frac{10}{5y} - \frac{y}{5y}$$

$$= \frac{10 - y}{5y}$$

45.
$$\frac{2}{x} - \frac{3}{4} = \frac{1}{x}$$

$$4x\left(\frac{2}{x} - \frac{3}{4}\right) = 4x\left(\frac{1}{x}\right)$$

$$4x\left(\frac{2}{x}\right) - 4x\left(\frac{3}{4}\right) = 4$$

$$8 - 3x = 4$$

$$-3x = -4$$

$$x = \frac{4}{3}$$

$$\left\{\frac{4}{3}\right\}$$

47.
$$\frac{1}{x + 1} - \frac{2}{x} = 0$$

$$x(x + 1)\left(\frac{1}{x + 1} - \frac{2}{x}\right) = x(x + 1)(0)$$

$$x(x + 1)\left(\frac{1}{x + 1}\right) - x(x + 1)\left(\frac{2}{x}\right) = 0$$

$$x - 2(x + 1) = 0$$

$$x - 2x - 2 = 0$$

$$-x - 2 = 0$$

$$-2 = x$$

$$\{-2\}$$

49.
$$\frac{4}{2 - x} - 1 = \frac{3}{2 - x}$$

$$(2 - x)\left(\frac{4}{2 - x} - 1\right) = (2 - x)\left(\frac{3}{2 - x}\right)$$

$$(2 - x)\left(\frac{4}{2 - x}\right) - (2 - x)(1) = 3$$

$$4 - 2 + x = 3$$

$$2 + x = 3$$

$$x = 1$$

$$\{1\}$$

51.
$$\frac{3}{x} = \frac{5}{x + 4}$$

$$x(x + 4)\left(\frac{3}{x}\right) = x(x + 4)\left(\frac{5}{x + 4}\right)$$

$$3x + 12 = 5x$$

$$12 = 2x$$

$$6 = x$$

$$\{6\}$$

53.

$$\frac{14.3}{w} - \frac{3.1}{2w} = 1.1$$

$$2w\left(\frac{14.3}{w} - \frac{3.1}{2w}\right) = 2w(1.1)$$

$$2w\left(\frac{14.3}{w}\right) - 2w\left(\frac{3.1}{2w}\right) = 2.2w$$

$$28.6 - 3.1 = 2.2w$$

$$25.5 = 2.2w$$

$$11.5\overline{90} = w$$

$$\{11.5\overline{90}\}$$

55.

$$\frac{1}{x-5} + \frac{x}{25 - x^2} = 0$$

$$\frac{1}{x-5} + \frac{x}{-(x^2 - 25)} = 0$$

$$\frac{1}{x-5} - \frac{x}{x^2 - 25} = 0$$

$$\frac{1}{x-5} - \frac{x}{(x-5)(x+5)} = 0$$

$$(x-5)(x+5)\left[\frac{1}{x-5} - \frac{x}{(x-5)(x+5)}\right] = (x-5)(x+5)(0)$$

$$(x-5)(x+5)\left(\frac{1}{x-5}\right) - (x-5)(x+5)\left[\frac{x}{(x-5)(x+5)}\right] = 0$$

$$x + 5 - x = 0$$

$$5 = 0$$

$$\emptyset$$

57.

$$\frac{1}{x+3} = \frac{x}{x+1} - 1$$

$$(x+3)(x+1)\left(\frac{1}{x+3}\right) = (x+3)(x+1)\left(\frac{x}{x+1} - 1\right)$$

$$x + 1 = (x+3)(x+1)\left(\frac{x}{x+1}\right) - (x+3)(x+1)(1)$$

114

$$x + 1 = x^2 + 3x - (x^2 + 4x + 3)$$

$$x + 1 = x^2 + 3x - x^2 - 4x - 3$$

$$x + 1 = -x - 3$$

$$2x = -4$$

$$x = -2$$

$$\{-2\}$$

59.
$$\frac{x}{x - 3} = \frac{x + 1}{x - 1}$$

$$(x - 1)(x - 3)\left(\frac{x}{x - 3}\right) = (x - 1)(x - 3)\left(\frac{x + 1}{x - 1}\right)$$

$$x^2 - x = x^2 - 2x - 3$$

$$x = -3$$

$$\{-3\}$$

61.
$$\frac{x}{x + 4} - \frac{2}{x - 3} = 1$$

$$(x + 4)(x - 3)\left(\frac{x}{x + 4} - \frac{2}{x - 3}\right) = (x + 4)(x - 3)(1)$$

$$(x + 4)(x - 3)\left(\frac{x}{x + 4}\right) - (x + 4)(x - 3)\left(\frac{2}{x - 3}\right) = x^2 + x - 12$$

$$x^2 - 3x - (2x + 8) = x^2 + x - 12$$

$$x^2 - 3x - 2x - 8 = x^2 + x - 12$$

$$x^2 - 5x - 8 = x^2 + x - 12$$

$$-6x = -4$$

$$x = \frac{4}{6}$$

$$x = \frac{2}{3}$$

$$\left\{\frac{2}{3}\right\}$$

63.

$$\frac{3}{x-1} - 1 + \frac{x^2-5}{x^2+2x-3} = 0$$

$$\frac{3}{x-1} - 1 + \frac{x^2-5}{(x+3)(x-1)} = 0$$

$$(x+3)(x-1)\left[\frac{3}{x-1} - 1 + \frac{x^2-5}{(x+3)(x-1)}\right] = (x+3)(x-1)(0)$$

$$(x+3)(x-1)\left(\frac{3}{x-1}\right) - (x+3)(x-1)(1) + (x+3)(x-1)\left[\frac{x^2-5}{(x+3)(x-1)}\right] = 0$$

$$3x + 9 - (x^2 + 2x - 3) + x^2 - 5 = 0$$

$$3x + 9 - x^2 - 2x + 3 + x^2 - 5 = 0$$

$$x + 7 = 0$$

$$x = -7$$

$$\{-7\}$$

65.

$$\frac{2}{x+1} + \frac{3}{x+2} = \frac{-3}{x^2+3x+2}$$

$$\frac{2}{x+1} + \frac{3}{x+2} = \frac{-3}{(x+2)(x+1)}$$

$$(x+1)(x+2)\left(\frac{2}{x+1} + \frac{3}{x+2}\right) = (x+1)(x+2)\left[\frac{-3}{(x+2)(x+1)}\right]$$

$$(x+1)(x+2)\left(\frac{2}{x+1}\right) + (x+1)(x+2)\left(\frac{3}{x+2}\right) = -3$$

$$2x + 4 + 3x + 3 = -3$$

$$5x + 7 = -3$$

$$5x = -10$$

$$x = -2$$

$x = -2$ causes a denominator to equal zero in the original problem.

$\emptyset$

67.

$$\frac{3}{x-1} - \frac{1}{x+2} = \frac{9}{x^2+x-2}$$

$$\frac{3}{x-1} - \frac{1}{x+2} = \frac{9}{(x+2)(x-1)}$$

$$(x-1)(x+2)\left(\frac{3}{x-1} - \frac{1}{x+2}\right) = (x-1)(x+2)\left[\frac{9}{(x+2)(x-1)}\right]$$

116

$$(x - 1)(x + 2)\left(\frac{3}{x - 1}\right) - (x - 1)(x + 2)\left(\frac{1}{x + 2}\right) = 9$$

$$3x + 6 - (x - 1) = 9$$

$$3x + 6 - x + 1 = 9$$

$$2x + 7 = 9$$

$$2x = 2$$

$$x = 1$$

$$\emptyset$$

69. $\dfrac{x + 1}{x - 1} > 0$

Free boundary numbers: $x - 1 = 0$

$$x = 1$$

Other boundary numbers:

$$\frac{x + 1}{x - 1} = 0$$

$$(x - 1)\left(\frac{x + 1}{x - 1}\right) = (x - 1)(0)$$

$$x + 1 = 0$$

$$x = -1$$

$$\underset{-1}{\underline{\quad A \quad}} \Big| \underset{1}{\underline{\quad B \quad}} \Big| \underline{\quad C \quad}$$

A: $x = -2$; $\dfrac{-2 + 1}{-2 - 1} > 0$

$$\frac{1}{3} > 0 \quad T$$

B: $x = 0$; $\dfrac{0 + 1}{0 - 1} > 0$

$$-1 > 0 \quad F$$

C: $x = 2$; $\dfrac{2 + 1}{2 - 1} > 0$

$$3 > 0 \quad T$$

$$\{x \mid x < -1 \text{ or } x > 1\}$$

71. $\dfrac{x - 4}{x - 5} \leq 0$

Free boundary number: $x - 5 = 0$

$$x = 5$$

Other boundary numbers:

$$\frac{x - 4}{x - 5} = 0$$

$$(x - 5)\left(\frac{x - 4}{x - 5}\right) = (x - 5)(0)$$

$$x - 4 = 0$$

$$x = 4$$

$$\underset{4}{\underline{\quad A \quad}} \Big| \underset{5}{\underline{\quad B \quad}} \Big| \underline{\quad C \quad}$$

A: $x = 0$; $\dfrac{0 - 4}{0 - 5} \leq 0$

$$\frac{4}{5} \leq 0 \quad F$$

B: $x = \dfrac{9}{2}$; $\dfrac{\dfrac{9}{2} - 4}{\dfrac{9}{2} - 5} \leq 0$

$$-1 \leq 0 \quad T$$

C: $x = 6$; $\dfrac{6 - 4}{6 - 5} \leq 0$

$$2 \leq 0 \quad F$$

$$\{x \mid 4 \leq x < 5 \}$$

73. $\dfrac{x + 2}{x} < 0$

Free boundary number: $x = 0$

Other boundary numbers:

$$\dfrac{x + 2}{x} = 0$$

$$x\left(\dfrac{x + 2}{x}\right) = x(0)$$

$$x + 2 = 0$$

$$x = -2$$

$$\underset{\substack{ \\ -2 \qquad\ 0}}{\underline{\quad A \quad\Big|\quad B \quad\Big|\quad C \quad}}$$

A: $x = -3$; $\dfrac{-3 + 2}{-3} < 0$

$$\dfrac{1}{3} < 0 \qquad \text{F}$$

B: $x = -1$; $\dfrac{-1 + 2}{-1} < 0$

$$-1 < 0 \qquad \text{T}$$

C: $x = 1$; $\dfrac{1 + 2}{1} < 0$

$$3 < 0 \qquad \text{F}$$

$$\{x \mid -2 < x < 0\}$$

75. $\dfrac{r - 5}{r + 2} \leq 2$

Free boundary numbers: $r + 2 = 0$

$$r = -2$$

Other boundary numbers:

$$\dfrac{r - 5}{r + 2} = 2$$

$$(r + 2)\left(\dfrac{r - 5}{r + 2}\right) = (r + 2)(2)$$

$$r - 5 = 2r + 4$$

$$-9 = r$$

$$\underset{\substack{ \\ -9 \qquad\ -2}}{\underline{\quad A \quad\Big|\quad B \quad\Big|\quad C \quad}}$$

A: $r = -10$; $\dfrac{-10 - 5}{-10 + 2} \leq 2$

$$\dfrac{15}{8} \leq 2 \qquad \text{T}$$

B: $r = -3$; $\dfrac{-3 - 5}{-3 + 2} \leq 2$

$$8 \leq 2 \qquad \text{F}$$

C: $r = 0$; $\dfrac{0 - 5}{0 + 2} \leq 2$

$$-\dfrac{5}{2} \leq 2 \qquad \text{T}$$

$$(-\infty, -9] \cup (-2, \infty)$$

77. $\dfrac{2x + 3}{x + 4} \geq 1$

Free boundary number:

$$x + 4 = 0$$

$$x = -4$$

Other boundary numbers:

$$\dfrac{2x + 3}{x + 4} = 1$$

$$(x + 4)\left(\dfrac{2x + 3}{x + 4}\right) = (x + 4)(1)$$

$$2x + 3 = x + 4$$

$$x = 1$$

$$\underset{\substack{ \\ -4 \qquad\ 1}}{\underline{\quad A \quad\Big|\quad B \quad\Big|\quad C \quad}}$$

A: $x = -5$; $\dfrac{2(-5) + 3}{-5 + 4} \geq 1$

$$7 \geq 1 \qquad \text{T}$$

118

B: $x = 0$; $\quad \dfrac{2(0) + 3}{0 + 4} \geq 1$

$$\dfrac{3}{4} \geq 1 \quad F$$

C: $x = 2$; $\quad \dfrac{2(2) + 3}{2 + 4} \geq 1$

$$\dfrac{7}{6} \geq 1 \quad T$$

$(-\infty, -4) \cup [1, \infty)$

79. $\dfrac{1}{x - 1} \leq \dfrac{1}{x + 1}$

Free boundary numbers:

$x - 1 = 0 \qquad x + 1 = 0$

$x = 1 \qquad\quad x = -1$

Other boundary numbers:

$$\dfrac{1}{x - 1} = \dfrac{1}{x + 1}$$

$$(x - 1)(x + 1)\left(\dfrac{1}{x - 1}\right) = (x - 1)(x + 1)\left(\dfrac{1}{x + 1}\right)$$

$$x + 1 = x - 1$$

$$0 = -2$$

None

$$\underset{-1}{\underline{\quad A \quad | \quad B \quad | \quad C \quad}}_{1}$$

A: $x = -2$; $\quad \dfrac{1}{-2 - 1} \leq \dfrac{1}{-2 + 1}$

$$-\dfrac{1}{3} \leq -1 \quad F$$

B: $x = 0$; $\quad \dfrac{1}{0 - 1} \leq \dfrac{1}{0 + 1}$

$$-1 \leq 1 \quad T$$

C: $x = 2$; $\quad \dfrac{1}{2 - 1} \leq \dfrac{1}{2 + 1}$

$$1 \leq \dfrac{1}{3} \quad F$$

$(-1, 1)$

81.

	D	T	T = D/R
Lila	8	r	8/r
Jessye	6	r - 2	6/(r - 2)

Lila's time = Jessye's time

$$\dfrac{8}{r} = \dfrac{6}{r - 2}$$

$$r(r - 2)\left(\dfrac{8}{r}\right) = r(r - 2)\left(\dfrac{6}{r - 2}\right)$$

$$8r - 16 = 6r$$

$$-16 = -2r$$

$$8 = r$$

Lila's rate is 8 mph.

83.

	D	R	T = D/R
To work	15	r	15/r
To home	15 + 10 = 25	r + 20	25/(r + 20)

time to work = time to home

$$\dfrac{15}{r} = \dfrac{25}{r + 20}$$

$$r(r + 20)\left(\dfrac{15}{r}\right) = r(r + 20)\left(\dfrac{25}{r + 20}\right)$$

$$15r + 300 = 25r$$

$$300 = 10r$$

$$30 = r$$

His speed going to work is 30 mph.

119

85. $\dfrac{\text{Democrats}}{\text{Republicans}}$: $\dfrac{3}{8} = \dfrac{45}{x}$

$$8x\left(\dfrac{3}{8}\right) = 8x\left(\dfrac{45}{x}\right)$$

$$3x = 360$$

$$x = 120$$

There are 120 Republicans.

87. $\dfrac{\text{gallons}}{\text{liters}}$: $\dfrac{1}{3.8} = \dfrac{5}{x}$

$$3.8x\left(\dfrac{1}{3.8}\right) = 3.8x\left(\dfrac{5}{x}\right)$$

$$x = 19$$

There are 19 liters in 5 gallons.

89. 1st number: x

2nd number: $3x$

$$\dfrac{1}{x} + \dfrac{1}{3x} = \dfrac{1}{4}$$

$$12x\left(\dfrac{1}{x} + \dfrac{1}{3x}\right) = 12x\left(\dfrac{1}{4}\right)$$

$$12x\left(\dfrac{1}{x}\right) + 12x\left(\dfrac{1}{3x}\right) = 3x$$

$$12 + 4 = 3x$$

$$16 = 3x$$

$$\dfrac{16}{3} = x$$

$$3x = 3\left(\dfrac{16}{3}\right) = 16$$

The two numbers are $\dfrac{16}{3}$ and 16.

91. number: x

$$\dfrac{1}{3x} = \dfrac{1}{x + 2}$$

$$3x(x + 2)\left(\dfrac{1}{3x}\right) = 3x(x + 2)\left(\dfrac{1}{x + 2}\right)$$

$$x + 2 = 3x$$

$$2 = 2x$$

$$1 = x$$

The number is 1.

93. number: x

$$\dfrac{1}{x - 1} > 0$$

Free boundary numbers: $x - 1 = 0$

$$x = 1$$

Other boundary numbers:

$$\dfrac{1}{x - 1} = 0$$

$$(x - 1)\left(\dfrac{1}{x - 1}\right) = (x - 1)(0)$$

$$1 = 0$$

None

$$\overset{A}{\underline{}}\,\overset{B}{\underset{1}{|}}\,\underline{}$$

A: $x = 0$: $\quad \dfrac{1}{0 - 1} > 0$

$$-1 > 0 \quad F$$

B: $x = 2$; $\quad \dfrac{1}{2 - 1} > 0$

$$1 > 0 \quad T$$

All numbers greater than 1.

95.

$$\frac{a}{x} + \frac{b}{2x} = c$$

$$2x\left(\frac{a}{x} + \frac{b}{2x}\right) = 2x(c)$$

$$2x\left(\frac{a}{x}\right) + 2x\left(\frac{b}{2x}\right) = 2cx$$

$$2a + b = 2cx$$

$$\frac{2a + b}{2c} = x$$

$$\left\{\frac{2a + b}{2c}\right\}$$

97. $\dfrac{x - a}{x - b} \le 0;\qquad a > 0,\ \ b < 0$

Free boundary numbers: $\ x - b = 0$

$$x = b$$

Other boundary numbers:

$$\frac{x - a}{x - b} = 0$$

$$(x - b)\left(\frac{x - a}{x - b}\right) = (x - b)(0)$$

$$x - a = 0$$

$$x = a$$

$$\underline{\quad A\quad}\underset{b}{|}\underline{\quad B\quad}\underset{a}{|}\underline{\quad C\quad}$$

A: $x = b - a$;

$$\frac{b - a - a}{b - a - b} \le 0$$

$$\frac{b - 2a}{-a} \le 0$$

$$(-a)\left(\frac{b - 2a}{-a}\right) \ge (-a)(0)$$

$$b - 2a \ge 0$$

$$b \ge 2a \quad \text{F}$$

B: $x = 0$;

$$\frac{0 - a}{0 - b} \le 0$$

$$\frac{a}{b} \le 0$$

$$b\left(\frac{a}{b}\right) \ge b(0)$$

$$a \ge 0 \quad \text{T}$$

C: $x = a - b$;

$$\frac{a - b - a}{a - b - b} \le 0$$

$$\frac{-b}{a - 2b} \le 0$$

$$(a - 2b)\left(\frac{-b}{a - 2b}\right) \le (a - 2b)$$

$$-b \le 0$$

$$b \ge 0 \quad \text{F}$$

$$\{x \mid b < x \le a\}$$

Problem Set 3.5

1.

$$\frac{\dfrac{2}{a}}{\dfrac{4}{a^2}} = \frac{2}{a} \cdot \frac{a^2}{4}$$

$$= \frac{2a^2}{4a}$$

$$= \frac{a}{2}$$

3.

$$\frac{\dfrac{-3r^3}{s^4}}{\dfrac{18r^4}{s^6}} = \frac{-3r^3}{s^4} \cdot \frac{s^6}{18r^4}$$

$$= \frac{-3r^3 s^6}{18r^4 s^4}$$

$$= -\frac{s^2}{6r}$$

5. $\dfrac{\dfrac{4s^2}{p-2}}{\dfrac{12s}{p-2}} = \dfrac{4s^2}{p-2} \cdot \dfrac{p-2}{12s}$

$= \dfrac{4s^2(p-2)}{12s(p-2)}$

$= \dfrac{s}{3}$

7. $\dfrac{\dfrac{z-2}{24}}{\dfrac{z-2}{36}} = \dfrac{z-2}{24} \cdot \dfrac{36}{z-2}$

$= \dfrac{36(z-2)}{24(z-2)}$

$= \dfrac{3}{2}$

9. $\dfrac{\dfrac{1}{y}-1}{\dfrac{1}{y}+1} = \dfrac{\left(\dfrac{1}{y}-1\right)(y)}{\left(\dfrac{1}{y}+1\right)(y)}$

$= \dfrac{\left(\dfrac{1}{y}\right)y - 1(y)}{\left(\dfrac{1}{y}\right)y + 1(y)}$

$= \dfrac{1-y}{1+y}$

11. $\dfrac{\dfrac{1}{m}+\dfrac{1}{n}}{\dfrac{1}{m}-\dfrac{1}{n}} = \dfrac{\left(\dfrac{1}{m}+\dfrac{1}{n}\right)(mn)}{\left(\dfrac{1}{m}-\dfrac{1}{n}\right)(mn)}$

$= \dfrac{\left(\dfrac{1}{m}\right)(mn) + \left(\dfrac{1}{n}\right)(mn)}{\left(\dfrac{1}{m}\right)(mn) - \left(\dfrac{1}{n}\right)(mn)}$

$= \dfrac{n+m}{n-m}$

13. $\dfrac{\dfrac{1}{b-1}+2}{3-\dfrac{1}{1-b}} = \dfrac{\dfrac{1}{b-1}+2}{3-\dfrac{1}{-(b-1)}}$

$= \dfrac{\dfrac{1}{b-1}+2}{3+\dfrac{1}{b-1}}$

$= \dfrac{\left(\dfrac{1}{b-1}+2\right)(b-1)}{\left(3+\dfrac{1}{b-1}\right)(b-1)}$

$= \dfrac{\left(\dfrac{1}{b-1}\right)(b-1) + 2(b-1)}{3(b-1) + \left(\dfrac{1}{b-1}\right)(b-1)}$

$= \dfrac{1+2b-2}{3b-3+1}$

$= \dfrac{2b-1}{3b-2}$

15. $\dfrac{a^{-2}+b^{-1}}{(ab)^{-1}} = \dfrac{\dfrac{1}{a^2}+\dfrac{1}{b}}{\dfrac{1}{ab}}$

$= \dfrac{\left(\dfrac{1}{a^2}+\dfrac{1}{b}\right)(a^2b)}{\left(\dfrac{1}{ab}\right)(a^2b)}$

$= \dfrac{\left(\dfrac{1}{a^2}\right)(a^2b) + \left(\dfrac{1}{b}\right)(a^2b)}{a}$

$= \dfrac{b+a^2}{a}$

17. $\dfrac{\dfrac{3}{a}}{\dfrac{9}{a^3}} = \dfrac{3}{a} \cdot \dfrac{a^3}{9}$

$\qquad = \dfrac{3a^3}{9a}$

$\qquad = \dfrac{a^2}{3}$

19. $\dfrac{\dfrac{xy}{x-2}}{\dfrac{ax}{x-2}} = \dfrac{xy}{x-2} \cdot \dfrac{x-2}{ax}$

$\qquad = \dfrac{xy(x-2)}{ax(x-2)}$

$\qquad = \dfrac{y}{a}$

21. $\dfrac{\dfrac{z-2}{64}}{\dfrac{z-2}{48}} = \dfrac{z-2}{64} \cdot \dfrac{48}{z-2}$

$\qquad = \dfrac{48(z-2)}{64(z-2)}$

$\qquad = \dfrac{3}{4}$

23. $\dfrac{\dfrac{-14}{3r+3s}}{\dfrac{21}{r+s}} = \dfrac{-14}{3r+3s} \cdot \dfrac{r+s}{21}$

$\qquad = \dfrac{-14}{3(r+s)} \cdot \dfrac{r+s}{21}$

$\qquad = \dfrac{-14(r+s)}{63(r+s)}$

$\qquad = -\dfrac{2}{9}$

25. $\dfrac{\dfrac{m+1}{m^2-1}}{\dfrac{3}{m-1}} = \dfrac{m+1}{m^2-1} \cdot \dfrac{m-1}{3}$

$\qquad = \dfrac{m+1}{(m+1)(m-1)} \cdot \dfrac{m-1}{3}$

$\qquad = \dfrac{(m+1)(m-1)}{3(m+1)(m-1)}$

$\qquad = \dfrac{1}{3}$

27. $\dfrac{\dfrac{24}{x-5}}{\dfrac{128}{125-x^3}} = \dfrac{24}{x-5} \cdot \dfrac{125-x^3}{128}$

$\qquad = \dfrac{24}{x-5} \cdot \dfrac{(5-x)(25+5x+x^2)}{128}$

$\qquad = \dfrac{24}{x-5} \cdot \dfrac{-(x-5)(25+5x+x^2)}{128}$

$\qquad = \dfrac{-24(x-5)(25+5x+x^2)}{128(x-5)}$

$\qquad = -\dfrac{3(25+5x+x^2)}{16}$

29. $\dfrac{1 - \dfrac{1}{r-1}}{1 + \dfrac{1}{r-1}} = \dfrac{\left(1 - \dfrac{1}{r-1}\right)(r-1)}{\left(1 + \dfrac{1}{r-1}\right)(r-1)}$

$\qquad = \dfrac{1(r-1) - \left(\dfrac{1}{r-1}\right)(r-1)}{1(r-1) + \left(\dfrac{1}{r-1}\right)(r-1)}$

$\qquad = \dfrac{r-1-1}{r-1+1}$

$\qquad = \dfrac{r-2}{r}$

31. $\dfrac{3y^{-1} - 2}{y^{-1} + 4} = \dfrac{\dfrac{3}{y} - 2}{\dfrac{1}{y} + 4}$

$= \dfrac{\left(\dfrac{3}{y} - 2\right)y}{\left(\dfrac{1}{y} + 4\right)y}$

$= \dfrac{\left(\dfrac{3}{y}\right)y - 2y}{\left(\dfrac{1}{y}\right)y + 4y}$

$= \dfrac{3 - 2y}{1 + 4y}$

33. $\dfrac{\dfrac{1}{2 - b} + 1}{1 - \dfrac{1}{b - 2}} = \dfrac{\dfrac{1}{2 - b} + 1}{1 - \dfrac{1}{-(2 - b)}}$

$= \dfrac{\dfrac{1}{2 - b} + 1}{1 + \dfrac{1}{2 - b}}$

$= \dfrac{\left(\dfrac{1}{2 - b} + 1\right)(2 - b)}{\left(1 + \dfrac{1}{2 - b}\right)(2 - b)}$

$= \dfrac{\left(\dfrac{1}{2 - b}\right)(2 - b) + 1(2 - b)}{1(2 - b) + \left(\dfrac{1}{2 - b}\right)(2 - b)}$

$= \dfrac{1 + 2 - b}{2 - b + 1}$

$= \dfrac{3 - b}{3 - b}$

$= 1$

35. $\dfrac{\dfrac{3t^2 + 2st - s^2}{st}}{\dfrac{3}{s} - \dfrac{1}{t}} = \dfrac{\dfrac{(3t^2 + 2st - s^2)}{st}(st)}{\left(\dfrac{3}{s} - \dfrac{1}{t}\right)(st)}$

$= \dfrac{3t^2 + 2st - s^2}{\left(\dfrac{3}{s}\right)(st) - \left(\dfrac{1}{t}\right)st}$

$= \dfrac{(3t - s)(t + s)}{3t - s}$

$= t + s$

37. $\dfrac{2a^{-1} + 2b^{-1}}{\dfrac{a^3 + b^3}{ab}} = \dfrac{\dfrac{2}{a} + \dfrac{2}{b}}{\dfrac{a^3 + b^3}{ab}}$

$= \dfrac{\left(\dfrac{2}{a} + \dfrac{2}{b}\right)(ab)}{\left(\dfrac{a^3 + b^3}{ab}\right)(ab)}$

$= \dfrac{\left(\dfrac{2}{a}\right)ab + \left(\dfrac{2}{b}\right)ab}{a^3 + b^3}$

$= \dfrac{2b + 2a}{a^3 + b^3}$

$= \dfrac{2(b + a)}{(a + b)(a^2 - ab + b^2)}$

$= \dfrac{2}{a^2 - ab + b^2}$

39. $\dfrac{\dfrac{t+5}{t^2-16}}{1+\dfrac{1}{t+4}}$

$= \dfrac{\dfrac{t+5}{(t+4)(t-4)}}{1+\dfrac{1}{t+4}}$

$= \dfrac{\left[\dfrac{t+5}{(t+4)(t-4)}\right](t+4)(t-4)}{\left(1+\dfrac{1}{t+4}\right)(t+4)(t-4)}$

$= \dfrac{t+5}{1(t+4)(t-4)+\left(\dfrac{1}{t+4}\right)(t+4)(t-4)}$

$= \dfrac{t+5}{t^2-16+t-4}$

$= \dfrac{t+5}{t^2+t-20}$

$= \dfrac{t+5}{(t+5)(t-4)}$

$= \dfrac{1}{t-4}$

41. $\dfrac{\dfrac{x}{x^2-x-12}}{\dfrac{x}{x-4}} = \dfrac{x}{x^2-x-12}\cdot\dfrac{x-4}{x}$

$= \dfrac{x}{(x-4)(x+3)}\cdot\dfrac{x-4}{x}$

$= \dfrac{x(x-4)}{x(x-4)(x+3)}$

$= \dfrac{1}{x+3}$

43. $\dfrac{\dfrac{2x+1}{x^2+x}}{\dfrac{2x}{x+1}-\dfrac{1}{x}}$

$= \dfrac{\dfrac{2x+1}{x(x+1)}}{\dfrac{2x}{x+1}-\dfrac{1}{x}}$

$= \dfrac{\left[\dfrac{2x+1}{x(x+1)}\right]x(x+1)}{\left(\dfrac{2x}{x+1}-\dfrac{1}{x}\right)x(x+1)}$

$= \dfrac{2x+1}{\left(\dfrac{2x}{x+1}\right)x(x+1)-\left(\dfrac{1}{x}\right)x(x+1)}$

$= \dfrac{2x+1}{2x^2-(x+1)}$

$= \dfrac{2x+1}{2x^2-x-1}$

$= \dfrac{2x+1}{(2x+1)(x-1)} = \dfrac{1}{x-1}$

45. $\dfrac{\dfrac{m+3}{m-3}-\dfrac{m+3}{m-3}}{\dfrac{m+3}{m-3}+\dfrac{m+3}{m-3}} = \dfrac{\dfrac{m+3-(m+3)}{m-3}}{\dfrac{m+3+m+3}{m-3}}$

$= \dfrac{\dfrac{m+3-m-3}{m-3}}{\dfrac{2m+6}{m-3}}$

$= \dfrac{\dfrac{0}{m-3}}{\dfrac{2m+6}{m-3}}$

$= \dfrac{0}{m-3}\cdot\dfrac{m-3}{2m+6}$

$= \dfrac{0(m-3)}{(m-3)(2m+6)}$

$= 0$

47.

$$\frac{1 + \dfrac{1}{x} - \dfrac{1}{x+1}}{\dfrac{x^2+1}{x+1} - \dfrac{1}{x}}$$

$$= \frac{\left(1 + \dfrac{1}{x} - \dfrac{1}{x+1}\right)(x)(x+1)}{\left(\dfrac{x^2+1}{x+1} - \dfrac{1}{x}\right)(x)(x+1)}$$

$$= \frac{1(x)(x+1) + \left(\dfrac{1}{x}\right)(x)(x+1) - \left(\dfrac{1}{x+1}\right)(x)(x+1)}{\left(\dfrac{x^2+1}{x+1}\right)(x)(x+1) - \left(\dfrac{1}{x}\right)(x)(x+1)}$$

$$= \frac{x^2 + x + x + 1 - x}{x^3 + x - (x+1)}$$

$$= \frac{x^2 + x + 1}{x^3 + x - x - 1}$$

$$= \frac{x^2 + x + 1}{x^3 - 1}$$

$$= \frac{x^2 + x + 1}{(x-1)(x^2 + x + 1)}$$

$$= \frac{1}{x-1}$$

49.

$$\frac{\dfrac{3}{x} - \dfrac{2}{y} - \dfrac{4}{z}}{\dfrac{1}{x} - \dfrac{1}{y} - \dfrac{1}{z}} = \frac{\left(\dfrac{3}{x} - \dfrac{2}{y} - \dfrac{4}{z}\right)xyz}{\left(\dfrac{1}{x} - \dfrac{1}{y} - \dfrac{1}{z}\right)xyz}$$

$$= \frac{\left(\dfrac{3}{x}\right)xyz - \left(\dfrac{2}{y}\right)xyz - \left(\dfrac{4}{z}\right)xyz}{\left(\dfrac{1}{x}\right)xyz - \left(\dfrac{1}{y}\right)xyz) - \left(\dfrac{1}{z}\right)xyz}$$

$$= \frac{3yz - 2xz - 4xy}{yz - xz - xy}$$

51. $i = \dfrac{f}{1 - \dfrac{f}{7}}$

(a) $f = \dfrac{7}{2}$

$$i = \frac{\dfrac{7}{2}}{1 - \dfrac{\frac{7}{2}}{7}}$$

$$= \frac{\dfrac{7}{2}}{1 - \dfrac{7}{2} \cdot \dfrac{1}{7}}$$

$$= \frac{\dfrac{7}{2}}{1 - \dfrac{1}{2}}$$

$$= \frac{\dfrac{7}{2}}{\dfrac{1}{2}}$$

$$= \frac{7}{2} \cdot \frac{2}{1}$$

$$= 7 \text{ ft}$$

(b) $f = \dfrac{1}{2}$

$$i = \frac{\dfrac{1}{2}}{1 - \dfrac{\frac{1}{2}}{7}}$$

$$= \frac{\dfrac{1}{2}}{1 - \dfrac{1}{2} \cdot \dfrac{1}{7}} = \frac{\dfrac{1}{2}}{1 - \dfrac{1}{14}} = \frac{\dfrac{1}{2}}{\dfrac{13}{14}}$$

$$= \frac{1}{2} \cdot \frac{14}{13} = \frac{7}{13} \text{ ft}$$

53. $1 + \dfrac{1}{1 + \dfrac{1}{1 + 1}} = 1 + \dfrac{1}{1 + \dfrac{1}{2}}$

$\qquad\qquad\qquad = 1 + \dfrac{1}{\dfrac{3}{2}}$

$\qquad\qquad\qquad = 1 + \dfrac{1}{1} \cdot \dfrac{2}{3}$

$\qquad\qquad\qquad = 1 + \dfrac{2}{3}$

$\qquad\qquad\qquad = \dfrac{5}{3}$

55. $\dfrac{1 + \dfrac{1}{1 - \dfrac{1}{x}}}{1 - \dfrac{1}{1 + \dfrac{1}{x}}}$

$= \dfrac{1 + \left(\dfrac{1}{1 - \dfrac{1}{x}}\right)\left(\dfrac{x}{x}\right)}{1 - \left(\dfrac{1}{1 + \dfrac{1}{x}}\right)\left(\dfrac{x}{x}\right)}$

$= \dfrac{1 + \dfrac{x}{1(x) - \left(\dfrac{1}{x}\right)x}}{1 - \dfrac{x}{1(x) + \left(\dfrac{1}{x}\right)x}}$

$= \dfrac{1 + \dfrac{x}{x - 1}}{1 - \dfrac{x}{x + 1}}$

$= \dfrac{\left(1 + \dfrac{x}{x - 1}\right)(x - 1)(x + 1)}{\left(1 - \dfrac{x}{x + 1}\right)(x - 1)(x + 1)}$

$= \dfrac{1(x - 1)(x + 1) + \left(\dfrac{x}{x - 1}\right)(x - 1)(x + 1)}{1(x - 1)(x + 1) - \left(\dfrac{x}{x + 1}\right)(x - 1)(x + 1)}$

$= \dfrac{x^2 - 1 + x^2 + x}{x^2 - 1 - (x^2 - x)}$

$= \dfrac{2x^2 + x - 1}{x^2 - 1 - x^2 + x}$

$= \dfrac{2x^2 + x - 1}{x - 1}$

Chapter 3 Review Problems

1. $x = 0$

3. $67 = 0$

None

5. $\dfrac{x - 1}{x^2 + 4x + 3} = \dfrac{x - 1}{(x + 3)(x + 1)}$

$\qquad x + 3 = 0 \qquad x + 1 = 0$

$\qquad\qquad x = -3 \qquad\qquad x = -1$

7. $\dfrac{4}{3p} = \dfrac{4}{3p} \cdot \dfrac{(2p)(p - 2)}{(2p)(p - 2)}$

$\qquad = \dfrac{8p(p - 2)}{6p^2(p - 2)}$

$\qquad = \dfrac{8p^2 - 16p}{6p^2(p - 2)}$

9. $x^2 - 8x + 16 = (x - 4)^2$

$\qquad \dfrac{x + 3}{x - 4} = \dfrac{(x + 3)}{(x - 4)} \cdot \dfrac{(x - 4)}{(x - 4)}$

$\qquad\qquad = \dfrac{x^2 - x - 12}{(x - 4)^2}$

11. $\dfrac{m^5 n}{m^7 n^3} = \dfrac{1}{m^2 n^2}$

13. $\dfrac{x^2 - 4x - 5}{x^3 + 1} = \dfrac{(x - 5)(x + 1)}{(x + 1)(x^2 - x + 1)}$

$\qquad = \dfrac{x - 5}{x^2 - x + 1}$

$\qquad = \dfrac{a^2(a^2 - 3a + 9)(a + b)^2}{a(a + b)(a - b)(a + 3)(a^2 - 3a + 9)}$

$\qquad = \dfrac{a(a + b)}{(a - b)(a + 3)}$

15. $\dfrac{s^3t - st}{s^2 + s} = \dfrac{st(s^2 - 1)}{s(s + 1)}$

$\qquad = \dfrac{st(s + 1)(s - 1)}{s(s + 1)}$

$\qquad = t(s - 1)$

21. $\dfrac{r^2st^3}{u^4v} \cdot \dfrac{-t^5}{u} \cdot \dfrac{u^2v^3}{rst}$

$\qquad = \dfrac{-r^2st^8u^2v^3}{rstu^5v}$

$\qquad = -\dfrac{rt^7v^2}{u^3}$

17. $\dfrac{12 - 3x^2}{2x^2 + x - 15} \cdot \dfrac{2x^2 - 3x - 5}{x^2 - 4x + 4}$

$\qquad = \dfrac{3(4 - x^2)}{(2x - 5)(x + 3)} \cdot \dfrac{(2x - 5)(x + 1)}{(x - 2)^2}$

$\qquad = \dfrac{3(2 - x)(2 + x)}{(2x - 5)(x + 3)} \cdot \dfrac{(2x - 5)(x + 1)}{(x - 2)^2}$

$\qquad = \dfrac{-3(x - 2)(2 + x)}{(2x - 5)(x + 3)} \cdot \dfrac{(2x - 5)(x + 1)}{(x - 2)^2}$

$\qquad = \dfrac{-3(x - 2)(2 + x)(2x - 5)(x + 1)}{(2x - 5)(x + 3)(x - 2)^2}$

$\qquad = \dfrac{-3(2 + x)(x + 1)}{(x + 3)(x - 2)}$

$\qquad = -\dfrac{3(2 + x)(x + 1)}{(x + 3)(x - 2)}$

23. $\dfrac{u^2 - uv + 2uw - 2vw}{u^3 + 8w^3} \div \dfrac{u^2 + 4uw + 4w^2}{u^2 - 2uw + 4w^2}$

$\qquad = \dfrac{u^2 - uv + 2uw - 2vw}{u^3 + 8w^3} \cdot \dfrac{u^2 - 2uw + 4w^2}{u^2 + 4uw + 4w^2}$

$\qquad = \dfrac{u(u - v) + 2w(u - v)}{(u + 2w)(u^2 - 2uw + 4w^2)} \cdot \dfrac{u^2 - 2uw + 4w^2}{(u + 2w)^2}$

$\qquad = \dfrac{(u - v)(u + 2w)}{(u + 2w)(u^2 - 2uw + 4w^2)} \cdot \dfrac{u^2 - 2uw + 4w^2}{(u + 2w)^2}$

$\qquad = \dfrac{(u - v)(u + 2w)(u^2 - 2uw + 4w^2)}{(u + 2w)^3(u^2 - 2uw + 4w^2)}$

$\qquad = \dfrac{u - v}{(u + 2w)^2}$

19. $\dfrac{a^2 - 3a + 9}{a^3 - ab^2} \div \dfrac{a^3 + 27}{a^4 + 2a^3b + a^2b^2}$

$\qquad = \dfrac{a^2 - 3a + 9}{a^3 - ab^2} \cdot \dfrac{a^4 + 2a^3b + a^2b^2}{a^3 + 27}$

$\qquad = \dfrac{a^2 - 3a + 9}{a(a^2 - b^2)} \cdot \dfrac{a^2(a^2 + 2ab + b^2)}{(a + 3)(a^2 - 3a + 9)}$

$\qquad = \dfrac{a^2 - 3a + 9}{a(a + b)(a - b)} \cdot \dfrac{a^2(a + b)^2}{(a + 3)(a^2 - 3a + 9)}$

25. $\dfrac{1 - 2y + y^2}{6y^2 - y - 1} \cdot \dfrac{y^2 + 7y + 12}{3y^2 + y - 4} \cdot \dfrac{6y^2 + 5y - 4}{y^2 + 2y - 3}$

$\qquad = \dfrac{y^2 - 2y + 1}{6y^2 - y - 1} \cdot \dfrac{y^2 + 7y + 12}{3y^2 + y - 4} \cdot \dfrac{6y^2 + 5y - 4}{y^2 + 2y - 3}$

$\qquad = \dfrac{(y - 1)^2}{(3y + 1)(2y - 1)} \cdot \dfrac{(y + 4)(y + 3)}{(3y + 4)(y - 1)} \cdot \dfrac{(3y + 4)(2y - 1)}{(y + 3)(y - 1)}$

$\qquad = \dfrac{(y - 1)^2(y + 4)(y + 3)(3y + 4)(2y - 1)}{(3y + 1)(2y - 1)(3y + 4)(y - 1)(y + 3)(y - 1)}$

$\qquad = \dfrac{y + 4}{3y + 1}$

27. $\dfrac{12x^2 - 5x - 2}{x^2 + 2xy + y^2} \div \dfrac{12x^2 + x - 6}{x + y}$

$= \dfrac{12x^2 - 5x - 2}{x^2 + 2xy + y^2} \cdot \dfrac{x + y}{12x^2 + x - 6}$

$= \dfrac{(4x + 1)(3x - 2)}{(x + y)^2} \cdot \dfrac{x + y}{(4x + 3)(3x - 2)}$

$= \dfrac{(4x + 1)(3x - 2)(x + y)}{(x + y)^2(4x + 3)(3x - 2)}$

$= \dfrac{4x + 1}{(x + y)(4x + 3)}$

29. $\dfrac{12t^3 - 27t}{t^8 - 16} \cdot \dfrac{t^4 + 2t^2}{2t^2 - 3t - 9}$

$= \dfrac{3t(4t^2 - 9)}{(t^4 + 4)(t^4 - 4)} \cdot \dfrac{t^2(t^2 + 2)}{(2t + 3)(t - 3)}$

$= \dfrac{3t(2t + 3)(2t - 3)}{(t^4 + 4)(t^2 - 2)(t^2 + 2)} \cdot \dfrac{t^2(t^2 + 2)}{(2t + 3)(t - 3)}$

$= \dfrac{3t^3(2t + 3)(2t - 3)(t^2 + 2)}{(t^4 + 4)(t^2 - 2)(t^2 + 2)(2t + 3)(t - 3)}$

$= \dfrac{3t^3(2t - 3)}{(t^4 + 4)(t^2 - 2)(t - 3)}$

31. $\dfrac{96t^7}{375s^2} \div \dfrac{-72t^4}{125s^7}$

$= \dfrac{96t^7}{375s^2} \cdot \dfrac{125s^7}{-72t^4}$

$= \dfrac{96 \cdot 125t^7s^7}{-72 \cdot 375t^4s^2}$

$= -\dfrac{4t^3s^5}{9}$

33. $\dfrac{a^3 + a^2b + ab^2 + b^3}{32a^7b^3} \cdot \dfrac{-768a^2b}{a^4 - b^4}$

$= \dfrac{a^2(a + b) + b^2(a + b)}{32a^7b^3} \cdot \dfrac{-768a^2b}{(a^2 - b^2)(a^2 + b^2)}$

$= \dfrac{(a + b)(a^2 + b^2)}{32a^7b^3} \cdot \dfrac{-768a^2b}{(a + b)(a - b)(a^2 + b^2)}$

$= \dfrac{-768a^2b(a + b)(a^2 + b^2)}{32a^7b^3(a + b)(a - b)(a^2 + b^2)}$

$= -\dfrac{24}{a^5b^2(a - b)}$

35. $\dfrac{m^2 - 16n^2}{6m^4 + 3m^2} \div \dfrac{64n^3 - m^3}{12m^2}$

$= \dfrac{m^2 - 16n^2}{6m^4 + 3m^2} \cdot \dfrac{12m^2}{64n^3 - m^3}$

$= \dfrac{(m - 4n)(m + 4n)}{3m^2(2m^2 + 1)} \cdot \dfrac{12m^2}{(4n - m)(16n^2 + 4nm + m^2)}$

$= \dfrac{-(4n - m)(m + 4n)}{3m^2(2m^2 + 1)} \cdot \dfrac{12m^2}{(4n - m)(16n^2 + 4nm + m^2)}$

$= \dfrac{-12m^2(4n - m)(m + 4n)}{3m^2(2m^2 + 1)(4n - m)(16n^2 + 4nm + m^2)}$

$= -\dfrac{4(m + 4n)}{(2m + 1)(16n^2 + 4nm + m^2)}$

37. $\dfrac{\frac{x^2y}{x + 1}}{\frac{x}{x + 1}} = \dfrac{x^2y}{x + 1} \cdot \dfrac{x + 1}{x}$

$= \dfrac{x^2y(x + 1)}{x(x + 1)}$

$= xy$

39. $\dfrac{\frac{1}{x} - \frac{2}{xy}}{\frac{2}{x} + \frac{1}{xy}} = \dfrac{\left(\frac{1}{x} - \frac{2}{xy}\right)xy}{\left(\frac{2}{x} + \frac{1}{xy}\right)xy}$

$= \dfrac{\left(\frac{1}{x}\right)xy - \left(\frac{2}{xy}\right)xy}{\left(\frac{2}{x}\right)xy + \left(\frac{1}{xy}\right)xy}$

$= \dfrac{y - 2}{2y + 1}$

41.
$$\dfrac{\dfrac{3t^2 + 5t}{t^2 - 25}}{\dfrac{2}{t - 5} + \dfrac{1}{t + 5}}$$

$$= \dfrac{\dfrac{3t^2 + 5t}{(t - 5)(t + 5)}}{\dfrac{2}{t - 5} + \dfrac{1}{t + 5}}$$

$$= \dfrac{\left[\dfrac{3t^2 + 5t}{(t - 5)(t + 5)}\right](t - 5)(t + 5)}{\left(\dfrac{2}{t - 5} + \dfrac{1}{t + 5}\right)(t - 5)(t + 5)}$$

$$= \dfrac{3t^2 + 5t}{\left(\dfrac{2}{t - 5}\right)(t - 5)(t + 5) + \left(\dfrac{1}{t + 5}\right)(t - 5)(t + 5)}$$

$$= \dfrac{3t^2 + 5t}{2t + 10 + t - 5}$$

$$= \dfrac{3t^2 + 5t}{3t + 5}$$

$$= \dfrac{t(3t + 5)}{3t + 5}$$

$$= t$$

43.
$$\dfrac{\dfrac{1}{x - 5} - \dfrac{1}{x + 3}}{\dfrac{16x^2 + 16}{x^2 - 2x - 15}}$$

$$= \dfrac{\dfrac{1}{x - 5} - \dfrac{1}{x + 3}}{\dfrac{16(x^2 + 1)}{(x - 5)(x + 3)}}$$

$$= \dfrac{\left(\dfrac{1}{x - 5} - \dfrac{1}{x + 3}\right)(x - 5)(x + 3)}{\left[\dfrac{16(x^2 + 1)}{(x - 5)(x + 3)}\right](x - 5)(x + 3)}$$

$$= \dfrac{\left(\dfrac{1}{x - 5}\right)(x - 5)(x + 3) - \left(\dfrac{1}{x + 3}\right)(x - 5)(x + 3)}{16(x^2 + 1)}$$

$$= \dfrac{x + 3 - x + 5}{16(x^2 + 1)}$$

$$= \dfrac{8}{16(x^2 + 1)}$$

$$= \dfrac{1}{2(x^2 + 1)}$$

45.

$$\dfrac{\dfrac{1}{s-5} + \dfrac{s+5}{s^2+5s+25}}{\dfrac{2s+5}{s^3-125}}$$

$$= \dfrac{\dfrac{1}{s-5} + \dfrac{s+5}{s^2+5s+25}}{\dfrac{2s+5}{(s-5)(s^2+5s+25)}}$$

$$= \dfrac{\left(\dfrac{1}{s-5} + \dfrac{s+5}{s^2+5s+25}\right)(s-5)(s^2+5s+25)}{\left[\dfrac{2s+5}{(s-5)(s^2+5s+25)}\right](s-5)(s^2+5s+25)}$$

$$= \dfrac{\left(\dfrac{1}{s-5}\right)(s-5)(s^2+5s+25) + \left(\dfrac{s+5}{s^2+5s+25}\right)(s-5)(s^2+5s+25)}{2s+5}$$

$$= \dfrac{s^2+5s+25 + s^2-25}{2s+5}$$

$$= \dfrac{2s^2+5s}{2s+5}$$

$$= \dfrac{s(2s+5)}{2s+5}$$

$$= s$$

47. $\dfrac{x}{x+7} \leq 2$

Free boundary number:

$$x + 7 = 0$$

$$x = -7$$

Other boundary numbers:

$$\dfrac{x}{x+7} = 2$$

$$(x+7)\left(\dfrac{x}{x+7}\right) = (x+7)(2)$$

$$x = 2x + 14$$

$$-x = 14$$

$$x = -14$$

$$\underset{-14 \qquad -7}{\overset{\quad A \quad | \quad B \quad | \quad C}{\rule{5cm}{0.4pt}}}$$

131

A: $x = -15$; $\dfrac{-15}{-15 + 7} \leq 2$

$$\dfrac{15}{8} \leq 2 \quad T$$

B: $x = -8$; $\dfrac{-8}{-8 + 7} \leq 2$

$$8 \leq 2 \quad F$$

C: $x = 0$; $\dfrac{0}{0 + 7} \leq 2$

$$0 \leq 2 \quad T$$

$$(-\infty, -14] \cup (-7, \infty)$$

49. $\dfrac{1}{x} - \dfrac{1}{3x} = \dfrac{1}{2}$

$$6x\left(\dfrac{1}{x} - \dfrac{1}{3x}\right) = 6x\left(\dfrac{1}{2}\right)$$

$$6x\left(\dfrac{1}{x}\right) - 6x\left(\dfrac{1}{3x}\right) = 3x$$

$$6 - 2 = 3x$$

$$4 = 3x$$

$$\dfrac{4}{3} = x$$

$$\left\{\dfrac{4}{3}\right\}$$

51. $\dfrac{x - 3}{x + 2} \geq 6$

Free boundary number: $x + 2 = 0$

$$x = -2$$

Other boundary numbers:

$$\dfrac{x - 3}{x + 2} = 6$$

$$(x + 2)\left(\dfrac{x - 3}{x + 2}\right) = (x + 2)(6)$$

$$x - 3 = 6x + 12$$

$$-15 = 5x$$

$$-3 = x$$

$$\underline{\quad A \quad \overset{|}{\underset{-3}{}} \quad B \quad \overset{|}{\underset{-2}{}} \quad C \quad}$$

A: $x = -4$; $\dfrac{-4 - 3}{-4 + 2} \geq 6$

$$\dfrac{7}{2} \geq 6 \quad F$$

B: $x = -\dfrac{5}{2}$; $\dfrac{\dfrac{-5}{2} - 3}{-\dfrac{5}{2} + 2} \geq 6$

$$11 \geq 6 \quad T$$

C: $x = 0$; $\dfrac{0 - 3}{0 + 2} \geq 6$

$$-\dfrac{3}{2} \geq 6 \quad F$$

$$[-3, -2)$$

53. $\dfrac{1}{x - 2} = \dfrac{2}{x - 2}$

$$(x - 2)\left(\dfrac{1}{x - 2}\right) = (x - 2)\left(\dfrac{2}{x - 2}\right)$$

$$1 = 2$$

$$\emptyset$$

55. $\dfrac{1}{x + 2} < \dfrac{1}{x - 1}$

Free boundary numbers:

$x + 2 = 0 \qquad x - 1 = 0$
$ x = -2 \qquad x = 1$

Other boundary numbers:

$$\dfrac{1}{x + 2} = \dfrac{1}{x - 1}$$

$$(x + 2)(x - 1)\left(\dfrac{1}{x + 2}\right) = (x + 2)(x - 1)\left(\dfrac{1}{x - 1}\right)$$

$$x - 1 = x + 2$$

$$-1 = 2$$

None

$A: \quad x = -3; \qquad \dfrac{1}{-3 + 2} < \dfrac{1}{-3 - 1}$

$$-1 < -\dfrac{1}{4} \qquad \text{T}$$

$B: \quad x = 0; \qquad \dfrac{1}{0 + 2} < \dfrac{1}{0 - 1}$

$$\dfrac{1}{2} < -1 \qquad \text{F}$$

$C: \quad x = 2; \qquad \dfrac{1}{2 + 2} < \dfrac{1}{2 - 1}$

$$\dfrac{1}{4} < 1 \qquad \text{T}$$

$$(-\infty, -2) \cup (1, \infty)$$

57. $\qquad\qquad \dfrac{5x}{2x - 3} = \dfrac{3}{2}$

$$2(2x - 3)\left(\dfrac{5x}{2x - 3}\right) = 2(2x - 3)\left(\dfrac{3}{2}\right)$$

$$10x = 6x - 9$$

$$4x = -9$$

$$x = -\dfrac{9}{4}$$

$$\left\{-\dfrac{9}{4}\right\}$$

59.

	D	R	T = D/R
truck	675	$r - 6$	$675/(r - 6)$
car	750	r	$750/r$

truck's time = car's time

$$\dfrac{675}{r - 6} = \dfrac{750}{r}$$

$$r(r - 6)\left(\dfrac{675}{r - 6}\right) = r(r - 6)\left(\dfrac{750}{r}\right)$$

$$675r = 750r - 4500$$

$$-75r = -4500$$

$$r = 60$$

$$r - 6 = 60 - 6 = 54$$

The speed of the truck is 54 mph.

61. $\dfrac{\text{width}}{\text{length}}: \quad \dfrac{1}{1.6} = \dfrac{x}{x + 6}$

$$1.6(x + 6)\left(\dfrac{1}{1.6}\right) = 1.6(x + 6)\left(\dfrac{x}{x + 6}\right)$$

$$x + 6 = 1.6x$$

$$6 = 0.6x$$

$$10 = x$$

$$x + 6 = 10 + 6 = 16$$

The measurements will be 10 in. by 16 in.

133

63. $\dfrac{\text{width}}{\text{length}}$: $\dfrac{1}{1.9} = \dfrac{x}{x + 6}$

$$1.9(x + 6)\left(\dfrac{1}{1.9}\right) = 1.9(x + 6)\left(\dfrac{x}{x + 6}\right)$$

$$x + 6 = 1.9x$$

$$6 = 0.9x$$

$$6.7 = x$$

$$x + 6 = 6.7 + 6 = 12.7$$

The dimensions will be 6.7 in. by 12.7 in.

65. 1st number: x

2nd number: $2x$

$$\dfrac{1}{x} + \dfrac{1}{2x} = 1$$

$$2x\left(\dfrac{1}{x} + \dfrac{1}{2x}\right) = 2x(1)$$

$$2 + 1 = 2x$$

$$3 = 2x$$

$$\dfrac{3}{2} = x$$

$$2x = 2\left(\dfrac{3}{2}\right) = 3$$

The numbers are $\dfrac{3}{2}$ and 3.

67. number: x

$$\dfrac{x}{1 + x} < 0$$

Free boundary number:

$$1 + x = 0$$

$$x = -1$$

Other boundary numbers:

$$\dfrac{x}{1 + x} = 0$$

$$(1 + x)\left(\dfrac{x}{1 + x}\right) = (1 + x)(0)$$

$$x = 0$$

$$\underline{\quad A \quad}_{\substack{| \\ -1}}\underline{\quad B \quad}_{\substack{| \\ 0}}\underline{\quad C \quad}$$

A: $x = -2$; $\qquad \dfrac{-2}{1 + (-2)} < 0$

$$2 < 0 \qquad \text{F}$$

B: $x = -\dfrac{1}{2}$; $\qquad \dfrac{-\dfrac{1}{2}}{1 + \left(-\dfrac{1}{2}\right)} < 0$

$$-1 < 0 \qquad \text{T}$$

C: $x = 1$; $\qquad \dfrac{1}{1 + 1} < 0$

$$\dfrac{1}{2} < 0 \qquad \text{F}$$

The number must be between -1 and 0.

69. $27x^3 - 1$

$= (3x)^3 - 1^3$

$= (3x - 1)[(3x)^2 + (3x)(1) + 1^2]$

$= (3x - 1)(9x^2 + 3x + 1)$

71. Factored

73. 2 terms; $x^3 + 1 = (x + 1)(x^2 - x + 1)$

$x - 1$ is not a factor.

75. 3 terms; $x^2 - x - 2 = (x - 2)(x + 1)$

$x - 1$ is not a factor.

77. 2 terms; $3(x - 1) - (x - 1)^2$

$$= (x - 1)[3 - (x - 1)]$$
$$= (x - 1)(3 - x + 1)$$
$$= (x - 1)(4 - x)$$

79. $(x + 2y)^2 = x^2 + 2x(2y) + (2y)^2$

$$= x^2 + 4xy + 4y^2$$

81. $-2^2 = -1 \cdot 2 \cdot 2 = -4$

83. $(-2)^3 = -2 \cdot -2 \cdot -2 = -8$

85. $(-a)^5 = (-1)^5(a)^5 = -a^5$

87. $\dfrac{a(y + z) - b(y + z)}{(y + z)(y - z)} = \dfrac{(y + z)(a - b)}{(y + z)(y - z)}$

$$= \dfrac{a - b}{y - z}$$

89.
$$\dfrac{\dfrac{4}{x - 2} + 1}{\dfrac{2x}{x + 2} - 1}$$

$$= \dfrac{\left(\dfrac{4}{x - 2} + 1\right)(x - 2)(x + 2)}{\left(\dfrac{2x}{x + 2} - 1\right)(x - 2)(x + 2)}$$

$$= \dfrac{\left(\dfrac{4}{x - 2}\right)(x - 2)(x + 2) + 1(x - 2)(x + 2)}{\left(\dfrac{2x}{x + 2}\right)(x - 2)(x + 2) - 1(x - 2)(x + 2)}$$

$$= \dfrac{4x + 8 + x^2 - 4}{2x^2 - 4x - (x^2 - 4)}$$

$$= \dfrac{x^2 + 4x + 4}{2x^2 - 4x - x^2 + 4}$$

$$= \dfrac{x^2 + 4x + 4}{x^2 - 4x + 4}$$

$$= \dfrac{(x + 2)^2}{(x - 2)^2}$$

91. Inequality;

$$\dfrac{1}{x - 1} \le \dfrac{2}{x + 1}$$

Free boundary numbers:

$$x - 1 = 0 \qquad x + 1 = 0$$
$$x = 1 \qquad x = -1$$

Other boundary numbers:

$$\dfrac{1}{x - 1} = \dfrac{2}{x + 1}$$

$$(x - 1)(x + 1)\left(\dfrac{1}{x - 1}\right) = (x - 1)(x + 1)\left(\dfrac{2}{x + 1}\right)$$

$$x + 1 = 2x - 2$$

$$3 = x$$

$$\underline{\quad A \quad |\quad B \quad |\quad C \quad|\quad D \quad}$$
$$\qquad\quad -1 \qquad 1 \qquad 3$$

A: $x = -2$; $\qquad \dfrac{1}{-2 - 1} \le \dfrac{2}{-2 + 1}$

$$-\dfrac{1}{3} \le -2 \qquad \text{F}$$

B: $x = 0$; $\qquad \dfrac{1}{0 - 1} \le \dfrac{2}{0 + 1}$

$$-1 \le 2 \qquad \text{T}$$

C: $x = 2$; $\qquad \dfrac{1}{2 - 1} \le \dfrac{2}{2 + 1}$

$$1 \le \dfrac{2}{3} \qquad \text{F}$$

D: $x = 4$; $\qquad \dfrac{1}{4 - 1} \le \dfrac{2}{4 + 1}$

$$\dfrac{1}{3} \le \dfrac{2}{5} \qquad \text{T}$$

$$(-1, 1) \cup [3, \infty)$$

93. Equation;

$$2x - 5 = 3(1 - x)$$

$$2x - 5 = 3 - 3x$$

$$5x = 8$$

$$x = \frac{8}{5}$$

$$\left\{\frac{8}{5}\right\}$$

95. Expression;

$$-3^2 + 3^{-2} = -9 + \frac{1}{3^2}$$

$$= -9 + \frac{1}{9}$$

$$= \frac{-9(9)}{9} + \frac{1}{9}$$

$$= \frac{-81}{9} + \frac{1}{9}$$

$$= \frac{-81 + 1}{9}$$

$$= \frac{-80}{9}$$

$$\text{or } -\frac{80}{9}$$

Chapter 3 Test

1. $x(x + 3) = 0$

$$x = 0 \qquad x + 3 = 0$$

$$x = -3$$

C

2. $x^2 - 14x + 49 = (x - 7)^2$

$$\frac{3x}{x - 7} = \frac{3x(x - 7)}{(x - 7)(x - 7)}$$

$$= \frac{3x^2 - 21x}{x^2 - 14x + 49}$$

B

3. $\dfrac{x^3 - 8y^3}{x^2 - 4y^2}$

$$= \frac{(x - 2y)(x^2 + 2xy + 4y^2)}{(x - 2y)(x + 2y)}$$

$$= \frac{x^2 + 2xy + 4y^2}{x + 2y}$$

A

4. $\dfrac{x}{x + 4} - \dfrac{x - 3}{x - 2}$

$$= \frac{x(x - 2)}{(x + 4)(x - 2)} - \frac{(x - 3)(x + 4)}{(x - 2)(x + 4)}$$

$$= \frac{x^2 - 2x}{(x + 4)(x - 2)} - \frac{x^2 + x - 12}{(x - 2)(x + 4)}$$

$$= \frac{x^2 - 2x - (x^2 + x - 12)}{(x + 4)(x - 2)}$$

$$= \frac{x^2 - 2x - x^2 - x + 12}{(x + 4)(x - 2)}$$

$$= \frac{-3x + 12}{(x + 4)(x - 2)}$$

$$= \frac{-3(x - 4)}{(x + 4)(x - 2)}$$

C

5. $\dfrac{\dfrac{3}{x} + \dfrac{1}{y}}{\dfrac{9y}{x} - \dfrac{x}{y}} = \dfrac{\left(\dfrac{3}{x} + \dfrac{1}{y}\right)(xy)}{\left(\dfrac{9y}{x} - \dfrac{x}{y}\right)(xy)}$

$$= \frac{3y + x}{9y^2 - x^2}$$

$$= \frac{3y + x}{(3y + x)(3y - x)}$$

$$= \frac{1}{3y - x}$$

B

6. $$\frac{a}{ab - b^2} + \frac{b}{a^2 - ab}$$

$$= \frac{a}{b(a - b)} + \frac{b}{a(a - b)}$$

$$= \frac{a(a)}{b(a - b)(a)} + \frac{b(b)}{a(a - b)(b)}$$

$$= \frac{a^2}{ab(a - b)} + \frac{b^2}{ab(a - b)}$$

$$= \frac{a^2 + b^2}{ab(a - b)}$$

7. $$\frac{x^3 + 3x^2}{x^3 - 6x^2 + 9x} \div \frac{x^2 + 2x - 3}{x^2 - 9}$$

$$= \frac{x^3 + 3x^2}{x^3 - 6x^2 + 9x} \cdot \frac{x^2 - 9}{x^2 + 2x - 3}$$

$$= \frac{x^2(x + 3)}{x(x - 3)^2} \cdot \frac{(x + 3)(x - 3)}{(x + 3)(x - 1)}$$

$$= \frac{x^2(x + 3)^2(x - 3)}{x(x - 3)^2(x + 3)(x - 1)}$$

$$= \frac{x(x + 3)}{(x - 3)(x - 1)}$$

8. $$\frac{y + 1}{y - 2} - \frac{y^2 + 5y + 1}{y^2 + y - 6}$$

$$= \frac{y + 1}{y - 2} - \frac{y^2 + 5y + 1}{(y + 3)(y - 2)}$$

$$= \frac{(y + 1)(y + 3)}{(y - 2)(y + 3)} - \frac{y^2 + 5y + 1}{(y + 3)(y - 2)}$$

$$= \frac{y^2 + 4y + 3}{(y - 2)(y + 3)} - \frac{y^2 + 5y + 1}{(y + 3)(y - 2)}$$

$$= \frac{y^2 + 4y + 3 - (y^2 + 5y + 1)}{(y - 2)(y + 3)}$$

$$= \frac{y^2 + 4y + 3 - y^2 - 5y - 1}{(y - 2)(y + 3)}$$

$$= \frac{-y + 2}{(y - 2)(y + 3)}$$

$$= \frac{-(y - 2)}{(y - 2)(y + 3)}$$

$$= -\frac{1}{y + 3}$$

9. $$\frac{6x^2 + 13x + 6}{9x^2 - 4} \cdot \frac{9x^2 - 12x + 4}{2x^2 + x - 3}$$

$$= \frac{(3x + 2)(2x + 3)}{(3x + 2)(3x - 2)} \cdot \frac{(3x - 2)(3x - 2)}{(2x + 3)(x - 1)}$$

$$= \frac{(3x + 2)(2x + 3)(3x - 2)^2}{(3x + 2)(3x - 2)(2x + 3)(x - 1)}$$

$$= \frac{3x - 2}{x - 1}$$

10. $$\frac{\dfrac{2t}{s \cdot t} + 1}{\dfrac{4t}{s + t} - 1} = \frac{\left(\dfrac{2t}{s + t} + 1\right)(s + t)}{\left(\dfrac{4t}{s + t} - 1\right)(s + t)}$$

$$= \frac{2t + s + t}{4t - s - t}$$

$$= \frac{3t + s}{3t - s}$$

11. $$\frac{x - 3}{x + 1} \geq 2$$

Free boundary number: $x + 1 = 0$

$$x = -1$$

Other boundary numbers:

$$\frac{x - 3}{x + 1} = 2$$

$$(x + 1)\left(\frac{x - 3}{x + 1}\right) = (x + 1)(2)$$

$$x - 3 = 2x + 2$$

$$-5 = x$$

$$\begin{array}{c c c} \underline{A} & \underline{B} & \underline{C} \\ {}^{|}_{-5} & {}^{|}_{-1} & \end{array}$$

A: $x = -6$; $\quad \dfrac{-6 - 3}{-6 + 1} \geq 2$

$$\dfrac{9}{5} \geq 2 \quad F$$

B: $x = -2$; $\quad \dfrac{-2 - 3}{-2 + 1} \geq 2$

$$5 > 2 \quad T$$

C: $x = 0$; $\quad \dfrac{0 - 3}{0 + 1} \geq 2$

$$-3 \geq 2 \quad F$$

$$[-5, -1)$$

12. $\quad \dfrac{1}{3x} - \dfrac{3}{2x} = \dfrac{1}{3}$

$$6x\left(\dfrac{1}{3x} - \dfrac{3}{2x}\right) = 6x\left(\dfrac{1}{3}\right)$$

$$2 - 9 = 2x$$

$$-7 = 2x$$

$$-\dfrac{7}{2} = x$$

$$\left\{-\dfrac{7}{2}\right\}$$

13. $\quad \dfrac{2}{x - 1} - \dfrac{1}{x + 2} = \dfrac{8}{x^2 + x - 2}$

$$\dfrac{2}{x - 1} - \dfrac{1}{x + 2} = \dfrac{8}{(x + 2)(x - 1)}$$

$$(x - 1)(x + 2)\left(\dfrac{2}{x - 1} - \dfrac{1}{x + 2}\right) = (x - 1)(x + 2)\left(\dfrac{8}{(x + 2)(x - 1)}\right)$$

$$2x + 4 - x + 1 = 8$$

$$x + 5 = 8$$

$$x = 3$$

$$\{3\}$$

14. number: x

$$\dfrac{1}{x - 1} \leq \dfrac{1}{x + 1}$$

Free boundary numbers:

$x - 1 = 0 \qquad x + 1 = 0$
$\quad x = 1 \qquad\qquad x = -1$

Other boundary numbers:

$$\dfrac{1}{x - 1} = \dfrac{1}{x + 1}$$

$$(x + 1)(x - 1)\left(\dfrac{1}{x - 1}\right) = (x + 1)(x - 1)\left(\dfrac{1}{x + 1}\right)$$

$$x + 1 = x - 1$$

$$0 = -2$$

None

$$\begin{array}{c c c} \underline{A} & \underline{B} & \underline{C} \\ {}^{|}_{-1} & {}^{|}_{1} & \end{array}$$

A: $x = -2$; $\qquad \dfrac{1}{-2 - 1} \leq \dfrac{1}{-2 + 1}$

$$-\dfrac{1}{3} \leq -1 \quad F$$

B: $x = 0$; $\qquad \dfrac{1}{0 - 1} \leq \dfrac{1}{0 + 1}$

$$-1 \leq 1 \quad T$$

C: $x = 2$; $\qquad \dfrac{1}{2 - 1} \leq \dfrac{1}{2 + 1}$

$$1 \leq \dfrac{1}{3} \quad F$$

All numbers between -1 and 1.

15.

	D	R	T = D/R
Jane	10	$r + 3$	$10/(r + 3)$
Bob	6	r	$6/r$

$$\text{Jane's time} = \text{Bob's time}$$

$$\frac{10}{r + 3} = \frac{6}{r}$$

$$r(r + 3)\left(\frac{10}{r + 3}\right) = r(r + 3)\left(\frac{6}{r}\right)$$

$$10r = 6r + 18$$

$$4r = 18$$

$$r = \frac{9}{2} = 4\frac{1}{2}$$

$$r + 3 = \frac{15}{2} = 7\frac{1}{2}$$

Bob's rate is $4\frac{1}{2}$ mph and

Jane's rate is $7\frac{1}{2}$ mph.

CHAPTER 4

Problem Set 4.1

1. $2\sqrt{49} = 2 \cdot 7$
$\qquad = 14$

3. $\sqrt[3]{125} = \sqrt[3]{5^3}$
$\qquad = 5$

5. $2\sqrt{36} - \sqrt{16} = 2(6) - 4$
$\qquad\qquad\qquad = 8$

7. $\sqrt{\dfrac{256}{625}} = \dfrac{\sqrt{256}}{\sqrt{625}}$
$\qquad\qquad = \dfrac{16}{25}$

9. $-\sqrt{4x^2} = -2x$

11. $\sqrt{100z^6} = 10z^3$

13. $\sqrt{\dfrac{121y^4}{x^2z^2}} = \dfrac{\sqrt{121y^4}}{\sqrt{x^2z^2}}$
$\qquad\qquad = \dfrac{11y^2}{xz}$

15. $\sqrt{50} = \sqrt{5^2 \cdot 2}$
$\qquad = \sqrt{5^2} \cdot \sqrt{2}$
$\qquad = 5\sqrt{2}$

17. $\sqrt{252} = \sqrt{2^2 \cdot 3^2 \cdot 7}$
$\qquad = 2 \cdot 3\sqrt{7}$
$\qquad = 6\sqrt{7}$

19. $\sqrt[3]{40} = \sqrt[3]{2^3 \cdot 5}$
$\qquad = \sqrt[3]{2^3} \cdot \sqrt[3]{5}$
$\qquad = 2\sqrt[3]{5}$

21. $\sqrt[3]{27x^7y^3} = \sqrt[3]{3^3 \cdot x^7y^3}$
$\qquad\qquad = \sqrt[3]{3^3x^6y^3} \cdot \sqrt[3]{x}$
$\qquad\qquad = 3x^2y\sqrt[3]{x}$

23. (a) $\sqrt{4y^2} = 2y$

(b) $\sqrt{4 + y^2}$

(c) $\sqrt{(2 + y)^2} = 2 + y$

25. $3\sqrt{36} = 3(6)$
$\qquad = 18$

27. $-\sqrt{100} = -10$

29. $\sqrt[3]{216} = \sqrt[3]{6^3}$
$\qquad = 6$

31. $-\sqrt{16x^4y^2} = -4x^2y$

33. $\sqrt{72} = \sqrt{6^2 \cdot 2}$
$\qquad = \sqrt{6^2} \cdot \sqrt{2}$
$\qquad = 6\sqrt{2}$

35. $\sqrt{288} = \sqrt{12^2 \cdot 2}$
$\qquad = 12\sqrt{2}$

37. $\sqrt[3]{54} = \sqrt[3]{3^3 \cdot 2}$
$\qquad = \sqrt[3]{3^3} \cdot \sqrt[3]{2}$
$\qquad = 3\sqrt[3]{2}$

140

39. $\sqrt[5]{-486} = \sqrt[5]{-1 \cdot 3^5 \cdot 2}$

$\qquad = \sqrt[5]{-1} \cdot \sqrt[5]{3^5} \cdot \sqrt[5]{2}$

$\qquad = -3\sqrt[5]{2}$

41. $\sqrt{\dfrac{16}{81}} = \dfrac{\sqrt{16}}{\sqrt{81}}$

$\qquad = \dfrac{4}{9}$

43. $\sqrt{\dfrac{27}{49}} = \dfrac{\sqrt{27}}{\sqrt{49}}$

$\qquad = \dfrac{\sqrt{3^2 \cdot 3}}{\sqrt{7^2}}$

$\qquad = \dfrac{\sqrt{3^2} \cdot \sqrt{3}}{7}$

$\qquad = \dfrac{3\sqrt{3}}{7}$

45. $-3\sqrt{\dfrac{8}{9}} = \dfrac{-3\sqrt{2^2 \cdot 2}}{\sqrt{3^2}}$

$\qquad = \dfrac{-3\sqrt{2^2} \cdot \sqrt{2}}{3}$

$\qquad = -2\sqrt{2}$

47. $\sqrt{9x^3y^2} = \sqrt{3^2x^2y^2 \cdot x}$

$\qquad = \sqrt{3^2x^2y^2} \cdot \sqrt{x}$

$\qquad = 3xy\sqrt{x}$

49. $\sqrt[3]{\dfrac{64}{27}} = \dfrac{\sqrt[3]{64}}{\sqrt[3]{27}}$

$\qquad = \dfrac{\sqrt[3]{4^3}}{\sqrt[3]{3^3}}$

$\qquad = \dfrac{4}{3}$

51. $\sqrt{\dfrac{75}{49}} = \dfrac{\sqrt{75}}{\sqrt{49}}$

$\qquad = \dfrac{\sqrt{5^2 \cdot 3}}{\sqrt{7^2}}$

$\qquad = \dfrac{\sqrt{5^2} \cdot \sqrt{3}}{7}$

$\qquad = \dfrac{5\sqrt{3}}{7}$

53. $\sqrt[3]{\dfrac{56}{27}} = \dfrac{\sqrt[3]{56}}{\sqrt[3]{27}}$

$\qquad = \dfrac{\sqrt[3]{2^3 \cdot 7}}{\sqrt[3]{3^3}}$

$\qquad = \dfrac{\sqrt[3]{2^3} \cdot \sqrt[3]{7}}{3}$

$\qquad = \dfrac{2\sqrt[3]{7}}{3}$

55. $3\sqrt{7x^2y^3z} = 3\sqrt{x^2y^2 \cdot 7yz}$

$\qquad = 3\sqrt{x^2y^2} \cdot \sqrt{7yz}$

$\qquad = 3xy\sqrt{7yz}$

57. $-\sqrt{450x^3y^9} = -\sqrt{2 \cdot 15^2 \cdot x^2 \cdot x \cdot y^8 \cdot y}$

$\qquad = -\sqrt{15^2x^2y^8} \cdot \sqrt{2xy}$

$\qquad = -15xy^4\sqrt{2xy}$

59. $\sqrt{\dfrac{196}{p^4q^2}} = \dfrac{\sqrt{196}}{\sqrt{p^4q^2}}$

$\qquad = \dfrac{\sqrt{14^2}}{\sqrt{p^4q^2}}$

$\qquad = \dfrac{14}{p^2q}$

61. $\sqrt[3]{16z^3} = \sqrt[3]{2^3z^3 \cdot 2}$

$\qquad = \sqrt[3]{2^3z^3} \cdot \sqrt[3]{2}$

$\qquad = 2z \sqrt[3]{2}$

63. $\dfrac{\sqrt[3]{216a^6b^{12}c^4}}{d^{27}} = \dfrac{\sqrt[3]{216a^6b^{12}c^4}}{\sqrt[3]{d^{27}}}$

$\qquad = \dfrac{\sqrt[3]{6^3a^6b^{12}c^3 \cdot c}}{\sqrt[3]{d^{27}}}$

$\qquad = \dfrac{\sqrt[3]{6^3a^6b^{12}c^3}\,\sqrt[3]{c}}{\sqrt[3]{d^{27}}}$

$\qquad = \dfrac{6a^2b^4c \sqrt[3]{c}}{d^9}$

65. $\sqrt[5]{-96x^{15}z^{26}} = \sqrt[5]{(-2)^5 \cdot x^{15}z^{25} \cdot 3z}$

$\qquad = -2x^3z^5 \sqrt[5]{3z}$

67. $\sqrt{(13 + 2x)^2} = 13 + 2x$

69. $\sqrt{36 + 9v^2} = \sqrt{9(4 + r^2)}$

$\qquad = \sqrt{9}\sqrt{4 + v^2}$

$\qquad = 3\sqrt{4 + v^2}$

71. Let $x = -2$.

$\qquad \sqrt{x^2} = x \qquad ?$

$\qquad \sqrt{(-2)^2} = -2 \qquad ?$

$\qquad \sqrt{4} = -2 \qquad ?$

$\qquad 2 \ne -2$

73. $\sqrt{x^2} = \begin{cases} x, & \text{if } x \ge 0 \\ -x, & \text{if } x < 0 \end{cases}$

$\qquad \sqrt{x^2} = |x|$

75. $\sqrt{b^2} = |b|$

$\qquad = -b$

(Since $b < 0$)

77. $\sqrt{(5 + a)^2} = |5 + a|$

$\qquad = 5 + a$

79. $\sqrt{(a + b)^2} = |a + b|$

<u>Problem Set 4.2</u>

1. $3\sqrt{13} + 5\sqrt{13} = 8\sqrt{13}$

3. $11\sqrt{3} - 19\sqrt{3} = -8\sqrt{3}$

5. $\sqrt{54} + \sqrt{6} = \sqrt{3^2 \cdot 3 \cdot 2} + \sqrt{6}$

$\qquad = 3\sqrt{6} + \sqrt{6}$

$\qquad = 4\sqrt{6}$

7. $\sqrt[3]{-8} + \sqrt[3]{81} = \sqrt[3]{(-2)^3} + \sqrt[3]{3^3 \cdot 3}$

$\qquad = -2 + 3\sqrt[3]{3}$

9. $\sqrt{8t^3} + 3t\sqrt{2t} = \sqrt{2^2 \cdot 2t^2 \cdot t} + 3t\sqrt{2t}$

$\qquad = 2t\sqrt{2t} + 3t\sqrt{2t}$

$\qquad = 5t\sqrt{2t}$

11. $3\sqrt[3]{8x^4y^2} - 5x\sqrt[3]{-xy^2}$

$\qquad = 3\sqrt[3]{2^3x^3 \cdot xy^2} - 5x\sqrt[3]{(-1)^3xy^2}$

$\qquad = 3 \cdot 2x\sqrt[3]{xy^2} - 5x \cdot (-1)\sqrt[3]{xy^2}$

$\qquad = 6x\sqrt[3]{xy^2} + 5x\sqrt[3]{xy^2}$

$\qquad = 11x\sqrt[3]{xy^2}$

13. $3\sqrt{13} + 8\sqrt{13} = 11\sqrt{13}$

142

15. $4\sqrt{18} - 5\sqrt{18} = -\sqrt{18}$

$$= -\sqrt{3^2 \cdot 2}$$

$$= -3\sqrt{2}$$

17. $\sqrt{75} + 6\sqrt{12}$

$$= \sqrt{5^2 \cdot 3} + 6\sqrt{2^2 \cdot 3}$$

$$= 5\sqrt{3} + 6 \cdot 2\sqrt{3}$$

$$= 5\sqrt{3} + 12\sqrt{3}$$

$$= 17\sqrt{3}$$

19. $\sqrt[3]{128} - \sqrt[3]{16}$

$$= \sqrt[3]{4^3 \cdot 2} - \sqrt[3]{2^3 \cdot 2}$$

$$= 4\sqrt[3]{2} - 2\sqrt[3]{2}$$

$$= 2\sqrt[3]{2}$$

21. $9\sqrt{3x} + \sqrt{27x}$

$$= 9\sqrt{3x} + \sqrt{3^2 \cdot 3x}$$

$$= 9\sqrt{3x} + 3\sqrt{3x}$$

$$= 12\sqrt{3x}$$

23. $2\sqrt{24} + 3\sqrt{54} - \sqrt{6}$

$$= 2\sqrt{2^2 \cdot 6} + 3\sqrt{3^2 \cdot 6} - \sqrt{6}$$

$$= 2 \cdot 2\sqrt{6} + 3 \cdot 3\sqrt{6} - \sqrt{6}$$

$$= 4\sqrt{6} + 9\sqrt{6} - \sqrt{6}$$

$$= 12\sqrt{6}$$

25. $t^2\sqrt{8t} + 2t\sqrt{2t^3} - \sqrt{2t^5}$

$$= t^2\sqrt{2^2 \cdot 2t} + 2t\sqrt{2t^2 \cdot t} - \sqrt{2t^4 \cdot t}$$

$$= t^2 \cdot 2\sqrt{2t} + 2t \cdot t\sqrt{2t} - t^2\sqrt{2t}$$

$$= 2t^2\sqrt{2t} + 2t^2\sqrt{2t} - t^2\sqrt{2t}$$

$$= 3t^2\sqrt{2t}$$

27. $\dfrac{3}{8}\sqrt{48x^3} + \dfrac{1}{2}x\sqrt{3x}$

$$= \dfrac{3}{8}\sqrt{4^2 \cdot 3x^2 \cdot x} + \dfrac{1}{2}x\sqrt{3x}$$

$$= \dfrac{3}{8} \cdot 4x\sqrt{3x} + \dfrac{1}{2}x\sqrt{3x}$$

$$= \dfrac{3}{2}x\sqrt{3x} + \dfrac{1}{2}x\sqrt{3x}$$

$$= 2x\sqrt{3x}$$

29. $2\sqrt[3]{54} - \sqrt[3]{250} + 7\sqrt[3]{2}$

$$= 2\sqrt[3]{3^3 \cdot 2} - \sqrt[3]{5^3 \cdot 2} + 7\sqrt[3]{2}$$

$$= 2 \cdot 3\sqrt[3]{2} - 5\sqrt[3]{2} + 7\sqrt[3]{2}$$

$$= 6\sqrt[3]{2} - 5\sqrt[3]{2} + 7\sqrt[3]{2}$$

$$= 8\sqrt[3]{2}$$

31. $\sqrt[3]{-192} - 2\sqrt[3]{-375}$

$$= \sqrt[3]{(-4)^3 \cdot 3} - 2\sqrt[3]{(-5)^3 \cdot 3}$$

$$= -4\sqrt[3]{3} - 2 \cdot (-5)\sqrt[3]{3}$$

$$= -4\sqrt[3]{3} + 10\sqrt[3]{3}$$

$$= 6\sqrt[3]{3}$$

33. $\dfrac{2}{9}\sqrt{18x} + \dfrac{1}{3}\sqrt{8x}$

$$= \dfrac{2}{9}\sqrt{3^2 \cdot 2x} + \dfrac{1}{3}\sqrt{2^2 \cdot 2x}$$

$$= \dfrac{2}{9} \cdot 3\sqrt{2x} + \dfrac{1}{3} \cdot 2\sqrt{2x}$$

$$= \dfrac{2}{3}\sqrt{2x} + \dfrac{2}{3}\sqrt{2x}$$

$$= \dfrac{4}{3}\sqrt{2x}$$

$35.\quad 6\sqrt{\dfrac{8x^3y}{81}} - 5x\sqrt{\dfrac{32xy}{225}}$

$\quad = 6\dfrac{\sqrt{2^2 \cdot 2x^2 \cdot xy}}{\sqrt{9^2}} - \dfrac{5x\sqrt{4^2 \cdot 2xy}}{\sqrt{15^2}}$

$\quad = \dfrac{6 \cdot 2x\sqrt{2xy}}{9} - \dfrac{5x \cdot 4\sqrt{2xy}}{15}$

$\quad = \dfrac{4x\sqrt{2xy}}{3} - \dfrac{4x\sqrt{2xy}}{3}$

$\quad = 0$

$37.\quad x^m\sqrt{x^{2n}} + x^n\sqrt{x^{2m}}$

$\quad = x^m\sqrt{(x^n)^2} + x^m\sqrt{(x^m)^2}$

$\quad = x^m \cdot x^n + x^n \cdot x^m$

$\quad = x^{m+n} + x^{n+m}$

$\quad = 2x^{m+n}$

$39.\quad 3x^n\sqrt[n]{x^{mn}} - x^m\sqrt[m]{x^{mn}}$

$\quad = 3x^n\sqrt[n]{(x^m)^n} - x^m\sqrt[m]{(x^n)^m}$

$\quad = 3x^n \cdot x^m - x^m \cdot x^n$

$\quad = 3x^{n+m} - x^{m+n}$

$\quad = 2x^{m+n}$

Problem Set 4.3

$1.\quad 2\sqrt{z} \cdot \sqrt{z} = 2(\sqrt{z})^2$

$\quad\quad\quad\quad\quad = 2z$

$3.\quad \sqrt[3]{24xy^2} \cdot \sqrt[3]{9xy} = \sqrt[3]{216x^2y^3}$

$\quad\quad\quad\quad\quad\quad\quad\quad = 6y\sqrt[3]{x^2}$

$5.\quad (3\sqrt{7})^2 = 3^2(\sqrt{7})^2$

$\quad\quad\quad\quad\quad = 9 \cdot 7$

$\quad\quad\quad\quad\quad = 63$

$7.\quad \sqrt{3}(\sqrt{2} - \sqrt{3}) = \sqrt{3} \cdot \sqrt{2} - \sqrt{3} \cdot \sqrt{3}$

$\quad\quad\quad\quad\quad\quad\quad = \sqrt{6} - (\sqrt{3})^2$

$\quad\quad\quad\quad\quad\quad\quad = \sqrt{6} - 3$

$9.\quad 2\sqrt{3}(3\sqrt{3} + \sqrt{6}) = 2\sqrt{3} \cdot 3\sqrt{3} + 2\sqrt{3} \cdot \sqrt{6}$

$\quad\quad\quad\quad\quad\quad\quad\quad = 6(\sqrt{3})^2 + 2\sqrt{18}$

$\quad\quad\quad\quad\quad\quad\quad\quad = 6 \cdot 3 + 2 \cdot 3\sqrt{2}$

$\quad\quad\quad\quad\quad\quad\quad\quad = 18 + 6\sqrt{2}$

$11.\quad (2 + \sqrt{3})(3 - \sqrt{3}) = 2 \cdot 3 - 2\sqrt{3} + 3\sqrt{3} - (\sqrt{3})^2$

$\quad\quad\quad\quad\quad\quad\quad\quad = 6 + \sqrt{3} - 3$

$\quad\quad\quad\quad\quad\quad\quad\quad = 3 + \sqrt{3}$

$13.\quad (2\sqrt{3} - 3\sqrt{2})^2 = (2\sqrt{3})^2 - 2(2\sqrt{3})(3\sqrt{2}) + (3\sqrt{2})^2$

$\quad\quad\quad\quad\quad\quad\quad\quad = 2^2(\sqrt{3})^2 - 12\sqrt{6} + 3^2(\sqrt{2})^2$

$\quad\quad\quad\quad\quad\quad\quad\quad = 4 \cdot 3 - 12\sqrt{6} + 9 \cdot 2$

$\quad\quad\quad\quad\quad\quad\quad\quad = 12 - 12\sqrt{6} + 18$

$\quad\quad\quad\quad\quad\quad\quad\quad = 30 - 12\sqrt{6}$

$15.\quad \sqrt{x} \cdot \sqrt{x} \cdot \sqrt{x} = (\sqrt{x})^2 \cdot \sqrt{x}$

$\quad\quad\quad\quad\quad\quad\quad = x\sqrt{x}$

$17.\quad \sqrt{7xy^3} \cdot 2\sqrt{14xy} = 2\sqrt{7^2 \cdot 2x^2y^4}$

$\quad\quad\quad\quad\quad\quad\quad\quad = 2 \cdot 7xy^2\sqrt{2}$

$\quad\quad\quad\quad\quad\quad\quad\quad = 14xy^2\sqrt{2}$

$19.\quad \sqrt{7}(2\sqrt{2} - \sqrt{3}) = \sqrt{7} \cdot 2\sqrt{2} - \sqrt{7} \cdot \sqrt{3}$

$\quad\quad\quad\quad\quad\quad\quad\quad = 2\sqrt{14} - \sqrt{21}$

$21.\quad \sqrt[3]{3s^2t^4} \cdot \sqrt[3]{18s} = \sqrt[3]{3^3 \cdot 2s^3t^4}$

$\quad\quad\quad\quad\quad\quad\quad\quad = 3st\sqrt[3]{2t}$

23. $(3\sqrt{3})^2 = 3^2 \cdot (\sqrt{3})^2$

$= 9 \cdot 3$

$= 27$

25. $(1 - \sqrt{2})(1 - 3\sqrt{2})$

$= 1 \cdot 1 - 1 \cdot 3\sqrt{2} - 1 \cdot \sqrt{2} + 3(\sqrt{2})^2$

$= 1 - 3\sqrt{2} - \sqrt{2} + 3 \cdot 2$

$= 1 - 4\sqrt{2} + 6$

$= 7 - 4\sqrt{2}$

27. $(s - \sqrt{2t})^2 = (s)^2 - 2(s)(\sqrt{2t}) + (\sqrt{2t})^2$

$= s^2 - 2s\sqrt{2t} + 2t$

29. $\sqrt{6}(2\sqrt{2} - \sqrt{3}) = \sqrt{6} \cdot 2\sqrt{2} - \sqrt{6} \cdot \sqrt{3}$

$= 2\sqrt{12} - \sqrt{18}$

$= 2 \cdot 2\sqrt{3} - 3\sqrt{2}$

$= 4\sqrt{3} - 3\sqrt{2}$

31. $2\sqrt{3}(2\sqrt{5} + 4\sqrt{3}) = 2\sqrt{3} \cdot 2\sqrt{5} + 2\sqrt{3} \cdot 4\sqrt{3}$

$= 4\sqrt{15} + 8(\sqrt{3})^2$

$= 4\sqrt{15} + 8 \cdot 3$

$= 4\sqrt{15} + 24$

33. $\sqrt[3]{2x^2y^2}(\sqrt[3]{4xy^4} + 2y\sqrt[3]{12xy})$

$= \sqrt[3]{2x^2y^2} \cdot \sqrt[3]{4xy^4} + \sqrt[3]{2x^2y^2} \cdot 2y\sqrt[3]{12xy}$

$= \sqrt[3]{8x^3y^6} + 2y\sqrt[3]{24x^3y^3}$

$= 2xy^2 + 4xy^2\sqrt[3]{3}$

35. $\sqrt[5]{4k^9}\left(2\sqrt[5]{8k} - \sqrt[5]{\dfrac{1}{4}k}\right)$

$= \sqrt[5]{4k^9} \cdot 2\sqrt[5]{8k} - \sqrt[5]{4k^9} \cdot \sqrt[5]{\dfrac{1}{4}k}$

$= 2\sqrt[5]{32k^{10}} - \sqrt[5]{k^{10}}$

$= 2 \cdot 2k^2 - k^2$

$= 4k^2 - k^2$

$= 3k^2$

37. $(2\sqrt{3} + 3)(2\sqrt{3} - 2)$

$= (2\sqrt{3})^2 - 2 \cdot 2\sqrt{3} + 3 \cdot 2\sqrt{3} - 3 \cdot 2$

$= 2^2(\sqrt{3})^2 - 4\sqrt{3} + 6\sqrt{3} - 6$

$= 4 \cdot 3 + 2\sqrt{3} - 6$

$= 12 + 2\sqrt{3} - 6$

$= 6 + 2\sqrt{3}$

39. $(x\sqrt{2} - y\sqrt{3})(x\sqrt{2} + y\sqrt{3})$

$= (x\sqrt{2})^2 - (y\sqrt{3})^2$

$= x^2(\sqrt{2})^2 - y^2(\sqrt{3})^2$

$= 2x^2 - 3y^2$

41. $(2\sqrt{2x} + \sqrt{3})^2 = (2\sqrt{2x})^2 + 2(2\sqrt{2x})(\sqrt{3}) + (\sqrt{3})^2$

$= 2^2(\sqrt{2x})^2 + 4\sqrt{6x} + 3$

$= 4(2x) + 4\sqrt{6x} + 3$

$= 8x + 4\sqrt{6x} + 3$

43. $(\sqrt{3} + 2\sqrt{6})(2\sqrt{3} - 3\sqrt{6})$

$= \sqrt{3} \cdot 2\sqrt{3} - \sqrt{3} \cdot 3\sqrt{6} + 2\sqrt{6} \cdot 2\sqrt{3} - 2\sqrt{6} \cdot 3\sqrt{6}$

$= 2(\sqrt{3})^2 - 3\sqrt{18} + 4\sqrt{18} - 6(\sqrt{6})^2$

$= 2 \cdot 3 + \sqrt{18} - 6 \cdot 6$

$= 6 + 3\sqrt{2} - 36$

$= 3\sqrt{2} - 30$

45. $(\sqrt{2} + 1)(\sqrt{2} - 2)(\sqrt{2} - 1)$

$= [(\sqrt{2} + 1)(\sqrt{2} - 1)](\sqrt{2} - 2)$

$= [(\sqrt{2})^2 - 1^2](\sqrt{2} - 2)$

$= (2 - 1)(\sqrt{2} - 2)$

$= 1(\sqrt{2} - 2)$

$= \sqrt{2} - 2$

47. $\left(\dfrac{1}{2} - \sqrt{2}\right)\left(2 - \dfrac{1}{2}\sqrt{2}\right)$

$= \dfrac{1}{2} \cdot 2 - \dfrac{1}{2} \cdot \dfrac{1}{2}\sqrt{2} - 2\sqrt{2} + \dfrac{1}{2}(\sqrt{2})^2$

$= 1 - \dfrac{1}{4}\sqrt{2} - 2\sqrt{2} + \dfrac{1}{2} \cdot 2$

$= 1 - \dfrac{9}{4}\sqrt{2} + 1$

$= 2 - \dfrac{9}{4}\sqrt{2}$

49. $\left(\sqrt[3]{2} + 1\right)^3$

$= \left(\sqrt[3]{2}\right)^3 + 3\left(\sqrt[3]{2}\right)^2(1) + 3\left(\sqrt[3]{2}\right)(1)^2 + 1^3$

$= 2 + 3\sqrt[3]{4} + 3\sqrt[3]{2} + 1$

$= 3 + 3\sqrt[3]{4} + 3\sqrt[3]{2}$

51. $\left(\sqrt[3]{5} + 1\right)^2\left(\sqrt[3]{5} + 2\right)$

$= \left[\left(\sqrt[3]{5}\right)^2 + 2\left(\sqrt[3]{5}\right)(1) + 1^2\right]\left(\sqrt[3]{5} + 2\right)$

$= \left(\sqrt[3]{25} + 2\sqrt[3]{5} + 1\right)\left(\sqrt[3]{5} + 2\right)$

$= \sqrt[3]{25} \cdot \sqrt[3]{5} + 2\sqrt[3]{5} \cdot \sqrt[3]{5} + 1 \cdot \sqrt[3]{5}$

$\quad + 2\sqrt[3]{25} + 2 \cdot 2\sqrt[3]{5} + 2 \cdot 1$

$= \sqrt[3]{125} + 2\sqrt[3]{25} + \sqrt[3]{5} + 2\sqrt[3]{25} + 4\sqrt[3]{5} + 2$

$= 7 + 4\sqrt[3]{25} + 5\sqrt[3]{5}$

53. $\left(\sqrt[3]{2} - \sqrt[3]{3}\right)^3$

$= \left(\sqrt[3]{2}\right)^3 - 3\left(\sqrt[3]{2}\right)^2\left(\sqrt[3]{3}\right) + 3\left(\sqrt[3]{2}\right)\left(\sqrt[3]{3}\right)^2 - \left(\sqrt[3]{3}\right)^3$

$= 2 - 3\sqrt[3]{12} + 3\sqrt[3]{18} - 3$

$= -1 - 3\sqrt[3]{12} + 3\sqrt[3]{18}$

$= 3\sqrt[3]{18} - 3\sqrt[3]{12} - 1$

55. $\sqrt{10} + \sqrt{6} = \sqrt{2 \cdot 5} + \sqrt{2 \cdot 3}$

$\qquad\qquad\quad = \sqrt{2} \cdot \sqrt{5} + \sqrt{2} \cdot \sqrt{3}$

$\qquad\qquad\quad = \sqrt{2}(\sqrt{5} + \sqrt{3})$

57. $\sqrt{6} - \sqrt{3} = \sqrt{2 \cdot 3} - \sqrt{3}$

$\qquad\qquad\quad = \sqrt{2} \cdot \sqrt{3} - \sqrt{3}$

$\qquad\qquad\quad = \sqrt{3}(\sqrt{2} - 1)$

59. $\sqrt{12} + \sqrt{20} = 2\sqrt{3} + 2\sqrt{5}$

$\qquad\qquad\quad\ = 2(\sqrt{3} + \sqrt{5})$

61. $\left(\sqrt[3]{3} - \sqrt[3]{2}\right)\left(\sqrt[3]{9} + \sqrt[3]{6} + \sqrt[3]{4}\right)$

$= \sqrt[3]{3} \cdot \sqrt[3]{9} + \sqrt[3]{3} \cdot \sqrt[3]{6} + \sqrt[3]{3} \cdot \sqrt[3]{4}$

$\quad - \sqrt[3]{2} \cdot \sqrt[3]{9} - \sqrt[3]{2} \cdot \sqrt[3]{6} - \sqrt[3]{2} \cdot \sqrt[3]{4}$

$= \sqrt[3]{27} + \sqrt[3]{18} + \sqrt[3]{12} - \sqrt[3]{18} - \sqrt[3]{12} - \sqrt[3]{8}$

$= \sqrt[3]{27} - \sqrt[3]{8}$

$= 3 - 2$

$= 1$

63. $\left(\sqrt[3]{a} - \sqrt[3]{b}\right)\left(\sqrt[3]{a^2} + \sqrt[3]{ab} + \sqrt[3]{b^2}\right)$

$= \sqrt[3]{a} \cdot \sqrt[3]{a^2} + \sqrt[3]{a} \cdot \sqrt[3]{ab} + \sqrt[3]{a} \cdot \sqrt[3]{b^2}$

$\quad - \sqrt[3]{b} \cdot \sqrt[3]{a^2} - \sqrt[3]{b} \cdot \sqrt[3]{ab} - \sqrt[3]{b} \cdot \sqrt[3]{b^2}$

$= \sqrt[3]{a^3} + \sqrt[3]{a^2b} + \sqrt[3]{ab^2} - \sqrt[3]{a^2b} - \sqrt[3]{ab^2} - \sqrt[3]{b^3}$

$$= \sqrt[3]{a^3} - \sqrt[3]{b^3}$$

$$= a - b$$

Difference of two cubes.

Problem Set 4.4

1. $\dfrac{\sqrt{48}}{\sqrt{6}} = \sqrt{\dfrac{48}{6}}$

$$= \sqrt{8}$$

$$= 2\sqrt{2}$$

3. $\dfrac{\sqrt{90}}{\sqrt{5}} = \sqrt{\dfrac{90}{5}}$

$$= \sqrt{18}$$

$$= 3\sqrt{2}$$

5. $\dfrac{\sqrt{6} - \sqrt{15}}{\sqrt{3}} = \dfrac{\sqrt{6}}{\sqrt{3}} - \dfrac{\sqrt{15}}{\sqrt{3}}$

$$= \sqrt{\dfrac{6}{3}} - \sqrt{\dfrac{15}{3}}$$

$$= \sqrt{2} - \sqrt{5}$$

7. $\dfrac{1}{\sqrt{5}} = \dfrac{1 \cdot \sqrt{5}}{\sqrt{5} \cdot \sqrt{5}}$

$$= \dfrac{\sqrt{5}}{5}$$

9. $\dfrac{6}{2\sqrt{6}} = \dfrac{6 \cdot \sqrt{6}}{2\sqrt{6} \cdot \sqrt{6}}$

$$= \dfrac{6\sqrt{6}}{2 \cdot 6}$$

$$= \dfrac{\sqrt{6}}{2}$$

11. $\dfrac{1}{\sqrt[3]{7}} = \dfrac{1 \cdot \sqrt[3]{7^2}}{\sqrt[3]{7} \cdot \sqrt[3]{7^2}}$

$$= \dfrac{\sqrt[3]{49}}{\sqrt[3]{7^3}}$$

$$= \dfrac{\sqrt[3]{49}}{7}$$

13. $\dfrac{1}{\sqrt[4]{4}} = \dfrac{1 \cdot \sqrt[4]{2^2}}{\sqrt[4]{2^2} \cdot \sqrt[4]{2^2}}$

$$= \dfrac{\sqrt[4]{2^2}}{\sqrt[4]{2^4}}$$

$$= \dfrac{\sqrt[4]{4}}{2}$$

15. $\dfrac{5}{\sqrt{5} + 2} = \dfrac{5 \cdot (\sqrt{5} - 2)}{(\sqrt{5} + 2)(\sqrt{5} - 2)}$

$$= \dfrac{5(\sqrt{5} - 2)}{(\sqrt{5})^2 - 2^2}$$

$$= \dfrac{5(\sqrt{5} - 2)}{5 - 4}$$

$$= \dfrac{5\sqrt{5} - 10}{1}$$

$$= 5\sqrt{5} - 10$$

17. $\dfrac{2\sqrt{3}}{\sqrt{2} - \sqrt{3}} = \dfrac{2\sqrt{3}(\sqrt{2} + \sqrt{3})}{(\sqrt{2} - \sqrt{3})(\sqrt{2} + \sqrt{3})}$

$$= \dfrac{2\sqrt{3} \cdot \sqrt{2} + 2\sqrt{3} \cdot \sqrt{3}}{(\sqrt{2})^2 - (\sqrt{3})^2}$$

$$= \dfrac{2\sqrt{6} + 2 \cdot 3}{2 - 3}$$

$$= \dfrac{2\sqrt{6} + 6}{-1}$$

$$= -2\sqrt{6} - 6$$

19. $\dfrac{\sqrt{3}}{\sqrt{3} + 3\sqrt{2}} = \dfrac{\sqrt{3} \cdot \sqrt{3}}{(\sqrt{3} + 3\sqrt{2}) \cdot \sqrt{3}}$

$\qquad = \dfrac{3}{\sqrt{3} \cdot \sqrt{3} + 3\sqrt{2} \cdot \sqrt{3}}$

$\qquad = \dfrac{3}{3 + 3\sqrt{6}}$

$\qquad = \dfrac{3}{3(1 + \sqrt{6})}$

$\qquad = \dfrac{1}{1 + \sqrt{6}}$

21. $\dfrac{\sqrt{72}}{\sqrt{24}} = \sqrt{\dfrac{72}{24}}$

$\qquad = \sqrt{3}$

23. $\dfrac{\sqrt[3]{32}}{\sqrt[3]{2}} = \sqrt[3]{16}$

$\qquad = 2\sqrt[3]{2}$

25. $\dfrac{\sqrt{120}}{\sqrt{10}} = \sqrt{\dfrac{120}{10}}$

$\qquad = \sqrt{12}$

$\qquad = 2\sqrt{3}$

27. $\dfrac{\sqrt{6} - \sqrt{3}}{\sqrt{3}} = \dfrac{\sqrt{6}}{\sqrt{3}} - \dfrac{\sqrt{3}}{\sqrt{3}}$

$\qquad = \sqrt{\dfrac{6}{3}} - 1$

$\qquad = \sqrt{2} - 1$

29. $\dfrac{\sqrt[3]{-56}}{\sqrt[3]{-7}} = \sqrt[3]{\dfrac{-56}{-7}}$

$\qquad = \sqrt[3]{8}$

$\qquad = 2$

31. $\dfrac{\sqrt{15x}}{\sqrt{3x}} = \sqrt{\dfrac{15x}{3x}}$

$\qquad = \sqrt{5}$

33. $\dfrac{6\sqrt{7} + 9\sqrt{21}}{3\sqrt{7}} = \dfrac{6\sqrt{7}}{3\sqrt{7}} + \dfrac{9\sqrt{21}}{3\sqrt{7}}$

$\qquad = 2\sqrt{\dfrac{7}{7}} + 3\sqrt{\dfrac{21}{7}}$

$\qquad = 2\sqrt{1} + 3\sqrt{3}$

$\qquad = 2 + 3\sqrt{3}$

35. $\dfrac{x\sqrt{6xy^3}}{\sqrt{3x^3}} = x\sqrt{\dfrac{6xy^3}{3x^3}}$

$\qquad = x\sqrt{\dfrac{2y^3}{x^2}}$

$\qquad = \dfrac{xy\sqrt{2y}}{x}$

$\qquad = y\sqrt{2y}$

37. $\dfrac{3\sqrt{12}}{2\sqrt{3}} = \dfrac{3}{2}\sqrt{\dfrac{12}{3}}$

$\qquad = \dfrac{3}{2}\sqrt{4}$

$\qquad = \dfrac{3}{2}(2)$

$\qquad = 3$

39. $\dfrac{10\sqrt{3} + 4\sqrt{15}}{2\sqrt{3}} = \dfrac{10\sqrt{3}}{2\sqrt{3}} + \dfrac{4\sqrt{15}}{2\sqrt{3}}$

$\qquad = 5\sqrt{\dfrac{3}{3}} + 2\sqrt{\dfrac{15}{3}}$

$\qquad = 5(1) + 2\sqrt{5}$

$\qquad = 5 + 2\sqrt{5}$

41.
$$\frac{2}{\sqrt{11}} = \frac{2 \cdot \sqrt{11}}{\sqrt{11} \cdot \sqrt{11}}$$

$$= \frac{2\sqrt{11}}{11}$$

43.
$$\frac{1}{2\sqrt{5}} = \frac{1 \cdot \sqrt{5}}{2\sqrt{5} \cdot \sqrt{5}}$$

$$= \frac{\sqrt{5}}{2 \cdot 5}$$

$$= \frac{\sqrt{5}}{10}$$

45.
$$\frac{\sqrt{3}}{\sqrt{5}} = \frac{\sqrt{3} \cdot \sqrt{5}}{\sqrt{5} \cdot \sqrt{5}}$$

$$= \frac{\sqrt{15}}{5}$$

47.
$$\frac{3}{3 + \sqrt{2}} = \frac{3(3 - \sqrt{2})}{(3 + \sqrt{2})(3 - \sqrt{2})}$$

$$= \frac{9 - 3\sqrt{2}}{3^2 - (\sqrt{2})^2}$$

$$= \frac{9 - 3\sqrt{2}}{9 - 2}$$

$$= \frac{9 - 3\sqrt{2}}{7}$$

49.
$$\frac{2\sqrt{5}}{5\sqrt{2}} = \frac{2\sqrt{5} \cdot \sqrt{2}}{5\sqrt{2} \cdot \sqrt{2}}$$

$$= \frac{2\sqrt{10}}{5 \cdot 2}$$

$$= \frac{\sqrt{10}}{5}$$

51.
$$\frac{\sqrt{3}}{2 + \sqrt{3}} = \frac{\sqrt{3}(2 - \sqrt{3})}{(2 + \sqrt{3})(2 - \sqrt{3})}$$

$$= \frac{2\sqrt{3} - (\sqrt{3})^2}{2^2 - (\sqrt{3})^2}$$

$$= \frac{2\sqrt{3} - 3}{4 - 3}$$

$$= \frac{2\sqrt{3} - 3}{1}$$

$$= 2\sqrt{3} - 3$$

53.
$$\frac{3\sqrt{2}}{\sqrt{6} - 2\sqrt{2}} = \frac{3\sqrt{2}(\sqrt{6} + 2\sqrt{2})}{(\sqrt{6} - 2\sqrt{2})(\sqrt{6} + 2\sqrt{2})}$$

$$= \frac{3\sqrt{2} \cdot \sqrt{6} + 3 \cdot 2 \cdot (\sqrt{2})^2}{(\sqrt{6})^2 - (2\sqrt{2})^2}$$

$$= \frac{3\sqrt{12} + 6 \cdot 2}{6 - 4 \cdot 2}$$

$$= \frac{3\sqrt{12} + 12}{-2}$$

$$= \frac{6\sqrt{3} + 12}{-2}$$

$$= -3\sqrt{3} - 6$$

55.
$$\frac{1 - \sqrt{5}}{1 + \sqrt{5}} = \frac{(1 - \sqrt{5})(1 - \sqrt{5})}{(1 + \sqrt{5})(1 - \sqrt{5})}$$

$$= \frac{1^2 - 2(1)(\sqrt{5}) + (\sqrt{5})^2}{1^2 - (\sqrt{5})^2}$$

$$= \frac{1 - 2\sqrt{5} + 5}{1 - 5}$$

$$= \frac{6 - 2\sqrt{5}}{-4}$$

$$= \frac{2(3 - \sqrt{5})}{-4}$$

$$= -\frac{3 - \sqrt{5}}{2}$$

or $\dfrac{\sqrt{5} - 3}{2}$

57. $\dfrac{1}{\sqrt[3]{2}} = \dfrac{1 \cdot \sqrt[3]{2^2}}{\sqrt[3]{2} \cdot \sqrt[3]{2^2}}$

$= \dfrac{\sqrt[3]{4}}{\sqrt[3]{2^3}}$

$= \dfrac{\sqrt[3]{4}}{2}$

59. $\dfrac{4}{\sqrt[4]{8}} = \dfrac{4 \cdot \sqrt[4]{2}}{\sqrt[4]{2^3} \cdot \sqrt[4]{2}}$

$= \dfrac{4\sqrt[4]{2}}{\sqrt[4]{2^4}}$

$= \dfrac{4\sqrt[4]{2}}{2}$

$= 2\sqrt[4]{2}$

61. $\dfrac{\sqrt{6}}{5} = \dfrac{\sqrt{6} \cdot \sqrt{6}}{5 \cdot \sqrt{6}}$

$= \dfrac{6}{5\sqrt{6}}$

63. $\dfrac{\sqrt{2}}{1 - \sqrt{2}} = \dfrac{\sqrt{2} \cdot \sqrt{2}}{\left(1 - \sqrt{2}\right) \cdot \sqrt{2}}$

$= \dfrac{2}{\sqrt{2} - \left(\sqrt{2}\right)^2}$

$= \dfrac{2}{\sqrt{2} - 2}$

65. $\dfrac{-5\sqrt{3}}{3 + \sqrt{6}} = \dfrac{-5\sqrt{3} \cdot \sqrt{3}}{\left(3 + \sqrt{6}\right) \cdot \sqrt{3}}$

$= \dfrac{-5(3)}{3\sqrt{3} + \sqrt{18}}$

$= \dfrac{-15}{3\sqrt{3} + 3\sqrt{2}}$

$= \dfrac{-15}{3\left(\sqrt{3} + \sqrt{2}\right)}$

$= -\dfrac{5}{\sqrt{3} + \sqrt{2}}$

67. $\dfrac{3 + \sqrt{5}}{5 - \sqrt{5}} = \dfrac{\left(3 + \sqrt{5}\right)\left(3 - \sqrt{5}\right)}{\left(5 - \sqrt{5}\right)\left(3 - \sqrt{5}\right)}$

$= \dfrac{3^2 - \left(\sqrt{5}\right)^2}{15 - 5\sqrt{5} - 3\sqrt{5} + \left(\sqrt{5}\right)^2}$

$= \dfrac{9 - 5}{15 - 8\sqrt{5} + 5}$

$= \dfrac{4}{20 - 8\sqrt{5}}$

$= \dfrac{4}{4\left(5 - 2\sqrt{5}\right)}$

$= \dfrac{1}{5 - 2\sqrt{5}}$

69. $\dfrac{2\sqrt{2} + \sqrt{3}}{\sqrt{2} - 3\sqrt{3}} = \dfrac{\left(2\sqrt{2} + \sqrt{3}\right)\left(2\sqrt{2} - \sqrt{3}\right)}{\left(\sqrt{2} - 3\sqrt{3}\right)\left(2\sqrt{2} - \sqrt{3}\right)}$

$= \dfrac{\left(2\sqrt{2}\right)^2 - \left(\sqrt{3}\right)^2}{2\left(\sqrt{2}\right)^2 - \sqrt{6} - 6\sqrt{6} + 3\left(\sqrt{3}\right)^2}$

$= \dfrac{4 \cdot 2 - 3}{2(2) - 7\sqrt{6} + 3(3)}$

$= \dfrac{5}{4 - 7\sqrt{6} + 9}$

$= \dfrac{5}{13 - 7\sqrt{6}}$

71. $\dfrac{\sqrt{x+h}-\sqrt{x}}{h} = \dfrac{(\sqrt{x+h}-\sqrt{x})(\sqrt{x+h}+\sqrt{x})}{h(\sqrt{x+h}+\sqrt{x})}$

$\qquad\qquad = \dfrac{(\sqrt{x+h})^2 - (\sqrt{x})^2}{h(\sqrt{x+h}+\sqrt{x})}$

$\qquad\qquad = \dfrac{x+h-x}{h(\sqrt{x+h}+\sqrt{x})}$

$\qquad\qquad = \dfrac{h}{h(\sqrt{x+h}+\sqrt{x})}$

$\qquad\qquad = \dfrac{1}{\sqrt{x+h}+\sqrt{x}}$

73. $\dfrac{1}{\sqrt[3]{3}-\sqrt[3]{2}} = \dfrac{1 \cdot \left[(\sqrt[3]{3})^2 + \sqrt[3]{3}\cdot\sqrt[3]{2} + (\sqrt[3]{2})^2 \right]}{(\sqrt[3]{3}-\sqrt[3]{2})\cdot\left[(\sqrt[3]{3})^2 + \sqrt[3]{3}\cdot\sqrt[3]{2} + (\sqrt[3]{2})^2 \right]}$

$\qquad\qquad = \dfrac{\sqrt[3]{9} + \sqrt[3]{6} + \sqrt[3]{4}}{(\sqrt[3]{3})^3 - (\sqrt[3]{2})^3}$

$\qquad\qquad = \dfrac{\sqrt[3]{9} + \sqrt[3]{6} + \sqrt[3]{4}}{3 - 2}$

$\qquad\qquad = \dfrac{\sqrt[3]{9} + \sqrt[3]{6} + \sqrt[3]{4}}{1}$

$\qquad\qquad = \sqrt[3]{9} + \sqrt[3]{6} + \sqrt[3]{4}$

Problem Set 4.5

1. $\sqrt{x+1} = 2$

$\qquad (\sqrt{x+1})^2 = 2^2$

$\qquad\qquad x + 1 = 4$

$\qquad\qquad\quad x = 3$

Check:

LS: $\sqrt{3+1} = \sqrt{4} = 2$

RS: 2

Solution: $\{3\}$

3. $\sqrt{2t+3} = 5$

$\qquad (\sqrt{2t+3})^2 = 5^2$

$\qquad\qquad 2t + 3 = 25$

$\qquad\qquad\quad 2t = 22$

$\qquad\qquad\quad\ t = 11$

Check:

LS: $\sqrt{2(11)+3} = \sqrt{25} = 5$

RS: 5

Solution: $\{11\}$

5. $\sqrt{2-x} = 3$

$\qquad (\sqrt{2-x})^2 = 3^2$

$\qquad\qquad 2 - x = 9$

$\qquad\qquad\ -x = 7$

$\qquad\qquad\quad x = -7$

Check:

LS: $\sqrt{2-(-7)} = \sqrt{9} = 3$

RS: 3

Solution: $\{-7\}$

7. $\sqrt{x} = 0$

$\qquad (\sqrt{x})^2 = 0^2$

$\qquad\qquad x = 0$

Check:

LS: $\sqrt{0} = 0$

RS: 0

Solution: $\{0\}$

9. $\sqrt{x - 3} = -8$

$\left(\sqrt{x - 3}\right)^2 = (-8)^2$

$x - 3 = 64$

$x = 67$

Check:

LS: $\sqrt{67 - 3} = \sqrt{64} = 8$

RS: -8

Solution: $\emptyset$

11. $\sqrt{w + 1} = \sqrt{3w - 1}$

$\left(\sqrt{w + 1}\right)^2 = \left(\sqrt{3w - 1}\right)^2$

$w + 1 = 3w - 1$

$2 = 2w$

$1 = w$

Check:

LS: $\sqrt{1 + 1} = \sqrt{2}$

RS: $\sqrt{3(1) - 1} = \sqrt{2}$

Solution: $\{1\}$

13. $\sqrt{3x + 2} + \sqrt{2x + 7} = 0$

$\sqrt{3x + 2} = -\sqrt{2x + 7}$

$\left(\sqrt{3x + 2}\right)^2 = \left(-\sqrt{2x + 7}\right)^2$

$3x + 2 = 2x + 7$

$x = 5$

Check:

LS: $\sqrt{3(5) + 2} + \sqrt{2(5) + 7}$

$= \sqrt{17} + \sqrt{17}$

$= 2\sqrt{17}$

RS: 0

Solution: $\emptyset$

15. $\sqrt{x + 1} = \sqrt{2x - 2}$

$\left(\sqrt{x + 1}\right)^2 = \left(\sqrt{2x - 2}\right)^2$

$x + 1 = 2x - 2$

$3 = x$

Check:

LS: $\sqrt{3 + 1} = \sqrt{4} = 2$

RS: $\sqrt{2(3) - 2} = \sqrt{4} = 2$

Solution: $\{3\}$

17. $\sqrt{x - 1} = \sqrt{x + 1}$

$\left(\sqrt{x - 1}\right)^2 = \left(\sqrt{x + 1}\right)^2$

$x - 1 = x + 1$

$0 = 2$

Solution: $\emptyset$

19. $12 = 3\sqrt{x - 1}$

$4 = \sqrt{x - 1}$

$4^2 = \left(\sqrt{x - 1}\right)^2$

$16 = x - 1$

$17 = x$

Check:

LS: 12

RS: $3\sqrt{17 - 1} = 3\sqrt{16} = 3(4) = 12$

Solution: $\{17\}$

21. $\sqrt{2x + 3} - 3\sqrt{x - 2} = 0$

$$\sqrt{2x + 3} = 3\sqrt{x - 2}$$

$$\left(\sqrt{2x + 3}\right)^2 = \left(3\sqrt{x - 2}\right)^2$$

$$2x + 3 = 9(x - 2)$$

$$2x + 3 = 9x - 18$$

$$21 = 7x$$

$$3 = x$$

Check:

LS: $\sqrt{2(3) + 3} - 3\sqrt{3 - 2}$

$= \sqrt{9} - 3\sqrt{1}$

$= 3 - 3$

$= 0$

RS: 0

Solution: {3}

23. $\sqrt{x^2 + 3} - x = 1$

$$\sqrt{x^2 + 3} = x + 1$$

$$\left(\sqrt{x^2 + 3}\right)^2 = (x + 1)^2$$

$$x^2 + 3 = x^2 + 2x + 1$$

$$2 = 2x$$

$$1 = x$$

Check:

LS: $\sqrt{1^2 + 3} - 1 = \sqrt{4} - 1$

$= 2 - 1$

$= 1$

RS: 1

Solution: {1}

25. $\sqrt{y^2 - 2y} = -y$

$$\left(\sqrt{y^2 - 2y}\right)^2 = (-y)^2$$

$$y^2 - 2y = y^2$$

$$-2y = 0$$

$$y = 0$$

Check:

LS: $\sqrt{0^2 - 2(0)} = \sqrt{0} = 0$

RS: $-0 = 0$

Solution: {0}

27. $\sqrt{x^2 + x} = x$

$$\left(\sqrt{x^2 + x}\right)^2 = x^2$$

$$x^2 + x = x^2$$

$$x = 0$$

Check:

LS: $\sqrt{0^2 + 0} = \sqrt{0} = 0$

RS: 0

Solution: {0}

29. $\sqrt{x^2 - 4} = x + 2$

$$\left(\sqrt{x^2 - 4}\right)^2 = (x + 2)^2$$

$$x^2 - 4 = x^2 + 4x + 4$$

$$-8 = 4x$$

$$-2 = x$$

Check:

LS: $\sqrt{(-2)^2 - 4} = \sqrt{4 - 4}$

$= \sqrt{0}$

$= 0$

RS: $-2 + 2 = 0$

Solution: {-2}

31. $\sqrt{4x^2 + 9} + 1 = 2x$

$\sqrt{4x^2 + 9} = 2x - 1$

$\left(\sqrt{4x^2 + 9}\right)^2 = (2x - 1)^2$

$4x^2 + 9 = 4x^2 - 4x + 1$

$8 = -4x$

$-2 = x$

Check:

LS: $\sqrt{4(-2)^2 + 9} + 1 = \sqrt{25} + 1$

$= 5 + 1$

$= 6$

RS: $2(-2) = 4$

Solution: $\emptyset$

33. $\sqrt{z - 5} = -7$

$\left(\sqrt{z - 5}\right)^2 = (-7)^2$

$z - 5 = 49$

$z = 54$

Check:

LS: $\sqrt{54 - 5} = \sqrt{49} = 7$

RS: -7

Solution: $\emptyset$

35. $\sqrt{3x + 4} = 4$

$\left(\sqrt{3x + 4}\right)^2 = 4^2$

$3x + 4 = 16$

$3x = 12$

$x = 4$

Check:

LS: $\sqrt{3(4) + 4} = \sqrt{16} = 4$

RS: 4

Solution: $\{4\}$

37. $\sqrt{1 - x} = 5$

$\left(\sqrt{1 - x}\right)^2 = 5^2$

$1 - x = 25$

$-x = 24$

$x = -24$

Check:

LS: $\sqrt{1 - (-24)} = \sqrt{25} = 5$

RS: 5

Solution: $\{-24\}$

39. $\sqrt{3 - 5x} = -2$

$\left(\sqrt{3 - 5x}\right)^2 = (-2)^2$

$3 - 5x = 4$

$-5x = 1$

$x = -\dfrac{1}{5}$

Check:

LS: $\sqrt{3 - 5\left(-\dfrac{1}{5}\right)} = \sqrt{4} = 2$

RS: -2

Solution: $\emptyset$

41. $\sqrt{2 - t} = 0$

$\left(\sqrt{2 - t}\right)^2 = 0^2$

$2 - t = 0$

$2 = t$

Check:

LS: $\sqrt{2 - 2} = \sqrt{0} = 0$

RS: 0

Solution: $\{2\}$

154

43. $\sqrt{2x - 1} = 3$

$\left(\sqrt{2x - 1}\right)^2 = 3^2$

$2x - 1 = 9$

$2x = 10$

$x = 5$

Check:

LS: $\sqrt{2(5) - 1} = \sqrt{9} = 3$

RS: 3

Solution: $\{5\}$

45. $\sqrt{2 - x} = \sqrt{x + 3}$

$\left(\sqrt{2 - x}\right)^2 = \left(\sqrt{x + 3}\right)^2$

$2 - x = x + 3$

$-1 = 2x$

$-\dfrac{1}{2} = x$

Check:

LS: $\sqrt{2 - \left(-\dfrac{1}{2}\right)} = \sqrt{\dfrac{5}{2}}$

RS: $\sqrt{-\dfrac{1}{2} + 3} = \sqrt{\dfrac{5}{2}}$

Solution: $\left\{-\dfrac{1}{2}\right\}$

47. $\sqrt{1 - 3s} + \sqrt{2s + 11} = 0$

$\sqrt{1 - 3s} = -\sqrt{2s + 11}$

$\left(\sqrt{1 - 3s}\right)^2 = \left(-\sqrt{2s + 11}\right)^2$

$1 - 3s = 2s + 11$

$-10 = 5s$

$-2 = s$

Check:

LS: $\sqrt{1 - 3(-2)} + \sqrt{2(-2) + 11}$

$= \sqrt{7} + \sqrt{7}$

$= 2\sqrt{7}$

RS: 0

Solution: $\emptyset$

49. $\sqrt{1 + 3x} = \sqrt{3x}$

$\left(\sqrt{1 + 3x}\right)^2 = \left(\sqrt{3x}\right)^2$

$1 + 3x = 3x$

$1 = 0$

Solution: $\emptyset$

51. $3\sqrt{6 - x} = 2\sqrt{x + 46}$

$\left(3\sqrt{6 - x}\right)^2 = \left(2\sqrt{x + 46}\right)^2$

$9(6 - x) = 4(x + 46)$

$54 - 9x = 4x + 184$

$-130 = 13x$

$-10 = x$

Check:

LS: $3\sqrt{6 - (-10)} = 3\sqrt{16}$

$= 3(4)$

$= 12$

RS: $2\sqrt{-10 + 46} = 2\sqrt{36}$

$= 2(6)$

$= 12$

Solution: $\{-10\}$

53. $2\sqrt{x} = \sqrt{3x + 1}$

$\left(2\sqrt{x}\right)^2 = \left(\sqrt{3x + 1}\right)^2$

$4x = 3x + 1$

$x = 1$

Check:

LS: $2\sqrt{1} = 2$

RS: $\sqrt{3(1) + 1} = \sqrt{4} = 2$

Solution: $\{1\}$

55. $\sqrt{x^2 - 1} = 1 - x$

$\left(\sqrt{x^2 - 1}\right)^2 = (1 - x)^2$

$x^2 - 1 = 1 - 2x + x^2$

$-2 = -2x$

$1 = x$

Check:

LS: $\sqrt{1^2 - 1} = \sqrt{0} = 0$

RS: $1 - 1 = 0$

Solution: $\{1\}$

57. $\sqrt{x^2 - 3x} = x - 3$

$\left(\sqrt{x^2 - 3x}\right)^2 = (x - 3)^2$

$x^2 - 3x = x^2 - 6x + 9$

$3x = 9$

$x = 3$

Check:

LS: $\sqrt{3^2 - 3(3)} = \sqrt{0} = 0$

RS: $3 - 3 = 0$

Solution: $\{3\}$

59. $\sqrt{x^2 + 11} = x + 1$

$\left(\sqrt{x^2 + 11}\right)^2 = (x + 1)^2$

$x^2 + 11 = x^2 + 2x + 1$

$10 = 2x$

$5 = x$

Check:

LS: $\sqrt{5^2 + 11} = \sqrt{36} = 6$

RS: $5 + 1 = 6$

Solution: $\{5\}$

61. $\sqrt{x^2 - 1} = x + 1$

$\left(\sqrt{x^2 - 1}\right)^2 = (x + 1)^2$

$x^2 - 1 = x^2 + 2x + 1$

$-2 = 2x$

$-1 = x$

Check:

LS: $\sqrt{(-1)^2 - 1} = \sqrt{0} = 0$

RS: $-1 + 1 = 0$

Solution: $\{-1\}$

63. $22.822 = \sqrt{x - 61.190}$

$(22.822)^2 = \left(\sqrt{x - 61.190}\right)^2$

$520.844 = x - 61.190$

$582.034 = x$

Check:

LS: 22.822

RS: $\sqrt{582.034 - 61.190} = 22.822$

Solution: $\{582.034\}$

65. $8.631 = 15.889\sqrt{6.331 - x}$

$0.543 = \sqrt{6.331 - x}$

$(0.543)^2 = \left(\sqrt{6.331 - x}\right)^2$

$0.295 = 6.331 - x$

$-6.036 = -x$

$6.036 = x$

Check:

LS: 8.631

RS: $15.889\sqrt{6.331 - 6.036} = 8.628$

Difference between LS and RS is due to rounding.

Solution: $\{6.036\}$

67. $\sqrt{2x + 1.133} - \sqrt{x + 4} = 0$

$\sqrt{2x + 1.133} = \sqrt{x + 4}$

$\left(\sqrt{2x + 1.133}\right)^2 = \left(\sqrt{x + 4}\right)^2$

$2x + 1.133 = x + 4$

$x = 2.867$

Check:

LS: $\sqrt{2(2.867) + 1.133} - \sqrt{2.867 + 4}$

$= 2.620 - 2.620$

$= 0$

RS: 0

Solution: $\{2.867\}$

69. $v = \sqrt{32h}$

$4 = \left(\sqrt{32h}\right)$

$4^2 = \left(\sqrt{32h}\right)^2$

$16 = 32h$

$\dfrac{1}{2} = h$

Check:

LS: 4

RS: $\sqrt{32\left(\dfrac{1}{2}\right)} = \sqrt{16} = 4$

The wave height is $\dfrac{1}{2}$ ft.

71. $r = \sqrt{1.9h}$

$2 = \sqrt{1.9h}$

$2^2 = \left(\sqrt{1.9h}\right)^2$

$4 = 1.9h$

$2.1 = h$

Check:

LS: 2

RS: $\sqrt{1.9(2.1)} = 2.0$

The height would be 2.1 in.

Problem Set 4.6

1. $49^{1/2} = \sqrt{49} = 7$

3. $169^{1/2} = \sqrt{169} = 13$

5. $(-100)^{1/2} = \sqrt{-100}$

Not a real number.

7. $81^{1/4} = \sqrt[4]{81} = 3$

9. $1024^{1/10} = \sqrt[10]{1024}$

$= \sqrt[10]{2^{10}}$

$= 2$

11. $64^{5/6} = 64^{(1/6)5}$

$= \left(64^{1/6}\right)^5$

$= \left(\sqrt[6]{64}\right)^5$

$= 2^5$

$= 32$

157

13. $4^{7/2} = 4^{(1/2)7}$

$\quad = (4^{1/2})^7$

$\quad = (\sqrt{4})^7$

$\quad = 2^7$

$\quad = 128$

15. $27^{-2/3} = 27^{(1/3)(-2)}$

$\quad = (27^{1/3})^{-2}$

$\quad = (\sqrt[3]{27})^{-2}$

$\quad = 3^{-2}$

$\quad = \dfrac{1}{3^2}$

$\quad = \dfrac{1}{9}$

17. $(-49)^{-3/2} = (\sqrt{-49})^{-3}$

Not a real number.

19. $\left(\dfrac{8}{125}\right)^{2/3} = \left(\sqrt[3]{\dfrac{8}{125}}\right)^2$

$\quad = \left(\dfrac{\sqrt[3]{8}}{\sqrt[3]{125}}\right)^2$

$\quad = \left(\dfrac{2}{5}\right)^2$

$\quad = \dfrac{2^2}{5^2}$

$\quad = \dfrac{4}{25}$

21. $\left(\dfrac{196}{169}\right)^{3/2} = \left(\sqrt{\dfrac{196}{169}}\right)^3$

$\quad = \left(\dfrac{\sqrt{196}}{\sqrt{169}}\right)^3$

$\quad = \left(\dfrac{14}{13}\right)^3$

$\quad = \dfrac{14^3}{13^3}$

$\quad = \dfrac{2744}{2197}$

23. $\left(\dfrac{100}{121}\right)^{-3/2} = \left(\sqrt{\dfrac{100}{121}}\right)^{-3}$

$\quad = \left(\dfrac{\sqrt{100}}{\sqrt{121}}\right)^{-3}$

$\quad = \left(\dfrac{10}{11}\right)^{-3}$

$\quad = \dfrac{10^{-3}}{11^{-3}}$

$\quad = \dfrac{11^3}{10^3}$

$\quad = \dfrac{1331}{1000}$

25. $2^{1/2} \cdot 2^{5/2} = 2^{1/2 + 5/2}$

$\quad = 2^3$

$\quad = 8$

27. $(25^{3/8})^{4/3} = 25^{3/8 \cdot 4/3}$

$\quad = 25^{1/2}$

$\quad = \sqrt{25}$

$\quad = 5$

29. $\dfrac{2^{7/3}}{2^{1/3}}$ $2^{7/3\,-\,1/3}$

       $= 2^2$

       $= 4$

31. $(36x^2)^{1/2} = 36^{1/2}(x^2)^{1/2}$

              $= \sqrt{36}\,x^{2\cdot 1/2}$

              $= 6x$

33. $(49x^2y^{-2})^{1/2} = 49^{1/2}(x^2)^{1/2}(y^{-2})^{1/2}$

                  $= \sqrt{49}\,x^{2(1/2)}y^{-2(1/2)}$

                  $= 7xy^{-1}$

                  $= \dfrac{7x}{y}$

35. $(-64x^{-6}z)^{-1/3} = (-64)^{-1/3}(x^{-6})^{-1/3}z^{-1/3}$

                    $= \left(\sqrt[3]{-64}\right)^{-1}x^{-6(-1/3)}z^{-1/3}$

                    $= (-4)^{-1}x^2z^{-1/3}$

                    $= \dfrac{1}{-4}x^2z^{-1/3}$

                    $= \dfrac{x^2}{-4z^{1/3}}$

                    $= -\dfrac{x^2}{4z^{1/3}}$

37. $(8x^{-2}y)^{1/2} = 8^{1/2}(x^{-2})^{1/2}y^{1/2}$

               $= \sqrt{8}\,x^{(-2)(1/2)}y^{1/2}$

               $= 2\sqrt{2}\,x^{-1}y^{1/2}$

               $= \dfrac{2\sqrt{2}\,y^{1/2}}{x}$

39. $(3a^{2/3}x^{-1/3})^3 = 3^3(a^{2/3})^3(x^{-1/3})^3$

                 $= 27a^{(2/3)3}x^{(-1/3)3}$

                 $= 27a^2x^{-1}$

                 $= \dfrac{27a^2}{x}$

41. $2x^2(3x^{1/6} - 5x^{-5/6})$

     $= 2x^2 \cdot 3x^{1/6} - 2x^2 \cdot 5x^{-5/6}$

     $= 6x^{2+1/6} - 10x^{2+(-5/6)}$

     $= 6x^{13/6} - 10x^{7/6}$

43. $(x^{1/2} + x^{-1/2})^2 = (x^{1/2})^2 + 2(x^{1/2})(x^{-1/2}) + (x^{-1/2})^2$

                     $= x + 2x^0 + x^{-1}$

                     $= x + 2 + \dfrac{1}{x}$

45. $36^{1/2} = \sqrt{36} = 6$

47. $225^{1/2} = \sqrt{225} = 15$

49. $-(100)^{1/2} = -\sqrt{100} = -10$

51. $81^{1/2} = \sqrt{81} = 9$

53. $-64^{1/6} = -\sqrt[6]{64} = -2$

55. $216^{1/3} = \sqrt[3]{216} = 6$

57. $(-81)^{1/4} = \sqrt[4]{-81}$ Not a real number.

59. $1024^{3/10} = 1024^{(1/10)3}$

           $= \left(\sqrt[10]{1024}\right)^3$

           $= 2^3$

           $= 8$

61. $32^{7/5} = 32^{(1/5)7}$

       $= (32^{1/5})^7$

       $= \left(\sqrt[5]{32}\right)^7$

       $= 2^7$

       $= 128$

63. $9^{5/2} = 9^{(1/2)5}$

$\qquad = (9^{1/2})^5$

$\qquad = (\sqrt{9})^5$

$\qquad = 3^5$

$\qquad = 243$

65. $64^{-2/3} = 64^{(1/3)(-2)}$

$\qquad = (64^{1/3})^{-2}$

$\qquad = (\sqrt[3]{64})^{-2}$

$\qquad = 4^{-2}$

$\qquad = \dfrac{1}{4^2}$

$\qquad = \dfrac{1}{16}$

67. $(-27)^{-2/3} = (-27)^{(1/3)(-2)}$

$\qquad = ((-27)^{1/3})^{-2}$

$\qquad = (\sqrt[3]{-27})^{-2}$

$\qquad = (-3)^{-2}$

$\qquad = \dfrac{1}{(-3)^2}$

$\qquad = \dfrac{1}{9}$

69. $\left(\dfrac{64}{343}\right)^{2/3} = \left(\sqrt[3]{\dfrac{64}{343}}\right)^2$

$\qquad = \left(\dfrac{\sqrt[3]{64}}{\sqrt[3]{343}}\right)^2$

$\qquad = \left(\dfrac{4}{7}\right)^2$

$\qquad = \dfrac{4^2}{7^2}$

$\qquad = \dfrac{16}{49}$

71. $\left(\dfrac{256}{225}\right)^{3/2} = \left(\sqrt{\dfrac{256}{225}}\right)^3$

$\qquad = \left(\dfrac{\sqrt{256}}{\sqrt{225}}\right)^3$

$\qquad = \left(\dfrac{16}{15}\right)^3$

$\qquad = \dfrac{16^3}{15^3}$

$\qquad = \dfrac{4096}{3375}$

73. $\left(\dfrac{169}{324}\right)^{-3/2} = \left(\sqrt{\dfrac{169}{324}}\right)^{-3}$

$\qquad = \left(\dfrac{\sqrt{169}}{\sqrt{324}}\right)^{-3}$

$\qquad = \left(\dfrac{13}{18}\right)^{-3}$

$\qquad = \dfrac{13^{-3}}{18^{-3}}$

$\qquad = \dfrac{18^3}{13^3}$

$\qquad = \dfrac{5832}{2197}$

75. $(25y^2)^{1/2} = 25^{1/2}(y^2)^{1/2}$

$\qquad = \sqrt{25}\,y^{2 \cdot 1/2}$

$\qquad = 5y$

77. $(16x^4y^{-6})^{1/2} = 16^{1/2}(x^4)^{1/2}(y^{-6})^{1/2}$

$\qquad = \sqrt{16}\,x^{4(1/2)}y^{(-6)(1/2)}$

$\qquad = 4x^2y^{-3}$

$\qquad = \dfrac{4x^2}{y^3}$

79. $(64x^{-3}z)^{-1/3} = 64^{-1/3}(x^{-3})^{-1/3}z^{-1/3}$

$$= \left(\sqrt[3]{64}\right)^{-1}x^{(-3)(-1/3)}z^{-1/3}$$

$$= (4)^{-1}xz^{-1/3}$$

$$= \frac{1}{4}xz^{-1/3}$$

$$= \frac{x}{4z^{1/3}}$$

81. $(18x^2y^{-2})^{1/2} = 18^{1/2}(x^2)^{1/2}(y^{-2})^{1/2}$

$$= \sqrt{18}x^{2(1/2)}y^{(-2)(1/2)}$$

$$= 3\sqrt{2}xy^{-1}$$

$$= \frac{3\sqrt{2}x}{y}$$

83. $-3(3a^{3/2}x^{-1/2})^2 = -3(3)^2(a^{3/2})^2(x^{-1/2})^2$

$$= -3 \cdot 9a^{(3/2)(2)}x^{(-1/2)(2)}$$

$$= -27a^3x^{-1}$$

$$= -\frac{27a^3}{x}$$

85. $\dfrac{5x^{-1/3}y^{1/3}}{y^{2/3}} = 5x^{-1/3}y^{1/3-2/3}$

$$= 5x^{-1/3}y^{-1/3}$$

$$= \frac{5}{x^{1/3}y^{1/3}}$$

87. $\left(\dfrac{2x^{1/3}y}{3z^{2/3}}\right)^{-3} = \dfrac{(2x^{1/3}y)^{-3}}{(3z^{2/3})^{-3}}$

$$= \frac{2^{-3}(x^{1/3})^{-3}y^{-3}}{3^{-3}(z^{2/3})^{-3}}$$

$$= \frac{2^{-3}x^{-1}y^{-3}}{3^{-3}z^{-2}}$$

$$= \frac{3^3z^2}{2^3xy^3}$$

$$= \frac{27z^2}{8xy^3}$$

89. $\left(\dfrac{25x^{-2/3}}{y^{2/3}}\right)^{-3/2} = \dfrac{(25x^{-2/3})^{-3/2}}{(y^{2/3})^{-3/2}}$

$$= \frac{25^{-3/2}(x^{-2/3})^{-3/2}}{y^{-1}}$$

$$= \frac{\left(\sqrt{25}\right)^{-3}x}{y^{-1}}$$

$$= \frac{xy}{\left(\sqrt{25}\right)^3}$$

$$= \frac{xy}{5^3}$$

$$= \frac{xy}{125}$$

91. $3x^3(2x^{1/3} - 5x^{-5/3}) = 3x^3 \cdot 2x^{1/3} - 3x^3 \cdot 5x^{-5/3}$

$$= 6x^{3+1/3} - 15x^{3+(-5/3)}$$

$$= 6x^{10/3} - 15x^{4/3}$$

93. $2a^{-1/2}b^{3/2}(3a^{1/2}b^{-3/2} + 1)$

$$= 2a^{-1/2}b^{3/2} \cdot 3a^{1/2}b^{-3/2} + 2a^{-1/2}b^{3/2} \cdot 1$$

$$= 6a^{-1/2+1/2}b^{3/2+(-3/2)} + 2a^{-1/2}b^{3/2}$$

$$= 6a^0b^0 + 2a^{-1/2}b^{3/2}$$

$$= 6 \cdot 1 \cdot 1 + \frac{2b^{3/2}}{a^{1/2}}$$

$$= 6 + \frac{2b^{3/2}}{a^{1/2}}$$

95. $(x^{3/2} + x^{-3/2})(x^{3/2} - x^{-3/2})$

$$= (x^{3/2})^2 - (x^{-3/2})^2$$

$$= x^3 - x^{-3}$$

$$= x^3 - \frac{1}{x^3}$$

97. $(a^{1/3} + b^{1/3})^3$

$$= (a^{1/3})^3 + 3(a^{1/3})^2(b^{1/3}) + 3(a^{1/3})(b^{1/3})^2 + (b^{1/3})^3$$

$$= a + 3a^{2/3}b^{1/3} + 3a^{1/3}b^{2/3} + b$$

99. $(x^{1/3} - x^{-1/3})^3$

$= (x^{1/3})^3 - 3(x^{1/3})^2(x^{-1/3}) + 3(x^{1/3})(x^{-1/3})^2 - (x^{-1/3})$

$= x - 3x^{2/3}x^{-1/3} + 3x^{1/3}x^{-2/3} - x^{-1}$

$= x - 3x^{2/3+(-1/3)} + 3x^{1/3+(-2/3)} - x^{-1}$

$= x - 3x^{1/3} + 3x^{-1/3} - x^{-1}$

$= x - 3x^{1/3} + \dfrac{3}{x^{1/3}} - \dfrac{1}{x}$

101. $x^{2/3} - x^{1/3} - 6 = (x^{1/3})^2 - x^{1/3} - 6$

$= (x^{1/3} - 3)(x^{1/3} + 2)$

103. $(-1)^{2/2}$: If you are taking a square root as the denominator of the power indicates, the base must be nonnegative.

105. $\sqrt[8]{x^4 y^4} = (x^4 y^4)^{1/8}$

$= (x^4)^{1/8}(y^4)^{1/8}$

$= x^{1/2} y^{1/2}$

$= (xy)^{1/2}$

$= \sqrt{xy}$

Problem Set 4.7

1. $\sqrt{-9} = \sqrt{9}i = 3i = 0 + 3i$

3. $-\sqrt{-121} = -\sqrt{121}i = -11i = 0 + (-11)i$

5. $3 - \sqrt{-4} = 3 - \sqrt{4}i = 3 - 2i = 3 + (-2)i$

7. $\sqrt{50} = 5\sqrt{2} = 5\sqrt{2} + 0i$

9. $(\sqrt{-2})^2 = (\sqrt{2}i)^2 = (\sqrt{2})^2 i^2 = 2(-1) = -2 + 0i$

11. $i^{11} = i^3 = -i$

13. $i^{48} = i^0 = 1$

15. $i^{-235} = \dfrac{1}{i^{235}}$

$= \dfrac{1}{i^3}$

$= \dfrac{1}{-i}$

$= \dfrac{1(i)}{-i(i)}$

$= \dfrac{i}{-i^2}$

$= \dfrac{i}{-(-1)}$

$= i$

17. $(3 + 5i) - (2 + 3i) = (3 + 5i) + (-2 - 3i)$

$= (3 - 2) + (5i - 3i)$

$= 1 + 2i$

19. $(-4 - i) + (1 + 5i) = (-4 + 1) + (-i + 5i)$

$= -3 + 4i$

21. $i(14i) = 14i^2 = 14(-1) = -14$

23. $3i(-2 - 5i) = (3i)(-2) - (3i)(5i)$

$= -6i - 15i^2$

$= -6i - 15(-1)$

$= -6i + 15$

$= 15 - 6i$

25. $(1 - i)(5 + 2i) = 5 + 2i - 5i - 2i^2$

$= 5 - 3i - 2(-1)$

$= 5 - 3i + 2$

$= 7 - 3i$

27. $(2 - 3i)^2 = 2^2 - 2(2)(3i) + (3i)^2$

$= 4 - 12i + 9i^2$

$= 4 - 12i + 9(-1)$

$= 4 - 12i - 9$

$= -5 - 12i$

29. $\left(\sqrt{8} - \sqrt{-12}\right)^2 = \left(2\sqrt{2} - \sqrt{12}i\right)^2$

$$= \left(2\sqrt{2} - 2\sqrt{3}i\right)^2$$

$$= \left(2\sqrt{2}\right)^2 - 2\left(2\sqrt{2}\right)\left(2\sqrt{3}i\right) + \left(2\sqrt{3}i\right)^2$$

$$= 8 - 8\sqrt{6}i + 12i^2$$

$$= 8 - 8\sqrt{6}i + 12(-1)$$

$$= 8 - 8\sqrt{6}i - 12$$

$$= -4 - 8\sqrt{6}i$$

31. $(5 + 2i)(5 - 2i) = 5^2 - (2i)^2$

$$= 25 - 4i^2$$

$$= 25 - 4(-1)$$

$$= 25 + 4$$

$$= 29$$

33. $\left(\sqrt{7} + \sqrt{-5}\right)\left(\sqrt{7} - \sqrt{-5}\right) = \left(\sqrt{7} + \sqrt{5}i\right)\left(\sqrt{7} - \sqrt{5}i\right)$

$$= \left(\sqrt{7}\right)^2 - \left(\sqrt{5}i\right)^2$$

$$= 7 - 5i^2$$

$$= 7 - 5(-1)$$

$$= 7 + 5$$

$$= 12$$

35. $\dfrac{1}{2 - i} = \dfrac{1(2 + i)}{(2 - i)(2 + i)}$

$$= \dfrac{2 + i}{2^2 - i^2}$$

$$= \dfrac{2 + i}{4 - (-1)}$$

$$= \dfrac{2 + i}{5}$$

$$= \dfrac{2}{5} + \dfrac{1}{5}i$$

37. $\dfrac{1 + 2i}{1 + 3i} = \dfrac{(1 + 2i)(1 - 3i)}{(1 + 3i)(1 - 3i)}$

$$= \dfrac{1 - 3i + 2i - 6i^2}{1^2 - (3i)^2}$$

$$= \dfrac{1 - i - 6(-1)}{1 - 9i^2}$$

$$= \dfrac{1 - i + 6}{1 - 9(-1)}$$

$$= \dfrac{7 - i}{10}$$

$$= \dfrac{7}{10} - \dfrac{1}{10}i$$

39. $(2 - i) \div (2 + i) = \dfrac{2 - i}{2 + i}$

$$= \dfrac{(2 - i)(2 - i)}{(2 + i)(2 - i)}$$

$$= \dfrac{4 - 2i - 2i + i^2}{2^2 - i^2}$$

$$= \dfrac{4 - 4i - 1}{4 - (-1)}$$

$$= \dfrac{3 - 4i}{5}$$

$$= \dfrac{3}{5} - \dfrac{4}{5}i$$

41. $\sqrt{-25} = \sqrt{25}i = 5i = 0 + 5i$

43. $2 - \sqrt{-5} = 2 - \sqrt{5}i$

45. $\sqrt{-7} = \sqrt{7}i = 0 + \sqrt{7}i$

47. $2\sqrt{-2} = 2\sqrt{2}i = 0 + 2\sqrt{2}i$

49. $i^7 = i^3 = -i$

51. $i^{30} = i^2 = -1$

53. $i^{-3} = \dfrac{1}{i^3}$

$\qquad = \dfrac{1}{-i}$

$\qquad = \dfrac{1(i)}{-i(i)}$

$\qquad = \dfrac{i}{-i^2}$

$\qquad = \dfrac{i}{-(-1)}$

$\qquad = i$

55. $(1 - i) + (2 - 3i) = (1 + 2) + (-i - 3i)$

$\qquad\qquad\qquad\quad = 3 - 4i$

57. $(5 + 2i) + (3 + 5i) = (5 + 3) + (2i + 5i)$

$\qquad\qquad\qquad\quad = 8 + 7i$

59. $(3 + 3i) - (5 + 5i) = (3 + 3i) + (-5 - 5i)$

$\qquad\qquad\qquad\quad = (3 - 5) + (3i - 5i)$

$\qquad\qquad\qquad\quad = -2 - 2i$

61. $i(9i) = 9i^2 = 9(-1) = -9$

63. $i(2 - 3i) = 2i - 3i^2$

$\qquad\qquad = 2i - 3(-1)$

$\qquad\qquad = 2i + 3$

$\qquad\qquad = 3 + 2i$

65. $(1 - i)(3 + 2i) = 3 + 2i - 3i - 2i^2$

$\qquad\qquad\qquad = 3 - i - 2(-1)$

$\qquad\qquad\qquad = 3 - i + 2$

$\qquad\qquad\qquad = 5 - i$

67. $(1 - 3i)(2 - 3i) = 2 - 3i - 6i + 9i^2$

$\qquad\qquad\qquad = 2 - 9i + 9(-1)$

$\qquad\qquad\qquad = 2 - 9i - 9$

$\qquad\qquad\qquad = -7 - 9i$

69. $(-2 - 3i)(1 - 2i) = -2 + 4i - 3i + 6i^2$

$\qquad\qquad\qquad\quad = -2 + i + 6(-1)$

$\qquad\qquad\qquad\quad = -2 + i - 6$

$\qquad\qquad\qquad\quad = -8 + i$

71. $(1 - i)^2 = 1^2 - 2(1)(i) + i^2$

$\qquad\qquad = 1 - 2i - 1$

$\qquad\qquad = -2i$

73. $(3 + 4i)^2 = 3^2 + 2(3)(4i) + (4i)^2$

$\qquad\qquad = 9 + 24i + 16i^2$

$\qquad\qquad = 9 + 24i + 16(-1)$

$\qquad\qquad = 9 + 24i - 16$

$\qquad\qquad = -7 + 24i$

75. $(\sqrt{3} + 3i)^2 = (\sqrt{3})^2 + 2(\sqrt{3})(3i) + (3i)^2$

$\qquad\qquad = 3 + 6\sqrt{3}i + 9i^2$

$\qquad\qquad = 3 + 6\sqrt{3}i + 9(-1)$

$\qquad\qquad = 3 + 6\sqrt{3}i - 9$

$\qquad\qquad = -6 + 6\sqrt{3}i$

77. $(\sqrt{3} + \sqrt{2}i)^2 = (\sqrt{3})^2 + 2(\sqrt{3})(\sqrt{2}i) + (\sqrt{2}i)^2$

$\qquad\qquad = 3 + 2\sqrt{6}i + 2i^2$

$\qquad\qquad = 3 + 2\sqrt{6}i + 2(-1)$

$\qquad\qquad = 3 + 2\sqrt{6}i - 2$

$\qquad\qquad = 1 + 2\sqrt{6}i$

79. $(3 - i)(3 + i) = 3^2 - i^2$

$\qquad\qquad = 9 - (-1)$

$\qquad\qquad = 10$

81. $(7 - 10i)(7 + 10i) = 7^2 - (10i)^2$

$\qquad\qquad\qquad = 49 - 100i^2$

$\qquad\qquad\qquad = 49 - 100(-1)$

$\qquad\qquad\qquad = 49 + 100$

$\qquad\qquad\qquad = 149$

83. $\left(\sqrt{3} - 2i\right)\left(\sqrt{3} + 2i\right) = \left(\sqrt{3}\right)^2 - (2i)^2$

$$= 3 - 4i^2$$
$$= 3 - 4(-1)$$
$$= 3 + 4$$
$$= 7$$

85. $\left(\sqrt{2} - \sqrt{3}i\right)\left(\sqrt{2} + \sqrt{3}i\right) = \left(\sqrt{2}\right)^2 - \left(\sqrt{3}i\right)^2$

$$= 2 - 3i^2$$
$$= 2 - 3(-1)$$
$$= 2 + 3$$
$$= 5$$

87. $\dfrac{1}{1 - i} = \dfrac{1(1 + i)}{(1 - i)(1 + i)}$

$$= \dfrac{1 + i}{1^2 - i^2}$$
$$= \dfrac{1 + i}{1 - (-1)}$$
$$= \dfrac{1 + i}{2}$$
$$= \dfrac{1}{2} + \dfrac{1}{2}i$$

89. $\dfrac{3}{1 - 2i} = \dfrac{3(1 + 2i)}{(1 - 2i)(1 + 2i)}$

$$= \dfrac{3 + 6i}{1^2 - (2i)^2}$$
$$= \dfrac{3 + 6i}{1 - 4i^2}$$
$$= \dfrac{3 + 6i}{i - 4(-1)}$$
$$= \dfrac{3 + 6i}{1 + 4}$$
$$= \dfrac{3 + 6i}{5}$$
$$= \dfrac{3}{5} + \dfrac{6}{5}i$$

91. $\dfrac{i}{4 + 4i} = \dfrac{i(4 - 4i)}{(4 + 4i)(4 - 4i)}$

$$= \dfrac{4i - 4i^2}{4^2 - (4i)^2}$$
$$= \dfrac{4i - 4(-1)}{16 - 16i^2}$$
$$= \dfrac{4i + 4}{16 - 16(-1)}$$
$$= \dfrac{4i + 4}{16 + 16}$$
$$= \dfrac{4i + 4}{32}$$
$$= \dfrac{4}{32}i + \dfrac{4}{32}$$
$$= \dfrac{1}{8} + \dfrac{1}{8}i$$

93. $\dfrac{2 + i}{1 - 3i} = \dfrac{(2 + i)(1 + 3i)}{(1 - 3i)(1 + 3i)}$

$$= \dfrac{2 + 6i + i + 3i^2}{1^2 - (3i)^2}$$
$$= \dfrac{2 + 7i + 3(-1)}{1 - 9i^2}$$
$$= \dfrac{2 + 7i - 3}{1 - 9(-1)}$$
$$= \dfrac{-1 + 7i}{10}$$
$$= -\dfrac{1}{10} + \dfrac{7}{10}i$$

95. $(5 - i) \div (5 + i) = \dfrac{5 - i}{5 + i}$

$$= \dfrac{(5 - i)(5 - i)}{(5 + i)(5 - i)}$$
$$= \dfrac{5^2 - 2(5)(i) + i^2}{5^2 - i^2}$$
$$= \dfrac{25 - 10i - 1}{25 - (-1)}$$

$$= \frac{24 - 10i}{26}$$

$$= \frac{24}{26} - \frac{10}{26}i$$

$$= \frac{12}{13} - \frac{5}{13}i$$

97. $(2 + 3i)^3$

$$= 2^3 + 3(2)^2(3i) + 3(2)(3i)^2 + (3i)^3$$

$$= 8 + 36i + 54i^2 + 27i^3$$

$$= 8 + 36i + 54(-1) + 27(-i)$$

$$= 8 + 36i - 54 - 27i$$

$$= -46 + 9i$$

99. $x^2 + 1 = 0$

Consider $x = i$

LS: $i^2 + 1 = -1 + 1 = 0$

RS: 0

$x = i$ is a solution.

Consider $x = -i$

LS: $(-i)^2 + 1 = i^2 + 1 = -1 + 1 = 0$

RS: 0

$x = -i$ is a solution.

101. $x^2 + 9 = x^2 - (-9)$

$$= x^2 - (3i)^2$$

$$= (x - 3i)(x + 3i)$$

Chapter 4 Review Problems

1. $\sqrt{75} = \sqrt{25 \cdot 3}$

$$= \sqrt{25} \cdot \sqrt{3}$$

$$= 5\sqrt{3}$$

3. $\sqrt{8x^3yz^4} = \sqrt{4 \cdot 2x^2 \cdot xyz^4}$

$$= \sqrt{4x^2z^4} \cdot \sqrt{2xy}$$

$$= 2xz^2\sqrt{2xy}$$

5. $\sqrt[3]{\frac{-8x^{-2}}{xy^3}} = \sqrt[3]{\frac{-8}{x^3y^3}}$

$$= \frac{\sqrt[3]{-8}}{\sqrt[3]{x^3y^3}}$$

$$= \frac{-2}{xy}$$

$$= -\frac{2}{xy}$$

7. $-49^{1/2} = -\sqrt{49} = -7$

9. $8^{-2/3} = (8^{1/3})^{-2}$

$$= \left(\sqrt[3]{8}\right)^{-2}$$

$$= 2^{-2}$$

$$= \frac{1}{2^2}$$

$$= \frac{1}{4}$$

11. $\left(\frac{2x^{1/2}y^{-1/3}}{z^{1/3}}\right)^6 = \frac{(2x^{1/2}y^{-1/3})^6}{(z^{1/3})^6}$

$$= \frac{2^6(x^{1/2})^6(y^{-1/3})^6}{(z^{1/3})^6}$$

$$= \frac{64x^3y^{-2}}{z^2}$$

$$= \frac{64x^3}{y^2z^2}$$

13. $\dfrac{x^{1/3}y^{-2/3}}{x^{-2/3}y^{1/3}} = x^{1/3-(-2/3)}y^{(-2/3)-1/3}$

$\qquad = xy^{-1}$

$\qquad = \dfrac{x}{y}$

15. $\dfrac{4^{-1/2} + 1}{4^{1/2}} = \dfrac{(\sqrt{4})^{-1} + 1}{\sqrt{4}}$

$\qquad = \dfrac{2^{-1} + 1}{2}$

$\qquad = \dfrac{\frac{1}{2} + 1}{2}$

$\qquad = \dfrac{\frac{3}{2}}{2}$

$\qquad = \dfrac{3}{4}$

17. $\sqrt{432} - 2\sqrt{147} + 3\sqrt{3}$

$\quad = \sqrt{144 \cdot 3} - 2\sqrt{49 \cdot 3} + 3\sqrt{3}$

$\quad = 12\sqrt{3} - 2 \cdot 7\sqrt{3} + 3\sqrt{3}$

$\quad = 12\sqrt{3} - 14\sqrt{3} + 3\sqrt{3}$

$\quad = \sqrt{3}$

19. $\sqrt{2}(\sqrt{6} - 1) = \sqrt{2} \cdot \sqrt{6} - \sqrt{2} \cdot 1$

$\qquad = \sqrt{12} - \sqrt{2}$

$\qquad = 2\sqrt{3} - \sqrt{2}$

21. $(\sqrt{2} - \sqrt{3})^2 = (\sqrt{2})^2 - 2(\sqrt{2})(\sqrt{3}) + (\sqrt{3})^2$

$\qquad = 2 - 2\sqrt{6} + 3$

$\qquad = 5 - 2\sqrt{6}$

23. $x^{1/2}(x^{1/2}y - x^{-1/2}) = x^{1/2} \cdot x^{1/2}y - x^{1/2} \cdot x^{-1/2}$

$\qquad = x^{1/2+1/2}y - x^{1/2+(-1/2)}$

$\qquad = xy - x^0$

$\qquad = xy - 1$

25. $(2x^{1/2} - 3y^{-1/2})(2x^{1/2} + 3y^{-1/2})$

$\quad = (2x^{1/2})^2 - (3y^{-1/2})^2$

$\quad = 2^2(x^{1/2})^2 - 3^2(y^{-1/2})^2$

$\quad = 4x - 9y^{-1}$

$\quad = 4x - \dfrac{9}{y}$

27. $\dfrac{11}{7\sqrt{11}} = \dfrac{11 \cdot \sqrt{11}}{7\sqrt{11} \cdot \sqrt{11}}$

$\qquad = \dfrac{11\sqrt{11}}{7 \cdot 11}$

$\qquad = \dfrac{\sqrt{11}}{7}$

29. $\dfrac{1}{\sqrt[3]{6}} = \dfrac{1 \cdot \sqrt[3]{6^2}}{\sqrt[3]{6} \cdot \sqrt[3]{6^2}}$

$\qquad = \dfrac{\sqrt[3]{36}}{\sqrt[3]{6^3}}$

$\qquad = \dfrac{\sqrt[3]{36}}{6}$

31. $\dfrac{\sqrt[3]{4} + \sqrt[3]{2}}{\sqrt[3]{2}} = \dfrac{(\sqrt[3]{4} + \sqrt[3]{2}) \cdot \sqrt[3]{2^2}}{\sqrt[3]{2} \cdot \sqrt[3]{2^2}}$

$\qquad = \dfrac{\sqrt[3]{4} \cdot \sqrt[3]{4} + \sqrt[3]{2} \cdot \sqrt[3]{4}}{\sqrt[3]{2^3}}$

$\qquad = \dfrac{\sqrt[3]{16} + \sqrt[3]{8}}{2}$

$$= \frac{2\sqrt[3]{2} + 2}{2}$$

$$= \frac{2\left(\sqrt[3]{2} + 1\right)}{2}$$

$$= \sqrt[3]{2} + 1$$

33. $\dfrac{\sqrt{2}}{\sqrt{2} + \sqrt{6}} = \dfrac{\sqrt{2}\left(\sqrt{2} - \sqrt{6}\right)}{\left(\sqrt{2} + \sqrt{6}\right)\left(\sqrt{2} - \sqrt{6}\right)}$

$$= \frac{\sqrt{2} \cdot \sqrt{2} - \sqrt{2} \cdot \sqrt{6}}{\left(\sqrt{2}\right)^2 - \left(\sqrt{6}\right)^2}$$

$$= \frac{2 - \sqrt{12}}{2 - 6}$$

$$= \frac{2 - 2\sqrt{3}}{-4}$$

$$= \frac{2\left(1 - \sqrt{3}\right)}{-4}$$

$$= -\frac{1 - \sqrt{3}}{2}$$

$$\text{or } \frac{\sqrt{3} - 1}{2}$$

35. $\sqrt{3x + 2} = 2$

$$\left(\sqrt{3x + 2}\right)^2 = 2^2$$

$$3x + 2 = 4$$

$$3x = 2$$

$$x = \frac{2}{3}$$

Check:

LS: $\sqrt{3\left(\dfrac{2}{3}\right) + 2} = \sqrt{4} = 2$

RS: 2

Solution: $\left\{\dfrac{2}{3}\right\}$

37. $\sqrt{6s - 2} = \sqrt{3s + 1}$

$$\left(\sqrt{6s - 2}\right)^2 = \left(\sqrt{3s + 1}\right)^2$$

$$6s - 2 = 3s + 1$$

$$3s = 3$$

$$s = 1$$

Check:

LS: $\sqrt{6(1) - 2} = \sqrt{4} = 2$

RS: $\sqrt{3(1) + 1} = \sqrt{4} = 2$

Solution: $\{1\}$

39. $\sqrt{x + 2} = \sqrt{x - 5}$

$$\left(\sqrt{x + 2}\right)^2 = \left(\sqrt{x - 5}\right)^2$$

$$x + 2 = x - 5$$

$$0 = -7$$

Solution: $\emptyset$

41. $\sqrt{-225} = \sqrt{225}\,i = 15i = 0 + 15i$

43. $\sqrt{-2}\left(\sqrt{2} - \sqrt{-2}\right) = \sqrt{2}\,i\left(\sqrt{2} - \sqrt{2}\,i\right)$

$$= \left(\sqrt{2}\,i\right)\left(\sqrt{2}\right) - \left(\sqrt{2}\,i\right)\left(\sqrt{2}\,i\right)$$

$$= 2i - 2i^2$$

$$= 2i - 2(-1)$$

$$= 2i + 2$$

$$= 2 + 2i$$

45. $(5 - 4i) - (2 - 6i) = (5 - 4i) + (-2 + 6i)$

$$= (5 - 2) + (-4i + 6i)$$

$$= 3 + 2i$$

47. $(1 + 2i)(5 - 3i) = 5 - 3i + 10i - 6i^2$

$$= 5 + 7i - 6(-1)$$

$$= 5 + 7i + 6$$

$$= 11 + 7i$$

49. $(6 - 7i)(6 + 7i) = 6^2 - (7i)^2$

$\qquad = 36 - 49i^2$

$\qquad = 36 - 49(-1)$

$\qquad = 36 + 49$

$\qquad = 85$

51. $(2 - 3i)^3 = 2^3 - 3(2)^2(3i) + 3(2)(3i)^2 - (3i)^3$

$\qquad = 8 - 36i + 54i^2 - 27i^3$

$\qquad = 8 - 36i + 54(-1) - 27(-i)$

$\qquad = 8 - 36i - 54 + 27i$

$\qquad = -46 - 9i$

53. $r^2(x - y) + t^2(x - y)$

$\quad = (x - y)(r^2 + t^2)$

55. Factored form

57. $4x^2 - 12xy + 9y^2 = (2x)^2 - 2(2x)(3y) + (3y)^2$

$\qquad = (2x - 3y)^2$

59. 2 terms; $1 - x^3 = (1 - x)(1 + x + x^2)$

$\quad 1 - x$ is a factor.

61. 1 term; $1 - x$ is a factor.

63. $\dfrac{8x^3 - 1}{1 - 4x^2} = \dfrac{(2x - 1)(4x^2 + 2x + 1)}{(1 - 2x)(1 + 2x)}$

$\qquad = \dfrac{-(1 - 2x)(4x^2 + 2x + 1)}{(1 - 2x)(1 + 2x)}$

$\qquad = -\dfrac{4x^2 + 2x + 1}{1 + 2x}$

65. $-4^{-1/2} = -\left(\sqrt{4}\right)^{-1}$

$\qquad = -2^{-1}$

$\qquad = -\dfrac{1}{2}$

67. $\dfrac{2^{-1} + 3^{-1}}{2^{-2}} = \dfrac{\dfrac{1}{2} + \dfrac{1}{3}}{\dfrac{1}{2^2}}$

$\qquad = \dfrac{\dfrac{5}{6}}{\dfrac{1}{4}}$

$\qquad = \dfrac{5}{6} \cdot \dfrac{4}{1}$

$\qquad = \dfrac{10}{3}$

69. $\sqrt{(x^2 + y^2)^2} = x^2 + y^2$

71. $\left(x + \sqrt{2}\right)^2 = x^2 + 2x \cdot \sqrt{2} + \left(\sqrt{2}\right)^2$

$\qquad = x^2 + 2\sqrt{2}x + 2$

73. $\dfrac{r^2(x - a) - s^2(x - a)}{(x - a)(r + s)}$

$\quad = \dfrac{(x - a)(r^2 - s^2)}{(x - a)(r + s)}$

$\quad = \dfrac{(x - a)(r - s)(r + s)}{(x - a)(r + s)}$

$\quad = r - s$

75.

$$\frac{1}{x+1} + \frac{1}{x-2} = \frac{1}{x^2 - x - 2}$$

$$\frac{1}{x+1} + \frac{1}{x-2} = \frac{1}{(x-2)(x+1)}$$

$$(x+1)(x-2)\left(\frac{1}{x+1} + \frac{1}{x-2}\right) = (x+1)(x-2)\left[\frac{1}{(x-2)(x+1)}\right]$$

$$(x+1)(x-2)\left(\frac{1}{x+1}\right) + (x+1)(x-2)\left(\frac{1}{x-2}\right) = 1$$

$$x - 2 + x + 1 = 1$$

$$2x - 1 = 1$$

$$2x = 2$$

$$x = 1$$

$$\{1\}$$

77.

$$\frac{1 - \dfrac{1}{x+1}}{\dfrac{1}{x-2} + 1} = \frac{\left(1 - \dfrac{1}{x+1}\right)(x+1)(x-2)}{\left(\dfrac{1}{x-2} + 1\right)(x+1)(x-2)}$$

$$= \frac{1(x+1)(x-2) - \left(\dfrac{1}{x+1}\right)(x+1)(x-2)}{\left(\dfrac{1}{x-2}\right)(x+1)(x-2) + 1(x+1)(x-2)}$$

$$= \frac{x^2 - 2x + x - 2 - (x-2)}{x+1 + x^2 - 2x + x - 2}$$

$$= \frac{x^2 - x - 2 - x + 2}{x^2 - 1}$$

$$= \frac{x^2 - 2x}{x^2 - 1}$$

79. Equation;

$$\sqrt{x+3} = 2$$

$$\left(\sqrt{x+3}\right)^2 = 2^2$$

$$x + 3 = 4$$

$$x = 1$$

79. (con′t)

Check:

LS: $\sqrt{1+3} = \sqrt{4} = 2$

RS: 2

Solution: $\{1\}$

170

81. Equation;

$$\frac{3(x + 1)}{2} - 1 = \frac{x + 2}{5}$$

$$10\left[\frac{3(x + 1)}{2} - 1\right] = 10\left(\frac{x + 2}{5}\right)$$

$$10\left[\frac{3(x + 1)}{2}\right] - 10 \cdot 1 = 2(x + 2)$$

$$15(x + 1) - 10 = 2x + 4$$

$$15x + 15 - 10 = 2x + 4$$

$$15x + 5 = 2x + 4$$

$$13x = -1$$

$$x = -\frac{1}{13}$$

83. Expression;

$$(3 - 2i)(1 + i) = 3 + 3i - 2i - 2i^2$$

$$= 3 + i - 2(-1)$$

$$= 3 + i + 2$$

$$= 5 + i$$

Chapter 4 Test

1. $(6^{-1/2})^{-2} = 6^{(-1/2)(-2)}$

$$= 6$$

   A

2. $\dfrac{(b^{1/2})^6}{(b^{-1})^2} = \dfrac{b^3}{b^{-2}}$

$$= b^{3-(-2)}$$

$$= b^5$$

   B

3. $\sqrt[3]{-8x^4y^2} = \sqrt[3]{-8x^3 \cdot xy^2}$

$$= \sqrt[3]{-8x^3} \cdot \sqrt[3]{xy^2}$$

$$= -2x\sqrt[3]{xy^2}$$

   C

4. $\sqrt{32} + \sqrt{18} - \sqrt{50}$

$$= \sqrt{16 \cdot 2} + \sqrt{9 \cdot 2} - \sqrt{25 \cdot 2}$$

$$= 4\sqrt{2} + 3\sqrt{2} - 5\sqrt{2}$$

$$= 2\sqrt{2}$$

   B

5. $i^{117} = i^1 = i$

   B

6. $(\sqrt{5} - 2)^2 = (\sqrt{5})^2 - 2\sqrt{5}(2) + 2^2$

$$= 5 - 4\sqrt{5} + 4$$

$$= 9 - 4\sqrt{5}$$

   D

7. $\left(-\dfrac{729}{64}\right)^{2/3} = \left(\sqrt[3]{\dfrac{-729}{64}}\right)^2$

$$= \left(\frac{\sqrt[3]{-729}}{\sqrt[3]{64}}\right)^2$$

$$= \left(\frac{-9}{4}\right)^2$$

$$= \frac{(-9)^2}{4^2}$$

$$= \frac{81}{16}$$

   A

8. $\sqrt{-3}(2 + \sqrt{-3}) = \sqrt{3}i(2 + \sqrt{3}i)$

$$= 2\sqrt{3}i + (\sqrt{3})^2i^2$$

$$= 2\sqrt{3}i + 3(-1)$$

$$= 2\sqrt{3}i - 3$$

$$= -3 + 2\sqrt{3}i$$

   C

9. $(5 - 2i)^2 = 5^2 - 2(5)(2i) + (2i)^2$

$\qquad = 25 - 20i + 4i^2$

$\qquad = 25 - 20i + 4(-1)$

$\qquad = 25 - 20i - 4$

$\qquad = 21 - 20i$

D

10. $\sqrt{x + 1} = \sqrt{2x - 2}$

$\left(\sqrt{x + 1}\right)^2 = \left(\sqrt{2x - 2}\right)^2$

$x + 1 = 2x - 2$

$3 = x$

Check:

LS: $\sqrt{3 + 1} = \sqrt{4} = 2$

RS: $\sqrt{2(3) - 2} = \sqrt{4} = 2$

Solution: $\{3\}$

11. $\sqrt{x + 3} + \sqrt{2x - 3} = 0$

$\sqrt{x + 3} = -\sqrt{2x - 3}$

$\left(\sqrt{x + 3}\right)^2 = \left(-\sqrt{2x - 3}\right)^2$

$x + 3 = 2x - 3$

$6 = x$

Check:

LS: $\sqrt{6 + 3} + \sqrt{2(6) - 3}$

$\qquad = \sqrt{9} + \sqrt{9}$

$\qquad = 3 + 3$

$\qquad = 6$

RS: 0

Solution: $\emptyset$

12. $\left(\dfrac{36a^4b^{-12}}{4c^{-6}}\right)^{-1/2} = \left(\dfrac{9a^4b^{-12}}{c^{-6}}\right)^{-1/2}$

$\qquad = \dfrac{(9a^4b^{-12})^{-1/2}}{(c^{-6})^{-1/2}}$

$\qquad = \dfrac{9^{-1/2}(a^4)^{-1/2}(b^{-12})^{-1/2}}{c^3}$

$\qquad = \dfrac{\left(\sqrt{9}\right)^{-1}a^{-2}b^6}{c^3}$

$\qquad = \dfrac{3^{-1}a^{-2}b^6}{c^3}$

$\qquad = \dfrac{b^6}{3a^2c^3}$

13. $\dfrac{i}{1 + i} = \dfrac{i(1 - i)}{(1 + i)(1 - i)}$

$\qquad = \dfrac{i - i^2}{1^2 - i^2}$

$\qquad = \dfrac{i - (-1)}{1 - (-1)}$

$\qquad = \dfrac{i + 1}{2}$

$\qquad = \dfrac{1}{2} + \dfrac{1}{2}i$

14. $\dfrac{3}{2\sqrt{6}} = \dfrac{3 \cdot \sqrt{6}}{2\sqrt{6} \cdot \sqrt{6}}$

$\qquad = \dfrac{3\sqrt{6}}{2 \cdot 6}$

$\qquad = \dfrac{\sqrt{6}}{4}$

15. $\dfrac{7 - \sqrt{2}}{1 + \sqrt{2}} = \dfrac{(7 - \sqrt{2})(1 - \sqrt{2})}{(1 + \sqrt{2})(1 - \sqrt{2})}$

$\qquad = \dfrac{7 - 7\sqrt{2} - \sqrt{2} + 2}{1^2 - \left(\sqrt{2}\right)^2}$

$\qquad = \dfrac{9 - 8\sqrt{2}}{1 - 2}$

$\qquad = \dfrac{9 - 8\sqrt{2}}{-1}$

$\qquad = -9 + 8\sqrt{2}$

172

CHAPTER 5

1. $(x + 7)(x + 3) = 0$

$x + 7 = 0$ or $x + 3 = 0$

$x = -7$ or $x = -3$

$\{-7, -3\}$

3. $(x + 5)(x + 5) = 0$

$x + 5 = 0$ or $x - 5 = 0$

$x = -5$ or $x = 5$

$\{-5, 5\}$

5. $2x(x + 5) = 0$

$2x = 0$ or $x + 5 = 0$

$x = 0$ or $x = -5$

$\{0, -5\}$

7. $-2x(3x - 1) = 0$

$-2x = 0$ or $3x - 1 = 0$

$x = 0$ or $3x = 1$

$x = 0$ or $x = \dfrac{1}{3}$

$\left\{0, \dfrac{1}{3}\right\}$

9. $(x + 4)(x + 4) = 0$

$x + 4 = 0$ or $x + 4 = 0$

$x = -4$ or $x = -4$

$\{-4\}$

11. $-2(t - 3)(2t - 9) = 0$

$(t - 3)(2t - 9) = 0$

$t - 3 = 0$ or $2t - 9 = 0$

$t = 3$ or $2t = 9$

$t = 3$ or $t = \dfrac{9}{2}$

$\left\{3, \dfrac{9}{2}\right\}$

13. $(z - 2i)(z + 2i) = 0$

$z - 2i = 0$ or $z + 2i = 0$

$z = 2i$ or $z = -2i$

$\{2i, -2i\}$

15. $x^2 - x - 2 = 0$

$(x - 2)(x + 1) = 0$

$x - 2 = 0$ or $x + 1 = 0$

$x = 2$ or $x = -1$

$\{2, -1\}$

17. $x^2 - 4x + 3 = 0$

$(x - 3)(x - 1) = 0$

$x - 3 = 0$ or $x - 1 = 0$

$x = 3$ or $x = 1$

$\{3, 1\}$

19. $4v^2 + 5v + 1 = 0$

$(4v + 1)(v + 1) = 0$

$4v + 1 = 0$ or $v + 1 = 0$

$4v = -1$ or $v = -1$

$v = -\dfrac{1}{4}$ or $v = -1$

$\left\{-\dfrac{1}{4}, -1\right\}$

21. $x^2 - 2x = 0$

$x(x - 2) = 0$

$x = 0$ or $x - 2 = 0$

$x = 0$ or $x = 2$

$\{0, 2\}$

23. $(x + 6)(x - 1) = -10$

$x^2 + 5x - 6 = -10$

$x^2 + 5x + 4 = 0$

$x + 4 = 0$ or $x + 1 = 0$

$x = -4$ or $x = -1$

$\{-4, -1\}$

25. $\dfrac{2}{5}x^2 - \dfrac{3}{5}x - 1 = 0$

$5\left(\dfrac{2}{5}x^2 - \dfrac{3}{5}x - 1\right) = 5(0)$

$2x^2 - 3x - 5 = 0$

$(2x - 5)(x + 1) = 0$

$2x - 5 = 0 \quad \text{or} \quad x + 1 = 0$

$2x = 5 \quad \text{or} \qquad x = -1$

$x = \dfrac{5}{2} \quad \text{or} \qquad x = -1$

$\left\{\dfrac{5}{2}, \ -1\right\}$

27. $\dfrac{3}{2}t^2 + \dfrac{5}{2}t = 1$

$2\left(\dfrac{3}{2}t^2 + \dfrac{5}{2}t\right) = 2(1)$

$3t^2 + 5t = 2$

$3t^2 + 5t - 2 = 0$

$(3t - 1)(t + 2) = 0$

$3t - 1 = 0 \quad \text{or} \quad t + 2 = 0$

$3t = 1 \quad \text{or} \qquad t = -2$

$t = \dfrac{1}{3} \quad \text{or} \qquad t = -2$

$\left\{\dfrac{1}{3}, \ -2\right\}$

29. $x^2 - 4 = 0$

$(x + 2)(x - 2) = 0$

$x + 2 = 0 \quad \text{or} \quad x - 2 = 0$

$x = -2 \quad \text{or} \qquad x = 2$

$\{-2, \ 2\}$

31. $2x^2 - 6x = 0$

$2x(x - 3) = 0$

$2x = 0 \quad \text{or} \quad x - 3 = 0$

$x = 0 \quad \text{or} \qquad x = 3$

$\{0, \ 3\}$

33. $-6x^2 + 5x + 1 = 0$

$6x^2 - 5x - 1 = 0$

$(6x + 1)(x - 1) = 0$

$6x + 1 = 0 \quad \text{or} \quad x - 1 = 0$

$6x = -1 \quad \text{or} \qquad x = 1$

$x = -\dfrac{1}{6} \quad \text{or} \qquad x = 1$

$\left\{-\dfrac{1}{6}, \ 1\right\}$

35. $x^2 - 16x + 64 = 0$

$(x - 8)(x - 8) = 0$

$x - 8 = 0$

$x = 8$

37. $(x + 1)(x - 2) < 0$

Boundary numbers:

$(x + 1)(x - 2) = 0$

$x = -1, \quad x = 2$

$$\underset{\substack{\quad\ \ \ -1 \qquad\ \ \ 2}}{\dfrac{\quad A \quad}{}\Big|\dfrac{\quad B \quad}{}\Big|\dfrac{\quad C \quad}{}}$$

A: $x = -2$; $\quad (-2 + 1)(-2 - 1) < 0$

$\qquad\qquad\qquad\qquad\qquad 3 < 0 \qquad$ F

B: $x = 0$; $\qquad (0 + 1)(0 - 2) < 0$

$\qquad\qquad\qquad\qquad\qquad -2 < 0 \qquad$ T

C: $x = 3$; $\qquad (3 + 1)(3 - 2) < 0$

$\qquad\qquad\qquad\qquad\qquad 4 < 0 \qquad$ F

$\{x \mid -1 < x < 2\}$

39. $(x - 3)(x + 1) \le 0$

Boundary numbers:

$(x - 3)(x + 1) = 0$

$x = 3, \quad x = -1$

$$\underset{\substack{\quad\ \ \ -1 \qquad\ \ \ 3}}{\dfrac{\quad A \quad}{}\Big|\dfrac{\quad B \quad}{}\Big|\dfrac{\quad C \quad}{}}$$

A: $x = -2$; $(-2 - 3)(-2 + 1) \le 0$

$5 \le 0$ F

B: $x = 0$; $(0 - 3)(0 + 1) \le 0$

$-3 \le 0$ T

C: $x = 4$; $(4 - 3)(4 + 1) \le 0$

$5 \le 0$ F

$\{x \mid -1 \le x \le 3\}$

41. $x^2 - 4x + 3 < 0$

Boundary numbers:

$x^2 - 4x + 3 = 0$

$(x - 3)(x - 1) = 0$

$x = 3, \ x = 1$

$$\underset{1 \qquad 3}{\underline{\quad A \quad | \quad B \quad | \quad C \quad}}$$

A: $x = 0$; $0^2 - 4(0) + 3 < 0$

$3 < 0$ F

B: $x = 2$; $2^2 - 4(2) + 3 < 0$

$-1 < 0$ T

C: $x = 4$; $4^2 - 4(4) + 3 < 0$

$3 < 0$ F

$(1, \ 3)$

43. $x^2 + 6x > -5$

Boundary numbers:

$x^2 + 6x = -5$

$x^2 + 6x + 5 = 0$

$(x + 5)(x + 1) = 0$

$x = -5, \ x = -1$

$$\underset{-5 \qquad -1}{\underline{\quad A \quad | \quad B \quad | \quad C \quad}}$$

A: $x = -6$; $(-6)^2 + 6(-6) > -5$

$0 > -5$ T

B: $x = -2$; $(-2)^2 + 6(-2) > -5$

$-8 > -5$ F

C: $x = 0$; $0^2 + 6(0) > -5$

$0 > -5$ T

$(-\infty, \ -5) \cup (-1, \ \infty)$

45. $-x^2 - x + 2 > 0$

Boundary numbers:

$-x^2 - x + 2 = 0$

$x^2 + x - 2 = 0$

$(x + 2)(x - 1) = 0$

$x = -2, \ x = 1$

$$\underset{-2 \qquad 1}{\underline{\quad A \quad | \quad B \quad | \quad C \quad}}$$

A: $x = -3$; $-(-3)^2 - (-3) + 2 > 0$

$-4 > 0$ F

B: $x = 0$; $-0^2 - 0 + 2 > 0$

$2 > 0$ T

C: $x = 2$; $-2^2 - 2 + 2 > 0$

$-4 > 0$ F

$(-2, \ 1)$

47. $x^2 - 5x - 6 = 0$

$(x - 6)(x + 1) = 0$

$x - 6 = 0$ or $x + 1 = 0$

$x = 6$ or $x = -1$

$\{6, \ -1\}$

49. $x^2 + 7x + 12 = 0$

$(x + 4)(x + 3) = 0$

$x + 4 = 0$ or $x + 3 = 0$

$x = -4$ or $x = -3$

$\{-4, \ -3\}$

51. $x^2 - 6x - 7 = 0$

$(x - 7)(x + 1) = 0$

$x - 7 = 0$ or $x + 1 = 0$

$x = 7$ or $x = -1$

$\{7, \ -1\}$

53. $x^2 - 6x - 16 = 0$

$(x - 8)(x + 2) = 0$

$x - 8 = 0$ or $x + 2 = 0$

$x = 8$ or $x = -2$

$\{8, \ -2\}$

55. $(x - 7)(x - 8) = 6$

$x^2 - 15x + 56 = 6$

$x^2 - 15x + 50 = 0$

$(x - 5)(x - 10) = 0$

$(x - 5) = 0$ or $x - 10 = 0$

$x = 5$ or $x = 10$

$\{5, \ 10\}$

57. $x^2 - 24x + 144 = 0$

$(x - 12)(x - 12) = 0$

$x - 12 = 0$

$x = 12$

$\{12\}$

59. $5x^2 - 11x + 2 = 0$

$(5x - 1)(x - 2) = 0$

$5x - 1 = 0$ or $x - 2 = 0$

$5x = 1$ or $x = 2$

$x = \dfrac{1}{5}$ or $x = 2$

$\left\{\dfrac{1}{5}, \ 2\right\}$

61. $3x^2 + 5x + 2 = 0$

$(3x + 2)(x + 1) = 0$

$3x + 2 = 0$ or $x + 1 = 0$

$3x = -2$ or $x = -1$

$x = -\dfrac{2}{3}$ or $x = -1$

$\left\{-\dfrac{2}{3}, \ -1\right\}$

63. $2x^2 - 7x + 6 = 0$

$(2x - 3)(x - 2) = 0$

$2x - 3 = 0$ or $x - 2 = 0$

$2x = 3$ or $x = 2$

$x = \dfrac{3}{2}$ or $x = 2$

$\left\{\dfrac{3}{2}, \ 2\right\}$

65. $5x^2 - 6x - 8 = 0$

$(5x + 4)(x - 2) = 0$

$5x + 4 = 0$ or $x - 2 = 0$

$5x = -4$ or $x = 2$

$x = -\dfrac{4}{5}$ or $x = 2$

$\left\{-\dfrac{4}{5}, \ 2\right\}$

67. $(x - 5)(x - 4) = 12$

$x^2 - 9x + 8 = 0$

$(x - 8)(x - 1) = 0$

$x - 8 = 0$ or $x - 1 = 0$

$x = 8$ or $x = 1$

$\{8, 1\}$

69. $(x - 6)(x - 2) = -3$

$x^2 - 8x + 12 = -3$

$x^2 - 8x + 15 = 0$

$(x - 5)(x - 3) = 0$

$x - 5 = 0$ or $x - 3 = 0$

$x = 5$ or $x = 3$

$\{5, \ 3\}$

71. $6x^2 + x = 12$

$6x^2 + x - 12 = 0$

$(2x + 3)(3x - 4) = 0$

$2x + 3 = 0$ or $3x - 4 = 0$

$2x = -3$ or $3x = 4$

$x = -\dfrac{3}{2}$ or $x = \dfrac{4}{3}$

$\left\{ -\dfrac{3}{2}, \ \dfrac{4}{5} \right\}$

73. $4x^2 + 4x - 3 = 0$

$(2x + 3)(2x - 1) = 0$

$2x + 3 = 0$ or $2x - 1 = 0$

$2x = -3$ or $2x = 1$

$x = -\dfrac{3}{2}$ or $x = \dfrac{1}{2}$

$\left\{ -\dfrac{3}{2}, \ \dfrac{1}{2} \right\}$

75. $2s^2 - 2s - 4 = 0$

$2(s^2 - s - 2) = 0$

$s^2 - s - 2 = 0$

$(s - 2)(s + 1) = 0$

$s - 2 = 0$ or $s + 1 = 0$

$s = 2$ or $s = -1$

$\{2, \ -1\}$

77. $-x^2 + x + 12 = 0$

$x^2 - x - 12 = 0$

$(x - 4)(x + 3) = 0$

$x - 4 = 0$ or $x + 3 = 0$

$x = 4$ or $x = -3$

$\{4, \ -3\}$

79. $-2t^2 + 24t + 26 = 0$

$-2(t^2 - 12t - 13) = 0$

$t^2 - 12t - 13 = 0$

$(t - 13)(t + 1) = 0$

$t - 13 = 0$ or $t + 1 = 0$

$t = 13$ or $t = -1$

$\{13, \ -1\}$

81. $5x^2 + \dfrac{1}{3}x = 2$

$3\left(5x^2 + \dfrac{1}{3}x \right) = 3(2)$

$15x^2 + x = 6$

$15x^2 + x - 6 = 0$

$(3x + 2)(5x - 3) = 0$

$3x + 2 = 0$ or $5x - 3 = 0$

$3x = -2$ or $5x = 3$

$x = -\dfrac{2}{3}$ or $x = \dfrac{3}{5}$

$\left\{ -\dfrac{2}{3}, \ \dfrac{3}{5} \right\}$

83.
$$5x^2 + \frac{7}{4}x - \frac{3}{2} = 0$$

$$4\left(5x^2 + \frac{7}{4}x - \frac{3}{2}\right) = 4(0)$$

$$20x^2 + 7x - 6 = 0$$

$$(4x + 3)(5x - 2) = 0$$

$$4x + 3 = 0 \quad \text{or} \quad 5x - 2 = 0$$

$$4x = -3 \quad \text{or} \quad 5x = 2$$

$$x = -\frac{3}{4} \quad \text{or} \quad x = \frac{2}{5}$$

$$\left\{-\frac{3}{4}, \frac{2}{5}\right\}$$

85. $(3x + 1)(x - 4) < 0$

Boundary numbers:

$$(3x + 1)(x - 4) = 0$$
$$x = -\frac{1}{3}, \quad x = 4$$

$$\frac{\quad A \quad | \quad B \quad | \quad C \quad}{\quad -\frac{1}{3} \quad\quad 4 \quad}$$

A: $x = -1;$ $\quad [3(-1) + 1](-1 - 4) < 0$

$$10 < 0 \quad \text{F}$$

B: $x = 0;$ $\quad [3(0) + 1](0 - 4) < 0$

$$-4 < 0 \quad \text{T}$$

C: $x = 5;$ $\quad [3(5) + 1](5 - 4) < 0$

$$16 < 0 \quad \text{F}$$

$$\left\{x \mid -\frac{1}{3} < x < 4\right\}$$

87. $(x + 1)(x + 1) < 0$

Boundary numbers:

$$(x + 1)(x + 1) = 0$$

$$x = -1$$

$$\frac{\quad A \quad | \quad B \quad}{\quad -1 \quad}$$

A: $x = -2;$ $\quad (-2 + 1)(-2 + 1) < 0$

$$1 < 0 \quad \text{F}$$

B: $x = 0;$ $\quad (0 + 1)(0 + 1) < 0$

$$1 < 0 \quad \text{F}$$

$$\emptyset$$

89. $(5 + x)(5 + x) \le 0$

Boundary numbers:

$$(5 + x)(5 + x) = 0$$

$$x = -5$$

$$\frac{\quad A \quad | \quad B \quad}{\quad -5 \quad}$$

A: $x = -6;$ $\quad (5 - 6)(5 - 6) \le 0$

$$1 \le 0 \quad \text{F}$$

B: $x = 0;$ $\quad (5 + 0)(5 + 0) \le 0$

$$25 \le 0 \quad \text{F}$$

The only solution is the boundary number -5.

$$\{-5\}$$

91. $x^2 - 7x + 6 > 0$

Boundary numbers:

$$x^2 - 7x + 6 = 0$$

$$(x - 6)(x - 1) = 0$$

$$x = 6, \quad x = 1$$

$$\frac{\quad A \quad | \quad B \quad | \quad C \quad}{\quad 1 \quad\quad 6 \quad}$$

A: $x = 0;$ $\quad 0^2 - 7(0) + 6 > 0$

$$6 > 0 \quad \text{T}$$

B: $x = 2;$ $\quad 2^2 - 7(2) + 6 > 0$

$$-4 > 0 \quad \text{F}$$

C: $x = 7;$ $\quad 7^2 - 7(7) + 6 > 0$

$$6 > 0 \quad \text{T}$$

$$\{x \mid x < 1 \text{ or } x > 6\}$$

93. $x^2 - 7x > 8$

Boundary numbers:

$$x^2 - 7x = 8$$

$$x^2 - 7x - 8 = 0$$

$$(x - 8)(x + 1) = 0$$

$$x = 8, \quad x = -1$$

$$\underset{-1 \qquad 8}{\overline{\quad A \quad | \quad B \quad | \quad C \quad}}$$

A: $x = -2$; $(-2)^2 - 7(-2) > 8$

$18 > 8$ T

B: $x = 0$; $0^2 - 7(0) > 8$

$0 > 8$ F

C: $x = 9$; $9^2 - 7(9) > 8$

$18 > 8$ T

$\{x \mid x < -1 \text{ or } x > 8\}$

95. $x^2 > 16$

Boundary numbers:

$$x^2 = 16$$

$$x^2 - 16 = 0$$

$$(x + 4)(x - 4) = 0$$

$$x = -4, \quad x = 4$$

$$\underset{-4 \qquad 4}{\overline{\quad A \quad | \quad B \quad | \quad C \quad}}$$

A: $x = -5$; $(-5)^2 > 16$

$25 > 16$ T

B: $x = 0$; $0^2 > 16$

$0 > 16$ F

C: $x = 5$; $5^2 > 16$

$25 > 16$ T

$\{x \mid x < -4 \text{ or } x > 4\}$

97. $-t^2 - 5t \geq 0$

Boundary numbers:

$$-t^2 - 5t = 0$$

$$t^2 + 5t = 0$$

$$t(t + 5) = 0$$

$$t = 0, \quad t = -5$$

$$\underset{-5 \qquad 0}{\overline{\quad A \quad | \quad B \quad | \quad C \quad}}$$

A: $t = -6$; $-(-6)^2 - 5(-6) \geq 0$

$-6 \geq 0$ F

B: $t = -1$; $-(-1)^2 - 5(-1) \geq 0$

$4 \geq 0$ T

C: $t = 1$; $-1^2 - 5(1) \geq 0$

$-6 \geq 0$ F

$\{3t \mid -5 \leq t \leq 0\}$

99. $s^2 + 3 < -4s$

Boundary numbers:

$$s^2 + 3 = -4s$$

$$s^2 + 4s + 3 = 0$$

$$(s + 3)(s + 1) = 0$$

$$s = -3, \quad s = -1$$

$$\underset{-3 \qquad -1}{\overline{\quad A \quad | \quad B \quad | \quad C \quad}}$$

A: $s = -4$; $(-4)^2 + 3 < -4(-4)$

$19 < 16$ F

B: $s = -2$; $(-2)^2 + 3 < -4(-2)$

$7 < 8$ T

C: $s = 0$; $0^2 + 3 < -4(0)$

$3 < 0$ F

$\{s \mid -3 < s < -1\}$

101. $5x + 14 \geq x^2$

Boundary numbers:

$5x + 14 = x^2$

$0 = x^2 - 5x - 14$

$0 = (x - 7)(x + 2)$

$x = 7, \quad x = -2$

$$\underline{\quad A \quad \underset{-2}{|} \quad B \quad \underset{7}{|} \quad C \quad}$$

A: $\quad x = -3; \quad 5(-3) + 14 \geq (-3)^2$

$-1 \geq 9 \quad$ F

B: $\quad x = 0; \qquad 5(0) + 14 \geq 0^2$

$14 \geq 0 \quad$ T

C: $\quad x = 8; \qquad 5(8) + 14 \geq 8^2$

$54 \geq 64 \quad$ F

$$\{x \mid -2 \leq x \leq 7\}$$

103. $10x^2 + x - 3 > 0$

Boundary numbers:

$10x^2 + x - 3 = 0$

$(5x + 3)(2x - 1) = 0$

$x = -\dfrac{3}{5}, \quad x = \dfrac{1}{2}$

$$\underline{\quad A \quad \underset{-\frac{3}{5}}{|} \quad B \quad \underset{\frac{1}{2}}{|} \quad C \quad}$$

A: $\quad x = -1; \quad 10(-1)^2 - 1 - 3 > 0$

$6 > 0 \quad$ T

B: $\quad x = 0; \qquad 10(0)^2 + 0 - 3 > 0$

$-3 > 0 \quad$ F

C: $\quad x = 1; \qquad 10(1)^2 + 1 - 3 > 0$

$8 > 0 \quad$ T

$$\left\{x \mid x < -\frac{3}{5} \text{ or } x > \frac{1}{2}\right\}$$

105. $21x^2 + 19x - 12 \leq 0$

Boundary numbers:

$21x^2 + 19 - 12 = 0$

$(7x - 3)(3x + 4) = 0$

$x = \dfrac{3}{7}, \quad x = -\dfrac{4}{3}$

$$\underline{\quad A \quad \underset{-\frac{4}{3}}{|} \quad B \quad \underset{\frac{3}{7}}{|} \quad C \quad}$$

A: $\quad x = -2; \quad 21(-2)^2 + 19(-2) - 12 \leq 0$

$34 \leq 0 \quad$ F

B: $\quad x = 0; \qquad 21(0)^2 + 19(0) - 12 \leq 0$

$-12 \leq 0 \quad$ T

C: $\quad x = 1; \qquad 21(1)^2 + 19(1) - 12 \leq 0$

$28 \leq 0 \quad$ F

$$\left[-\frac{4}{3}, \frac{3}{7}\right]$$

107. $12x^2 + 5x - 2 < 0$

Boundary numbers:

$12x^2 + 5x - 2 = 0$

$(4x - 1)(3x + 2) = 0$

$x = \dfrac{1}{4}, \quad x = -\dfrac{2}{3}$

$$\underline{\quad A \quad \underset{-\frac{2}{3}}{|} \quad B \quad \underset{\frac{1}{4}}{|} \quad C \quad}$$

A: $\quad x = -1; \quad 12(-1)^2 + 5(-1) - 2 < 0$

$5 < 0 \quad$ F

B: $\quad x = 0; \qquad 12(0)^2 + 5(0) - 2 < 0$

$-2 < 0 \quad$ T

C: $\quad x = 1; \qquad 12(1)^2 + 5(1) - 2 < 0$

$15 < 0 \quad$ F

$$\left(-\frac{2}{3}, \frac{1}{4}\right)$$

109. $10x^2 - 21x - 49 \geq 0$

Boundary numbers:

$10x^2 - 21x - 49 = 0$

$(5x + 7)(2x - 7) = 0$

$x = -\dfrac{7}{5}, \quad x = \dfrac{7}{2}$

$$\underline{\quad A \quad \mid \quad B \quad \mid \quad C \quad}$$
$$\qquad -\dfrac{7}{5} \qquad \dfrac{7}{2}$$

A: $x = -2$; $\quad 10(-2)^2 - 21(-2) - 49 \geq 0$

$$33 \geq 0 \quad T$$

B: $x = 0$; $\quad 10(0)^2 - 21(0) - 49 \geq 0$

$$-49 \geq 0 \quad F$$

C: $x = 4$; $\quad 10(4)^2 - 21(4) - 49 \geq 0$

$$27 \geq 0 \quad T$$

$$\left(-\infty, \ -\dfrac{7}{5}\right] \cup \left[\dfrac{7}{2}, \ \infty\right)$$

111. $x^2 - 5 = 0$

$x^2 - \left(\sqrt{5}\right)^2 = 0$

$\left(x - \sqrt{5}\right)\left(x + \sqrt{5}\right) = 0$

$x - \sqrt{5} = 0 \quad$ or $\quad x + \sqrt{5} = 0$

$x = \sqrt{5} \quad$ or $\quad x = -\sqrt{5}$

$$\left\{\sqrt{5}, \ -\sqrt{5}\right\}$$

113. $w^2 + 5 = 0$

$w^2 - (-5) = 0$

$w^2 - \left(\sqrt{5}i\right)^2 = 0$

$\left(w - \sqrt{5}i\right)\left(w + \sqrt{5}i\right) = 0$

$w - \sqrt{5}i = 0 \quad$ or $\quad w + \sqrt{5}i = 0$

$w = \sqrt{5}i \quad$ or $\quad w = -\sqrt{5}i$

$$\left\{\sqrt{5}i, \ -\sqrt{5}i\right\}$$

115. $4x^2 + 25 = 0$

$(2x)^2 - (-25) = 0$

$(2x)^2 - (5i)^2 = 0$

$(2x - 5i)(2x + 5i) = 0$

$2x - 5i = 0 \quad$ or $2x + 5i = 0$

$2x = 5i \quad$ or $\quad 2x = -5i$

$x = \dfrac{5}{2}i \quad$ or $\qquad x = -\dfrac{5}{2}i$

$$\left\{\dfrac{5}{2}i, \ -\dfrac{5}{2}i\right\}$$

117. $\{1, 2\}$

$x = 1$ or $\quad x = 2$

$x - 1 = 0$ or $x - 2 = 0$

$(x - 1)(x - 2) = 0$

$x^2 - 3x + 2 = 0$

119. $\{\pm 2\}$

$x = 2$ or $\qquad x = -2$

$x - 2 = 0$ or $x + 2 = 0$

$(x - 2)(x + 2) = 0$

$x^2 - 4 = 0$

121. $\{\pm 2i\}$

$x = 2i$ or $\qquad x = -2i$

$x - 2i = 0$ or $x + 2i = 0$

$(x - 2i)(x + 2i) = 0$

$x^2 - 4i^2 = 0$

$x^2 - 4(-1) = 0$

$x^2 + 4 = 0$

Problem Set 5.2

1. 1st girl's age: x
   2nd girl's age: $x + 2$

   $$x(x + 2) = 288$$

   $$x^2 + 2x = 288$$

   $$x^2 + 2x - 288 = 0$$

   $$(x + 18)(x - 16) = 0$$

   $$x = -18 \quad \text{or} \quad x = 16$$

   Since $x > 0$, $x = 16$

   $$x + 2 = 18$$

   Their ages are 16 and 18.

3. length of 1st trail: x
   length of 2nd trail: $20 - x$

   $$x(20 - x) = 91$$

   $$20x - x^2 = 91$$

   $$0 = x^2 - 20x + 91$$

   $$0 = (x - 7)(x - 13)$$

   $$x = 7 \quad \text{or} \quad x = 13$$

   $$20 - x = 13 \qquad 20 - x = 7$$

   The lengths are 7 miles and 13 miles.

5. 1st consecutive integer: x
   2nd consecutive integer: $x + 1$
   3rd consecutive integer: $x + 2$

   $$x^2 + (x + 1)^2 + (x + 2)^2 = 77$$

   $$x^2 + x^2 + 2x + 1 + x^2 + 4x + 4 = 77$$

   $$3x^2 + 6x - 72 = 0$$

   $$x^2 + 2x - 24 = 0$$

   $$(x + 6)(x - 4) = 0$$

   $$x = -6 \quad \text{or} \quad x = 4$$

   Since $x > 0$, $x = 4$, $x + 1 = 5$, $x + 2 = 6$

   The numbers are 4, 5 and 6.

7. $80t - 16t^2 > 96$

   Boundary numbers:

   $$80t - 16t^2 = 96$$

   $$0 = 16t^2 - 80t + 96$$

   $$0 = t^2 - 5t + 6$$

   $$0 = (t - 3)(t - 2)$$

   $$t = 3, \quad t = 2$$

   $$\underline{\quad A \quad \overset{|}{\underset{2}{}} \quad B \quad \overset{|}{\underset{3}{}} \quad C \quad}$$

   A: $t = 0$; $\quad 80(0) - 16(0)^2 > 96$

   $$0 > 96 \quad F$$

   B: $t = \dfrac{5}{2}$; $\quad 80\left(\dfrac{5}{2}\right) - 16\left(\dfrac{5}{2}\right)^2 > 96$

   $$100 > 96 \quad T$$

   C: $t = 4$; $\quad 80(4) - 16(4)^2 > 96$

   $$64 > 96 \quad F$$

   The time must be between 2 and 3 seconds.

9.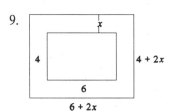

   $$(4 + 2x)(6 + 2x) = 48$$

   $$24 + 20x + 4x^2 = 48$$

   $$4x^2 + 20x - 24 = 0$$

   $$x^2 + 5x - 6 = 0$$

   $$(x + 6)(x - 1) = 0$$

   $$x = -6 \quad \text{or} \quad x = 1$$

   Since $x > 0$, $x = 1$.

   The width of the frame should be 1 foot.

11. altitude: x
base: $9 + x$

$$56 = \frac{1}{2}(9 + x)x$$

$$2(56) = 2\left(\frac{1}{2}\right)(9 + x)x$$

$$112 = 9x + x^2$$

$$0 = x^2 + 9x - 112$$

$$0 = (x + 16)(x - 7)$$

$x = -16$ or $x = 7$

Since $x > 0$, $x = 7$, $9 + x = 16$

The altitude is 7 ft and the base is 16 ft.

13. shorter leg: x
longer leg: $2x + 2$

$$x^2 + (2x + 2)^2 = 13^2$$

$$x^2 + 4x^2 + 8x + 4 = 169$$

$$5x^2 + 8x - 165 = 0$$

$$(5x + 33)(x - 5) = 0$$

$$x = -\frac{33}{5} \quad \text{or} \quad x = 5$$

Since $x > 0$, $x = 5$, $2x + 2 = 12$

The legs are 5 ft and 12 ft.

15. 1st consecutive integer: x
2nd consecutive integer: $x + 1$

$$x(x + 1) = 132$$

$$x^2 + x = 132$$

$$x^2 + x - 132 = 0$$

$$(x + 12)(x - 11) = 0$$

$$x = -12 \quad \text{or} \quad x = 11$$

Since $x < 0$, $x = -12$

$$x + 1 = -11$$

The integers are -12 and -11.

17. 1st number: x
2nd number: $9 - x$

$$x(9 - x) = 20$$

$$9x - x^2 = 20$$

$$0 = x^2 - 9x + 20$$

$$0 = (x - 5)(x - 4)$$

$x = 5$ or $x = 4$

$9 - x = 4$ $9 - x = 5$

The numbers are 4 and 5.

19. 1st odd, consecutive integer: x
2nd odd, consecutive integer: $x + 2$
3rd odd, consecutive integer: $x + 4$

$$x^2 + (x + 2)^2 + (x + 4)^2 = 35$$

$$x^2 + x^2 + 4x + 4 + x^2 + 8x + 16 = 35$$

$$3x^2 + 12x + 20 = 35$$

$$3x^2 + 12x - 15 = 0$$

$$x^2 + 4x - 5 = 0$$

$$(x + 5)(x - 1) = 0$$

$$x = -5 \quad \text{or} \quad x = 1$$

Since $x > 0$, $x = 1$, $x + 2 = 3$, $x + 4 = 5$

The integers are 1, 3 and 5.

21. width: x
length: $2x - 3$

$$x(2x - 3) = 65$$

$$2x^2 - 3x = 65$$

$$2x^2 - 3x - 65 = 0$$

$$(2x - 13)(x + 5) = 0$$

$$x = \frac{13}{2} \quad \text{or} \quad x = -5$$

Since $x > 0$, $x = \frac{13}{2}$, $2x - 3 = 10$

The dimensions are $6\frac{1}{2}$ m by 10 m.

23. $t(3t - 14) < 160$

Boundary numbers:

$$3t^2 - 14t = 160$$

$$3t^2 - 14t - 160 = 0$$

$$(3t + 16)(t - 10) = 0$$

$$t = -\frac{16}{3}, \quad t = 10$$

$$\underline{\quad A \quad}_{\,|}\underline{\quad B \quad}_{\,|}\underline{\quad C \quad}$$
$$\quad -\frac{16}{3} \quad 10$$

A: $t = -6$; $\qquad -6[3(-6) - 14] < 160$

$$192 < 160 \quad F$$

B: $t = 0$; $\qquad 0[3(0) - 14] < 160$

$$0 < 160 \quad T$$

C: $t = 11$; $\qquad 11[3(11) - 14] < 160$

$$209 < 160 \quad F$$

$$-\frac{16}{3} < t < 10$$

Since $t > 0$, the time must be between 0 and 10 minutes.

25.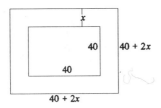

$$(40 + 2x)(40 + 2x) = 2025$$

$$1600 + 160x + 4x^2 = 2025$$

$$4x^2 + 160x - 425 = 0$$

$$(2x - 5)(2x + 85) = 0$$

$$x = \frac{5}{2} \quad \text{or} \quad x = -\frac{85}{2}$$

Since $x > 0$, $x = \frac{5}{2}$

The sidewalk should be $2\frac{1}{2}$ ft wide.

27.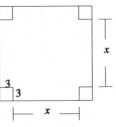

$$x \cdot x \cdot 3 = 192$$

$$3x^2 = 192$$

$$3x^2 - 192 = 0$$

$$x^2 - 64 = 0$$

$$(x + 8)(x - 8) = 0$$

$$x = -8 \quad \text{or} \quad x = 8$$

Since $x > 0$, $x = 8$

One side of the sheet was $8 + 3 + 3 = 14$ in.

29. width: x
length: $2x - 4$

$$x(2x - 4) = 70$$

$$2x^2 - 4x = 70$$

$$2x^2 - 4x - 70 = 0$$

$$x^2 - 2x - 35 = 0$$

$$(x - 7)(x + 5) = 0$$

$$x = 7 \quad \text{or} \quad x = -5$$

Since $x > 0$, $x = 7$, $2x - 4 = 10$

The poster is 7 in. by 10 in.

31. $\frac{1}{2}$ perimeter = $\frac{1}{2}(44) = 22$

width: x
length: $22 - x$

$x(22 - x) = 120$

$22x - x^2 = 120$

$0 = x^2 - 22x + 120$

$0 = (x - 12)(x - 10)$

$x = 12$ or $x = 10$

$22 - x = 10,$ $22 - x = 12$

The dimensions are 10 m by 12 m.

33. $x^2 + 20x - 200 > 600$

Boundary numbers:

$x^2 + 20x - 200 = 600$

$x^2 + 20x - 800 = 0$

$(x + 40)(x - 20) = 0$

$x = -40,\ x = 20$

$$\underline{\quad\ \ A\ \ \ |\ \ B\ \ |\ \ C\quad\quad}$$
$$\text{-40}\quad\ 20$$

A: $x = -50$; $(-50)^2 + 20(-50) - 200 > 600$

$1300 > 600$ T

B: $x = 0$; $0^2 + 20(0) - 200 > 600$

$-200 > 600$ F

C: $x = 30$; $30^2 + 20(30) - 200 > 600$

$1300 > 600$ T

$x < -40$ or $x > 20$

Since $x > 0$, $x > 20$ is the solution set.

They must sell at least 20 dresses.

35. A.) width: x
length: $2x - 10$

$x(2x - 10) = 600$

$2x^2 - 10x = 600$

$2x^2 - 10x - 600 = 0$

$x^2 - 5x - 300 = 0$

$(x - 20)(x + 15) = 0$

$x = 20$ or $x = -15$

Since $x > 0$, $x = 20$, $2x - 10 = 30$

The dimensions of the office are 20 ft by 30 ft.

B.)   20 ft by 30 ft = 240 in by 360 in

Area = (240)(360) = 86400 sq in

$\dfrac{\text{bottles}}{\text{sq in:}}\quad \dfrac{32}{1} = \dfrac{n}{86400}$

$n = (32)(86400)$

$n = 2,764,800$

2,764,800 bottles are needed.

37. height: x
base: $2x$

$\frac{1}{2}x(2x) = 49$

$x^2 = 49$

$x^2 - 49 = 0$

$(x + 7)(x - 7) = 0$

$x = -7$ or $x = 7$

Since $x > 0$, $x = 7$, $2x = 14$

The height is 7 yd and the base is 14 yd.

185

39. Shorter leg's length: x
longer leg's length: $2x + 4$

$$x^2 + (2x + 4)^2 = 26^2$$

$$x^2 + 4x^2 + 16x + 16 = 676$$

$$5x^2 + 16x - 660 = 0$$

$$(5x + 66)(x - 10) = 0$$

$$x = -\frac{66}{5} \quad \text{or} \quad x = 10$$

Since $x > 0$, $x = 10$, $2x + 4 = 24$

The legs are 10 m and 24 m.

41.

	T	R	D = RT
south train	2	$r + 70$	$2(r + 70)$
east train	2	r	$2r$

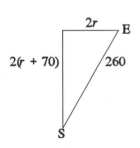

$$(2r)^2 + [2(r + 70)]^2 = 260^2$$

$$4r^2 + 4(r + 70)^2 = 67600$$

$$4r^2 + 4(r^2 + 140r + 4900) = 67600$$

$$4r^2 + 4r^2 + 560r + 19600 = 67600$$

$$8r^2 + 560r - 48000 = 0$$

$$r^2 + 70r - 6000 = 0$$

$$(r + 120)(r - 50) = 0$$

$$r = -120 \quad \text{or} \quad r = 50$$

Since $r > 0$, $r = 50$

$$r + 70 = 120$$

The southbound train is traveling at 120 mph and the eastbound train is traveling at 50 mph.

Problem Set 5.3

1. $x^2 - 25 = 0$

$$x^2 = 25$$

$$x = \pm\sqrt{25}$$

$$x = \pm 5$$

$$\{\pm 5\}$$

3. $x^2 - 5 = 0$

$$x^2 = 5$$

$$x = \pm\sqrt{5}$$

$$\{\pm\sqrt{5}\}$$

5. $x^2 - 48 = 0$

$$x^2 = 48$$

$$x = \pm\sqrt{48}$$

$$x = \pm 4\sqrt{3}$$

$$\{\pm 4\sqrt{3}\}$$

7. $2t^2 + 16 = 0$

$$2t^2 = -16$$

$$t^2 = -8$$

$$t = \pm\sqrt{-8}$$

$$t = \pm\sqrt{8}i$$

$$t = \pm 2\sqrt{2}i$$

$$\{\pm 2\sqrt{2}i\}$$

9. $(x + 3)^2 = 4$

$x + 3 = \pm 2$

$x = -3 \pm 2$

$x = -3 + 2$ or $x = -3 - 2$

$x = -1$ or $x = -5$

$\{-1, -5\}$

11. $(x - 5)^2 = -7$

$x - 5 = \pm\sqrt{-7}$

$x - 5 = \pm\sqrt{7}i$

$x = 5 \pm \sqrt{7}i$

$\{5 \pm \sqrt{7}i\}$

13. $x^2 + 4x$

Add: $\left[\frac{1}{2}(4)\right]^2 = 2^2 = 4$

15. $x^2 + 7x$

Add: $\left[\frac{1}{2}(7)\right]^2 = \left(\frac{7}{2}\right)^2 = \frac{49}{4}$

17. $x^2 + x$

Add: $\left[\frac{1}{2}(1)\right]^2 = \left(\frac{1}{2}\right)^2 = \frac{1}{4}$

19. $x^2 - 4x - 5 = 0$

$x^2 - 4x = 5$

$x^2 - 4x + 2^2 = 5 + 2^2$

$(x - 2)^2 = 9$

$x - 2 = \pm\sqrt{9}$

$x - 2 = \pm 3$

$x = 2 \pm 3$

$x = 2 + 3$ or $x = 2 - 3$

$x = 5$ or $x = -1$

$\{5, -1\}$

21. $x^2 + x - 6 = 0$

$x^2 + x = 6$

$x^2 + x + \left(\frac{1}{2}\right)^2 = 6 + \left(\frac{1}{2}\right)^2$

$\left(x + \frac{1}{2}\right)^2 = 6 + \frac{1}{4}$

$\left(x + \frac{1}{2}\right)^2 = \frac{25}{4}$

$x + \frac{1}{2} = \pm\sqrt{\frac{25}{4}}$

$x + \frac{1}{2} = \pm\frac{5}{2}$

$x = -\frac{1}{2} \pm \frac{5}{2}$

$x = -\frac{1}{2} + \frac{5}{2}$ or $x = -\frac{1}{2} - \frac{5}{2}$

$x = 2$ or $x = -3$

$\{2, -3\}$

23. $x^2 - 6x + 1 = 0$

$x^2 - 6x = -1$

$x^2 - 6x + (3)^2 = -1 + (3)^2$

$(x - 3)^2 = 8$

$x - 3 = \pm\sqrt{8}$

$x - 3 = \pm 2\sqrt{2}$

$x = 3 \pm 2\sqrt{2}$

$\{3 \pm 2\sqrt{2}\}$

25. $3w^2 - 6w - 9 = 0$

$w^2 - 2w - 3 = 0$

$w^2 - 2w = 3$

$w^2 - 2w + 1^2 = 3 + 1^2$

$(w - 1)^2 = 3 + 1$

$(w - 1)^2 = 4$

$w - 1 = \pm\sqrt{4}$

$w - 1 = \pm 2$

$= 1 \pm 2$

$w = 1 + 2 \quad \text{or} \quad w = 1 - 2$

$w = 3 \qquad \text{or} \quad w = -1$

$\{3, -1\}$

27. $4x^2 + 12x + 6 = 0$

$x^2 + 3x + \dfrac{3}{2} = 0$

$x^2 + 3x = -\dfrac{3}{2}$

$x^2 + 3x + \left(\dfrac{3}{2}\right)^2 = -\dfrac{3}{2} + \left(\dfrac{3}{2}\right)^2$

$\left(x + \dfrac{3}{2}\right)^2 = -\dfrac{3}{2} + \dfrac{9}{4}$

$\left(x + \dfrac{3}{2}\right)^2 = \dfrac{3}{4}$

$x + \dfrac{3}{2} = \pm\sqrt{\dfrac{3}{4}}$

$x + \dfrac{3}{2} = \pm\dfrac{\sqrt{3}}{2}$

$x = -\dfrac{3}{2} \pm \dfrac{\sqrt{3}}{2}$

$\left\{ -\dfrac{3}{2} \pm \dfrac{\sqrt{3}}{2} \right\}$

29. $x^2 - 2x - 8 = 0$

$x^2 - 2x = 8$

$x^2 - 2x + (1)^2 = 8 + (1)^2$

$(x - 1)^2 = 9$

$x - 1 = \pm\sqrt{9}$

$x - 1 = \pm 3$

$x = 1 \pm 3$

$x - 1 = 3 \quad \text{or} \quad x - 1 = -3$

$x = 4 \quad \text{or} \quad x = -2$

$\{4, \ -2\}$

31. $x^2 - 2x + 1 = 0$

$x^2 - 2x = -1$

$x^2 - 2x + (1)^2 = -1 + (1)^2$

$(x - 1)^2 = 0$

$x - 1 = \pm\sqrt{0}$

$x - 1 = 0$

$x = 1$

$\{1\}$

33. $3x^2 - 12x = -9$

$x^2 - 4x = -3$

$x^2 - 4x + (2)^2 = -3 + (2)^2$

$(x - 2)^2 = 1$

$x - 2 = \pm\sqrt{1}$

$x - 2 = \pm 1$

$x = 2 \pm 1$

$x = 2 + 1 \quad$ or $\quad x = 2 - 1$

$x = 3 \qquad$ or $\quad x = 1$

$\{3, \ 1\}$

35. $x^2 - 3x - 4 = 0$

$x^2 - 3x = 4$

$x^2 - 3x + \left(\dfrac{3}{2}\right)^2 = 4 + \left(\dfrac{3}{2}\right)^2$

$\left(x - \dfrac{3}{2}\right)^2 = 4 + \dfrac{9}{4}$

$\left(x - \dfrac{3}{2}\right)^2 = \dfrac{25}{4}$

$x - \dfrac{3}{2} = \pm\sqrt{\dfrac{25}{4}}$

$x - \dfrac{3}{2} = \pm\dfrac{5}{2}$

$x = \dfrac{3}{2} \pm \dfrac{5}{2}$

$x = \dfrac{3}{2} + \dfrac{5}{2} \quad$ or $\quad x = \dfrac{3}{2} - \dfrac{5}{2}$

$x = 4 \qquad$ or $\quad x = -1$

$\{4, \ -1\}$

37. $4x^2 - 20x + 16 = 0$

$x^2 - 5x + 4 = 0$

$x^2 - 5x = -4$

$x^2 - 5x + \left(\dfrac{5}{2}\right)^2 = -4 + \left(\dfrac{5}{2}\right)^2$

$\left(x - \dfrac{5}{2}\right)^2 = -4 + \dfrac{25}{4}$

$\left(x - \dfrac{5}{2}\right)^2 = \dfrac{9}{4}$

$x - \dfrac{5}{2} = \pm\sqrt{\dfrac{9}{4}}$

$x - \dfrac{5}{2} = \pm\dfrac{3}{2}$

$x = \dfrac{5}{2} \pm \dfrac{3}{2}$

$x = \dfrac{5}{2} + \dfrac{3}{2} \quad$ or $\quad x = \dfrac{5}{2} - \dfrac{3}{2}$

$x = 4 \qquad$ or $\quad x = 1$

$\{4, \ 1\}$

39. $2u^2 - 2u = 2$

$u^2 - u = 1$

$u^2 - u + \left(\dfrac{1}{2}\right)^2 = 1 + \left(\dfrac{1}{2}\right)^2$

$\left(u - \dfrac{1}{2}\right)^2 = 1 + \dfrac{1}{4}$

$\left(u - \dfrac{1}{2}\right)^2 = \dfrac{5}{4}$

$u - \dfrac{1}{2} = \pm\sqrt{\dfrac{5}{4}}$

$u - \dfrac{1}{2} = \pm\dfrac{\sqrt{5}}{2}$

$u = \dfrac{1}{2} \pm \dfrac{\sqrt{5}}{2}$

$\left\{\dfrac{1}{2} \pm \dfrac{\sqrt{5}}{2}\right\}$

41. $2x^2 + 5x - 4 = 0$

$$x^2 + \frac{5}{2}x - 2 = 0$$

$$x^2 + \frac{5}{2}x = 2$$

$$x^2 + \frac{5}{2}x + \left(\frac{5}{4}\right)^2 = 2 + \left(\frac{5}{4}\right)^2$$

$$\left(x + \frac{5}{4}\right)^2 = 2 + \frac{25}{16}$$

$$\left(x + \frac{5}{4}\right)^2 = \frac{57}{16}$$

$$x + \frac{5}{4} = \pm\sqrt{\frac{57}{16}}$$

$$x + \frac{5}{4} = \pm\frac{\sqrt{57}}{4}$$

$$x = \frac{5}{4} \pm \frac{\sqrt{57}}{4}$$

$$\left\{-\frac{5}{4} \pm \frac{\sqrt{57}}{4}\right\}$$

43. $3v^2 - 4v + 10 = 0$

$$v^2 - \frac{4}{3}v + \frac{10}{3} = 0$$

$$v^2 - \frac{4}{3}v = -\frac{10}{3}$$

$$v^2 - \frac{4}{3}v + \left(\frac{2}{3}\right)^2 = -\frac{10}{3} + \left(\frac{2}{3}\right)^2$$

$$\left(v - \frac{2}{3}\right)^2 = -\frac{10}{3} + \frac{4}{9}$$

$$\left(v - \frac{2}{3}\right)^2 = -\frac{26}{9}$$

$$v - \frac{2}{3} = \pm\sqrt{-\frac{26}{9}}$$

$$v - \frac{2}{3} = \pm\sqrt{\frac{26}{9}}\,i$$

$$v - \frac{2}{3} = \pm\frac{\sqrt{26}}{3}\,i$$

$$v = \frac{2}{3} \pm \frac{\sqrt{26}}{3}\,i$$

$$\left\{\frac{2}{3} \pm \frac{\sqrt{26}}{3}\,i\right\}$$

45. $2x^2 - 7x + 5 = 0$

$$x^2 - \frac{7}{2}x + \frac{5}{2} = 0$$

$$x^2 - \frac{7}{2}x = -\frac{5}{2}$$

$$x^2 - \frac{7}{2}x + \left(\frac{7}{4}\right)^2 = -\frac{5}{2} + \left(\frac{7}{4}\right)^2$$

$$\left(x - \frac{7}{4}\right)^2 = -\frac{5}{2} + \frac{49}{16}$$

$$\left(x - \frac{7}{4}\right)^2 = \frac{9}{16}$$

$$x - \frac{7}{4} = \pm\sqrt{\frac{9}{16}}$$

$$x - \frac{7}{4} = \pm\frac{3}{4}$$

$$x = \frac{7}{4} \pm \frac{3}{4}$$

$$x = \frac{7}{4} + \frac{3}{4} \quad \text{or} \quad x = \frac{7}{4} - \frac{3}{4}$$

$$x = \frac{5}{2} \quad \text{or} \quad x = 1$$

$$\left\{\frac{5}{2},\ 1\right\}$$

47. $4x^2 - 6x + 10 = 0$

$$x^2 - \frac{3}{2}x + \frac{5}{2} = 0$$

$$x^2 - \frac{3}{2}x = -\frac{5}{2}$$

$$x^2 - \frac{3}{2}x + \left(\frac{3}{4}\right)^2 = -\frac{5}{2} + \left(\frac{3}{4}\right)^2$$

$$\left(x - \frac{3}{4}\right)^2 = -\frac{5}{2} + \frac{9}{16}$$

$$\left(x - \frac{3}{4}\right)^2 = -\frac{31}{16}$$

$$x - \frac{3}{4} = \pm\sqrt{-\frac{31}{16}}$$

$$x - \frac{3}{4} = \pm\sqrt{\frac{31}{16}}i$$

$$x - \frac{3}{4} = \pm\frac{\sqrt{31}}{4}i$$

$$x = \frac{3}{4} \pm \frac{\sqrt{31}}{4}i$$

$$\left\{\frac{3}{4} \pm \frac{\sqrt{31}}{4}i\right\}$$

49. $x^2 = 444$

$$x = \pm\sqrt{444}$$

$$x = \pm 21.071$$

$$\{\pm 21.071\}$$

51. $3t^2 = -18$

$$t^2 = -6$$

$$t = \pm\sqrt{-6}$$

$$t = \pm\sqrt{6}i$$

$$t = \pm 2.449i$$

$$\{\pm 2.449i\}$$

53. $4x^2 - 3.613 = 0$

$$4x^2 = 3.613$$

$$x^2 = 0.903$$

$$x = \pm\sqrt{0.903}$$

$$x = \pm 0.950$$

$$\{\pm 0.950\}$$

55. width: x
length: $3x$

A.) $x(3x) = 4.6875$

$$3x^2 = 4.6875$$

$$x^2 = 1.5625$$

$$x = \pm\sqrt{1.5625}$$

$$x = \pm 1.25$$

Since $x > 0$, $x = 1.25$, $3x = 3.75$

The ad is 1.25 in by 3.75 in.

B.) $\dfrac{\text{cost}}{\text{sq in}}$: $\dfrac{27.90}{1} = \dfrac{x}{4.6875}$

$$x = 130.78$$

The ad will cost $130.78.

C.) $\dfrac{\text{cost}}{\text{sq in}}$: $\dfrac{42.63}{1} = \dfrac{x}{4.6875}$

$$x = 199.83$$

The ad will cost $199.83.

D.) $2(130.78) + 2(199.83) = 661.22 \nleq 600$

No, Sam cannot run weekly ads with 2 of those ads in the Sunday paper for $600 or less.

57. distance between opposite corners: x

$$x^2 = 12.5^2 + 12.5^2$$

$$x^2 = 312.5$$

$$x = \pm\sqrt{312.5}$$

$$x = \pm17.7$$

Since $x > 0$, $x = 17.7$

The distance is approximately 17.7 in.

59. side length: x

$$x^2 = 43560$$

$$x = \pm\sqrt{43560}$$

$$x = \pm209$$

Since $x > 0$, $x = 209$

One side would be approximately 209 ft.

61. leg length: x

$$x^2 + (5\sqrt{2})^2 = 10^2$$

$$x^2 + 50 = 100$$

$$x^2 = 50$$

$$x = \pm\sqrt{50}$$

$$x = \pm5\sqrt{2}$$

Since $x > 0$, $x = 5\sqrt{2}$

The other leg is $5\sqrt{2}$ m.

63. width: x
length: $2x$

$$x(2x) = 20$$

$$2x^2 = 20$$

$$x^2 = 10$$

$$x = \pm\sqrt{10}$$

$$x = \pm3.16$$

Since $x > 0$, $x = 3.16$

The width would be approximately 3.1 ft or less

65. $x^2 = a + b$

$$x = \pm\sqrt{a + b}$$

$$\{\pm\sqrt{a + b}\}$$

67. $(x + 1)^2 = a$

$$x + 1 = \pm\sqrt{a}$$

$$x = -1 \pm \sqrt{a}$$

$$\{-1 \pm \sqrt{a}\}$$

69. $(x + b)^2 = 8$

$$x + b = \pm\sqrt{8}$$

$$x + b = \pm2\sqrt{2}$$

$$x = -b \pm 2\sqrt{2}$$

$$\{-b \pm 2\sqrt{2}\}$$

71. $4x^2 + 4ax - a = 0$

$$x^2 + ax - \frac{a}{4} = 0$$

$$x^2 + ax = \frac{a}{4}$$

$$x^2 + ax + \left(\frac{a}{2}\right)^2 = \frac{a}{4} + \left(\frac{a}{2}\right)^2$$

$$\left(x + \frac{a}{2}\right)^2 = \frac{a}{4} + \frac{a^2}{4}$$

$$\left(x + \frac{a}{2}\right)^2 = \frac{a + a^2}{4}$$

$$x + \frac{a}{2} = \pm\frac{\sqrt{a + a^2}}{4}$$

$$x + \frac{a}{2} = \pm\frac{\sqrt{a + a^2}}{2}$$

$$x = -\frac{a}{2} \pm \frac{\sqrt{a + a^2}}{2}$$

$$\left\{-\frac{a}{2} \pm \frac{\sqrt{a + a^2}}{2}\right\}$$

73. $0.8x^2 - 0.48x + 0.01 = 0$

$x^2 - 0.6x + 0.0125 = 0$

$x^2 - 0.6x = -0.0125$

$x^2 - 0.6x + (0.3)^2 = -0.0125 + (0.3)^2$

$(x - 0.3)^2 = -0.0125 + 0.09$

$(x - 0.3)^2 = 0.0775$

$x - 0.3 = \pm\sqrt{0.0775}$

$x - 0.3 = \pm 0.2784$

$x = 0.3 \pm 0.2784$

$x = 0.3 + 0.2784$ or $x = 0.3 - 0.2784$

$x = 0.5784$ or $x = 0.0216$

$\{0.5784, \ 0.0216\}$

75. $ax^2 + bx + c = 0; \ a \neq 0$

$x^2 + \frac{b}{a}x + \frac{c}{a} = 0$

$x^2 + \frac{b}{a}x = -\frac{c}{a}$

$x^2 + \frac{b}{a}x + \left(\frac{b}{2a}\right)^2 = -\frac{c}{a} + \left(\frac{b}{2a}\right)^2$

$\left(x + \frac{b}{2a}\right)^2 = -\frac{c}{a} + \frac{b^2}{4a^2}$

$\left(x + \frac{b}{2a}\right)^2 = -\frac{4ac}{4a^2} + \frac{b^2}{4a^2}$

$\left(x + \frac{b}{2a}\right)^2 = \frac{-4ac + b^2}{4a^2}$

$\left(x + \frac{b}{2a}\right)^2 = \frac{b^2 - 4ac}{4a^2}$

$x + \frac{b}{2a} = \pm\sqrt{\frac{b^2 - 4ac}{4a^2}}$

$x + \frac{b}{2a} = \pm\frac{\sqrt{b^2 - 4ac}}{2a}$

$x = -\frac{b}{2a} \pm \frac{\sqrt{b^2 - 4ac}}{2a}$

$x = \frac{-b \pm \sqrt{b^2 - 4ac}}{2a}$

$\left\{\frac{-b \pm \sqrt{b^2 - 4ac}}{2a}; \ a \neq 0\right\}$

Problem Set 5.4

1. $x^2 + 2x - 8 = 0$

$a = 1, \ b = 2, \ c = -8$

$x = \frac{-2 \pm \sqrt{2^2 - 4(1)(-8)}}{2(1)}$

$= \frac{-2 \pm \sqrt{4 + 32}}{2}$

$= \frac{-2 \pm \sqrt{36}}{2}$

$= \frac{-2 \pm 6}{2}$

$= -\frac{2}{2} \pm \frac{6}{2}$

$= -1 \pm 3$

$x = -1 + 3$ or $x = -1 - 3$

$x = 2$ or $x = -4$

$\{2, -4\}$

3. $x^2 + 2x = 0$

$a = 1, \ b = 2, \ c = 0$

$x = \frac{-2 \pm \sqrt{2^2 - 4(1)(0)}}{2(1)}$

$= \frac{-2 \pm \sqrt{4 - 0}}{2}$

$= \frac{-2 \pm \sqrt{4}}{2}$

$= \frac{-2 \pm 2}{2}$

$= \frac{2}{2} \pm \frac{2}{2}$

193

$$= -1 \pm 1$$

$x = -1 + 1 \quad \text{or} \quad x = -1 - 1$

$x = 0 \qquad \text{or} \quad x = -2$

$$\{0, \ -2\}$$

5. $x^2 + x - 1 = 0$

   $a = 1, \ b = 1, \ c = -1$

   $$x = \frac{-1 \pm \sqrt{1^2 - 4(1)(-1)}}{2(1)}$$

   $$= \frac{-1 \pm \sqrt{1 + 4}}{2}$$

   $$= \frac{-1 \pm \sqrt{5}}{2}$$

   $$= -\frac{1}{2} \pm \frac{\sqrt{5}}{2}$$

   $$\left\{ -\frac{1}{2} \pm \frac{\sqrt{5}}{2} \right\}$$

7. $x^2 - x + 1 = 0$

   $a = 1, \ b = -1, \ c = 1$

   $$x = \frac{-(-1) \pm \sqrt{(-1)^2 - 4(1)(1)}}{2(1)}$$

   $$= \frac{1 \pm \sqrt{1 - 4}}{2}$$

   $$= \frac{1 \pm \sqrt{-3}}{2}$$

   $$= \frac{1 \pm \sqrt{3}i}{2}$$

   $$= \frac{1}{2} \pm \frac{\sqrt{3}}{2}i$$

   $$\left\{ \frac{1}{2} \pm \frac{\sqrt{3}}{2}i \right\}$$

9. $x^2 + 1 = 0$

   $a = 1, \ b = 0, \ c = 1$

   $$x = \frac{-0 \pm \sqrt{0^2 - 4(1)(1)}}{2(1)}$$

   $$= \frac{0 \pm \sqrt{0 - 4}}{2}$$

   $$= \frac{\pm \sqrt{-4}}{2}$$

   $$= \frac{\pm 2i}{2}$$

   $$= \pm i$$

   $$\{\pm i\}$$

11. $x^2 + 5x + 9 = 0$

   $a = 1, \ b = 5, \ c = 9$

   $$x = \frac{-5 \pm \sqrt{5^2 - 4(1)(9)}}{2(1)}$$

   $$= \frac{-5 \pm \sqrt{25 - 36}}{2}$$

   $$= \frac{-5 \pm \sqrt{-11}}{2}$$

   $$= \frac{-5 \pm \sqrt{11}i}{2}$$

   $$= -\frac{5}{2} \pm \frac{\sqrt{11}}{2}i$$

   $$\left\{ -\frac{5}{2} \pm \frac{\sqrt{11}}{2}i \right\}$$

13.
$$x^2 - 2x = 1$$
$$x^2 - 2x - 1 = 0$$

$$a = 1, \quad b = -2, \quad c = -1$$

$$x = \frac{-(-2) \pm \sqrt{(-2)^2 - 4(1)(-1)}}{2(1)}$$

$$= \frac{2 \pm \sqrt{4 + 4}}{2}$$

$$= \frac{2 \pm \sqrt{8}}{2}$$

$$= \frac{2 \pm 2\sqrt{2}}{2}$$

$$= \frac{2}{2} \pm \frac{2\sqrt{2}}{2}$$

$$= 1 \pm \sqrt{2}$$

$$\{1 \pm \sqrt{2}\}$$

15. $x^2 + 4x - 2 = 0$

$$a = 1, \quad b = 4, \quad c = -2$$

$$x = \frac{-4 \pm \sqrt{4^2 - 4(1)(-2)}}{2(1)}$$

$$= \frac{-4 \pm \sqrt{16 + 8}}{2}$$

$$= \frac{-4 \pm \sqrt{24}}{2}$$

$$= \frac{-4 \pm 2\sqrt{6}}{2}$$

$$= -\frac{4}{2} \pm \frac{2\sqrt{6}}{2}$$

$$= -2 \pm \sqrt{6}$$

$$\{-2 \pm \sqrt{6}\}$$

17. $x^2 + 2x + 2 = 0$

$$a = 1, \quad b = 2, \quad c = 2$$

$$x = \frac{-2 \pm \sqrt{2^2 - 4(1)(2)}}{2(1)}$$

$$= \frac{-2 \pm \sqrt{4 - 8}}{2}$$

$$= \frac{-2 \pm \sqrt{-4}}{2}$$

$$= \frac{-2 \pm 2i}{2}$$

$$= -\frac{2}{2} \pm \frac{2}{2}i$$

$$= -1 \pm i$$

$$\{-1 \pm i\}$$

19.
$$\frac{1}{2}x - 1 = \frac{1}{4}x^2$$
$$4\left(\frac{1}{2}x - 1\right) = 4\left(\frac{1}{4}x^2\right)$$
$$2x - 4 = x^2$$
$$0 = x^2 - 2x + 4$$

$$a = 1, \quad b = -2, \quad c = 4$$

$$x = \frac{-(-2) \pm \sqrt{(-2)^2 - 4(1)(4)}}{2(1)}$$

$$= \frac{2 \pm \sqrt{4 - 16}}{2}$$

$$= \frac{2 \pm \sqrt{-12}}{2}$$

$$= \frac{2 \pm 2\sqrt{3}i}{2}$$

$$= \frac{2}{2} \pm \frac{2\sqrt{3}}{2}i$$

$$= 1 \pm \sqrt{3}i$$

$$\{1 \pm \sqrt{3}i\}$$

21. $\dfrac{2}{3}x^2 + \dfrac{11}{3}x - 7 = 0$

$3\left(\dfrac{2}{3}x^2 + \dfrac{11}{3}x - 7\right) = 3(0)$

$2x^2 + 11x - 21 = 0$

$a = 2, \quad b = 11, \quad c = -21$

$x = \dfrac{-11 \pm \sqrt{11^2 - 4(2)(-21)}}{2(2)}$

$ = \dfrac{-11 \pm \sqrt{121 + 168}}{4}$

$ = \dfrac{-11 \pm \sqrt{289}}{4}$

$ = \dfrac{-11 \pm 17}{4}$

$x = \dfrac{-11 + 17}{4} \quad \text{or} \quad x = \dfrac{-11 - 17}{4}$

$x = \dfrac{3}{2} \quad \text{or} \quad x = -7$

$$\left\{\dfrac{3}{2}, \ -7\right\}$$

23. $r(r - 6) - (2r - 1) = 0$

$r^2 - 6r - 2r + 1 = 0$

$r^2 - 8r + 1 = 0$

$a = 1, \quad b = -8, \quad c = 1$

$r = \dfrac{-(-8) \pm \sqrt{(-8)^2 - 4(1)(1)}}{2(1)}$

$ = \dfrac{8 \pm \sqrt{64 - 4}}{2}$

$ = \dfrac{8 \pm \sqrt{60}}{2}$

$ = \dfrac{8 \pm 2\sqrt{15}}{2}$

$ = \dfrac{8}{2} \pm \dfrac{2\sqrt{15}}{2}$

$ = 4 \pm \sqrt{15}$

$$\left\{4 \pm \sqrt{15}\right\}$$

25. $(2y + 3)(y - 1) = 3$

$2y^2 + y - 3 = 3$

$2y^2 + y - 6 = 0$

$a = 2, \quad b = 1, \quad c = -6$

$y = \dfrac{-1 \pm \sqrt{1^2 - 4(2)(-6)}}{2(2)}$

$ = \dfrac{-1 \pm \sqrt{1 + 48}}{4}$

$ = \dfrac{-1 \pm \sqrt{49}}{4}$

$ = \dfrac{-1 \pm 7}{4}$

$y = \dfrac{-1 + 7}{4} \quad \text{or} \quad y = \dfrac{-1 - 7}{4}$

$y = \dfrac{3}{2} \quad \text{or} \quad y = -2$

$$\left\{\dfrac{3}{2}, \ -2\right\}$$

27. $x^2 - x + 4 = 0$

$a = 1, \quad b = -1, \quad c = 4$

$b^2 - 4ac = (-1)^2 - 4(1)(4)$

$ = 1 - 16$

$ = -15$

Since the discriminant is negative there are two complex solutions.

29. $2x^2 + 3x = 4$

$2x^2 + 3x - 4 = 0$

$a = 2, \quad b = 3, \quad c = -4$

$b^2 - 4ac = 3^2 - 4(2)(-4)$

$ = 9 + 32$

$ = 41$

Since the discriminant is positive there are two real solutions.

31. $x^2 - 6x + 4 \leq 0$

Boundary numbers:

$x^2 - 6x + 4 = 0$

$$x = \frac{6 \pm \sqrt{36 - 4(1)(4)}}{2}$$

$$= \frac{6 \pm \sqrt{20}}{2}$$

$$= \frac{6 \pm 2\sqrt{5}}{2}$$

$$= 3 \pm \sqrt{5}$$

$$\underline{\quad A \quad}_{\Big|} \underline{\quad B \quad}_{\Big|} \underline{\quad C \quad}$$
$$\quad 3-\sqrt{5} \quad 3+\sqrt{5}$$

A: $x = 0$; $\quad 0^2 - 6(0) + 4 \leq 0$

$\qquad\qquad\qquad\qquad 4 \leq 0 \quad$ F

B: $x = 3$; $\quad 3^2 - 6(3) + 4 \leq 0$

$\qquad\qquad\qquad\quad -5 \leq 0 \quad$ T

C: $x = 6$; $\quad 6^2 - 6(6) + 4 \leq 0$

$\qquad\qquad\qquad\qquad 4 \leq 0 \quad$ F

$$\left[3 - \sqrt{5}, \ 3 + \sqrt{5} \right]$$

33. $x^2 - 6x + 10 < 0$

Boundary numbers:

$x^2 - 6x + 10 = 0$

$$x = \frac{6 \pm \sqrt{36 - 4(1)(10)}}{2}$$

$$= \frac{6 \pm \sqrt{-4}}{2}$$

None

$$\underline{\qquad\qquad A \qquad\qquad}$$

A: $x = 0$; $0^2 - 6(0) + 10 < 0$

$\qquad\qquad\qquad\quad 10 < 0 \quad$ F

$\qquad\qquad \emptyset$

35. $x^2 + 6x + 9 = 0$

$(x + 3)^2 = 0$

$x + 3 = \pm\sqrt{0}$

$x + 3 = 0$

$x = -3$

$\{-3\}$

37. $\quad 2x^2 - x - 1 = 0$

$(2x + 1)(x - 1) = 0$

$2x + 1 = 0 \quad$ or $\quad x - 1 = 0$

$x = -\dfrac{1}{2} \quad$ or $\qquad x = 1$

$$\left\{ -\frac{1}{2}, \ 1 \right\}$$

39. $x^2 - 2(x + 4) = 0$

$x^2 - 2x - 8 = 0$

$(x - 4)(x + 2) = 0$

$x - 4 = 0 \quad$ or $\quad x + 2 = 0$

$x = 4 \quad$ or $\qquad x = -2$

41. $\qquad\qquad 5x^2 = 13x + 6$

$5x^2 - 13x - 6 = 0$

$(5x + 2)(x - 3) \quad = 0$

$5x + 2 = 0 \quad$ or $\quad x - 3 = 0$

$x = -\dfrac{2}{5} \quad$ or $\qquad x = 3$

$$\left\{ -\frac{2}{5}, \ 3 \right\}$$

43.
$$6x^2 + x - 2 = 0$$
$$(3x + 2)(2x - 1) = 0$$
$$3x + 2 = 0 \quad \text{or} \quad 2x - 1 = 0$$
$$x = -\frac{2}{3} \quad \text{or} \quad x = \frac{1}{2}$$
$$\left\{-\frac{2}{3}, \ \frac{1}{2}\right\}$$

45. $x(x - 1) = 0$
$$x = 0 \quad \text{or} \quad x - 1 = 0$$
$$x = 0 \quad \text{or} \quad x = 1$$
$$\{0, \ 1\}$$

47. $x^2 + 25 = 0$
$$x^2 = -25$$
$$x = \pm\sqrt{-25}$$
$$x = \pm\sqrt{25}\,i$$
$$x = \pm 5i$$
$$\{\pm 5i\}$$

49. $x^2 + x - 4 = 0$
$$a = 1, \ b = 1, \ c = -4$$
$$x = \frac{-1 \pm \sqrt{1^2 - 4(1)(-4)}}{2(1)}$$
$$= \frac{-1 \pm \sqrt{1 + 16}}{2}$$
$$= \frac{-1 \pm \sqrt{17}}{2}$$
$$\left\{-\frac{1}{2} \pm \frac{\sqrt{17}}{2}\right\}$$

51. $\left(y - \frac{3}{2}\right)^2 = -\frac{11}{4}$
$$y - \frac{3}{2} = \pm\sqrt{-\frac{11}{4}}$$
$$y - \frac{3}{2} = \pm\sqrt{\frac{11}{4}}\,i$$

$$y - \frac{3}{2} = \pm\frac{\sqrt{11}}{2}\,i$$
$$y = \frac{3}{2} \pm \frac{\sqrt{11}}{2}\,i$$

53. $x^2 - 2x + 2 = 0$
$$a = 1, \ b = -2, \ c = 2$$
$$x = \frac{-(-2) \pm \sqrt{(-2)^2 - 4(1)(2)}}{2(1)}$$
$$= \frac{2 \pm \sqrt{4 - 8}}{2}$$
$$= \frac{2 \pm \sqrt{-4}}{2}$$
$$= \frac{2 \pm \sqrt{4}\,i}{2}$$
$$= \frac{2 \pm 2i}{2}$$
$$= \frac{2}{2} \pm \frac{2}{2}\,i$$
$$= 1 \pm i$$
$$\{1 \pm i\}$$

55. $x^2 - 6x + 10 = 0$
$$a = 1, \ b = -6, \ c = 10$$
$$x = \frac{-(-6) \pm \sqrt{(-6)^2 - 4(1)(10)}}{2(1)}$$
$$= \frac{6 \pm \sqrt{36 - 40}}{2}$$
$$= \frac{6 \pm \sqrt{-4}}{2}$$
$$= \frac{6 \pm \sqrt{4}\,i}{2}$$
$$= \frac{6 \pm 2i}{2}$$
$$= \frac{6}{2} \pm \frac{2}{2}\,i$$
$$= 3 \pm i$$
$$\{3 \pm i\}$$

57. $(2x + 1)(x + 3) = 2$

$2x^2 + 7x + 3 = 2$

$2x^2 + 7x + 1 = 0$

$a = 2, \quad b = 7, \quad c = 1$

$x = \dfrac{-7 \pm \sqrt{7^2 - 4(2)(1)}}{2(2)}$

$= \dfrac{-7 \pm \sqrt{49 - 8}}{4}$

$= \dfrac{-7 \pm \sqrt{41}}{4}$

$\left\{ -\dfrac{7}{4} \pm \dfrac{\sqrt{41}}{4} \right\}$

59. $y^2 + 0.1y - 0.06 = 0$

$(y + 0.3)(y - 0.2) = 0$

$y + 0.3 = 0 \quad$ or $\quad y - 0.2 = 0$

$y = -0.3$ or $\quad y = 0.2$

$\{-0.3, \ 0.2\}$

61. $x^2 + 8x + 7 = 0$

$(x + 7)(x + 1) = 0$

$x + 7 = 0 \quad$ or $\quad x + 1 = 0$

$x = -7 \quad$ or $\quad x = -1$

$\{-7, \ -1\}$

63. $2x^2 - 6 = 0$

$2x^2 = 6$

$x^2 = 3$

$x = \pm\sqrt{3}$

$\{\pm\sqrt{3}\}$

65. $3x^2 - 2x - 1 = 0$

$(3x + 1)(x - 1) = 0$

$3x + 1 = 0 \quad$ or $\quad x - 1 = 0$

$x = -\dfrac{1}{3} \quad$ or $\quad x = 1$

67. $2x^2 - x = 0$

$x(2x - 1) = 0$

$x = 0 \quad$ or $\quad 2x - 1 = 0$

$x = 0 \quad$ or $\quad x = \dfrac{1}{2}$

$\left\{ 0, \ \dfrac{1}{2} \right\}$

69. $z^2 = 4$

$z = \pm\sqrt{4}$

$z = \pm 2$

$\{\pm 2\}$

71. $2x^2 - 2x + 1 = 0$

$a = 2, \quad b = -2, \quad c = 1$

$x = \dfrac{-(-2) \pm \sqrt{(-2)^2 - 4(2)(1)}}{2(2)}$

$= \dfrac{2 \pm \sqrt{4 - 8}}{4}$

$= \dfrac{2 \pm \sqrt{-4}}{4}$

$= \dfrac{2 \pm \sqrt{4}\,i}{4}$

$= \dfrac{2 \pm 2i}{4}$

$= \dfrac{2}{4} \pm \dfrac{2}{4}i$

$= \dfrac{1}{2} \pm \dfrac{1}{2}i$

$\left\{ \dfrac{1}{2} \pm \dfrac{1}{2}i \right\}$

199

73. $9x^2 + 12x + 4 = 0$

$$(3x + 2)^2 = 0$$

$$3x + 2 = \pm\sqrt{0}$$

$$3x + 2 = 0$$

$$x = -\frac{2}{3}$$

$$\left\{-\frac{2}{3}\right\}$$

75. $x^2 - 6x + 6 \le 0$

Boundary numbers:

$x^2 - 6x + 6 = 0$

$$x = \frac{6 \pm \sqrt{36 - 4(1)(6)}}{2}$$

$$= \frac{6 \pm \sqrt{12}}{2}$$

$$= \frac{6 \pm 2\sqrt{3}}{2}$$

$$= 3 \pm \sqrt{3}$$

$$\frac{A \quad | \quad B \quad | \quad C}{3 - \sqrt{3} \quad 3 + \sqrt{3}}$$

A: $x = 0$; $0^2 - 6(0) + 6 \le 0$

$$6 \le 0 \quad F$$

B: $x = 3$; $3^2 - 6(3) + 6 \le 0$

$$-3 \le 0 \quad T$$

C: $x = 5$; $5^2 - 6(5) + 6 \le 0$

$$1 \le 0 \quad F$$

$$\left[3 - \sqrt{3}, \ 3 + \sqrt{3}\right]$$

77. $x^2 - 3x + 3 > 0$

Boundary numbers:

$x^2 - 3x + 3 = 0$

$$x = \frac{3 \pm \sqrt{9 - 4(1)(3)}}{2}$$

$$= \frac{3 \pm \sqrt{-3}}{2}$$

None

$$\underline{\qquad A \qquad}$$

A: $x = 0$; $0^2 - 3(0) + 3 > 0$

$$3 > 0 \quad T$$

$$(-\infty, \ \infty)$$

79. $x^2 + 4x - 2 > 0$

Boundary numbers:

$x^2 + 4x - 2 = 0$

$$x = \frac{-4 \pm \sqrt{16 - 4(1)(-2)}}{2}$$

$$= \frac{-4 \pm \sqrt{24}}{2}$$

$$= \frac{-4 \pm 2\sqrt{6}}{2}$$

$$= -2 \pm \sqrt{6}$$

$$\frac{A \quad | \quad B \quad | \quad C}{-2 - \sqrt{6} \quad -2 + \sqrt{6}}$$

A: $x = -6$; $(-6)^2 + 4(-6) - 2 > 0$

$$10 > 0 \quad T$$

B: $x = -2$; $(-2)^2 + 4(-2) - 2 > 0$

$$-6 > 0 \quad F$$

C: $x = 2$; $2^2 + 4(2) - 2 > 0$

$$10 > 0 \quad T$$

$$\left(-\infty, \ -2 - \sqrt{6}\right) \cup \left(-2 + \sqrt{6}, \ \infty\right)$$

81. $x^2 + 3x + 4 < 0$

Boundary numbers:

$x^2 + 3x + 4 = 0$

$$x = \frac{-3 \pm \sqrt{9 - 4(1)(4)}}{2}$$

$$= \frac{-3 \pm \sqrt{-7}}{2}$$

None

A

A: $x = 0$; $\quad 0^2 + 3(0) + 4 < 0$

$\qquad\qquad\qquad 4 < 0 \quad$ F

$\emptyset$

83. $ax^2 + bx = 0$

$\quad b^2 - 4ac = b^2 - 4(a)(0)$

$\qquad\qquad\quad = b^2$

Since b^2 is positive the solutions are real.

85. $x^2 - 2x + k = 0$

$\qquad b^2 - 4ac < 0$

$\quad (-2)^2 - 4(1)(k) < 0$

$\qquad\quad 4 - 4k < 0$

$\qquad\qquad\quad 1 < k$

$\qquad \{k \mid k > 1\}$

87. $\quad x^2 - 2x + k = 0$

$\qquad b^2 - 4ac = 0$

$\quad (-2)^2 - 4(1)(k) = 0$

$\qquad\quad 4 - 4k = 0$

$\qquad\qquad\quad 4 = 4k$

$\qquad\qquad\quad 1 = k$

$\qquad\qquad \{1\}$

89. $144t - 16t^2 = 300$
$\quad 0 = 16t^2 - 144t + 300$
$\quad 0 = 4t^2 - 36t + 75$

$$t = \frac{-(-36) \pm \sqrt{(-36)^2 - 4(4)(75)}}{2(4)}$$

$$= \frac{36 \pm \sqrt{1296 - 1200}}{8}$$

$$= \frac{36 \pm \sqrt{96}}{8}$$

$$= \frac{36 \pm 4\sqrt{6}}{8}$$

$$= \frac{36}{8} \pm \frac{4\sqrt{6}}{8}$$

$$= \frac{9}{2} \pm \frac{\sqrt{6}}{2}$$

$t = \dfrac{9 + \sqrt{6}}{2} \approx 5.7 \quad$ or $\quad t = \dfrac{9 - \sqrt{6}}{2} \approx 3.3$

It will first reach a height of 300 ft at approximately 3 sec.

91.
$$120 = 73I - 5I^2$$

$$5I^2 - 73I + 120 = 0$$

$$I = \frac{-(-73) \pm \sqrt{(-73)^2 - 4(5)(120)}}{2(5)}$$

$$= \frac{73 \pm \sqrt{5329 - 2400}}{10}$$

$$= \frac{73 \pm \sqrt{2929}}{10}$$

$$I = \frac{73 + \sqrt{2929}}{10} \quad \text{or} \quad I = \frac{73 - \sqrt{2929}}{10}$$

$$I = 12.71 \qquad \text{or} \quad 1.89$$

The current should be 1.89 amps or 12.71 amps.

93. $ax^2 + 3x - 2 = 0$

$a = a, \quad b = 3, \quad c = -2$

$$x = \frac{-3 \pm \sqrt{3^2 - 4(a)(-2)}}{2(a)}$$

$$= \frac{-3 \pm \sqrt{9 + 8a}}{2a}$$

$$\left\{ \frac{-3 \pm \sqrt{9 + 8a}}{2a} \right\}$$

95. $x^2 + (n - m)x - mn = 0; \quad (m > 0, \; n > 0)$

$a = 1, \quad b = n - m, \quad c = -mn$

$$x = \frac{-(n - m) \pm \sqrt{(n - m)^2 - 4(1)(-mn)}}{2(1)}$$

$$= \frac{m - n \pm \sqrt{n^2 - 2mn + m^2 + 4mn}}{2}$$

$$= \frac{m - n \pm \sqrt{n^2 + 2mn + m^2}}{2}$$

$$= \frac{m - n \pm \sqrt{(n + m)^2}}{2}$$

$$= \frac{m - n \pm (n + m)}{2}$$

$$x = \frac{m - n + (n + m)}{2} \quad \text{or} \quad x = \frac{m - n - (n + m)}{2}$$

$$x = \frac{2m}{2} \qquad \text{or} \quad x = \frac{m - n - n - m}{2}$$

$$x = m \qquad \text{or} \quad x = \frac{-2n}{2}$$

$$x = m \qquad \text{or} \quad x = -n$$

$$\{m, \; -n\}$$

97. $d = \frac{D}{L}(L^2 - a^2)$

$$d = DL - \frac{Da^2}{L}$$

$$dL = DL^2 - Da^2$$

$$0 = DL^2 - dL - Da^2$$

$a = D, \quad b = -d, \quad c = -Da^2$

$$L = \frac{-(-d) \pm \sqrt{(-d)^2 - 4(D)(-Da^2)}}{2(D)}$$

$$= \frac{d \pm \sqrt{d^2 + 4D^2a^2}}{2D}$$

$$\left\{ \frac{d \pm \sqrt{d^2 + 4D^2a^2}}{2D} \right\}$$

99. $L = \dfrac{2}{R^2} - d^2; \quad R > 0$

$d^2 = \dfrac{2}{R^2} - L$

$d^2 = \dfrac{2 - LR^2}{R^2}$

$d = \pm\sqrt{\dfrac{2 - LR^2}{R^2}}$

$d = \pm\dfrac{\sqrt{2 - LR^2}}{R}$

101. $R_1 = \dfrac{-b + \sqrt{b^2 - 4ac}}{2a}$

$R_2 = \dfrac{-b - \sqrt{b^2 - 4ac}}{2a}$

$R_1 + R_2 = \dfrac{-b + \sqrt{b^2 - 4ac}}{2a} + \dfrac{-b - \sqrt{b^2 - 4ac}}{2a}$

$= \dfrac{-b + \sqrt{b^2 - 4ac} - b - \sqrt{b^2 - 4ac}}{2a}$

$= \dfrac{-2b}{2a}$

$= -\dfrac{b}{a}$

103. $2x^2 + 3x + 4 = 0$

$a = 2, \quad b = 3, \quad c = 4$

$R_1 + R_2 = -\dfrac{b}{a} = -\dfrac{3}{2}$

$R_1 R_2 = \dfrac{c}{a} = \dfrac{4}{2} = 2$

105. $2x - 17 = 5x^2$

$0 = 5x^2 - 2x + 17$
$a = 5, \quad b = -2, \quad c = 17$

$R_1 + R_2 = -\dfrac{b}{a} = -\dfrac{-2}{5} = \dfrac{2}{5}$

$R_1 R_2 = \dfrac{c}{a} = \dfrac{17}{5}$

Problem Set 5.5

1. $(x - 5)(x + 4)(x + 3) = 0$

$x - 5 = 0 \quad$ or $\quad x + 4 = 0 \quad$ or $\quad x + 3 = 0$

$x = 5 \quad$ or $\qquad x = -4$ or $\qquad x = -3$

$\{5, -4, -3\}$

3. $x\left(x + \dfrac{3}{2}\right)(x + 17) = 0$

$x = 0 \quad$ or $\quad x + \dfrac{3}{2} = 0 \quad$ or $\quad x + 17 = 0$

$x = 0 \quad$ or $\qquad x = -\dfrac{3}{2}$ or $\qquad x = -17$

$\left\{0, -\dfrac{3}{2}, -17\right\}$

5. $x^4 - 10x^2 + 9 = 0$

$(x^2 - 9)(x^2 - 1) = 0$

$x^2 - 9 = 0 \quad$ or $\quad x^2 - 1 = 0$

$x^2 = 9 \quad$ or $\qquad x^2 = 1$

$x = \pm 3$ or $\qquad x = \pm 1$

$\{\pm 3, \pm 1\}$

7. $x^4 - 3x^2 - 4 = 0$

$(x^2 - 4)(x^2 + 1) = 0$

$x^2 - 4 = 0 \quad$ or $\quad x^2 + 1 = 0$

$x^2 = 4 \quad$ or $\qquad x^2 = -1$

$x = \pm 2$ or $\qquad x = \pm i$

$\{\pm 2, \pm i\}$

9.
$$x^3 = 27$$
$$x^3 - 27 = 0$$
$$(x - 3)(x^2 + 3x + 9) = 0$$
$$x - 3 = 0 \quad \text{or} \quad x^2 + 3x + 9 = 0$$
$$x = 3 \quad \text{or} \quad x = \frac{-3 \pm \sqrt{3^2 - 4(1)(9)}}{2(1)}$$
$$= \frac{-3 \pm \sqrt{-27}}{2}$$
$$= \frac{-3 \pm 3\sqrt{3}\,i}{2}$$
$$= -\frac{3}{2} \pm \frac{3\sqrt{3}}{2}\,i$$
$$\left\{ 3, \ -\frac{3}{2} \pm \frac{3\sqrt{3}}{2}\,i \right\}$$

11. $x(x + 2)(x - 1) > 0$

Boundary numbers:
$$x(x + 2)(x - 1) = 0$$
$$x = 0, \ x = -2, \ x = 1$$

$$\frac{A}{\underset{-2}{}} \Big| \frac{B}{\underset{0}{}} \Big| \frac{C}{\underset{1}{}} \Big| \frac{D}{}$$

A: $x = -3$; $\quad -3(-3 + 2)(-3 - 1) > 0$
$$-12 > 0 \quad F$$

B: $x = -1$; $\quad -1(-1 + 2)(-1 - 1) > 0$
$$2 > 0 \quad T$$

C: $x = \frac{1}{2}$; $\quad \frac{1}{2}\left(\frac{1}{2} + 2\right)\left(\frac{1}{2} - 1\right) > 0$
$$-\frac{5}{8} > 0 \quad F$$

D: $x = 2$; $\quad 2(2 + 2)(2 - 1) > 0$
$$8 > 0 \quad T$$
$$\{x \mid -2 < x < 0 \ \text{or} \ x > 1\}$$

13. $(x + 4)(x - 1)(x + 3) > 0$

Boundary numbers:
$$(x + 4)(x - 1)(x + 3) = 0$$
$$x = -4, \ x = 1, \ x = -3$$

$$\frac{A}{\underset{-4}{}} \Big| \frac{B}{\underset{-3}{}} \Big| \frac{C}{\underset{1}{}} \Big| \frac{D}{}$$

A: $x = -5$; $\quad (-5 + 4)(-5 - 1)(-5 + 3) > 0$
$$-12 > 0$$

B: $x = -\frac{7}{2}$; $\quad \left(-\frac{7}{2} + 4\right)\left(-\frac{7}{2} - 1\right)\left(-\frac{7}{2} + 3\right) > 0$
$$\frac{9}{8} > 0$$

C: $x = 0$; $\quad (0 + 4)(0 - 1)(0 + 3) > 0$
$$-12 > 0$$

D: $x = 2$; $\quad (2 + 4)(2 - 1)(2 + 3) > 0$
$$30 > 0 \quad T$$
$$\{x \mid -4 < x < -3 \ \text{or} \ x > 1\}$$

15.
$$x^4 + x^2 - 6 = 0$$
$$(x^2 + 3)(x^2 - 2) = 0$$
$$x^2 + 3 = 0 \quad \text{or} \quad x^2 - 2 = 0$$
$$x^2 = -3 \quad \text{or} \quad x^2 = 2$$
$$x = \pm\sqrt{3}\,i \quad \text{or} \quad x = \pm\sqrt{2}$$
$$\left\{ \pm\sqrt{3}\,i, \ \pm\sqrt{2} \right\}$$

17. $x^4 - 13x^2 + 12 = 0$

$(x^2 - 12)(x^2 - 1) = 0$

$x^2 - 12 = 0 \quad$ or $\quad x^2 - 1 = 0$

$x^2 = 12 \quad$ or $\quad x^2 = 1$

$x = \pm 2\sqrt{3} \quad$ or $\quad x = \pm 1$

$\{\pm 2\sqrt{3}, \ \pm 1\}$

19. $x^4 + 2x^2 = 0$

$x^2(x^2 + 2) = 0$

$x^2 = 0 \quad$ or $\quad x^2 + 2 = 0$

$x = \pm 0 \quad$ or $\quad x^2 = -2$

$x = 0 \quad$ or $\quad x = \pm\sqrt{2}i$

$\{0, \ \pm\sqrt{2}i\}$

21. $x^4 - 4 = 0$

$(x^2 + 2)(x^2 - 2) = 0$

$x^2 + 2 = 0 \quad$ or $\quad x^2 - 2 = 0$

$x^2 = -2 \quad$ or $\quad x^2 = 2$

$x = \pm\sqrt{2}i \quad$ or $\quad x = \pm\sqrt{2}$

$\{\pm\sqrt{2}i, \ \pm\sqrt{2}\}$

23. $4x^4 + 5x^2 - 6 = 0$

$(4x^2 - 3)(x^2 + 2) = 0$

$4x^2 - 3 = 0 \quad$ or $\quad x^2 + 2 = 0$

$4x^2 = 3 \quad$ or $\quad x^2 = -2$

$x^2 = \dfrac{3}{4} \quad$ or $\quad x = \pm\sqrt{2}i$

$x = \pm\dfrac{\sqrt{3}}{2}$

$\left\{\pm\dfrac{\sqrt{3}}{2}, \ \pm\sqrt{2}i\right\}$

25. $16x^4 - 8x^2 - 15 = 0$

$(4x^2 - 5)(4x^2 + 3) = 0$

$4x^2 - 5 = 0 \quad$ or $\quad 4x^2 + 3 = 0$

$4x^2 = 5 \quad$ or $\quad 4x^2 = -3$

$x^2 = \dfrac{5}{4} \quad$ or $\quad x^2 = -\dfrac{3}{4}$

$x = \pm\dfrac{\sqrt{5}}{2} \quad$ or $\quad x = \pm\dfrac{\sqrt{3}}{2}i$

$\left\{\pm\dfrac{\sqrt{5}}{2}, \ \pm\dfrac{\sqrt{3}}{2}i\right\}$

27. $\dfrac{1}{4}w^4 - 1 = 0$

$w^4 - 4 = 0$

$(w^2 + 2)(w^2 - 2) = 0$

$w^2 + 2 = 0 \quad$ or $\quad w^2 - 2 = 0$

$w^2 = -2 \quad$ or $\quad w^2 = 2$

$w = \pm\sqrt{2}i \quad$ or $\quad w = \pm\sqrt{2}$

$\{\pm\sqrt{2}i, \ \pm\sqrt{2}\}$

29. $x^3 + 1 = 0$

$(x + 1)(x^2 - x + 1) = 0$

$x + 1 = 0 \quad$ or $\quad x^2 - x + 1 = 0$

$x = -1 \quad$ or $\quad x = \dfrac{-(-1) \pm \sqrt{(-1)^2 - 4(1)(1)}}{2(1)}$

$= \dfrac{1 \pm \sqrt{1 - 4}}{2}$

$= \dfrac{1 \pm \sqrt{3}\,i}{2}$

$= \dfrac{1}{2} \pm \dfrac{\sqrt{3}}{2}\,i$

$\left\{-1, \ \dfrac{1}{2} \pm \dfrac{\sqrt{3}}{2}\,i\right\}$

$(r + 3)(r^2 - 3r + 9) = 0$

$r = -3 \quad$ or $\quad r = \dfrac{-(-3) \pm \sqrt{(-3)^2 - 4(1)(9)}}{2(1)}$

$= \dfrac{3 \pm \sqrt{-27}}{2}$

$= \dfrac{3 \pm 3\sqrt{3}\,i}{2}$

$= \dfrac{3}{2} \pm \dfrac{3\sqrt{3}}{2}\,i$

$\left\{0, -3, \ \dfrac{3}{2} \pm \dfrac{3\sqrt{3}}{2}\,i\right\}$

31. $\quad x^3 - 125 = 0$

$(x - \ )(x^2 + 5x + 25) = 0$

$x - \ = 0 \quad$ or $\quad x^2 + 5x + 25 = 0$

$x = 5 \quad$ or $\quad x = \dfrac{-5 \pm \sqrt{5^2 - 4(1)(25)}}{2(1)}$

$= \dfrac{-5 \pm \sqrt{-75}}{2}$

$= \dfrac{-5 \pm 5\sqrt{3}\,i}{2}$

$= -\dfrac{5}{2} \pm 5\dfrac{\sqrt{3}}{2}\,i$

$\left\{5, \ -\dfrac{5}{2} \pm 5\dfrac{\sqrt{3}}{2}\,i\right\}$

33. $\quad r^4 + 27r = 0$

$r(r^3 + 27) = 0$

$r = 0 \quad$ or $\quad r^3 + 27 = 0$

35. $\quad x^4 - 37x^2 + 36 = 0$

$(x^2 - 36)(x^2 - 1) = 0$

$x^2 - 36 = 0 \quad$ or $\quad x^2 - 1 = 0$

$x^2 = 36 \quad$ or $\quad x^2 = 1$

$x = \pm 6 \quad$ or $\quad x = \pm 1$

$\{\pm 6, \ \pm 1\}$

37. $\quad 2x^4 - 6x^2 = 0$

$2x^2(x^2 - 3) = 0$

$2x^2 = 0 \quad$ or $\quad x^2 - 3 = 0$

$x^2 = 0 \quad$ or $\quad x^2 = 3$

$x = 0 \quad$ or $\quad x = \pm\sqrt{3}$

$\{0, \ \pm\sqrt{3}\}$

39. $\quad x^4 - x^2 - 42 = 0$

$(x^2 - 7)(x^2 + 6) = 0$

$x^2 - 7 = 0 \quad$ or $\quad x^2 + 6 = 0$

$x^2 = 7 \quad$ or $\quad x^2 = -6$

$x = \pm\sqrt{7} \quad$ or $\quad x = \pm\sqrt{6}\,i$

$\{\pm\sqrt{7}, \ \pm\sqrt{6}\,i\}$

41. $x^4 - x^2 - 12 = 0$

$(x^2 - 4)(x^2 + 3) = 0$

$x^2 - 4 = 0 \quad$ or $\quad x^2 + 3 = 0$

$\qquad x^2 = 4 \qquad$ or $\qquad x^2 = -3$

$\qquad x = \pm 2 \qquad$ or $\qquad x = \pm\sqrt{3}\,i$

$$\left\{\pm 2, \ \pm\sqrt{3}\,i\right\}$$

43. $-x^4 + x^2 + 6 = 0$

$x^4 - x^2 - 6 = 0$

$(x^2 - 3)(x^2 + 2) = 0$

$x^2 - 3 = 0 \quad$ or $\quad x^2 + 2 = 0$

$\qquad x^2 = 3 \qquad$ or $\qquad x^2 = -2$

$\qquad x = \pm\sqrt{3} \qquad$ or $\qquad x = \pm\sqrt{2}\,i$

$$\left\{\pm\sqrt{3}, \ \pm\sqrt{2}\,i\right\}$$

45. $4x^4 - x^2 - 3 = 0$

$(4x^2 + 3)(x^2 - 1) = 0$

$4x^2 + 3 = 0 \qquad$ or $\quad x^2 - 1 = 0$

$\qquad x^2 = -\dfrac{3}{4} \qquad$ or $\qquad x^2 = 1$

$\qquad x = \pm\dfrac{\sqrt{3}}{2}\,i \quad$ or $\qquad x = \pm 1$

$$\left\{\pm\dfrac{\sqrt{3}}{2}\,i, \ \pm 1\right\}$$

47. $9x^4 + 11x^2 - 14 = 0$

$(9x^2 - 7)(x^2 + 2) = 0$

$9x^2 - 7 = 0 \qquad$ or $\quad x^2 + 2 = 0$

$\qquad x^2 = \dfrac{7}{9} \qquad$ or $\qquad x^2 = -2$

$\qquad x = \pm\dfrac{\sqrt{7}}{3} \quad$ or $\qquad x = \pm\sqrt{2}\,i$

$$\left\{\pm\dfrac{\sqrt{7}}{3}, \ \pm\sqrt{2}\,i\right\}$$

49. $\qquad x^4 = 16$

$x^4 - 16 = 0$

$(x^2 + 4)(x^2 - 4) = 0$

$x^2 + 4 = 0 \qquad$ or $\quad x^2 - 4 = 0$

$\qquad x^2 = -4 \quad$ or $\qquad x^2 = 4$

$\qquad x = \pm 2i \quad$ or $\qquad x = \pm 2$

$$\left\{\pm 2i, \ \pm 2\right\}$$

51. $\qquad 2x^3 + 54 = 0$

$x^3 + 27 = 0$

$(x + 3)(x^2 - 3x + 9) = 0$

207

$x + 3 = 0 \quad$ or $\quad x^2 - 3x + 9 = 0$

$x = -3 \quad$ or $\quad x = \dfrac{-(-3) \pm \sqrt{(-3)^2 - 4(1)(9)}}{2(1)}$

$\qquad\qquad\qquad = \dfrac{3 \pm \sqrt{-27}}{2}$

$\qquad\qquad\qquad = \dfrac{3 \pm 3\sqrt{3}\,i}{2}$

$\qquad\qquad\qquad = \dfrac{3}{2} \pm \dfrac{3\sqrt{3}}{2}\,i$

$\qquad\qquad \left\{ -3, \ \dfrac{3}{2} \pm \dfrac{3\sqrt{3}}{2}\,i \right\}$

$u^2 = 0 \ $ or $\ u - 1 = 0 \ $ or $\ u^2 + u + 1 = 0$

$u = 0 \ $ or $\qquad u = 1 \ $ or $\quad u = \dfrac{-1 \pm \sqrt{1^2 - 4(1)(1)}}{2(1)}$

$\qquad\qquad\qquad\qquad\qquad = \dfrac{-1 \pm \sqrt{-3}}{2}$

$\qquad\qquad\qquad\qquad\qquad = \dfrac{-1 \pm \sqrt{3}\,i}{2}$

$\qquad\qquad\qquad\qquad\qquad = -\dfrac{1}{2} \pm \dfrac{\sqrt{3}}{2}\,i$

$\qquad\qquad \left\{ 0, \ 1, \ -\dfrac{1}{2} \pm \dfrac{\sqrt{3}}{2}\,i \right\}$

53.

$2w^3 + 250 = 0$

$w^3 + 125 = 0$

$(w + 5)(w^2 - 5w + 25) = 0$

$w + 5 = 0 \quad$ or $\quad w^2 - 5w + 25 = 0$

$w = -5 \quad$ or $\quad w = \dfrac{-(-5) \pm \sqrt{(-5)^2 - 4(1)(25)}}{2(1)}$

$\qquad\qquad\qquad = \dfrac{5 \pm \sqrt{-75}}{2}$

$\qquad\qquad\qquad = \dfrac{5 \pm 5\sqrt{3}\,i}{2}$

$\qquad\qquad\qquad = \dfrac{5}{2} \pm \dfrac{5\sqrt{3}}{2}\,i$

$\qquad\qquad \left\{ -5, \ \dfrac{5}{2} \pm \dfrac{5\sqrt{3}}{2}\,i \right\}$

55.

$u^5 = u^2$

$u^5 - u^2 = 0$

$u^2(u^3 - 1) = 0$

$u^2(u - 1)(u^2 + u + 1) = 0$

57.

$\dfrac{1}{5}x^4 - 5 = 0$

$x^4 - 25 = 0$

$(x^2 + 5)(x^2 - 5) = 0$

$x^2 + 5 = 0 \qquad$ or $\quad x^2 - 5 = 0$

$x^2 = -5 \qquad$ or $\qquad x^2 = 5$

$x = \pm\sqrt{5}\,i \quad$ or $\qquad x = \pm\sqrt{5}$

$\qquad \left\{ \pm\sqrt{5}\,i, \ \pm\sqrt{5} \right\}$

59. $(x - 7)(x + 3)(x - 1) \geq 0$

Boundary numbers:

$(x - 7)(x + 3)(x - 1) = 0$

$x = 7, \quad x = -3, \quad x = 1$

$\underset{\qquad -3 \qquad\quad 1 \qquad\quad 7}{\underline{\quad A \quad | \quad B \quad | \quad C \quad | \quad D \quad}}$

A: $x = -4$;　$(-4 - 7)(-4 + 3)(-4 - 1) \geq 0$

$$-55 \geq 0 \quad F$$

B: $x = 0$;　　$(0 - 7)(0 + 3)(0 - 1) \geq 0$

$$21 \geq 0 \quad T$$

C: $x = 2$;　　$(2 - 7)(2 + 3)(2 - 1) \geq 0$

$$-25 \geq 0 \quad F$$

D: $x = 8$;　　$(8 - 7)(8 + 3)(8 - 1) \geq 0$

$$77 \geq 0 \quad T$$

$$[-3, 1] \cup [7, \infty)$$

61.　$x(x + 1)(x - 1) > 0$

Boundary numbers:

$$x(x + 1)(x - 1) = 0$$

$$x = 0, \quad x = -1, \quad x = 1$$

$$\underset{-1}{\overset{A}{\rule{3.5em}{0.4pt}}}|\underset{0}{\overset{B}{\rule{2.5em}{0.4pt}}}|\underset{1}{\overset{C}{\rule{2.5em}{0.4pt}}}\overset{D}{\rule{2.5em}{0.4pt}}$$

A: $x = -2$;　　$-2(-2 + 1)(-2 - 1) > 0$

$$-6 > 0$$

B: $x = -\dfrac{1}{2}$;　　$-\dfrac{1}{2}\left(-\dfrac{1}{2} + 1\right)\left(-\dfrac{1}{2} - 1\right) > 0$

$$\dfrac{3}{8} > 0$$

C: $x = \dfrac{1}{2}$;　　$\dfrac{1}{2}\left(\dfrac{1}{2} + 1\right)\left(\dfrac{1}{2} - 1\right) > 0$

$$-\dfrac{3}{8} > 0$$

D: $x = 2$;　　$2(2 + 1)(2 - 1) > 0$

$$6 > 0 \quad T$$

$$(-1, 0) \cup (1, \infty)$$

63.　$x^3 - 4x \geq 0$

Boundary numbers:

$$x^3 - 4x = 0$$

$$x(x^2 - 4) = 0$$

$$x = 0, \quad x^2 - 4 = 0$$

$$x^2 = 4$$

$$x = \pm 2$$

$$\underset{-2}{\overset{A}{\rule{3em}{0.4pt}}}|\underset{0}{\overset{B}{\rule{2.5em}{0.4pt}}}|\underset{2}{\overset{C}{\rule{2.5em}{0.4pt}}}|\overset{D}{\rule{2.5em}{0.4pt}}$$

A: $x = -3$;　　$(-3)^3 - 4(-3) \geq 0$

$$-15 \geq 0 \quad F$$

B: $x = -1$;　　$(-1)^3 - 4(-1) \geq 0$

$$3 \geq 0 \quad T$$

C: $x = 1$;　　$1^3 - 4(1) \geq 0$

$$-3 \geq 0 \quad F$$

D: $x = 3$;　　$3^3 - 4(3) \geq 0$

$$15 \geq 0 \quad T$$

$$[-2, 0] \cup [2, \infty)$$

65.　width: x

length: x^2

$$x(x^2) = 125$$

$$x^3 = 125$$

$$x^3 - 125 = 0$$

$$(x - 5)(x^2 + 5x + 25) = 0$$

$$x - 5 = 0 \quad \text{or} \quad x^2 + 5x + 25 = 0$$

$$x = 5 \quad \text{or} \quad x = \dfrac{-5 \pm \sqrt{5^2 - 4(1)(25)}}{2(1)}$$

$$= \dfrac{-5 \pm \sqrt{-75}}{2}$$

$$= \dfrac{-5 \pm 5\sqrt{3}\,i}{2}$$

Since x is a real number $x = 5$ and $x^2 = 25$.
The dimensions are 5 ft by 25 ft.

67. height: x
length: $x - 24$
width: $x - 24$

$$x(x - 24)^2 = 27648$$

$$x(x^2 - 48x + 576) = 27648$$

$$x^3 - 48x^2 + 576x = 27648$$

$$x^3 - 48x^2 + 576x - 27648 = 0$$

$$x^2(x - 48) + 576(x - 48) = 0$$

$$(x - 48)(x^2 + 576) = 0$$

$$x - 48 = 0 \quad \text{or} \quad x^2 + 576 = 0$$

$$x = 48 \quad \text{or} \quad x^2 = -576$$

$$x = \pm 24i$$

Since x is a real number, $x = 48$.
The height is 48 ft.

69. side of base: x
height: x

$$\frac{1}{3}(x \cdot x) \cdot x \leq 10$$

$$\frac{1}{3}x^3 \leq 10$$

Boundary numbers:

$$\frac{1}{3}x^3 = 10$$

$$x^3 = 30$$

$$x = \sqrt[3]{30} \approx 3.1$$

$$\underset{3.1}{\underline{A \qquad | \qquad B}}$$

A: $x = 0;\quad \dfrac{1}{3}(0)^3 \leq 10$

$$0 \leq 10 \quad T$$

B: $x = 4;\quad \dfrac{1}{3}(4)^3 \leq 10$

$$\frac{64}{3} \leq 10 \quad F$$

At most 3 ft high with a 3 ft square base.

71. length: x
width: $x - 1$
height: $x - 2$

$$x(x - 1)(x - 2) \geq 6$$

Boundary numbers:

$$x(x - 1)(x - 2) = 6$$

$$x(x^2 - 3x + 2) = 6$$

$$x^3 - 3x^2 + 2x = 6$$

$$x^3 - 3x^2 + 2x - 6 = 0$$

$$x^2(x - 3) + 2(x - 3) = 0$$

$$(x - 3)(x^2 + 2) = 0$$

$$x - 3 = 0 \quad \text{or} \quad x^2 + 2 = 0$$

$$x = 3 \quad \text{or} \quad x^2 = -2$$

$$x = \pm\sqrt{2}\,i$$

$$\underset{3}{\underline{A \qquad | \qquad B}}$$

A: $x = 0;\quad 0(0 - 1)(0 - 2) \geq 6$

$$0 \geq 6 \quad F$$

B: $x = 4;\quad 4(4 - 1)(4 - 2) \geq 6$

$$24 \geq 6 \quad T$$

$$\{x \mid x \geq 3\}$$

The length must be at least 3 ft.

73. $\quad x^6 - 64 = 0$

$\quad\quad (x^3)^2 - 8^2 = 0$

$\quad (x^3 - 8)(x^3 + 8) = 0$

$\quad\quad\quad x^3 - 8 = 0 \quad\quad\text{or}\quad\quad\quad\quad x^3 + 8 = 0$

$(x - 2)(x^2 + 2x + 4) = 0 \quad\quad\text{or}\quad\quad (x + 2)(x^2 - 2x + 4) = 0$

$x - 2 = 0 \quad\text{or}\quad x^2 + 2x + 4 = 0 \quad\quad\text{or}\quad x + 2 = 0 \quad\text{or}\quad x^2 - 2x + 4 = 0$

$x = 2 \quad\text{or}\quad x = \dfrac{-2 \pm \sqrt{2^2 - 4(1)(4)}}{2(1)} \quad\text{or}\quad\quad x = -2 \quad\text{or}\quad x = \dfrac{-(-2) \pm \sqrt{(-2)^2 - 4(1)(4)}}{2(1)}$

$\quad\quad\quad\quad = \dfrac{-2 \pm \sqrt{-12}}{2} \quad\quad\quad\quad\quad\quad\quad\quad\quad = \dfrac{2 \pm \sqrt{-12}}{2}$

$\quad\quad\quad\quad = \dfrac{-2 \pm 2\sqrt{3}\,i}{2} \quad\quad\quad\quad\quad\quad\quad\quad\quad = \dfrac{2 \pm 2\sqrt{3}\,i}{2}$

$\quad\quad\quad\quad = -1 \pm \sqrt{3}\,i \quad\quad\quad\quad\quad\quad\quad\quad\quad = 1 \pm \sqrt{3}\,i$

$\quad\quad\quad\quad \{2, \ \ -1 \pm \sqrt{3}i, \ \ -2, \ \ 1 \pm \sqrt{3}i\}$

75. $\quad x^3 - 2x^2 + 4x - 8 = 0$

$\quad x^2(x - 2) + 4(x - 2) = 0$

$\quad\quad\quad (x - 2)(x^2 + 4) = 0$

$\quad x - 2 = 0 \quad\text{or}\quad x^2 + 4 = 0$

$\quad\quad x = 2 \quad\text{or}\quad\quad x^2 = -4$

$\quad\quad\quad\quad\quad\quad\quad\quad\quad\quad x = \pm 2i$

$\quad\quad\quad\quad \{2, \ \pm 2i\}$

77. $\quad (x + 1)^4 - 5(x + 1)^2 + 4 = 0$

$\quad$ Let $u = x + 1$

$\quad\quad u^4 - 5u^2 + 4 = 0$

$\quad (u^2 - 4)(u^2 - 1) = 0$

$\quad u^2 - 4 = 0 \quad\text{or}\quad u^2 - 1 = 0$

$\quad\quad u^2 = 4 \quad\text{or}\quad\quad u^2 = 1$

$\quad\quad u = \pm 2 \quad\text{or}\quad\quad u = \pm 1$

77. (cont'd)

$x + 1 = \pm 2 \quad\quad\text{or}\quad x + 1 = \pm 1$

$\quad x = -1 \pm 2 \quad\quad\quad\quad x = -1 \pm 1$

$\quad x = -1 + 2 = 1 \quad\quad\quad x = -1 + 1 = 0$

$\quad\quad\quad\text{or} \quad\quad\quad\quad\quad\quad\quad\quad \text{or}$

$\quad x = -1 - 2 = -3 \quad\quad x = -1 - 1 = -2$

$\quad\quad\quad \{1, \ -3, \ 0, \ -2\}$

79. $\quad x = 4 \quad\text{or}\quad\quad x = 2 \quad\text{or}\quad\quad x = -1$

$\quad x - 4 = 0 \quad\text{or}\quad x - 2 = 0 \quad\text{or}\quad x + 1 = 0$

$\quad\quad\quad (x - 4)(x - 2)(x + 1) = 0$

$\quad\quad\quad\quad (x^2 - 6x + 8)(x + 1) = 0$

$\quad x^3 - 6x^2 + 8x + x^2 - 6x + 8 = 0$

$\quad\quad\quad x^3 - 5x^2 + 2x + 8 = 0$

81. $x^3 + 8 > 0$

Boundary numbers:

$$x^3 + 8 = 0$$

$$(x + 2)(x^2 - 2x + 4) = 0$$

$$x + 2 = 0 \quad \text{or} \quad x^2 - 2x + 4 = 0$$

$$x = -2 \qquad x = \frac{-(-2) \pm \sqrt{(-2)^2 - 4(1)(4)}}{2(1)}$$

$$= \frac{2 \pm \sqrt{-12}}{2}$$

$$= \frac{2 \pm 2\sqrt{3}i}{2}$$

$$= 1 \pm \sqrt{3}i$$

$$\underline{\quad A \quad}|\underline{\quad B \quad}$$
$${-2}$$

A: $x = -3;\quad (-3)^3 + 8 > 0$

$$-19 > 0 \quad \text{F}$$

B: $x = 0;\qquad 0^3 + 8 > 0$

$$8 > 0 \quad \text{T}$$

$$\{x \mid x > -2\}$$

83. $x^2(x - 2) \geq 0$

Boundary numbers:

$$x^2(x - 2) = 0$$

$$x = 0, \quad x = 2$$

$$\underline{\quad A \quad}|\underline{\quad B \quad}|\underline{\quad C \quad}$$
$${0}{2}$$

A: $x = -1;\qquad (-1)^2(-1 - 2) \geq 0$

$$-3 \geq 0 \quad \text{F}$$

B: $x = 1;\qquad 1^2(1 - 2) \geq 0$

$$-1 \geq 0 \quad \text{F}$$

C: $x = 3;\qquad 3^2(3 - 2) \geq 0$

$$9 \geq 0 \quad \text{T}$$

$$\{x \mid x \geq 2\}$$

Problem Set 5.6

1.
$$\frac{6}{x - 2} - \frac{3}{x} = 1$$

$$x(x - 2)\left(\frac{6}{x - 2} - \frac{3}{x}\right) = x(x - 2)(1)$$

$$6x - 3(x - 2) = x^2 - 2x$$

$$6x - 3x + 6 = x^2 - 2x$$

$$0 = x^2 - 5x - 6$$

$$0 = (x - 6)(x + 1)$$

$$x = 6, \quad \text{or} \quad x = -1$$

$$\{6, \ -1\}$$

3.
$$\frac{2}{x} - 1 = \frac{4}{x + 3}$$

$$x(x + 3)\left(\frac{2}{x} - 1\right) = x(x + 3)\left(\frac{4}{x + 3}\right)$$

$$2(x + 3) - x(x + 3) = 4x$$

$$2x + 6 - x^2 - 3x = 4x$$

$$-x^2 - 5x + 6 = 0$$

$$x^2 + 5x - 6 = 0$$

$$(x + 6)(x - 1) = 0$$

$$x = -6 \quad \text{or} \quad x = 1$$

$$\{-6, 1\}$$

5.
$$\frac{4}{x+5} - \frac{5}{x-2} = 3$$

$$(x+5)(x-2)\left(\frac{4}{x+5} - \frac{5}{x-2}\right) = (x+5)(x-2)(3)$$

$$4(x-2) - 5(x+5) = 3(x^2 + 3x - 10)$$

$$4x - 8 - 5x - 25 = 3x^2 + 9x - 30$$

$$0 = 3x^2 + 10x + 3$$

$$0 = (3x+1)(x+3)$$

$$x = -\frac{1}{3} \quad \text{or} \quad x = -3$$

$$\left\{-\frac{1}{3}, -3\right\}$$

7.
$$\frac{z}{z-2} + \frac{2}{z+1} = \frac{7z+1}{z^2 - z - 2}$$

$$\frac{z}{z-2} + \frac{2}{z+1} = \frac{7z+1}{(z-2)(z+1)}$$

$$(z-2)(z+1)\left(\frac{z}{z-2} + \frac{2}{z+1}\right) = (z-2)(z+1)\left[\frac{7z+1}{(z-2)(z+1)}\right]$$

$$z(z+1) + 2(z-2) = 7z+1$$

$$z^2 + z + 2z - 4 = 7z+1$$

$$z^2 - 4z - 5 = 0$$

$$(z-5)(z+1) = 0$$

$$z = 5 \quad \text{or} \quad z = -1$$

Since $z = -1$ causes a zero in a denominator, $z = 5$ is the only solution.

$$\{5\}$$

9.
$$\frac{x}{x+1} - \frac{2}{1-x} = \frac{8x-4}{x^2 - 1}$$

$$\frac{x}{x+1} + \frac{2}{x-1} = \frac{8x-4}{(x+1)(x-1)}$$

$$(x+1)(x-1)\left(\frac{x}{x+1} + \frac{2}{x-1}\right) = (x+1)(x-1)\left[\frac{8x-4}{(x+1)(x-1)}\right]$$

213

$$x(x - 1) + 2(x + 1) = 8x - 4$$

$$x^2 - x + 2x + 2 = 8x - 4$$

$$x^2 - 7x + 6 = 0$$

$$(x - 6)(x - 1) = 0$$

$$x = 6 \quad \text{or} \quad x = 1$$

Since $x = 1$ causes a zero in a denominator, $x = 6$ is the only solution.

$$\{6\}$$

11.
$$\frac{2x^2}{x^2 - 4x + 3} + \frac{1}{x - 1} = \frac{9}{x - 3}$$

$$\frac{2x^2}{(x - 1)(x - 3)} + \frac{1}{x - 1} = \frac{9}{x - 3}$$

$$(x - 1)(x - 3)\left[\frac{2x^2}{(x - 1)(x - 3)} + \frac{1}{x - 1}\right] = (x - 1)(x - 3)\left(\frac{9}{x - 3}\right)$$

$$2x^2 + x - 3 = 9x - 9$$

$$2x^2 - 8x + 6 = 0$$

$$x^2 - 4x + 3 = 0$$

$$(x - 3)(x - 1) = 0$$

$$x = 3 \quad \text{or} \quad x = 1$$

Since $x = 3$ and $x = 1$ both cause zero in a denominator, there is no solution.

$$\emptyset$$

13.
$$\frac{2}{x - 3} + \frac{x^2 - 1}{x^2 - 10x + 21} = \frac{12}{x - 7}$$

$$\frac{2}{x - 3} + \frac{x^2 - 1}{(x - 7)(x - 3)} = \frac{12}{x - 7}$$

$$(x - 3)(x - 7)\left[\frac{2}{x - 3} + \frac{x^2 - 1}{(x - 7)(x - 3)}\right] = (x - 3)(x - 7)\left(\frac{12}{x - 7}\right)$$

$$2x - 14 + x^2 - 1 = 12x - 36$$

$$x^2 - 10x + 21 = 0$$

$$(x - 7)(x - 3) = 0$$

$x = 7$ or $x = 3$

Since $x = 7$ and $x = 3$ both cause zero in a denominator, there is no solution.

$$\emptyset$$

15. $\dfrac{x}{12} - \dfrac{1}{x} + \dfrac{1}{12} \geq 0$

Free boundary number: $x = 0$

Other boundary numbers:

$$\dfrac{x}{12} - \dfrac{1}{x} + \dfrac{1}{12} = 0$$

$$12x\left(\dfrac{x}{12} - \dfrac{1}{x} + \dfrac{1}{12}\right) = 12x(0)$$

$$x^2 - 12 + x = 0$$

$$x^2 + x - 12 = 0$$

$$(x + 4)(x - 3) = 0$$

$$x = -4, \quad x = 3$$

$$\underline{\quad A \quad}\Big|\underline{\quad B \quad}\Big|\underline{\quad C \quad}\Big|\underline{\quad D \quad}$$
$$\qquad -4 \qquad 0 \qquad 3$$

A: $x = -5$; $\dfrac{-5}{12} - \dfrac{1}{-5} + \dfrac{1}{12} \geq 0$

$$-\dfrac{2}{15} \geq 0 \quad F$$

B: $x = -1$; $\dfrac{-1}{12} - \dfrac{1}{-1} + \dfrac{1}{12} \geq 0$

$$1 \geq 0 \quad T$$

C: $x = 1$; $\dfrac{1}{12} - \dfrac{1}{1} + \dfrac{1}{12} \geq 0$

$$-\dfrac{5}{6} \geq 0 \quad F$$

D: $x = 4$; $\dfrac{4}{12} - \dfrac{1}{4} + \dfrac{1}{12} \geq 0$

$$\dfrac{1}{6} \geq 0 \quad T$$

$[-4, 0) \cup [3, \infty)$

17. $x^{-2} + 2x^{-1} - 24 = 0$

$$\dfrac{1}{x^2} + \dfrac{2}{x} - 24 = 0$$

$$x^2\left(\dfrac{1}{x^2} + \dfrac{2}{x} - 24\right) = x^2(0)$$

$$1 + 2x - 24x^2 = 0$$

$$24x^2 - 2x - 1 = 0$$

$$(6x + 1)(4x - 1) = 0$$

$$x = -\dfrac{1}{6} \quad \text{or} \quad x = \dfrac{1}{4}$$

$$\left\{-\dfrac{1}{6}, \ \dfrac{1}{4}\right\}$$

215

19. $t^{-4} - 2t^{-2} + 1 = 0$

$$\frac{1}{t^4} - \frac{2}{t^2} + 1 = 0$$

$$t^4\left(\frac{1}{t^4} - \frac{2}{t^2} + 1\right) = t^4(0)$$

$$1 - 2t^2 + t^4 = 0$$

$$t^4 - 2t^2 + 1 = 0$$

$$(t^2 - 1)^2 = 0$$

$$t^2 - 1 = 0$$

$$t^2 = 1$$

$$t = \pm 1$$

$$\{\pm 1\}$$

21.

	D	R	T = D/R
actual	24	r	24/r
possible	24	r + 2	24/(r + 2)

$$\frac{24}{r + 2} = \frac{24}{r} - 1$$

$$r(r + 2)\left(\frac{24}{r + 2}\right) = r(r + 2)\left(\frac{24}{r} - 1\right)$$

$$24r = 24r + 48 - r^2 - 2r$$

$$r^2 + 2r - 48 = 0$$

$$(r + 8)(r - 6) = 0$$

$$r = -8 \quad \text{or} \quad r = 6$$

Since $r > 0$, $r = 6$

Its speed was 6 knots.

23.

	D	R	T = D/R
riding	22	r + 4	22/(r + 4)
walking	1	r	1/r

$$\frac{22}{r + 4} + \frac{1}{r} = 3$$

$$r(r + 4)\left(\frac{22}{r + 4} + \frac{1}{r}\right) = r(r + 4)(3)$$

$$22r + r + 4 = 3r^2 + 12r$$

$$0 = 3r^2 - 11r - 4$$

$$0 = (3r + 1)(r - 4)$$

$$r = -\frac{1}{3} \quad \text{or} \quad r = 4$$

Since $r > 0$, $r = 4$
His walking speed was 4 mph.

25.

	time for whole job	part done in 1 hr
smaller	x + 6	1/(x + 6)
larger	x	1/x
together	4	1/4

$$\frac{1}{x + 6} + \frac{1}{x} = \frac{1}{4}$$

$$4x(x + 6)\left(\frac{1}{x + 6} + \frac{1}{x}\right) = 4x(x + 6)\left(\frac{1}{4}\right)$$

$$4x + 4x + 24 = x^2 + 6x$$

$$0 = x^2 - 2x - 24$$

$$0 = (x - 6)(x + 4)$$

$$x = 6 \quad \text{or} \quad x = -4$$

Since $x > 0$, $x = 6$

$$x + 6 = 12$$

It would take the smaller pipe 12 hours and the larger pipe 6 hours.

27.
$$\frac{2x}{5} - \frac{2}{x} = \frac{1}{5}$$

$$5x\left(\frac{2x}{5} - \frac{2}{x}\right) = 5x\left(\frac{1}{5}\right)$$

$$2x^2 - 10 = x$$

$$2x^2 - x - 10 = 0$$

$$(2x - 5)(x + 2) = 0$$

$$x = \frac{5}{2} \quad \text{or} \quad x = -2$$

$$\left\{\frac{5}{2},\ -2\right\}$$

29.
$$\frac{-4}{3x - 1} = 2x + 3$$

$$(3x - 1)\left(\frac{-4}{3x - 1}\right) = (3x - 1)(2x + 3)$$

$$-4 = 6x^2 + 7x - 3$$

$$0 = 6x^2 + 7x + 1$$

$$0 = (6x + 1)(x + 1)$$

$$x = -\frac{1}{6} \quad \text{or} \quad x = -1$$

$$\left\{-\frac{1}{6},\ -1\right\}$$

31. $\dfrac{1}{x + 1} - \dfrac{10}{x - 2} = \dfrac{3}{2}$

$$2(x + 1)(x - 2)\left(\frac{1}{x + 1} - \frac{10}{x - 2}\right) = 2(x + 1)(x - 2)\left(\frac{3}{2}\right)$$

$$2(x - 2) - 20(x + 1) = 3(x + 1)(x - 2)$$

$$2x - 4 - 20x - 20 = 3x^2 - 3x - 6$$

$$0 = 3x^2 + 15x + 18$$

$$0 = x^2 + 5x + 6$$

$$0 = (x + 2)(x + 3)$$

$$x = -2 \quad \text{or} \quad x = -3$$

$$\{-2,\ -3\}$$

33.
$$\frac{2}{x^2 - 1} + \frac{3x + 6}{x^2 - x - 2} = \frac{8}{x^2 - 3x + 2}$$

$$\frac{2}{(x + 1)(x - 1)} + \frac{3x + 6}{(x - 2)(x + 1)} = \frac{8}{(x - 2)(x - 1)}$$

$$(x + 1)(x - 1)(x - 2)\left[\frac{2}{(x + 1)(x - 1)} + \frac{3x + 6}{(x - 2)(x + 1)}\right] = (x + 1)(x - 1)(x - 2)\left[\frac{8}{(x - 2)(x - 1)}\right]$$

$$2(x - 2) + (3x + 6)(x - 1) = 8(x + 1)$$

$$2x - 4 + 3x^2 + 3x - 6 = 8x + 8$$

$$3x^2 - 3x - 18 = 0$$

$$x^2 - x - 6 = 0$$

$$(x - 3)(x + 2) = 0$$

$$x = 3 \quad \text{or} \quad x = -2$$

$$\{3, \ -2\}$$

35.
$$\frac{5}{x + 2} - \frac{3}{x - 5} = \frac{x^2 - 8x - 6}{x^2 - 3x - 10}$$

$$\frac{5}{x + 2} - \frac{3}{x - 5} = \frac{x^2 - 8x - 6}{(x - 5)(x + 2)}$$

$$(x + 2)(x - 5)\left(\frac{5}{x + 2} - \frac{3}{x - 5}\right) = (x + 2)(x - 5)\left[\frac{x^2 - 8x - 6}{(x - 5)(x + 2)}\right]$$

$$5(x - 5) - 3(x + 2) = x^2 - 8x - 6$$

$$5x - 25 - 3x - 6 = x^2 - 8x - 6$$

$$0 = x^2 - 10x + 25$$

$$0 = (x - 5)^2$$

$$x = 5$$

Since $x = 5$ causes zero in a denominator, there is no solution.

$$\emptyset$$

37. $\dfrac{1}{x} + \dfrac{5}{3x} = \dfrac{x + 2}{3}$

$$3x\left(\dfrac{1}{x} + \dfrac{5}{3x}\right) = 3x\left(\dfrac{x + 2}{3}\right)$$

$$3 + 5 = x^2 + 2x$$

$$0 = x^2 + 2x - 8$$

$$0 = (x + 4)(x - 2)$$

$$x = -4 \quad \text{or} \quad x = 2$$

$$\{-4, \ 2\}$$

39.
$$\dfrac{2}{x + 7} + \dfrac{16}{x^2 + 6x - 7} + \dfrac{x}{x - 1} = 0$$

$$\dfrac{2}{x + 7} + \dfrac{16}{(x + 7)(x - 1)} + \dfrac{x}{x - 1} = 0$$

$$(x + 7)(x - 1)\left[\dfrac{2}{x + 7} + \dfrac{16}{(x + 7)(x - 1)} + \dfrac{x}{x - 1}\right] = (x + 7)(x - 1)(0)$$

$$2(x - 1) + 16 + x(x + 7) = 0$$

$$2x - 2 + 16 + x^2 + 7x = 0$$

$$x^2 + 9x + 14 = 0$$

$$(x + 7)(x + 2) = 0$$

$$x = -7 \quad \text{or} \quad x = -2$$

Since $x = -7$ causes zero in a denominator, the only solution is $x = -2$.

$$\{-2\}$$

41.
$$\dfrac{1}{5} - \dfrac{7}{15 - 5x} = \dfrac{x^2}{x - 3}$$

$$\dfrac{1}{5} - \dfrac{7}{5(3 - x)} = \dfrac{x^2}{x - 3}$$

$$\dfrac{1}{5} + \dfrac{7}{5(x - 3)} = \dfrac{x^2}{x - 3}$$

$$5(x - 3)\left[\dfrac{1}{5} + \dfrac{7}{5(x - 3)}\right] = 5(x - 3)\left(\dfrac{x^2}{x - 3}\right)$$

$$x - 3 + 7 = 5x^2$$

41. (con't)

$$0 = 5x^2 - x - 4$$

$$0 = (5x + 4)(x - 1)$$

$$x = -\dfrac{4}{5} \quad \text{or} \quad x = 1$$

$$\left\{-\dfrac{4}{5}, \ 1\right\}$$

43.

$$\frac{2}{t-3} - \frac{t}{2-t} = \frac{2}{t^2 - 5t + 6}$$

$$\frac{2}{6-3} + \frac{t}{t-2} = \frac{2}{(t-2)(t-3)}$$

$$(t-2)(t-3)\left(\frac{2}{t-3} + \frac{t}{t-2}\right) = (t-2)(t-3)\left[\frac{2}{(t-2)(t-3)}\right]$$

$$2(t-2) + t(t-3 = 2$$

$$2t - 4 + t^2 - 3t = 2$$

$$t^2 - t - 6 = 0$$

$$(t-3)(t+2) = 0$$

$$t = 3 \quad \text{or} \quad t = -2$$

Since $t = 3$ causes a denominator to equal zero, the only solution is $t = -2$.

$$\{-2\}$$

45.

$$\frac{1}{x} + \frac{x-3}{x+1} = \frac{16}{x^2 + x}$$

$$\frac{1}{x} + \frac{x-3}{x+1} = \frac{16}{x(x+1)}$$

$$x(x+1)\left(\frac{1}{x} + \frac{x-3}{x+1}\right) = x(x+1)\left[\frac{16}{x(x+1)}\right]$$

$$x + 1 + x(x-3) = 16$$

$$x + 1 + x^2 - 3x = 16$$

$$x^2 - 2x - 15 = 0$$

$$(x-5)(x+3) = 0$$

$$x = 5 \quad \text{or} \quad x = -3$$

$$\{5, \ -3\}$$

47.

$$\frac{1}{x^2} + \frac{1}{3} = \frac{4}{x^2}$$

$$3x^2\left(\frac{1}{x^2} + \frac{1}{3}\right) = 3x^2\left(\frac{4}{x^2}\right)$$

$$3 + x^2 = 12$$

$$x^2 = 9$$

$$x = \pm 3$$

$$\{\pm 3\}$$

49.

$$\frac{Y}{Y+5} = \frac{1}{Y-3} - \frac{8}{Y^2 + 2Y - 15}$$

$$\frac{Y}{Y+5} = \frac{1}{Y-3} - \frac{8}{(Y+5)(4-3)}$$

$$(Y+5)(Y-3)\left(\frac{Y}{Y+5}\right) = (Y+5)(Y-3)\left[\frac{1}{Y-3} - \frac{8}{(Y+5)(Y-3)}\right]$$

$$Y(Y-3) = Y+5-8$$

$$Y^2 - 3Y = Y - 3$$

$$Y^2 - 4Y + 3 = 0$$

$$(Y-3)(Y-1) = 0$$

$$Y = 3 \quad \text{or} \quad Y = 1$$

Since $Y = 3$ causes a denominator to equal zero, the only solution is $Y = 1$.

$$\{1\}$$

51.

$$\frac{-2}{x+1} - \frac{5}{3-x} = 1$$

$$(x+1)(3-x)\left(\frac{-2}{x+1} - \frac{5}{3-x}\right) = (x+1)(3-x)(1)$$

$$-2(3-x) - 5(x+1) = -x^2 + 2x + 3$$

$$-6 + 2x - 5x - 5 = -x^2 + 2x + 3$$

$$x^2 - 5x - 14 = 0$$

$$(x-7)(x+2) = 0$$

$$x = 7 \quad \text{or} \quad x = -2$$

$$\{7, \ -2\}$$

53.

$$\frac{6}{x} + \frac{x}{x+5} = \frac{30}{x^2 + 5x}$$

$$\frac{6}{x} + \frac{x}{x+5} = \frac{30}{x(x+5)}$$

$$x(x+5)\left(\frac{6}{x} + \frac{x}{x+5}\right) = x(x+5)\left[\frac{30}{x(x+5)}\right]$$

$$6(x+5) + x^2 = 30$$

$$6x + 30 + x^2 = 30$$

53. (con'*t*)

$$x^2 + 6x = 0$$

$$x(x+6) = 0$$

$$x = 0 \quad \text{or} \quad x = -6$$

Since $x = 0$ causes a denominator to equal 0, $x = -6$ is the only solution.

$$\{-6\}$$

55.

$$\frac{t + 3}{t - 1} + \frac{1}{t + 4} = \frac{2t^2 + 6t + 12}{t^2 + 3t - 4}$$

$$\frac{t + 3}{t - 1} + \frac{1}{t + 4} = \frac{2t^2 + 6t + 12}{(t + 4)(t - 1)}$$

$$(t - 1)(t + 4)\left(\frac{t + 3}{t - 1} + \frac{1}{t + 4}\right) = (t - 1)(t + 4)\left[\frac{2t^2 + 6t + 12}{(t + 4)(t - 1)}\right]$$

$$(t + 4)(t + 3) + t - 1 = 2t^2 + 6t + 12$$

$$t^2 + 7t + 12 + t - 1 = 2t^2 + 6t + 12$$

$$0 = t^2 - 2t + 1$$

$$0 = (t - 1)^2$$

$$t = 1$$

Since $t = 1$ causes a denominator to equal 0, there is no solution.

$$\emptyset$$

57.

$$\frac{1}{t^2 - 1} + \frac{2t}{2t^2 - 3t + 1} = \frac{1}{2t^2 + t - 1}$$

$$\frac{1}{(t + 1)(t - 1)} + \frac{2t}{(2t - 1)(t - 1)} = \frac{1}{(2t - 1)(t + 1)}$$

$$(t + 1)(t - 1)(2t - 1)\left[\frac{1}{(t + 1)(t - 1)} + \frac{2t}{(2t - 1)(t - 1)}\right] = (t + 1)(t - 1)(2t - 1)\left[\frac{1}{(2t - 1)(t + 1)}\right]$$

$$2t - 1 + 2t(t + 1) = t - 1$$

$$2t - 1 + 2t^2 + 2t = t - 1$$

$$2t^2 + 3t = 0$$

$$t(2t + 3) = 0$$

$$t = 0 \quad \text{or} \quad t = -\frac{3}{2}$$

$$\left\{0, \ -\frac{3}{2}\right\}$$

59.

$$\frac{18}{s^2 + s - 6} + \frac{s - 1}{s^2 + 5s + 6} = \frac{12}{s^2 - 4}$$

$$\frac{18}{(s + 3)(s - 2)} + \frac{s - 1}{(s + 2)(s + 3)} = \frac{12}{(s + 2)(s - 2)}$$

$$(s + 3)(s - 2)(s + 2)\left[\frac{18}{(s + 3)(s - 2)} + \frac{s - 1}{(s + 2)(s + 3)}\right] = (s + 3)(s - 2)(s + 2)\left[\frac{12}{(s + 2)(s - 2)}\right]$$

$$18(s + 2) + (s - 2)(s - 1) = 12(s + 3)$$

$$18s + 36 + s^2 - 3s + 2 = 12s + 36$$

$$s^2 + 3s + 2 = 0$$

$$(s + 2)(s + 1) = 0$$

$$s = -2 \quad \text{or} \quad s = -1$$

Since $s = -2$ causes a denominator to equal zero, the only solution is $s = -1$.

$$\{-1\}$$

61.

$$\frac{1}{x^2 - 1} + \frac{x}{2x^2 - 3x + 1} = \frac{1}{2x^2 + x - 1}$$

$$\frac{1}{(x + 1)(x - 1)} + \frac{x}{(2x - 1)(x - 1)} = \frac{1}{(2x - 1)(x + 1)}$$

$$(x + 1)(x - 1)(2x - 1)\left[\frac{1}{(x + 1)(x - 1)} + \frac{x}{(2x - 1)(x - 1)}\right] = (x + 1)(x - 1)(2x - 1)\left[\frac{1}{(2x - 1)(x + 1)}\right]$$

$$2x - 1 + x(x + 1) = x - 1$$

$$2x - 1 + x^2 + x = x - 1$$

$$x^2 + 2x = 0$$

$$x(x + 2) = 0$$

$$x = 0 \quad \text{or} \quad x = -2$$

$$\{0, \ -2\}$$

63.

$$\frac{2}{5y^2 - 15y - 50} + \frac{y + 1}{y^2 - y - 6} = \frac{-8}{5y^2 - 40y + 75}$$

$$\frac{2}{5(y - 5)(y + 2)} + \frac{y + 1}{(y - 3)(y + 2)} = \frac{-8}{5(y - 5)(y - 3)}$$

$$5(y - 5)(y + 2)(y - 3)\left[\frac{2}{5(y - 5)(y + 2)} + \frac{y + 1}{(y - 3)(y + 2)}\right] = 5(y - 5)(y + 2)(y - 3)\left[\frac{-8}{5(y - 5)(y - 3)}\right]$$

$$2(y - 3) + 5(y - 5)(y + 1) = -8(y + 2)$$

$$2y - 6 + 5(y^2 - 4y - 5) = -8y - 16$$

$$2y - 6 + 5y^2 - 20y - 25 = -8y - 16$$

$$5y^2 - 10y - 15 = 0$$

$$y^2 - 2y - 3 = 0$$

$$(y - 3)(y + 1) = 0$$

$$y = 3 \quad \text{or} \quad y = -1$$

Since $y = 3$ causes a denominator to equal zero the only solution is $y = -1$.

$$\{-1\}$$

65. $x^{-2} - 3x^{-1} - 10 = 0$

$$\frac{1}{x^2} - \frac{3}{x} - 10 = 0$$

$$x^2\left(\frac{1}{x^2} - \frac{3}{x} - 10\right) = x^2(0)$$

$$1 - 3x - 10x^2 = 0$$

$$10x^2 + 3x - 1 = 0$$

$$(5x - 1)(2x + 1) = 0$$

$$x = \frac{1}{5} \quad \text{or} \quad x = -\frac{1}{2}$$

$$\left\{\frac{1}{5}, \ -\frac{1}{2}\right\}$$

69. $4x^{-2} + 27 = 21x^{-1}$

$$\frac{4}{x^2} + 27 = \frac{21}{x}$$

$$x^2\left(\frac{4}{x^2} + 27\right) = x^2\left(\frac{21}{x}\right)$$

$$4 + 27x^2 = 21x$$

$$27x^2 - 21x + 4 = 0$$

$$(9x - 4)(3x - 1) = 0$$

$$x = \frac{4}{9} \quad \text{or} \quad x = \frac{1}{3}$$

$$\left\{\frac{4}{9}, \ \frac{1}{3}\right\}$$

67. $2x^{-2} - 17x^{-1} + 21 = 0$

$$\frac{2}{x^2} - \frac{17}{x} + 21 = 0$$

$$x^2\left(\frac{2}{x^2} - \frac{17}{x} + 21\right) = x^2(0)$$

$$2 - 17x + 21x^2 = 0$$

$$21x^2 - 17x + 2 = 0$$

$$(3x - 2)(7x - 1) = 0$$

$$x = \frac{2}{3} \quad \text{or} \quad x = \frac{1}{7}$$

$$\left\{\frac{2}{3}, \ \frac{1}{7}\right\}$$

71. $x^{-2} - 10x^{-1} + 25 = 0$

$$\frac{1}{x^2} - \frac{10}{x} + 25 = 0$$

$$x^2\left(\frac{1}{x^2} - \frac{10}{x} + 25\right) = x^2(0)$$

$$1 - 10x + 25x^2 = 0$$

$$25x^2 - 10x + 1 = 0$$

$$(5x - 1)^2 = 0$$

$$x = \frac{1}{5}$$

$$\left\{\frac{1}{5}\right\}$$

73. $2x^{-4} - 11x^{-2} + 9 = 0$

$$\frac{2}{x^4} - \frac{11}{x^2} + 9 = 0$$

$$x^4\left(\frac{2}{x^4} - \frac{11}{x^2} + 9\right) = x^4(0)$$

$$2 - 11x^2 + 9x^4 = 0$$

$$9x^4 - 11x^2 + 2 = 0$$

$$(9x^2 - 2)(x^2 - 1) = 0$$

$$x^2 = \frac{2}{9} \quad \text{or} \quad x^2 = 1$$

$$x = \pm\frac{\sqrt{2}}{3} \quad \text{or} \quad x = \pm1$$

$$\left\{\pm\frac{\sqrt{2}}{3}, \ \pm1\right\}$$

75. $\dfrac{-3}{x} - 1 < \dfrac{4}{1 - x}$

Free boundary numbers:

$x = 0, \ x = 1$

Other boundary numbers:

$$\frac{-3}{x} - 1 = \frac{4}{1 - x}$$

$$x(1 - x)\left(\frac{-3}{x} - 1\right) = x(1 - x)\left(\frac{4}{1 - x}\right)$$

$$-3(1 - x) - x(1 - x) = 4x$$

$$-3 + 3x - x + x^2 = 4x$$

$$x^2 - 2x - 3 = 0$$

$$(x - 3)(x + 1) = 0$$

$$x = 3, \ x = -1$$

$$\begin{array}{c|c|c|c|c}
A & B & C & D & E \\
\hline
-1 & 0 & 1 & 3 &
\end{array}$$

A: $x = -2$; $\quad \dfrac{-3}{-2} - 1 < \dfrac{4}{1 - (-2)}$

$$\frac{1}{2} < \frac{4}{3} \quad \text{T}$$

B: $x = -\dfrac{1}{2}$; $\quad \dfrac{-3}{-\dfrac{1}{2}} - 1 < \dfrac{4}{1 - \left(-\dfrac{1}{2}\right)}$

$$5 < \frac{8}{3} \quad \text{F}$$

C: $x = \dfrac{1}{2}$; $\quad \dfrac{-3}{\dfrac{1}{2}} - 1 < \dfrac{4}{1 - \dfrac{1}{2}}$

$$-7 < 8 \quad \text{T}$$

D: $x = 2$; $\quad \dfrac{-3}{2} - 1 < \dfrac{4}{1 - 2}$

$$\frac{-5}{2} < -4 \quad \text{F}$$

E: $x = 4$; $\quad \dfrac{-3}{4} - 1 < \dfrac{4}{1 - 4}$

$$\frac{-7}{4} < -\frac{4}{3} \quad \text{T}$$

$$(-\infty, -1) \cup (0, 1) \cup (3, \infty)$$

77. $\dfrac{14}{x - 2} - \dfrac{35}{x - 5} \le \dfrac{3}{2}$

Free boundary numbers:

$x = 2, \ x = 5$

Other boundary numbers:

$$\frac{14}{x - 2} - \frac{35}{x - 5} = \frac{3}{2}$$

$$2(x - 2)(x - 5)\left(\frac{14}{x - 2} - \frac{35}{x - 5}\right) = 2(x - 2)(x - 5)\left(\frac{3}{2}\right)$$

$$28(x - 5) - 70(x - 2) = 3(x - 2)(x - 5)$$

$$28x - 140 - 70x + 140 = 3(x^2 - 7x + 10)$$

$$-42x = 3x^2 - 21x + 30$$

$$0 = 3x^2 + 21x + 30$$

$$0 = x^2 + 7x + 10$$

$$0 = (x + 5)(x + 2)$$

$$x = -5, \quad x = -2$$

$$\underset{-5}{\underline{\quad A \quad}}\Big|\underset{-2}{\underline{\quad B \quad}}\Big|\underset{2}{\underline{\quad C \quad}}\Big|\underset{5}{\underline{\quad D \quad}}\underline{\quad E \quad}$$

A: $x = -6$; $\dfrac{14}{-6 - 2} - \dfrac{35}{-6 - 5} \le \dfrac{3}{2}$

$\dfrac{63}{44} \le \dfrac{3}{2}$ T

B: $x = -3$; $\dfrac{14}{-3 - 2} - \dfrac{35}{-3 - 5} \le \dfrac{3}{2}$

$\dfrac{63}{40} \le \dfrac{3}{2}$ F

C: $x = 0$; $\dfrac{14}{0 - 2} - \dfrac{35}{0 - 5} \le \dfrac{3}{2}$

$0 \le \dfrac{3}{2}$ T

D: $x = 3$; $\dfrac{14}{3 - 2} - \dfrac{35}{3 - 5} \le \dfrac{3}{2}$

$\dfrac{63}{2} \le \dfrac{3}{2}$ F

E: $x = 6$; $\dfrac{14}{6 - 2} - \dfrac{35}{6 - 5} \le \dfrac{3}{2}$

$\dfrac{63}{2} \le \dfrac{3}{2}$ T

$(-\infty, -5] \cup [-2, 2) \cup (5, \infty)$

79. $\dfrac{1}{x} + \dfrac{x + 3}{7} \ge \dfrac{5}{7x}$

Free boundary numbers:

$x = 0$

Other boundary numbers:

$$\dfrac{1}{x} + \dfrac{x + 3}{7} = \dfrac{5}{7x}$$

$$7x\left(\dfrac{1}{x} + \dfrac{x + 3}{7}\right) = 7x\left(\dfrac{5}{7x}\right)$$

$$7 + x(x + 3) = 5$$

$$7 + x^2 + 3x = 5$$

$$x^2 + 3x + 2 = 0$$

$$(x + 2)(x + 1) = 0$$

$$x = -2, \quad x = -1$$

$$\underset{-2}{\underline{\quad A \quad}}\Big|\underset{-1}{\underline{\quad B \quad}}\Big|\underset{0}{\underline{\quad C \quad}}\underline{\quad D \quad}$$

A: $x = -3$; $\dfrac{1}{-3} + \dfrac{-3 + 3}{7} \ge \dfrac{5}{7(-3)}$

$-\dfrac{1}{3} \ge -\dfrac{5}{21}$ F

B: $x = -\dfrac{3}{2}$; $\dfrac{1}{-\dfrac{3}{2}} + \dfrac{-\dfrac{3}{2} + 3}{7} \ge \dfrac{5}{7\left(-\dfrac{3}{2}\right)}$

$-\dfrac{19}{42} \ge -\dfrac{10}{21}$ T

C: $x = -\dfrac{1}{2}$; $\dfrac{1}{-\dfrac{1}{2}} + \dfrac{-\dfrac{1}{2} + 3}{7} \ge \dfrac{5}{7\left(-\dfrac{1}{2}\right)}$

$\dfrac{-23}{14} \ge -\dfrac{10}{7}$ F

D: $x = 1$; $\dfrac{1}{1} + \dfrac{1 + 3}{7} \ge \dfrac{5}{7(1)}$

$\dfrac{11}{7} \ge \dfrac{5}{7}$ T

$[-2, -1] \cup (0, \infty)$

81.

	D	R	T = D/R
walk	6	r - 2	6/(r - 2)
ride	3	r	3/r

$$\frac{6}{r - 2} = \frac{3}{r} + 1$$

$$r(r - 2)\left(\frac{6}{r - 2}\right) = r(r - 2)\left(\frac{3}{r} + 1\right)$$

$$6r = 3(r - 2) + r(r - 2)$$

$$6r = 3r - 6 + r^2 - 2r$$

$$0 = r^2 - 5r - 6$$

$$0 = (r - 6)(r + 1)$$

$$r = 6 \quad \text{or} \quad r = -1$$

Since $r > 0$, $r = 6$.

His speed on his bike was 6 mph.

83.

	D	R	T = D/R
1st part	64	r	64/r
2nd part	10	r - 6	10/(r - 6}

$$\frac{64}{r} + \frac{10}{r - 6} = 5$$

$$r(r - 6)\left(\frac{64}{r} + \frac{10}{r - 6}\right) = r(r - 6)(5)$$

$$64(r - 6) + 10r = 5r^2 - 30r$$

$$64r - 384 + 10r = 5r^2 - 30r$$

$$0 = 5r^2 - 104r + 384$$

$$0 = (5r - 24)(r - 16)$$

$$r = \frac{24}{5} \quad \text{or} \quad r = 16$$

$$r - 6 = -\frac{6}{5} \qquad r - 6 = 10$$

Since both rates must be positive, $r = 16$.
Her original speed was 16 mph.

85.

	time for whole job	part done in 1 hr
Cathy	x	1/x
Jim	x + 9	1/(x + 9)
together	6	1/6

$$\frac{1}{x} + \frac{1}{x + 9} = \frac{1}{6}$$

$$6x(x + 9)\left(\frac{1}{x} + \frac{1}{x + 9}\right) = 6x(x + 9)\left(\frac{1}{6}\right)$$

$$6x + 54 + 6x = x^2 + 9x$$

$$0 = x^2 - 3x - 54$$

$$0 = (x - 9)(x + 6)$$

$$x = 9 \quad \text{or} \quad x = -6$$

Since $x > 0$, $x = 9$

$$x + 9 = 18$$

It would take Cathy 9 hrs and Jim 18 hrs.

87. $4x^{-4} - 68x^{-2} + 225 = 0$

$$\frac{4}{x^4} - \frac{68}{x^2} + 225 = 0$$

$$x^4\left(\frac{4}{x^4} - \frac{68}{x^2} + 225\right) = x^4(0)$$

$$4 - 68x^2 + 225x^4 = 0$$

$$225x^4 - 68x^2 + 4 = 0$$

$$(25x^2 - 2)(9x^2 - 2) = 0$$

$$x^2 = \frac{2}{25} \quad \text{or} \quad x^2 = \frac{2}{9}$$

$$x = \pm\frac{\sqrt{2}}{5} \quad \text{or} \quad x = \pm\frac{\sqrt{2}}{3}$$

$$\left\{\pm\frac{\sqrt{2}}{5}, \ \pm\frac{\sqrt{2}}{3}\right\}$$

89. $9x^{-4} + 7x^{-2} - 2 = 0$

$$\frac{9}{x^4} + \frac{7}{x^2} - 2 = 0$$

$$x^4\left(\frac{9}{x^4} + \frac{7}{x^2} - 2\right) = x^4(0)$$

$$9 + 7x^2 - 2x^4 = 0$$

$$2x^4 - 7x^2 - 9 = 0$$

$$(2x^2 - 9)(x^2 + 1) = 0$$

$$x^2 = \frac{9}{2} \quad \text{or} \quad x^2 = -1$$

$$x = \pm\frac{3}{\sqrt{2}} \quad \text{or} \quad x = \pm i$$

$$x = \pm\frac{3\sqrt{2}}{2}$$

$$\left\{\pm\frac{3\sqrt{2}}{2}, \ \pm i\right\}$$

91. $16x^{-4} - 8x^{-2} - 15 = 0$

$$\frac{16}{x^4} - \frac{8}{x^2} - 15 = 0$$

$$x^4\left(\frac{16}{x^4} - \frac{8}{x^2} - 15\right) = x^4(0)$$

$$16 - 8x^2 - 15x^4 = 0$$

$$15x^4 + 8x^2 - 16 = 0$$

$$(5x^2 - 4)(3x^2 + 4) = 0$$

$$x^2 = \frac{4}{5} \quad \text{or} \quad x^2 = -\frac{4}{3}$$

$$x = \pm\frac{2}{\sqrt{5}} \quad \text{or} \quad x = \frac{2}{\sqrt{3}}i$$

$$x = \pm\frac{2\sqrt{5}}{5} \quad \text{or} \quad x = \pm\frac{2\sqrt{3}}{3}i$$

$$\left\{\pm\frac{2\sqrt{5}}{5}, \ \pm\frac{2\sqrt{3}}{3}i\right\}$$

93. $6x^{-2} - x^{-1} = 0$

$$\frac{6}{x^2} - \frac{1}{x} = 0$$

$$x^2\left(\frac{6}{x^2} - \frac{1}{x}\right) = x^2(0)$$

$$6 - x = 0$$

$$6 = x$$

$$\{6\}$$

95. $$\frac{3x}{x - 1} = \frac{x^2}{1 - x}$$

$$\frac{3x}{x - 1} = \frac{x^2}{-(x - 1)}$$

$$(x - 1)\left(\frac{3x}{x - 1}\right) = (x - 1)\left(-\frac{x^2}{x - 1}\right)$$

$$3x = -x^2$$

$$x^2 + 3x = 0$$

$$x(x + 3) = 0$$

$$x = 0 \quad \text{or} \quad x = -3$$
$$\{0, \; -3\}$$

97. $$\frac{5x}{x^2 + 3x - 4} + \frac{x}{x + 4} = \frac{x^2}{x - 1}$$

$$\frac{5x}{(x + 4)(x - 1)} + \frac{x}{x + 4} = \frac{x^2}{x - 1}$$

$$(x + 4)(x - 1)\left[\frac{5x}{(x + 4)(x - 1)} + \frac{x}{x + 4}\right] = (x + 4)(x - 1)\left(\frac{x^2}{x - 1}\right)$$

$$5x + x(x - 1) = x^2(x + 4)$$

$$5x + x^2 - x = x^3 + 4x^2$$

$$0 = x^3 + 3x^2 - 4x$$

$$0 = x(x^2 + 3x - 4)$$

$$0 = x(x + 4)(x - 1)$$

$$x = 0 \quad \text{or} \quad x = -4 \quad \text{or} \quad x = 1$$

Since $x = -4$ and $x = 1$ cause a denominator to equal zero, the only solution is $x = 0$.

$$\{0\}$$

99. $$\frac{a - 2}{x + 2} + \frac{x}{x - 1} = \frac{2a^2 - a + 2}{x^2 + x - 2}$$

$$\frac{a - 2}{x + 2} + \frac{x}{x - 1} = \frac{2a^2 - a + 2}{(x + 2)(x - 1)}$$

$$(x + 2)(x - 1)\left(\frac{a - 2}{x + 2} + \frac{x}{x - 1}\right) = (x + 2)(x - 1)\left[\frac{2a^2 - a + 2}{(x + 2)(x - 1)}\right]$$

$$(x - 1)(a - 2) + x(x + 2) = 2a^2 - a + 2$$

$$xa - 2x - a + 2 + x^2 + 2x = 2a^2 - a + 2$$

$$x^2 + ax - 2a^2 = 0$$

$$(x + 2a)(x - a) = 0$$

$$x = -2a \quad \text{or} \quad x = a$$

$$\{-2a, \; a\}$$

229

101. $(x + 1)^{-4} - 5(x + 1)^{-2} + 4 = 0$

$$\frac{1}{(x + 1)^4} - \frac{5}{(x + 1)^2} + 4 = 0$$

$$(x + 1)^4 \left[\frac{1}{(x + 1)^4} - \frac{5}{(x + 1)^2} + 4 \right] = (x + 1)^4(0)$$

$$1 - 5(x + 1)^2 + 4(x + 1)^4 = 0$$

$$4(x + 1)^4 - 5(x + 1)^2 + 1 = 0$$

Let $u = x + 1$

$$4u^4 - 5u^2 + 1 = 0$$

$$(4u^2 - 1)(u^2 - 1) = 0$$

$$u^2 = \frac{1}{4} \qquad \text{or} \qquad u^2 = 1$$

$$u = \pm\frac{1}{2} \qquad \text{or} \qquad u = \pm 1$$

$$x + 1 = \pm\frac{1}{2} \qquad \text{or} \quad x + 1 = \pm 1$$

$$x = -1 \pm \frac{1}{2} \quad \text{or} \qquad x = -1 \pm 1$$

$$x = -1 + \frac{1}{2} \text{ or } x = -1 - \frac{1}{2} \qquad \text{or} \quad x = -1 + 1 \quad \text{or} \quad x = -1 - 1$$

$$x = -\frac{1}{2} \qquad \text{or} \quad x = -\frac{3}{2} \qquad \text{or} \quad x = 0 \qquad \text{or} \quad x = -2$$

$$\left\{ -\frac{1}{2}, \ -\frac{3}{2}, \ 0, \ -2 \right\}$$

103. $\dfrac{2}{3\sqrt{x}} + \dfrac{\sqrt{x}}{3} = \dfrac{x^2}{3\sqrt{x}}$

$$3\sqrt{x}\left(\frac{2}{3\sqrt{x}} + \frac{\sqrt{x}}{3} \right) = 3\sqrt{x}\left(\frac{x^2}{3\sqrt{x}} \right)$$

$$2 + x = x^2$$

$$0 = x^2 - x - 2$$

$$0 = (x - 2)(x + 1)$$

$$x = 2 \quad \text{or} \quad x = -1$$

230

105. Ron's rate in still water: x

	D	R	T = D/R
downstream	30	$x + 5$	$30/(x + 5)$
upstream	30	$x - 5$	$30/(x - 5)$

$$\frac{30}{x + 5} + \frac{30}{x - 5} = 8$$

$$(x + 5)(x - 5)\left(\frac{30}{x + 5} + \frac{30}{x - 5}\right) = (x + 5)(x - 5)(8)$$

$$30(x - 5) + 30(x + 5) = 8(x^2 - 25)$$

$$30x - 150 + 30x + 150 = 8x^2 - 200$$

$$0 = 8x^2 - 60x - 200$$

$$0 = 2x^2 - 15x - 50$$

$$0 = (2x + 5)(x - 10)$$

$$x = -\frac{5}{2} \quad \text{or} \quad x = 10$$

Since $x > 0$, $x = 10$

His rate in still water is 10 mph.

Problem Set 5.7

1. $x = \sqrt{2x^2 - 3x + 2}$

$x^2 = \left(\sqrt{2x^2 - 3x + 2}\right)^2$

$x^2 = 2x^2 - 3x + 2$

$0 = x^2 - 3x + 2$

$0 = (x - 2)(x - 1)$

$x = 2 \quad \text{or} \quad x = 1$

Check: $x = 2$

LS: 2

RS: $\sqrt{2(2)^2 - 3(2) + 2} = \sqrt{4} = 2$

Checks.

Check: $x = 1$

LS: 1

RS: $\sqrt{2(1)^2 - 3(1) + 2} = \sqrt{1} = 1$

Checks.

$$\{2, \ 1\}$$

3. $x - 2 = \sqrt{6 - 3x}$

$(x - 2)^2 = \left(\sqrt{6 - 3x}\right)^2$

$x^2 - 4x + 4 = 6 - 3x$

$x^2 - x - 2 = 0$

$(x - 2)(x + 1) = 0$

$x = 2 \quad \text{or} \quad x = -1$

Check: $x = 2$

LS: $2 - 2 = 0$

RS: $\sqrt{6 - 3(2)} = \sqrt{0} = 0$

Checks.

Check: $x = -1$

LS: $-1 - 2 = -3$

RS: $\sqrt{6 - 3(-3)} = \sqrt{15}$

Does not check.

$$\{2\}$$

5. $x = \sqrt{2x + 6} - 3$

$x + 3 = \sqrt{2x + 6}$

$(x + 3)^2 = \left(\sqrt{2x + 6}\right)^2$

$x^2 + 6x + 9 = 2x + 6$

$x^2 + 4x + 3 = 0$

$(x + 3)(x + 1) = 0$

$x = -3 \quad \text{or} \quad x = -1$

Check: $x = -3$

LS: -3

RS: $\sqrt{2(-3) + 6} - 3 = \sqrt{0} - 3 = -3$

Checks.

Check: $x = -1$

LS: -1

RS: $\sqrt{2(-1) + 6} - 3 = \sqrt{4} - 3 = -1$

Checks.
$$\{-3, \ -1\}$$

7.
$$2x - 1 = \sqrt{2x^2 + 3x - 2}$$
$$(2x - 1)^2 = \left(\sqrt{2x^2 + 3x - 2}\right)^2$$
$$4x^2 - 4x + 1 = 2x^2 + 3x - 2$$
$$2x^2 - 7x + 3 = 0$$
$$(2x - 1)(x - 3) = 0$$
$$x = \frac{1}{2} \quad \text{or} \quad x = 3$$

Check: $x = \frac{1}{2}$

LS: $2\left(\frac{1}{2}\right) - 1 = 0$

RS: $\sqrt{2\left(\frac{1}{2}\right)^2 + 3\left(\frac{1}{2}\right) - 2} = \sqrt{0} = 0$

Checks.

Check: $x = 3$

LS: $2(3) - 1 = 5$

RS: $\sqrt{2(3)^2 + 3(3) - 2} = \sqrt{25} = 5$

Checks.
$$\left\{\frac{1}{2}, \ 3\right\}$$

9.
$$2x + 3 = \sqrt{8 + 7x - 2x^2}$$
$$(2x + 3)^2 = \left(\sqrt{8 + 7x - 2x^2}\right)^2$$

$$4x^2 + 12x + 9 = 8 + 7x - 2x^2$$
$$6x^2 + 5x + 1 = 0$$
$$(3x + 1)(2x + 1) = 0$$
$$x = -\frac{1}{3} \quad \text{or} \quad x = -\frac{1}{2}$$

Check: $x = -\frac{1}{3}$

LS: $2\left(-\frac{1}{3}\right) + 3 = \frac{7}{3}$

RS: $\sqrt{8 + 7\left(-\frac{1}{3}\right) - 2\left(-\frac{1}{3}\right)^2} = \frac{\sqrt{49}}{9} = \frac{7}{3}$

Checks.

Check: $x = -\frac{1}{2}$

LS: $2\left(-\frac{1}{2}\right) + 3 = 2$

RS: $\sqrt{8 + 7\left(-\frac{1}{2}\right) - 2\left(-\frac{1}{2}\right)^2} = \sqrt{4} = 2$

Checks.
$$\left\{-\frac{1}{3}, \ -\frac{1}{2}\right\}$$

11.
$$\sqrt{x - 3} - 1 = \sqrt{2x - 4}$$
$$\left(\sqrt{x - 3} - 1\right)^2 = \left(\sqrt{2x - 4}\right)^2$$
$$x - 3 - 2\sqrt{x - 3} + 1 = 2x - 4$$
$$-2\sqrt{x - 3} = x - 2$$
$$\left(-2\sqrt{x - 3}\right)^2 = (x - 2)^2$$
$$4(x - 3) = x^2 - 4x + 4$$
$$4x - 12 = x^2 - 4x + 4$$
$$0 = x^2 - 8x + 16$$
$$0 = (x - 4)^2$$
$$x = 4$$

Check:

LS: $\sqrt{4 - 3} - 1 = \sqrt{1} - 1 = 0$

RS: $\sqrt{2(4) - 4} = \sqrt{4} = 2$

$$\emptyset$$

13. $\sqrt{3x + 1} = 1 + \sqrt{2x - 1}$

$$\left(\sqrt{3x + 1}\right)^2 = \left(1 + \sqrt{2x - 1}\right)^2$$

$$3x + 1 = 1 + 2\sqrt{2x - 1} + 2x - 1$$

$$x + 1 = 2\sqrt{2x - 1}$$

$$(x + 1)^2 = \left(2\sqrt{2x - 1}\right)^2$$

$$x^2 + 2x + 1 = 4(2x - 1)$$

$$x^2 + 2x + 1 = 8x - 4$$

$$x^2 - 6x + 5 = 0$$

$$(x - 5)(x - 1) = 0$$

$$x = 5 \quad \text{or} \quad x = 1$$

Check: $x = 5$

LS: $\sqrt{3(5) + 1} = \sqrt{16} = 4$

RS: $1 + \sqrt{2(5) - 1} = 1 + \sqrt{9} = 1 + 3 = 4$

Checks.

Check: $x = 1$

LS: $\sqrt{3(1) + 1} = \sqrt{4} = 2$

RS: $1 + \sqrt{2(1) - 1} = 1 + \sqrt{1} = 1 + 1 = 2$

Checks.

$$\{5, \ 1\}$$

15. $\sqrt{2y - 5} = \sqrt{3y + 4} - 2$

$$\left(\sqrt{2y - 5}\right)^2 = \left(\sqrt{3y + 4} - 2\right)^2$$

$$2y - 5 = 3y + 4 - 4\sqrt{3y + 4} + 4$$

$$-y - 13 = -4\sqrt{3y + 4}$$

$$y + 13 = 4\sqrt{3y + 4}$$

$$y^2 + 26y + 169 = 16(3y + 4)$$

$$y^2 + 26y + 169 = 48y + 64$$

$$y^2 - 22y + 105 = 0$$

$$(y - 7)(y - 15) = 0$$

$$y = 7 \quad \text{or} \quad y = 15$$

Check: $y = 7$

LS: $\sqrt{2(7) - 5} = \sqrt{9} = 3$

RS: $\sqrt{3(7) + 4} - 2 = \sqrt{25} - 2 = 5 - 2 = 3$

Checks.

Check: $y = 15$

LS: $\sqrt{2(15) - 5} = \sqrt{25} = 5$

RS: $\sqrt{3(15) + 4} - 2 = \sqrt{49} - 2 = 7 - 2 = 5$

Checks.

$$\{7, \ 15\}$$

17. $\sqrt{5x + 21} = 1 - \sqrt{3x + 16}$

$$\left(\sqrt{5x + 21}\right)^2 = \left(1 - \sqrt{3x + 16}\right)^2$$

$$5x + 21 = 1 - 2\sqrt{3x + 16} + 3x + 16$$

$$2x + 4 = -2\sqrt{3x + 16}$$

$$(2x + 4)^2 = \left(-2\sqrt{3x + 16}\right)^2$$

$$4x^2 + 16x + 16 = 4(3x + 16)$$

$$4x^2 + 16x + 16 = 12x + 64$$

$$4x^2 + 4x - 48 = 0$$

$$x^2 + x - 12 = 0$$

$$(x + 4)(x - 3) = 0$$

$$x = -4 \quad \text{or} \quad x = 3$$

Check: $x = -4$

LS: $\sqrt{5(-4) + 21} = \sqrt{1} = 1$

RS: $1 - \sqrt{3(-4) + 16} = 1 - \sqrt{4} = 1 - 2 = -1$

Does not check.

Check: $x = 3$

LS: $\sqrt{5(3) + 21} = \sqrt{36} = 6$

RS: $1 - \sqrt{3(3) + 16} = 1 - \sqrt{25} = 1 - 5 = -4$

Does not check.

$$\emptyset$$

19. $x^{1/3} = 2$

$$(x^{1/3})^3 = 2^3$$

$$x = 8$$

$$\{8\}$$

21. $\sqrt[3]{2x - 1} - 3 = 0$

$$\sqrt[3]{2x - 1} = 3$$

$$\left(\sqrt[3]{2x - 1}\right)^3 = 3^3$$

$$2x - 1 = 27$$

$$2x = 28$$

$$x = 14$$

Check:

LS: $\sqrt[3]{2(14) - 1} - 3 = \sqrt[3]{27} - 3 = 3 - 3 = 0$

RS: $= 0$

$$\{14\}$$

23. $\sqrt{x + 2} = x$

$$\left(\sqrt{x + 2}\right)^2 = x^2$$

$$x + 2 = x^2$$

$$0 = x^2 - x - 2$$

$$0 = (x - 2)(x + 1)$$

$x = 2$ or $x = -1$

Check: $x = 2$

LS: $\sqrt{2 + 2} = \sqrt{4} = 2$

RS: $= 2$

Checks.

Check: $x = -1$

LS: $\sqrt{-1 + 2} = \sqrt{1} = 1$

RS: -1

Does not check.

$$\{2\}$$

25. $\sqrt{16x} = x + 3$

$$\left(\sqrt{16x}\right)^2 = (x + 3)^2$$

$$16x = x^2 + 6x + 9$$

$$0 = x^2 - 10x + 9$$

$$0 = (x - 9)(x - 1)$$

$x = 9$ or $x = 1$

Check: $x = 9$

LS: $\sqrt{16(9)} = \sqrt{144} = 12$

RS: $= 9 + 3 = 12$

Checks.

Check: $x = 1$

LS: $\sqrt{16(1)} = \sqrt{16} = 4$

RS: $1 + 3 = 4$

Checks.

$$\{9, \ 1\}$$

27. $\sqrt{4 - x} = x - 4$

$$\left(\sqrt{4 - x}\right)^2 = (x - 4)^2$$

$$4 - x = x^2 - 8x + 16$$

$$0 = x^2 - 7x + 12$$

$$0 = (x - 4)(x - 3)$$

$x = 4$ or $x = 3$

Check: $x = 4$

LS: $\sqrt{4 - 4} = \sqrt{0} = 0$

RS: $4 - 4 = 0$

Checks.

Check: $x = 3$

LS: $\sqrt{4 - 3} = \sqrt{1} = 1$

RS: $3 - 4 = -1$

Does not check.

$$\{4\}$$

29. $\sqrt{2x} = x - 4$

$$\left(\sqrt{2x}\right)^2 = (x - 4)^2$$

$$2x = x^2 - 8x + 16$$

$$0 = x^2 - 10x + 16$$

$$0 = (x - 8)(x - 2)$$

$$x = 8 \quad \text{or} \quad x = 2$$

Check: $x = 8$

LS: $\sqrt{2(8)} = \sqrt{16} = 4$

RS: $8 - 4 = 4$

Checks.

Check: $x = 2$

LS: $\sqrt{2(2)} = \sqrt{4} = 2$

RS: $2 - 4 = -2$

Does not check.

$$\{8\}$$

31. $\sqrt{8x + 1} - 2 = x$

$$\sqrt{8x + 1} = x + 2$$

$$\left(\sqrt{8x + 1}\right)^2 = (x + 2)^2$$

$$8x + 1 = x^2 + 4x + 4$$

$$0 = x^2 - 4x + 3$$

$$0 = (x - 3)(x - 1)$$

$$x = 3 \quad \text{or} \quad x = 1$$

Check: $x = 3$

LS: $\sqrt{8(3) + 1} - 2 = \sqrt{25} - 2 = 5 - 2 = 3$

RS: 3

Checks.

Check: $x = 1$

LS: $\sqrt{8(1) + 1} - 2 = \sqrt{9} - 2 = 3 - 2 = 1$

RS: 1

Checks.

$$\{3, \ 1\}$$

33. $\sqrt{z + 27} - 1 = \sqrt{2z + 20}$

$$\left(\sqrt{z + 27} - 1\right)^2 = \left(\sqrt{2z + 20}\right)^2$$

$$z + 27 - 2\sqrt{z + 27} + 1 = 2z + 20$$

$$-2\sqrt{z + 27} = z - 8$$

$$\left(-2\sqrt{z + 27}\right)^2 = (z - 8)^2$$

$$4(z + 27) = z^2 - 16z + 64$$

$$4z + 108 = z^2 - 16z + 64$$

$$0 = z^2 - 20z - 44$$

$$0 = (z - 22)(z + 2)$$

$$z = 22 \quad \text{or} \quad z = -2$$

Check: $z = 22$

LS: $\sqrt{22 + 27} - 1 = \sqrt{49} - 1 = 7 - 1 = 6$

RS: $\sqrt{2(22) + 20} = \sqrt{64} = 8$

Does not check.

Check: $z = -2$

LS: $\sqrt{-2 + 27} - 1 = \sqrt{25} - 1 = 5 - 1 = 4$

RS: $\sqrt{2(-2) + 20} = \sqrt{16} = 4$

Checks.

$$\{-2\}$$

35. $\sqrt{5w} - \sqrt{w-4} = 4$

$$\sqrt{5w} = 4 + \sqrt{w-4}$$

$$\left(\sqrt{5w}\right)^2 = \left(4 + \sqrt{w-4}\right)^2$$

$$5w = 16 + 8\sqrt{w-4} + w - 4$$

$$4w - 12 = 8\sqrt{w-4}$$

$$w - 3 = 2\sqrt{w-4}$$

$$(w-3)^2 = \left(2\sqrt{w-4}\right)^2$$

$$w^2 - 6w + 9 = 4(w-4)$$

$$w^2 - 6w + 9 = 4w - 16$$

$$w^2 - 10w + 25 = 0$$

$$(w-5)^2 = 0$$

$$w = 5$$

Check: $w = 5$

LS: $\sqrt{5(5)} - \sqrt{5-4}$

$= \sqrt{25} - \sqrt{1}$

$= 5 - 1$

$= 4$

RS: 4

Checks.

$$\{5\}$$

37. $\sqrt{3-2x} - \sqrt{x+4} = 2$

$$\sqrt{3-2x} = 2 + \sqrt{x+4}$$

$$\left(\sqrt{3-2x}\right)^2 = \left(2 + \sqrt{x+4}\right)^2$$

$$3 - 2x = 4 + 4\sqrt{x+4} + x + 4$$

$$-3x - 5 = 4\sqrt{x+4}$$

$$(-3x-5)^2 = \left(4\sqrt{x+4}\right)^2$$

$$9x^2 + 30x + 25 = 16(x+4)$$

$$9x^2 + 30x + 25 = 16x + 64$$

$$9x^2 + 14x - 39 = 0$$

$$(9x - 13)(x + 3) = 0$$

$$x = \frac{13}{9} \quad \text{or} \quad x = -3$$

Check: $x = \dfrac{13}{9}$

LS: $\sqrt{3 - 2\left(\dfrac{13}{9}\right)} - \sqrt{\dfrac{13}{9} + 4} = \sqrt{\dfrac{1}{9}} - \sqrt{\dfrac{49}{9}} = \dfrac{1}{3} - \dfrac{7}{3} = -2$

RS: 2

Does not check.

Check: $x = -3$

LS: $\sqrt{3 - 2(-3)} - \sqrt{-3+4} = \sqrt{9} - \sqrt{1} = 3 - 1 = 2$

RS: 2

Checks.

$$\{-3\}$$

39. $\sqrt{t+9} = 2 + \sqrt{2t+1}$

$$\left(\sqrt{t+9}\right)^2 = \left(2 + \sqrt{2t+1}\right)^2$$

$$t + 9 = 4 + 4\sqrt{2t+1} + 2t + 1$$

$$-t + 4 = 4\sqrt{2t+1}$$

$$(-t+4)^2 = \left(4\sqrt{2t+1}\right)^2$$

$$t^2 - 8t + 16 = 16(2t+1)$$

$$t^2 - 8t + 16 = 32t + 16$$

$$t^2 - 40t = 0$$

$$t(t-40) = 0$$

$$t = 0 \quad \text{or} \quad t = 40$$

Check: $t = 0$

LS: $\sqrt{0+9} = \sqrt{9} = 3$

RS: $2 + \sqrt{2(0)+1} = 2 + \sqrt{1} = 2 + 1 = 3$

Checks.

Check: $t = 40$

LS: $\sqrt{40 + 9} = \sqrt{49} = 7$

RS: $2 + \sqrt{2(40) + 1} = 2 + \sqrt{81} = 2 + 9 = 11$

Does not check.

$$\{0\}$$

41. $\sqrt{z - 2} + 2 = \sqrt{2z}$

$$\left(\sqrt{z - 2} + 2\right)^2 = \left(\sqrt{2z}\right)^2$$

$$z - 2 + 4\sqrt{z - 2} + 4 = 2z$$

$$4\sqrt{z - 2} = z - 2$$

$$\left(4\sqrt{z - 2}\right)^2 = (z - 2)^2$$

$$16(z - 2) = z^2 - 4z + 4$$

$$16z - 32 = z^2 - 4z + 4$$

$$0 = z^2 - 20z + 36$$

$$0 = (z - 2)(z - 18)$$

$$z = 2 \quad \text{or} \quad z = 18$$

Check: $z = 2$

LS: $\sqrt{2 - 2} + 2 = \sqrt{0} + 2 = 0 + 2 = 2$

RS: $\sqrt{2(2)} = \sqrt{4} = 2$

Checks.

Check: $z = 18$

LS: $\sqrt{18 - 2} + 2 = \sqrt{16} + 2 = 4 + 2 = 6$

RS: $\sqrt{2(18)} = \sqrt{36} = 6$

Checks.

$$\{2,\ 18\}$$

43. $\sqrt{2x + 1} + 1 = \sqrt{x + 4}$

$$\left(\sqrt{2x + 1} + 1\right)^2 = \left(\sqrt{x + 4}\right)^2$$

$$2x + 1 + 2\sqrt{2x + 1} + 1 = x + 4$$

$$2\sqrt{2x + 1} = -x + 2$$

$$\left(2\sqrt{2x + 1}\right)^2 = (-x + 2)^2$$

$$4(2x + 1) = x^2 - 4x + 4$$

$$8x + 4 = x^2 - 4x + 4$$

$$0 = x^2 - 12x$$

$$0 = x(x - 12)$$

$$x = 0 \quad \text{or} \quad x = 12$$

Check: $x = 0$

LS: $\sqrt{2(0) + 1} + 1 = \sqrt{1} + 1 = 1 + 1 = 2$

RS: $\sqrt{0 + 4} = \sqrt{4} = 2$

Checks.

Check: $x = 12$

LS: $\sqrt{2(12) + 1} + 1 = \sqrt{25} + 1 = 5 + 1 = 6$

RS: $\sqrt{12 + 4} = \sqrt{16} = 4$

Does not check.

$$\{0\}$$

45. $\sqrt[3]{x} = 1$

$$\left(\sqrt[3]{x}\right)^3 = 1^3$$

$$x = 1$$

$$\{1\}$$

47. $(3x + 1)^{1/3} = 2$

$$\left[(3x + 1)^{1/3}\right]^3 = 2^3$$

$$3x + 1 = 8$$

$$3x = 7$$

$$x = \frac{7}{3}$$

$$\left\{\frac{7}{3}\right\}$$

49.

$$S = D$$

$$\sqrt{11x} = \sqrt{121 - 2x^2}$$

$$\left(\sqrt{11x}\right)^2 = \left(\sqrt{121 - 2x^2}\right)^2$$

$$11x = 121 - 2x^2$$

$$2x^2 + 11x - 121 = 0$$

$$(2x - 11)(x + 11) = 0$$

$$x = \frac{11}{2} \quad \text{or} \quad x = -11$$

Since $x > 0$, $x = \dfrac{11}{2} = 5.5$.

The price is \$5.50.

51.

$$\sqrt{x} + \sqrt{x + 5} = \sqrt{5x + 5}$$

$$\left(\sqrt{x} + \sqrt{x + 5}\right)^2 = \left(\sqrt{5x + 5}\right)^2$$

$$x + 2\sqrt{x} \cdot \sqrt{x + 5} + x + 5 = 5x + 5$$

$$2\sqrt{x} \cdot \sqrt{x + 5} = 3x$$

$$\left(2\sqrt{x} \cdot \sqrt{x + 5}\right)^2 = (3x)^2$$

$$4x(x + 5) = 9x^2$$

$$4x^2 + 20x = 9x^2$$

$$0 = 5x^2 - 20x$$

$$0 = 5x(x - 4)$$

$$x = 0 \quad \text{or} \quad x = 4$$

Check: $x = 0$

LS: $\sqrt{0} + \sqrt{0 + 5} = 0 + \sqrt{5} = \sqrt{5}$

RS: $\sqrt{5(0) + 5} = \sqrt{5}$

Checks.

Check: $x = 4$

LS: $\sqrt{4} + \sqrt{4 + 5} = \sqrt{4} + \sqrt{9} = 2 + 3 = 5$

RS: $\sqrt{5(4) + 5} = \sqrt{25} = 5$

Checks.

$$\{0, \ 4\}$$

53.

$$\sqrt{4x + 2} = \sqrt{6x + 6} - \sqrt{2 - 2x}$$

$$\left(\sqrt{4x + 2}\right)^2 = \left(\sqrt{6x + 6} - \sqrt{2 - 2x}\right)^2$$

$$4x + 2 = 6x + 6 - 2\sqrt{6x + 6} \cdot \sqrt{2 - 2x} + 2 - 2x$$

$$-6 = -2\sqrt{6x + 6} \cdot \sqrt{2 - 2x}$$

$$(-6)^2 = \left(-2\sqrt{6x + 6} \cdot \sqrt{2 - 2x}\right)^2$$

$$36 = 4(6x + 6)(2 - 2x)$$

$$9 = (6x + 6)(2 - 2x)$$

$$9 = -12x^2 + 12$$

$$12x^2 = 3$$

$$x^2 = \frac{1}{4}$$

$$x = \pm\frac{1}{2}$$

Check: $x = \dfrac{1}{2}$

LS: $\sqrt{4\left(\dfrac{1}{2}\right) + 2} = \sqrt{4} = 2$

RS: $\sqrt{6\left(\dfrac{1}{2}\right) + 6} - \sqrt{2 - 2\left(\dfrac{1}{2}\right)} = \sqrt{9} - \sqrt{1} = 3 - 1 = 2$

Checks.

Check: $x = -\dfrac{1}{2}$

LS: $\sqrt{4\left(-\dfrac{1}{2}\right) + 2} = \sqrt{0} = 0$

RS: $\sqrt{6\left(-\dfrac{1}{2}\right) + 6} - \sqrt{2 - 2\left(-\dfrac{1}{2}\right)} = \sqrt{3} - \sqrt{3} = 0$

Checks.

$$\left\{\pm\frac{1}{2}\right\}$$

55.

$$x^{2/3} - x^{1/3} - 2 = 0$$

$$(x^{1/3})^2 - x^{1/3} - 2 = 0$$

$$(x^{1/3} - 2)(x^{1/3} + 1) = 0$$

$$x^{1/3} - 2 = 0 \quad \text{or} \quad x^{1/3} + 1 = 0$$

$$x^{1/3} = 2 \quad \text{or} \quad x^{1/3} = -1$$

$$(x^{1/3})^3 = 2^3 \quad \text{or} \quad (x^{1/3})^3 = (-1)^3$$

$$x = 8 \quad \text{or} \quad x = -1$$

$$\{8, \ -1\}$$

57.

$$x^{2/5} - 2x^{1/5} - 3 = 0$$

$$(x^{1/5})^2 - 2x^{1/5} - 3 = 0$$

$$(x^{1/5} - 3)(x^{1/5} + 1) = 0$$

$$x^{1/5} - 3 = 0 \quad \text{or} \quad x^{1/5} + 1 = 0$$

$$x^{1/5} = 3 \quad \text{or} \quad x^{1/5} = -1$$

$$(x^{1/5})^5 = 3^5 \quad \text{or} \quad (x^{1/5})^5 = (-1)^5$$

$$x = 243 \quad \text{or} \quad x = -1$$

$$\{243, \ -1\}$$

Chapter 5 Review Problems

1.

$$12x^2 + 5x - 2 = 0$$

$$(4x - 1)(3x + 2) = 0$$

$$4x - 1 = 0 \quad \text{or} \quad 3x + 2 = 0$$

$$4x = 1 \quad \text{or} \quad 3x = -2$$

$$x = \frac{1}{4} \quad \text{or} \quad x = -\frac{2}{3}$$

$$\left\{ \frac{1}{4}, \ -\frac{2}{3} \right\}$$

3. $x^2 - 7x > 0$

Boundary numbers:

$$x^2 - 7x = 0$$

$$x(x - 7) = 0$$

$$x = 0, \quad x = 7$$

$$\overset{A}{\underset{0}{\rule{1cm}{0.4pt}}} \overset{B}{\underset{7}{\rule{1cm}{0.4pt}}} \overset{C}{\rule{1cm}{0.4pt}}$$

A: $x = -1$; $\quad (-1)^2 - 7(-1) > 0$

$$8 > 0 \quad T$$

B: $x = 1$; $\quad 1^2 - 7(1) > 0$

$$-6 > 0 \quad F$$

C: $x = 8$; $\quad 8^2 - 7(8) > 0$

$$8 > 0 \quad T$$

$$(-\infty, 0) \cup (7, \infty)$$

5. $3x^2 + 4x - 3 = 0$

$$a = 3, \quad b = 4, \quad c = -3$$

$$x = \frac{-4 \pm \sqrt{4^2 - 4(3)(-3)}}{2(3)}$$

$$= \frac{-4 \pm \sqrt{16 + 36}}{6}$$

$$= \frac{-4 \pm \sqrt{52}}{6}$$

$$= \frac{-4 \pm 2\sqrt{13}}{6}$$

$$= \frac{2(-2 \pm \sqrt{13})}{6}$$

$$= \frac{-2 \pm \sqrt{13}}{3}$$

$$\left\{ \frac{-2 \pm \sqrt{13}}{3} \right\}$$

7.

$$\frac{3}{x - 5} + \frac{x}{3(x + 2)} = \frac{5}{x^2 - 3x - 10}$$

$$\frac{3}{x - 5} + \frac{x}{3(x + 2)} = \frac{5}{(x - 5)(x + 2)}$$

239

$$3(x+2)(x-5)\left[\frac{3}{x-5} + \frac{x}{3(x+2)}\right] = 3(x+2)(x-5)\left[\frac{5}{(x-5)(x+2)}\right]$$

$$9(x+2) + x(x-5) = 15$$

$$9x + 18 + x^2 - 5x = 15$$

$$x^2 + 4x + 3 = 0$$

$$(x+3)(x+1) = 0$$

$$x = -3 \quad \text{or} \quad x = -1$$

$$\{-3, \ -1\}$$

9. $4x^2 + 12x + 9 = 0$

$$(2x+3)^2 = 0$$

$$2x + 3 = 0$$

$$x = -\frac{3}{2}$$

$$\left\{-\frac{3}{2}\right\}$$

11. $4x^2 - x + 1 = 0$

$a = 4, \ b = -1, \ c = 1$

$$x = \frac{-(-1) \pm \sqrt{(-1)^2 - 4(4)(1)}}{2(4)}$$

$$= \frac{1 \pm \sqrt{1 - 16}}{8}$$

$$= \frac{1 \pm \sqrt{-15}}{8}$$

$$= \frac{1 \pm \sqrt{15}i}{8}$$

$$= \frac{1}{8} \pm \frac{\sqrt{15}}{8}i$$

$$\left\{\frac{1}{8} \pm \frac{\sqrt{15}}{8}i\right\}$$

13. $(3x+2)^2 = 9$

$$3x + 2 = \pm\sqrt{9}$$

$$3x + 2 = \pm 3$$

$$3x + 2 = 3 \quad \text{or} \quad 3x + 2 = -3$$

$$3x = 1 \quad \text{or} \quad 3x = -5$$

$$x = \frac{1}{3} \quad \text{or} \quad x = -\frac{5}{3}$$

$$\left\{\frac{1}{3}, \ -\frac{5}{3}\right\}$$

15.
$$\sqrt{6x+1} - 1 = \sqrt{3x+4}$$

$$\left(\sqrt{6x+1} - 1\right)^2 = \left(\sqrt{3x+4}\right)^2$$

$$6x + 1 - 2\sqrt{6x+1} + 1 = 3x + 4$$

$$-2\sqrt{6x+1} = -3x + 2$$

$$\left(-2\sqrt{6x+1}\right)^2 = (-3x+2)^2$$

$$4(6x+1) = 9x^2 - 12x + 4$$

$$24x + 4 = 9x^2 - 12x + 4$$

$$0 = 9x^2 - 36x$$

$$0 = 9x(x-4)$$

$$x = 0 \quad \text{or} \quad x = 4$$

Check: $x = 0$

LS: $\sqrt{6(0)+1} - 1 = \sqrt{1} - 1 = 1 - 1 = 0$

RS: $\sqrt{3(0)+4} = \sqrt{4} = 2$

Does not check.

240

Check: $x = 4$

LS: $\sqrt{6(4) + 1} - 1 = \sqrt{25} - 1 = 5 - 1 = 4$

RS: $\sqrt{3(4) + 4} = \sqrt{16} = 4$

Checks.

$$\{4\}$$

17. $18x^2 + 9x - 5 = 0$

$(6x + 5)(3x - 1) = 0$

$6x + 5 = 0 \quad \text{or} \quad 3x - 1 = 0$

$x = -\dfrac{5}{6} \quad \text{or} \quad x = \dfrac{1}{3}$

$$\left\{-\dfrac{5}{6}, \; \dfrac{1}{3}\right\}$$

19. $4x^2 - 9 = 0$

$4x^2 = 9$

$x^2 = \dfrac{9}{4}$

$x = \pm\sqrt{\dfrac{9}{4}}$

$x = \pm\dfrac{3}{2}$

$$\left\{\pm\dfrac{3}{2}\right\}$$

21. $\dfrac{1}{x} + \dfrac{x + 1}{x^2} = \dfrac{6}{x^3}$

$x^3\left(\dfrac{1}{x} + \dfrac{x + 1}{x^2}\right) + x^3\left(\dfrac{6}{x^3}\right)$

$x^2 + x(x + 1) = 6$

$x^2 + x^2 + x = 6$

$2x^2 + x - 6 = 0$

$(2x - 3)(x + 2) = 0$

$x = \dfrac{3}{2} \quad \text{or} \quad x = -2$

$$\left\{\dfrac{3}{2}, \; -2\right\}$$

23. $\dfrac{3}{x} + \dfrac{3}{x^2 - x} = \dfrac{x}{1 - x}$

$\dfrac{3}{x} + \dfrac{3}{x(x - 1)} = \dfrac{-x}{x - 1}$

$x(x - 1)\left[\dfrac{3}{x} + \dfrac{3}{x(x - 1)}\right] = x(x - 1)\left(-\dfrac{x}{x - 1}\right)$

$3(x - 1) + 3 = -x^2$

$3x - 3 + 3 = -x^2$

$x^2 + 3x = 0$

$x(x + 3) = 0$

$x = 0 \quad \text{or} \quad x = -3$

Since $x = 0$ causes a denominator to equal zero, the only solution is $x = -3$.

$$\{-3\}$$

25. $x^3 = \dfrac{1}{8}$

$x^3 - \dfrac{1}{8} = 0$

$8x^3 - 1 = 0$

$(2x - 1)(4x^2 + 2x + 1) = 0$

$2x - 1 = 0 \quad \text{or} \quad 4x^2 + 2x + 1 = 0$

$x = \dfrac{1}{2} \quad \text{or} \qquad x = \dfrac{-2 \pm \sqrt{2^2 - 4(4)(1)}}{2(4)}$

$= \dfrac{-2 \pm \sqrt{4 - 16}}{8}$

$= \dfrac{-2 \pm \sqrt{-12}}{8}$

$= \dfrac{-2 \pm 2\sqrt{3}i}{8}$

$= -\dfrac{2}{8} \pm \dfrac{2\sqrt{3}}{8}i$

$= -\dfrac{1}{4} \pm \dfrac{\sqrt{3}}{4}i$

$$\left\{\dfrac{1}{2}, \; -\dfrac{1}{4} \pm \dfrac{\sqrt{3}}{4}i\right\}$$

27. $3x^2 + 8x + 4 = 0$

$(3x + 2)(x + 2) = 0$

$3x + 2 = 0 \quad$ or $\quad x + 2 = 0$

$x = -\dfrac{2}{3} \quad$ or $\qquad x = -2$

$$\left\{-\dfrac{2}{3}, \ -2\right\}$$

29. $\sqrt{x + 2} - x = 0$

$\sqrt{x + 2} = x$

$\left(\sqrt{x + 2}\right)^2 = x^2$

$x + 2 = x^2$

$0 = x^2 - x - 2$

$0 = (x - 2)(x + 1)$

$x = 2 \quad$ or $\quad x = -1$

Check: $x = 2$

LS: $\sqrt{2 + 2} - 2 = \sqrt{4} - 2 = 2 - 2 = 0$

RS: 0

Checks.

Check: $x = -1$

LS: $\sqrt{-1 + 2} - (-1) = \sqrt{1} + 1 = 1 + 1 = 2$

RS: 0

Does not check.

$$\{2\}$$

31. $3x^2 - x - 2 \geq 0$

Boundary numbers:

$3x^2 - x - 2 = 0$

$(3x + 2)(x - 1) = 0$

$x = -\dfrac{2}{3}, \ x = 1$

A	B	C
$-\dfrac{2}{3}$	1	

A: $x = -1; \quad 3(-1)^2 - (-1) - 2 \geq 0$

$2 \geq 0 \quad$ T

B: $x = 0; \qquad 3(0)^2 - 0 - 2 \geq 0$

$-2 \geq 0 \quad$ F

C: $x = 2; \qquad 3(2)^2 - 2 - 2 \geq 0$

$8 \geq 0 \quad$ T

$$\left(-\infty, \ -\dfrac{2}{3}\right] \cup [1, \ \infty)$$

33. $x^4 - 50x^2 + 49 = 0$

$(x^2 - 49)(x^2 - 1) = 0$

$x^2 - 49 = 0 \quad$ or $\quad x^2 - 1 = 0$

$x^2 = 49 \quad$ or $\qquad x^2 = 1$

$x = \pm 7 \quad$ or $\qquad x = \pm 1$

$$\{\pm 7, \ \pm 1\}$$

35. $\sqrt{2x + 1} - \sqrt{x} = 1$

$\sqrt{2x + 1} = \sqrt{x} + 1$

$\left(\sqrt{2x + 1}\right)^2 = \left(\sqrt{x} + 1\right)^2$

$2x + 1 = x + 2\sqrt{x} + 1$

$x = 2\sqrt{x}$

$x^2 = \left(2\sqrt{x}\right)^2$

$x^2 = 4x$

$x^2 - 4x = 0$

$x(x - 4) = 0$

$x = 0 \quad$ or $\quad x = 4$

Check: $x = 0$

LS: $\sqrt{2(0) + 1} - \sqrt{0} = \sqrt{1} - 0 = 1$

RS: 1

Checks.

Check: $x = 4$

LS: $\sqrt{2(4) + 1} - \sqrt{4} = \sqrt{9} - \sqrt{4} = 3 - 2 = 1$

RS: 1

Checks.

$$\{0, \ 4\}$$

37. $\sqrt{2x + 5} - \sqrt{x - 1} = 2$

$$\sqrt{2x + 5} = 2 + \sqrt{x - 1}$$

$$\left(\sqrt{2x + 5}\right)^2 = \left(2 + \sqrt{x - 1}\right)^2$$

$$2x + 5 = 4 + 4\sqrt{x - 1} + x - 1$$

$$x + 2 = 4\sqrt{x - 1}$$

$$(x + 2)^2 = \left(4\sqrt{x - 1}\right)^2$$

$$x^2 + 4x + 4 = 16(x - 1)$$

$$x^2 + 4x + 4 = 16x - 16$$

$$x^2 - 12x + 20 = 0$$

$$(x - 2)(x - 10) = 0$$

$$x = 2 \quad \text{or} \quad x = 10$$

Check: $x = 2$

LS: $\sqrt{2(2) + 5} - \sqrt{2 - 1} = \sqrt{9} - \sqrt{1} = 3 - 1 = 2$

RS: 2

Checks.

Check: $x = 10$

LS: $\sqrt{2(10) + 5} - \sqrt{10 - 1} = \sqrt{25} - \sqrt{9} = 5 - 3 = 2$

RS: 2

Checks.

$$\{2, \ 10\}$$

243

39.
$$\frac{3}{x^2 + 6x + 8} + \frac{x + 4}{x^2 + 3x + 2} = \frac{5}{x^2 + 5x + 4}$$

$$\frac{3}{(x + 4)(x + 2)} + \frac{x + 4}{(x + 2)(x + 1)} = \frac{5}{(x + 4)(x + 1)}$$

$$(x + 4)(x + 2)(x + 1)\left[\frac{3}{(x + 4)(x + 2)} + \frac{x + 4}{(x + 2)(x + 1)}\right] = (x + 4)(x + 2)(x + 1)\left[\frac{5}{(x + 4)(x + 1)}\right]$$

$$3(x + 1) + (x + 4)(x + 4) = 5(x + 2)$$

$$3x + 3 + x^2 + 8x + 16 = 5x + 10$$

$$x^2 + 6x + 9 = 0$$

$$(x + 3)^2 = 0$$

$$x = -3$$

$$\{-3\}$$

41. $5x^2 + 5x + 2 = 0$

$a = 5, \quad b = 5, \quad c = 2$

$$x = \frac{-5 \pm \sqrt{5^2 - 4(5)(2)}}{2(5)}$$

$$= \frac{-5 \pm \sqrt{25 - 40}}{10}$$

$$= \frac{-5 \pm \sqrt{-15}}{10}$$

$$= \frac{-5 \pm \sqrt{15}i}{10}$$

$$= -\frac{5}{10} \pm \frac{\sqrt{15}}{10}i$$

$$= -\frac{1}{2} \pm \frac{\sqrt{15}}{10}i$$

$$\left\{-\frac{1}{2} \pm \frac{\sqrt{15}}{10}i\right\}$$

43.
$$\frac{1}{x + 1} + \frac{1}{3} = 2x$$

$$3(x + 1)\left(\frac{1}{x + 1} + \frac{1}{3}\right) = 3(x + 1)(2x)$$

43. (con't.)

$$3 + x + 1 = 6x^2 + 6x$$

$$0 = 6x^2 + 5x - 4$$

$$0 = (3x + 4)(2x - 1)$$

$$x = -\frac{4}{3} \quad \text{or} \quad x = \frac{1}{2}$$

$$\left\{-\frac{4}{3}, \frac{1}{2}\right\}$$

45. $(x - 2)^2 = 6$

$$x - 2 = \pm\sqrt{6}$$

$$x = 2 \pm \sqrt{6}$$

$$\{2 \pm \sqrt{6}\}$$

47. $\sqrt{2x + 7} - \sqrt{x} = 2$

$$\sqrt{2x + 7} = \sqrt{x} + 2$$

$$\left(\sqrt{2x + 7}\right)^2 = \left(\sqrt{x} + 2\right)^2$$

$$2x + 7 = x + 4\sqrt{x} + 4$$

$$x + 3 = 4\sqrt{x}$$

$$(x + 3)^2 = \left(4\sqrt{x}\right)^2$$

$$x^2 + 6x + 9 = 16x$$

$$x^2 - 10x + 9 = 0$$

$$(x - 9)(x - 1) = 0$$

$$x = 9 \quad \text{or} \quad x = 1$$

Check: $x = 9$

LS: $\sqrt{2(9) + 7} - \sqrt{9} = \sqrt{25} - \sqrt{9} = 5 - 3 = 2$

RS: 2

Checks.

Check: $x = 1$

LS: $\sqrt{2(1) + 7} - \sqrt{1} = \sqrt{9} - \sqrt{1} = 3 - 1 = 2$

RS: 2

Checks.

$$\{9, \; 1\}$$

49. $x^2(x^2 - 5) - 3(x^2 - 5) = 0$

$$(x^2 - 5)(x^2 - 3) = 0$$

$$x^2 - 5 = 0 \quad \text{or} \quad x^2 - 3 = 0$$

$$x^2 = 5 \quad \text{or} \quad x^2 = 3$$

$$x = \pm\sqrt{5} \quad \text{or} \quad x = \pm\sqrt{3}$$

$$\{\pm\sqrt{5}, \; \pm\sqrt{3}\}$$

51. $(x + 3)^2 = -5$

$$x + 3 = \pm\sqrt{-5}$$

$$x + 3 = \pm\sqrt{5}\,i$$

$$x = -3 \pm \sqrt{5}\,i$$

$$\left\{-3 \pm \sqrt{5}\,i\right\}$$

53. $16x^4 + 9x^2 - 7 = 0$

$$(16x^2 - 7)(x^2 + 1) = 0$$

$$16x^2 - 7 = 0 \quad \text{or} \quad x^2 + 1 = 0$$

$$x^2 = \frac{7}{16} \quad \text{or} \quad x^2 = -1$$

$$x = \pm\frac{\sqrt{7}}{4} \quad \text{or} \quad x = \pm i$$

$$\left\{\pm\frac{\sqrt{7}}{4}, \; \pm i\right\}$$

55. $\dfrac{9}{x - 1} + \dfrac{5}{3 + x} \geq 2$

Free boundary numbers:

$$x = 1, \quad x = -3$$

Other boundary numbers:

$$\frac{9}{x - 1} + \frac{5}{3 + x} = 2$$

$$(x - 1)(3 + x)\left(\frac{9}{x - 1} + \frac{5}{3 + x}\right) = (x - 1)(3 + x)(2)$$

$$9(3 + x) + 5(x - 1) = 2(x^2 + 2x - 3)$$

$$27 + 9x + 5x - 5 = 2x^2 + 4x - 6$$

$$0 = 2x^2 - 10x - 28$$

$$0 = x^2 - 5x - 14$$

$$0 = (x - 7)(x + 2)$$

$$x = 7, \quad x = -2$$

$$\overline{\underset{-3}{}\overset{A}{}\Big|\underset{-2}{}\overset{B}{}\Big|\underset{1}{}\overset{C}{}\Big|\underset{7}{}\overset{D}{}\Big|\overset{E}{}}$$

A: $x = -4$; $\qquad \dfrac{9}{-4 - 1} + \dfrac{5}{3 - 4} 1 \geq 2$

$$-\frac{34}{5} \geq 2 \quad \text{F}$$

B: $x = -\dfrac{5}{2}$; $\qquad \dfrac{9}{-\dfrac{5}{2} - 1} + \dfrac{5}{3 - \dfrac{5}{2}} \geq 2$

$$\frac{52}{7} \geq 2 \quad \text{T}$$

245

C: $x = 0$; $\qquad \dfrac{9}{0-1} + \dfrac{5}{3+0} \geq 2$

$$-\dfrac{22}{3} \geq 2 \quad \text{F}$$

D: $x = 2$; $\qquad \dfrac{9}{2-1} + \dfrac{5}{3+2} \geq 2$

$$10 \geq 2 \quad \text{T}$$

E: $x = 8$; $\qquad \dfrac{9}{8-1} + \dfrac{5}{3+8} \geq 2$

$$\dfrac{134}{77} \geq 2 \quad \text{F}$$

$$(-3, -2] \cup (1, 7]$$

57. $(x - 7)^2 = -8$

$x - 7 = \pm\sqrt{-8}$

$x - 7 = \pm 2\sqrt{2}\,i$

$x = 7 \pm 2\sqrt{2}\,i$

$\{7 \pm 2\sqrt{2}\,i\}$

59. 1st number: x

2nd number: $17 - x$

$x(17 - x) = -390$

$17x - x^2 = -390$

$0 = x^2 - 17x - 390$

$0 = (x + 13)(x - 30)$

$x = -13 \quad \text{or} \quad x = 30$

$17 - x = 30 \qquad 17 - x = -13$

The numbers are -13 and 30.

61. leg length: x

$x^2 + 2^2 = \left(3\sqrt{2}\right)^2$

$x^2 + 4 = 18$

$x^2 = 14$

$x = \pm\sqrt{14}$

Since $x > 0$, $x = \sqrt{14}$

The leg is $\sqrt{14}$ m.

63.

	D	R	T = D/R
1st part	15	$r + 6$	$15/(r + 6)$
2nd part	12	r	$12/r$

$$\dfrac{15}{r+6} + \dfrac{12}{r} = 1$$

$$r(r+6)\left(\dfrac{15}{r+6} + \dfrac{12}{r}\right) = r(r+6)(1)$$

$$15r + 12(r + 6) = r^2 + 6r$$

$$15r + 12r + 72 = r^2 + 6r$$

$$0 = r^2 - 21r - 72$$

$$0 = (r - 24)(r + 3)$$

$$r = 24 \quad \text{or} \quad r = -3$$

Since $r > 0$, $r = 24$

His speed on the last part of the trip was 24 mph.

65. $t^2 + 930t > 0.30(96000)$

$t^2 + 930t > 28800$

Boundary numbers:

$$t^2 + 930t = 28800$$

$t^2 + 930t - 28800 = 0$

$(t + 960)(t - 30) = 0$

$t = -960, \quad t = 30$

$$\underset{-960}{\overset{A}{\rule{0pt}{0pt}}} \bigg|\ \underset{30}{\overset{B}{\rule{0pt}{0pt}}}\ \bigg| \overset{C}{\rule{0pt}{0pt}}$$

A: $x = -1000$; $(-1000)^2 + 930(-1000) > 28800$

$$70000 > 28800 \quad \text{T}$$

246

B: $x = 0$; $\qquad 0^2 + 930(0) > 28800$

$\qquad\qquad\qquad\qquad 0 > 28800$ F

C: $x = 40$; $\qquad 40^2 + 930(40) > 28800$

$\qquad\qquad\qquad\qquad 38800 > 28800$ T

$\qquad (-\infty, -960) \cup (30, \infty)$

Since $t > 0$, $\ t > 30$

The maintenance cost will be greater than 30% of the purchase price after 30 years.

67. Factored

69. $16 - (x + 1)^2 = 4^2 - (x + 1)^2$

$\qquad = [4 - (x + 1)][4 + (x + 1)]$

$\qquad = (4 - x - 1)(4 + x + 1)$

$\qquad = (3 - x)(5 + x)$

71. $y(c - d) + x(c - d) = (c - d)(y + x)$

73. 3 terms; $\ x^2 - xy + y^2$ prime

$\quad x - y$ is not a factor.

75. 2 terms; $\ x - y(a + b)$ prime
$\quad x - y$ is not a factor.

77. $\dfrac{-3x^{-2}}{x^{-1}y^3} = \dfrac{-3x}{x^2y^3}$

$\qquad = -\dfrac{3}{xy^3}$

79. $a^{-1}b^2(a^{-2} + b) = a^{-1}b^2 \cdot a^{-2} + a^{-1}b^2 \cdot b$

$\qquad\qquad = a^{-3}b^2 + a^{-1}b^3$

$\qquad\qquad = \dfrac{b^2}{a^3} + \dfrac{b^3}{a}$

81. $\sqrt[5]{a^5 + b^5}\qquad$ Cannot be simplified.

83. $(3y - 1)^3 = (3y)^3 - 3(3y)^2(1) + 3(3y)(1)^2 - 1^3$

$\qquad\qquad = 27y^3 - 27y^2 + 9y - 1$

85. $\dfrac{x(b + c) - y(b + c)}{(b + c)(b - c)} = \dfrac{(b + c)(x - y)}{(b + c)(b - c)}$

$\qquad\qquad = \dfrac{x - y}{b - c}$

87. $\dfrac{3}{x + 4} - \dfrac{x}{x + 3} = \dfrac{3(x + 3)}{(x + 4)(x + 3)} - \dfrac{x(x + 4)}{(x + 3)(x + 4)}$

$\qquad = \dfrac{3x + 9}{(x + 4)(x + 3)} - \dfrac{x^2 + 4x}{(x + 3)(x + 4)}$

$\qquad = \dfrac{3x + 9 - (x^2 + 4x)}{(x + 4)(x + 3)}$

$\qquad = \dfrac{3x + 9 - x^2 - 4x}{(x + 4)(x + 3)}$

$\qquad = \dfrac{-x^2 - x + 9}{(x + 4)(x + 3)}$

89. Expression;

$\quad (1 - 2i)(1 + i) = 1 + i - 2i - 2i^2$

$\qquad\qquad = 1 - i - 2(-1)$

$\qquad\qquad = 1 - i + 2$

$\qquad\qquad = 3 - i$

91. Expression;

$\qquad \dfrac{x}{x^2 - 5x - 14} - \dfrac{1}{x - 7}$

$\quad = \dfrac{x}{(x - 7)(x + 2)} - \dfrac{1}{x - 7}$

$\quad = \dfrac{x}{(x - 7)(x + 2)} - \dfrac{1(x + 2)}{(x - 7)(x + 2)}$

$\quad = \dfrac{x - (x + 2)}{(x - 7)(x + 2)}$

$\quad = \dfrac{x - x - 2}{(x - 7)(x + 2)}$

$\quad = \dfrac{-2}{(x - 7)(x + 2)}$

93. Equation;

$$\left| \frac{1}{4}(x - 3) \right| = 3$$

$$\frac{1}{4}(x - 3) = 3 \quad \text{or} \quad \frac{1}{4}(x - 3) = -3$$

$$x - 3 = 12 \quad \text{or} \quad x - 3 = -12$$

$$x = 15 \quad \text{or} \quad x = -9$$

$$\{15, \ -9\}$$

95. Equation;

$$\frac{3}{5}(x + 2) - 1 = \frac{1}{3}(1 - x)$$

$$15\left[\frac{3}{5}(x + 2) - 1 \right] = 15\left[\frac{1}{3}(1 - x) \right]$$

$$9(x + 2) - 15 = 5(1 - x)$$

$$9x + 18 - 15 = 5 - 5x$$

$$14x = 2$$

$$x = \frac{2}{14}$$

$$x = \frac{1}{7}$$

$$\left\{ \frac{1}{7} \right\}$$

Chapter 5 Test

1. $x^2 - 2x - 3 = 5$

$$x^2 - 2x - 8 = 0$$

$$(x - 4)(x + 2) = 0$$

$$x = 4 \quad \text{or} \quad x = -2$$

B

2. $\dfrac{-3 \pm \sqrt{12}}{12} = \dfrac{-3 \pm 2\sqrt{3}}{12}$

$$= \frac{-3}{12} \pm \frac{2\sqrt{3}}{12}$$

$$= -\frac{1}{4} \pm \frac{\sqrt{3}}{6}$$

A

3. $2x^2 - 2x + 3 = 0$

$$a = 2, \quad b = -2, \quad c = 3$$

$$x = \frac{-(-2) \pm \sqrt{(-2)^2 - 4(2)(3)}}{2(2)}$$

$$= \frac{2 \pm \sqrt{4 - 24}}{4}$$

$$= \frac{2 \pm \sqrt{-20}}{4}$$

$$= \frac{2 \pm 2\sqrt{5}\,i}{4}$$

$$= \frac{1}{2} \pm \frac{\sqrt{5}}{2}\,i$$

D

4. $\qquad x = \sqrt{7 - x} + 1$

$$x - 1 = \sqrt{7 - x}$$

$$(x - 1)^2 = \left(\sqrt{7 - x} \right)^2$$

$$x^2 - 2x + 1 = 7 - x$$

$$x^2 - x - 6 = 0$$

$$(x - 3)(x + 2) = 0$$

$$x = 3 \quad \text{or} \quad x = -2$$

Check: $x = 3$

LS: 3

RS: $\sqrt{7 - 3} + 1 = \sqrt{4} + 1 = 3$

Checks.

Check: $x = -2$

LS: -2

RS: $\sqrt{7 - (-2)} + 1 = \sqrt{9} + 1 = 4$

Does not check.

$\{3\}$

C

5. $4x^4 - 17x^2 + 4 = 0$

$(4x^2 - 1)(x^2 - 4) = 0$

$4x^2 - 1 = 0 \quad \text{or} \quad x^2 - 4 = 0$

$x^2 = \dfrac{1}{4} \quad \text{or} \quad x^2 = 4$

$x = \pm\dfrac{1}{2} \quad \text{or} \quad x = \pm 2$

C

6. $3x^2 - 4x - 4 < 0$

Boundary numbers:

$3x^2 - 4x - 4 = 0$

$(3x + 2)(x - 2) = 0$

$x = -\dfrac{2}{3}, \quad x = 2$

$$\underbrace{}_{} \overset{A}{} \Big| \overset{B}{} \Big| \overset{C}{}$$

$$\begin{array}{c|c|c} A & B & C \\ \hline & & \\ -\frac{2}{3} & 2 & \end{array}$$

A: $x = -1$; $3(-1)^2 - 4(-1) - 4 < 0$

$3 < 0 \quad \text{F}$

B: $x = 0$; $3(0)^2 - 4(0) - 4 < 0$

$-4 < 0 \quad \text{T}$

C: $x = 3$; $3(3)^2 - 4(3) - 4 < 0$

$11 < 0 \quad \text{F}$

$\left(-\dfrac{2}{3}, \ 2\right)$

7. $\dfrac{3}{x + 1} + \dfrac{2}{x + 3} = 2$

$(x + 1)(x + 3)\left(\dfrac{3}{x + 1} + \dfrac{2}{x + 3}\right) = (x + 1)(x + 3)(2)$

$3(x + 3) + 2(x + 1) = (x^2 + 4x + 3)(2)$

$3x + 9 + 2x + 2 = 2x^2 + 8x + 6$

$0 = 2x^2 + 3x - 5$

$0 = (2x + 5)(x - 1)$

$x = -\dfrac{5}{2} \quad \text{or} \quad x = 1$

$\left\{-\dfrac{5}{2}, \ 1\right\}$

8. $(x - 3)^2 = 9$

$\sqrt{(x - 3)^2} = \pm\sqrt{9}$

$x - 3 = \pm 3$

$x = 3 \pm 3$

$x = 3 + 3 \quad \text{or} \quad x = 3 - 3$

$x = 6 \qquad \text{or} \quad x = 0$

$\{6, \ 0\}$

9. $x^2 \geq 3x$

Boundary numbers:

$x^2 = 3x$

$x^2 - 3x = 0$

$x(x - 3) = 0$

$x = 0, \quad x = 3$

$$\begin{array}{c|c|c} A & B & C \\ \hline & & \\ 0 & 3 & \end{array}$$

A: $x = -1$; $(-1)^2 \geq 3(-1)$

$1 \geq -3$ T

B: $x = 1$; $1^2 \geq 3(1)$

$1 \geq 3$ F

C: $x = 4$; $4^2 \geq 3(4)$

$16 \geq 12$ T

$(-\infty, 0] \cup [3, \infty)$

10. $x^3 - 8 = 0$

$(x - 2)(x^2 + 2x + 4) = 0$

$x - 2 = 0$ or $x^2 + 2x + 4 = 0$

$x = 2$ or $x = \dfrac{-2 \pm \sqrt{2^2 - 4(1)(4)}}{2(1)}$

$= \dfrac{-2 \pm \sqrt{4 - 16}}{2}$

$= \dfrac{-2 \pm \sqrt{-12}}{2}$

$= \dfrac{-2 \pm 2\sqrt{3}i}{2}$

$= \dfrac{-2}{2} \pm \dfrac{2\sqrt{3}}{2}i$

$= -1 \pm \sqrt{3}i$

$\{2, \; -1 \pm \sqrt{3}i\}$

11. $\dfrac{1}{x^4} + \dfrac{3}{x^2} - 4 = 0$

$x^4\left(\dfrac{1}{x^4} + \dfrac{3}{x^2} - 4\right) = x^4(0)$

$1 + 3x^2 - 4x^4 = 0$

$4x^4 - 3x^2 - 1 = 0$

$(4x^2 + 1)(x^2 - 1) = 0$

$4x^2 + 1 = 0$ or $x^2 - 1 = 0$

$x^2 = -\dfrac{1}{4}$ or $x^2 = 1$

$x = \pm\dfrac{1}{2}i$ or $x = \pm 1$

$\left\{\pm\dfrac{1}{2}i, \;\; \pm 1\right\}$

12. $\sqrt{x + 1} + 2 = \sqrt{2x + 10}$

$\left(\sqrt{x + 1} + 2\right)^2 = \left(\sqrt{2x + 10}\right)^2$

$x + 1 + 4\sqrt{x + 1} + 4 = 2x + 10$

$4\sqrt{x + 1} = x + 5$

$\left(4\sqrt{x + 1}\right)^2 = (x + 5)^2$

$16(x + 1) = x^2 + 10x + 25$

$16x + 16 = x^2 + 10x + 25$

$0 = x^2 - 6x + 9$

$0 = (x - 3)^2$

$x = 3$

Check: $x = 3$

LS: $\sqrt{3 + 1} + 2 = \sqrt{4} + 2 = 2 + 2 = 4$

RS: $\sqrt{2(3) + 10} = \sqrt{16} = 4$

Checks.

$\{3\}$

13.　$3x^2 - 6x - 3 = 0$

$$x^2 - 2x - 1 = 0$$

$$x^2 - 2x = 1$$

$$x^2 - 2x + (1)^2 = 1 + (1)^2$$

$$(x - 1)^2 = 2$$

$$x - 1 = \pm\sqrt{2}$$

$$x = 1 \pm \sqrt{2}$$

$$\{1 \pm \sqrt{2}\}$$

14.　$2x^2 - x - 1 = 0$

$$a = 2, \ b = -1, \ c = -1$$

$$b^2 - 4ac = (-1)^2 - 4(2)(-1)$$

$$= 1 + 8$$

$$= 9$$

B, since the discriminant is positive.

15.

	D	R	T = D/R
to Ventura	20	r	20/r
to S. B.	30	$r - 10$	30/($r - 10$)

$$\frac{20}{r} + \frac{30}{r - 10} = 4$$

$$r(r - 10)\left(\frac{20}{r} + \frac{30}{r - 10}\right) = r(r - 10)(4)$$

$$20(r - 10) + 30r = 4r^2 - 40r$$

$$20r - 200 + 30r = 4r^2 - 40r$$

$$0 = 4r^2 - 90r + 200$$

$$0 = 2r^2 - 45r + 100$$

$$0 = (2r - 5)(r - 20)$$

$$r = \frac{5}{2} \qquad \text{or} \quad r = 20$$

$$r - 10 = -\frac{15}{2} \qquad r - 10 = 10$$

Since both rates must be positive, $r = 20$.
Her original speed was 20 mph.

251

CHAPTER 6

Problem Set 6.1

1. (2, 5)   QI

3. (11, -1)   QIV

5. (-3, 0)   *x*-axis

7. (2.4, -0.01)   QIV

9. (0, 0)   *x*- and *y*-axis

11.

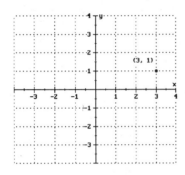

13.

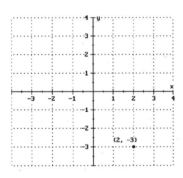

15.

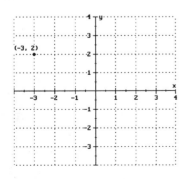

17.

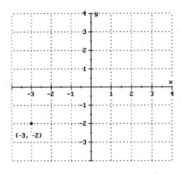

19.

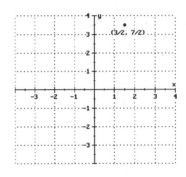

21.

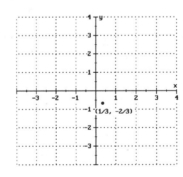

23.

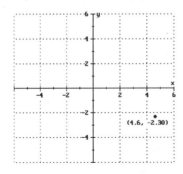

25. $(8, 6)$, $(5, 2)$

$$d = \sqrt{(8 - 5)^2 + (6 - 2)^2}$$
$$= \sqrt{3^2 + 4^2}$$
$$= \sqrt{9 + 16}$$
$$= \sqrt{25}$$
$$= 5$$

27. $(-7, 3)$, $(1, -3)$

$$d = \sqrt{(-7 - 1)^2 + [3 - (-3)]^2}$$
$$= \sqrt{(-8)^2 + 6^2}$$
$$= \sqrt{64 + 36}$$
$$= \sqrt{100}$$
$$= 10$$

29. $(-5, 1)$, $(-5, 14)$

$$d = \sqrt{[-5 - (-5)]^2 + (1 - 14)^2}$$
$$= \sqrt{0^2 + (-13)^2}$$
$$= \sqrt{0 + 169}$$
$$= \sqrt{169}$$
$$= 13$$

31. $(-1, -2)$, $(-3, -4)$

$$d = \sqrt{[-1 - (-3)]^2 + [-2 - (-4)]^2}$$
$$= \sqrt{2^2 + 2^2}$$
$$= \sqrt{4 + 4}$$
$$= \sqrt{8}$$
$$= 2\sqrt{2}$$

33. $(7, 0)$, $(0, 7)$

$$d = \sqrt{(7 - 0)^2 + (0 - 7)^2}$$
$$= \sqrt{7^2 + (-7)^2}$$
$$= \sqrt{49 + 49}$$
$$= \sqrt{98}$$
$$= 7\sqrt{2}$$

35. $\left(\dfrac{1}{2}, \dfrac{5}{2}\right)$, $\left(\dfrac{3}{2}, \dfrac{1}{2}\right)$

$$d = \sqrt{\left(\dfrac{1}{2} - \dfrac{3}{2}\right)^2 + \left(\dfrac{5}{2} - \dfrac{1}{2}\right)^2}$$
$$= \sqrt{(-1)^2 + 2^2}$$
$$= \sqrt{1 + 4}$$
$$= \sqrt{5}$$

37. $\left(\dfrac{1}{2}, \dfrac{1}{6}\right)$, $\left(\dfrac{1}{3}, 1\right)$

$$d = \sqrt{\left(\dfrac{1}{2} - \dfrac{1}{3}\right)^2 + \left(\dfrac{1}{6} - 1\right)^2}$$
$$= \sqrt{\left(\dfrac{1}{6}\right)^2 + \left(-\dfrac{5}{6}\right)^2}$$
$$= \sqrt{\dfrac{1}{36} + \dfrac{25}{36}}$$
$$= \sqrt{\dfrac{26}{36}}$$
$$= \dfrac{\sqrt{26}}{6}$$

39. (5, 1), (-3, -1)

$$x = \frac{5 + (-3)}{2} = \frac{2}{2} = 1$$

$$y = \frac{1 + (-1)}{2} = \frac{0}{2} = 0$$

midpoint: (1, 0)

41. (11, 8), (3, 2)

$$d = \sqrt{(11 - 3)^2 + (8 - 2)^2}$$

$$= \sqrt{8^2 + 6^2}$$

$$= \sqrt{64 + 36}$$

$$= \sqrt{100}$$

$$= 10$$

43. (2, 3), (14, 8)

$$d = \sqrt{(2 - 14)^2 + (3 - 8)^2}$$

$$= \sqrt{(-12)^2 + (-5)^2}$$

$$= \sqrt{144 + 25}$$

$$= \sqrt{169}$$

$$= 13$$

45. (10, 2), (2, 2)

$$d = \sqrt{(10 - 2)^2 + (2 - 2)^2}$$

$$= \sqrt{8^2 + 0^2}$$

$$= \sqrt{64}$$

$$= 8$$

47. (2, 3), (-4, 6)

$$d = \sqrt{[2 - (-4)]^2 + (3 - 6)^2}$$

$$= \sqrt{6^2 + (-3)^2}$$

$$= \sqrt{36 + 9}$$

$$= \sqrt{45}$$

$$= 3\sqrt{5}$$

49. (-2, -3), (1, 1)

$$d = \sqrt{(-2 - 1)^2 + (-3 - 1)^2}$$

$$= \sqrt{(-3)^2 + (-4)^2}$$

$$= \sqrt{9 + 16}$$

$$= \sqrt{25}$$

$$= 5$$

51. (-5, 12), (-4, 11)

$$d = \sqrt{[-5 - (-4)]^2 + (12 - 11)^2}$$

$$= \sqrt{(-1)^2 + 1^2}$$

$$= \sqrt{1 + 1}$$

$$= \sqrt{2}$$

53. (-2, 1), (-2, 10)

$$d = \sqrt{[-2 - (-2)]^2 + (1 - 10)^2}$$

$$= \sqrt{0^2 + (-9)^2}$$

$$= \sqrt{81}$$

$$= 9$$

55. (-4, -7), (5, -6)

$$d = \sqrt{(-4 - 5)^2 + [-7 - (-6)]^2}$$

$$= \sqrt{(-9)^2 + (-1)^2}$$

$$= \sqrt{81 + 1}$$

$$= \sqrt{82}$$

57. (0, 7), (-7, 0)

$$d = \sqrt{[0 - (-7)]^2 + (7 - 0)^2}$$

$$= \sqrt{7^2 + 7^2}$$

$$= \sqrt{49 + 49}$$

$$= \sqrt{98}$$

$$= 7\sqrt{2}$$

59. $\left(\dfrac{7}{2}, \dfrac{3}{2}\right), \left(-\dfrac{5}{2}, \dfrac{1}{2}\right)$

$$d = \sqrt{\left[\dfrac{7}{2} - \left(-\dfrac{5}{2}\right)\right]^2 + \left(\dfrac{3}{2} - \dfrac{1}{2}\right)^2}$$

$$= \sqrt{6^2 + 1^2}$$

$$= \sqrt{36 + 1}$$

$$= \sqrt{37}$$

61. $\left(\dfrac{1}{3}, \dfrac{1}{3}\right), \left(\dfrac{1}{4}, -\dfrac{1}{4}\right)$

$$d = \sqrt{\left(\dfrac{1}{3} - \dfrac{1}{4}\right)^2 + \left[\dfrac{1}{3} - \left(-\dfrac{1}{4}\right)\right]^2}$$

$$= \sqrt{\left(\dfrac{1}{12}\right)^2 + \left(\dfrac{7}{12}\right)^2}$$

$$= \sqrt{\dfrac{1}{144} + \dfrac{49}{144}}$$

$$= \sqrt{\dfrac{50}{144}}$$

$$= \dfrac{\sqrt{50}}{\sqrt{144}}$$

$$= \dfrac{5\sqrt{2}}{12}$$

63. $(-5, -4), (-1, -10)$

$$x = \dfrac{-5 + (-1)}{2} = \dfrac{-6}{2} = -3$$

$$y = \dfrac{-4 + (-10)}{2} = \dfrac{-14}{2} = -7$$

midpoint: $(-3, -7)$

65. $(\sqrt{5}, 0), (0, 3)$

$$d = \sqrt{(\sqrt{5} - 0)^2 + (0 - 3)^2}$$

$$= \sqrt{(\sqrt{5})^2 + (-3)^2}$$

$$= \sqrt{5 + 9}$$

$$= \sqrt{14}$$

$$\approx 3.74$$

$$x = \dfrac{\sqrt{5} + 0}{2} = \dfrac{\sqrt{5}}{2} \approx 1.1$$

$$y = \dfrac{0 + 3}{2} = \dfrac{3}{2} = 1.5$$

midpoint: $\left(\dfrac{\sqrt{5}}{2}, \dfrac{3}{2}\right)$

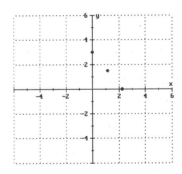

67. $(-3, \pi), \left(\dfrac{7}{4}, 1\right)$

$$d = \sqrt{\left(-3 - \dfrac{7}{4}\right)^2 + (\pi - 1)^2}$$

$$= \sqrt{\left(-\dfrac{19}{4}\right)^2 + (\pi - 1)^2}$$

$$\approx \sqrt{\dfrac{361}{16} + (2.14)^2}$$

$$\approx 5.21$$

67. (con't.)

$$x = \frac{-3 + \frac{7}{4}}{2} = \frac{-\frac{5}{4}}{2} = -\frac{5}{8}$$

$$y = \frac{\pi + 1}{2} \approx 2.1$$

midpoint: $\left(-\frac{5}{8}, \frac{\pi + 1}{2}\right)$

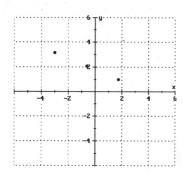

69. $(a, b); \quad a > 0, \ b > 0 \quad$ QI

71. $(c, b); \quad c < 0, \ b > 0 \quad$ QII

73. $(-a, c); \quad a > 0, \ c < 0 \qquad$ QIII
$$-a < 0$$

75. $(-b, -a); \quad b > 0, \quad a > 0 \qquad$ QIII
$$-b < 0, \quad -a < 0$$

77. $(-c, -d); \quad c < 0, \quad d < 0 \qquad$ QI
$$-c > 0, \quad -d > 0$$

79. $(-d, 0); \quad x$-axis

81. $P_1(2, 1), \ P_2(-2, -1), \ P_3(0, 0)$

$$d(P_1, P_2) = \sqrt{[2 - (-2)]^2 + [1 - (-1)]^2}$$
$$= \sqrt{4^2 + 2^2}$$
$$= \sqrt{16 + 4}$$
$$= \sqrt{20}$$
$$= 2\sqrt{5}$$

$$d(P_1, P_3) = \sqrt{(2 - 0)^2 + (1 - 0)^2}$$
$$= \sqrt{2^2 + 1^2}$$
$$= \sqrt{4 + 1}$$
$$= \sqrt{5}$$

$$d(P_2, P_3) = \sqrt{(-2 - 0)^2 + (-1 - 0)^2}$$
$$= \sqrt{(-2)^2 + (-1)^2}$$
$$= \sqrt{4 + 1}$$
$$= \sqrt{5}$$

Since $d(P_1, P_3) + d(P_2, P_3) = d(P_1, P_2)$, the three points are collinear.

83. $P_1(-1, 1), \ P_2(-2, 5), \ P_3(4, 2)$

$$d(P_1, P_2) = \sqrt{[-1 - (-2)]^2 + (1 - 5)^2}$$
$$= \sqrt{1^2 + (-4)^2}$$
$$= \sqrt{1 + 16}$$
$$= \sqrt{17}$$

$$d(P_1, P_3) = \sqrt{(-1 - 4)^2 + (1 - 2)^2}$$
$$= \sqrt{(-5)^2 + (-1)^2}$$
$$= \sqrt{25 + 1}$$
$$= \sqrt{26}$$

$$d(P_2, P_3) = \sqrt{(-2 - 4)^2 + (5 - 2)^2}$$
$$= \sqrt{(-6)^2 + 3^2}$$
$$= \sqrt{36 + 9}$$
$$= \sqrt{45}$$

$[d(P_1, P_2)]^2 + [d(P_1, P_3)]^2 \neq [d(P_2, P_3)^2]$, hence, it is not a right triangle.

Problem Set 6.2

1. $3x + 2y = 6$

If $x = 0, y = 3$.
(0, 3)
If $y = 0, x = 2$.
(2, 0)

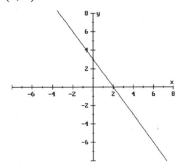

3. $-3x + 2y = 6$

If $x = 0, y = 3$.
(0, 3)
If $y = 0, x = -2$.
(-2, 0)

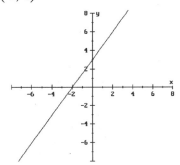

5. $x + y = 1$

If $x = 0, y = 1$.
(0, 1)
If $y = 0, x = 1$.
(1, 0)

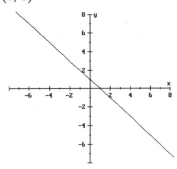

7. $4x + y = 4$

If $x = 0, y = 4$.
(0, 4)
If $y = 0, x = 1$.
(1, 0)

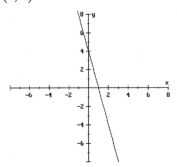

9. $3x - 4y = 4$

If $x = 0, y = -1$.
(0, -1)
If $y = 0, x = 4/3$.
(4/3, 0)

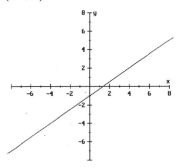

11. $2x - y - 1 = 0$

If $x = 0, y = -1$.
(0, -1)
If $y = 0, x = 1/2$.
(1/2, 0)

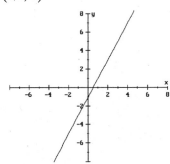

257

13. $10x - 15y + 25 = 0$

If $x = 0, y = 5/3$.
(0, 5/3)
If $y = 0, x = -5/2$.
(-5/2, 0)

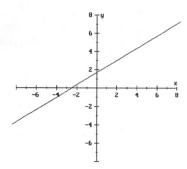

15. $x + y = 0$

(0, 0), (1, -1)

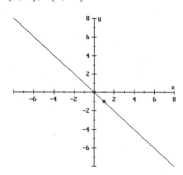

17. $3x = -2y$

(0, 0), (1, -3/2)

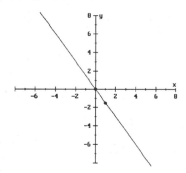

19. $x = 3$

(3, 0), (3, 1)

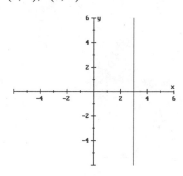

21. $x = 0$

(0, 0), (0, 1)

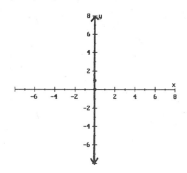

23. $2y = 7$

$y = 7/2$

(0, 7/2), (1, 7/2)

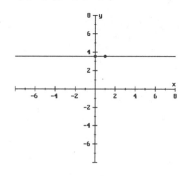

25. $d - 10t = 0$

(a) $(0, 0)$, $(1, 10)$

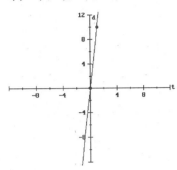

(b) Let $d = 70$: $70 - 10t = 0$

$$70 = 10t$$

$$7 = t$$

It will take 7 hours.

27. $5x - P = 10$

(a) $(0, -10)$, $(2, 0)$

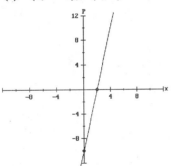

(b) Let $P = 0$, $x = 2$.
Two shirts must be sold to make no profit.

(c) Let $x = 100$: $5(100) - P = 10$

$$-P = -490$$

$$P = 490$$

The profit is $490.

(d) Let $x = 1$: $5(1) - P = 10$

$$-P = 5$$

$$P = -5$$

There is a loss of $5.

29. $4x + 3y = 12$

If $x = 0$, $y = 4$.
$(0, 4)$
If $y = 0$, $x = 3$.
$(3, 0)$

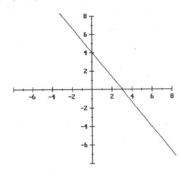

31. $-4x + 3y = 12$

If $x = 0$, $y = 4$.
$(0, 4)$
If $y = 0$, $x = -3$.
$(-3, 0)$

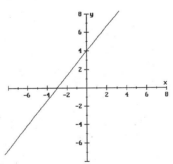

33. $4x + 3y = 0$

If $x = 0$, $y = 0$.
$(0, 0)$

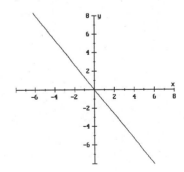

259

35. $x + 2y = 2$

If $x = 0$, $y = 1$.
(0, 1)
If $y = 0$, $x = 2$.
(2, 0)

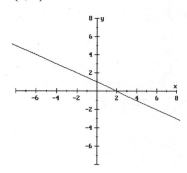

37. $x + 2y = 0$

If $x = 0$, $y = 0$.
(0, 0)

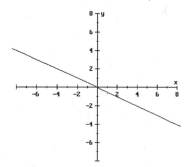

39. $7x + 2y = 14$

If $x = 0$, $y = 7$.
(0, 7)
If $y = 0$, $x = 2$.
(2, 0)

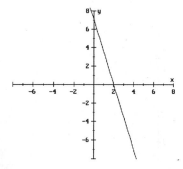

41. $x = -5$

If $y = 0$, $x = -5$.
(-5, 0)

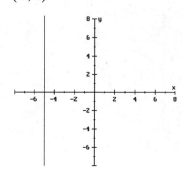

43. $3x + 4y = 6$

If $x = 0$, $y = 3/2$.
(0, 3/2)
If $y = 0$, $x = 2$.
(2, 0)

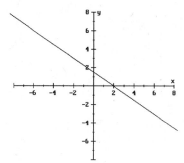

45. $4x + y = 1$

If $x = 0$, $y = 1$.
(0, 1)
If $y = 0$, $x = 1/4$.
(1/4, 0)

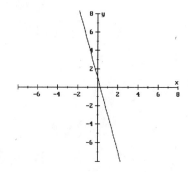

260

47. $3 + 2y = 0$

$\qquad 2y = -3$

$\qquad y = -3/2$

If $x = 0$, $y = -3/2$.
(0, −3/2)

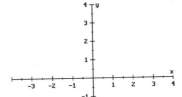

49. $2x + 5 = 3y$

If $x = 0$, $y = 5/3$.
(0, 5/3)
If $y = 0$, $x = -5/2$.
(−5/2, 0)

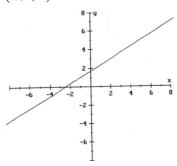

51. $12x - 15y + 20 = 0$

If $x = 0$, $y = 4/3$.
(0, 4/3)
If $y = 0$, $x = -5/3$.
(−5/3, 0)

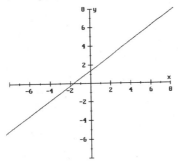

53. $0 = 1 + x$

$\qquad x = -1$

If $y = 0$, $x = -1$.
(−1, 0)

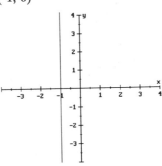

55. $4x + 3 = 0$

$\qquad 4x = -3$

$\qquad x = -3/4$

If $y = 0$, $x = -3/4$.
(−3/4, 0)

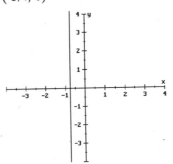

57. $A - 5L = 0$

(a)

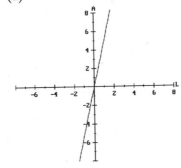

57. (con't.)

(b) Let $A = 105$: $105 - 5L = 0$

$$105 = 5L$$

$$21 = L$$

The length will be 21 m.

59. $C - 2x = 3$

(a)

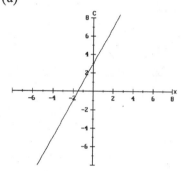

(b) Let $x = 15$: $C - 2(15) = 3$

$$C - 30 = 3$$

$$C = 33$$

The cost is $33.

61. $D - 3Y = 1$

(a)

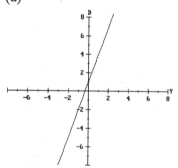

(b) Let $Y = 4$: $D - 3(4) = 1$

$$D = 13$$

The depreciation is $13,000.

63. $\dfrac{1}{2}x - \dfrac{1}{3}y = \dfrac{1}{12}$

If $x = 0, y = -\dfrac{1}{4}$.

$\left(0, -\dfrac{1}{4}\right)$

If $y = 0, x = \dfrac{1}{6}$.

$\left(\dfrac{1}{6}, 0\right)$

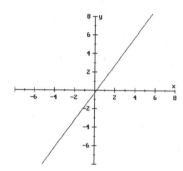

65. $\dfrac{x}{4} + \dfrac{y}{3} = \dfrac{1}{2}$

If $x = 0, y = \dfrac{3}{2}$.

$\left(0, \dfrac{3}{2}\right)$

If $y = 0, x = 2$.
$(2, 0)$

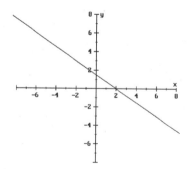

67. $y - 2x = \dfrac{6}{7}$

If $x = 0, y = \dfrac{6}{7}$.

$\left(0, \dfrac{6}{7}\right)$

If $y = 0, x = -\dfrac{3}{7}$.

$\left(-\dfrac{3}{7}, 0\right)$

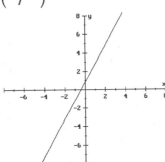

69. $Bx + Dy = D$

$Bx + \left(\dfrac{1}{2}B\right)y = \dfrac{1}{2}B$

$x + \dfrac{1}{2}y = \dfrac{1}{2}$

If $x = 0, y = 1$.
$(0, 1)$

If $y = 0, x = \dfrac{1}{2}$.

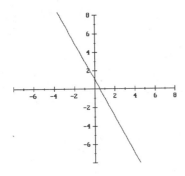

71. $Ax - By = CD$

$Ax - By = (2A)\left(\dfrac{1}{2}B\right)$

$Ax - By = AB$

If $x = 0, y = -A$.
$(0, -A)$ $A > 0$, $-A < 0$
If $y = 0, x = B$.
$(B, 0)$ $B < 0$

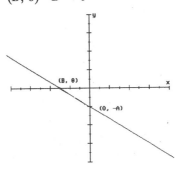

73. $C + Dy = 0$

$y = \dfrac{-C}{D}$

$C = 2A > 0$ $\qquad D = \dfrac{1}{2}B < 0$

$-C < 0$

$\dfrac{-C}{D} > 0$

If $x = 0, y = \dfrac{-C}{D}$.

$\left(0, \dfrac{-C}{D}\right)$

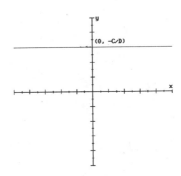

263

Problem Set 6.3

1. (2, 2), (3, 5)

$$m = \frac{5 - 2}{3 - 2} = \frac{3}{1} = 3$$

3. (2, 2), (5, 3)

$$m = \frac{3 - 2}{5 - 2} = \frac{1}{3}$$

5. $\left(\frac{5}{2}, \frac{1}{2}\right)$, $\left(\frac{7}{2}, \frac{5}{2}\right)$

$$m = \frac{\frac{5}{2} - \frac{1}{2}}{\frac{7}{2} - \frac{5}{2}} = \frac{2}{1} = 2$$

7. (4, −2), (2, 2)

$$m = \frac{-2-(-2)}{4 - 2} = \frac{4}{2} = -2 \quad 0$$

9. (−7, 4), (−3, 2)

$$m = \frac{4 - 2}{-7 - (-3)} = \frac{2}{-4} = -\frac{1}{2}$$

11. (−2, −6), (−4, −1)

$$m = \frac{-6 - (-1)}{-2 - (-4)} = \frac{-5}{2} = -\frac{5}{2}$$

13. (0, −9), (−4, 0)

$$m = \frac{-9 - 0}{0 - (-4)} = \frac{-9}{4} = -\frac{9}{4}$$

15. (−2, 3), (−5, 3)

$$m = \frac{3 - 3}{-2 - (-5)} = \frac{0}{3} = 0$$

17. (−2, 0), (0, 0)

$$m = \frac{0 - 0}{-2 - 0} = \frac{0}{-2} = 0$$

19. (−3, −6), (−3, −8)

$$m = \frac{-6 - (-8)}{-3 - (-3)} = \frac{2}{0} \text{ undefined}$$

21. $2x + 3y = 6$

$$3y = -2x + 6$$

$$y = -\frac{2}{3}x + 2$$

$$m = -\frac{2}{3}$$

23. $x + 7y = 14$

$$7y = -x + 14$$

$$y = -\frac{1}{7}x + 2$$

$$m = -\frac{1}{7}$$

25. $5x - y = 7$

$$-y = -5x + 7$$

$$y = 5x - 7$$

$$m = 5$$

27. $4x - 7y = 0$

$$-7y = -4x$$

$$y = \frac{4}{7}x$$

$$m = \frac{4}{7}$$

29. $y = x$

$$m = 1$$

31. $2x = 7$

$$x = \frac{7}{2}$$

Vertical line: slope undefined

264

33. $y = x - 1$

$m = 1, \quad b = -1$

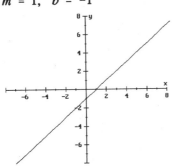

35. $y = \dfrac{1}{2}x + \dfrac{3}{2}$

$m = \dfrac{1}{2}, \quad b = \dfrac{3}{2}$

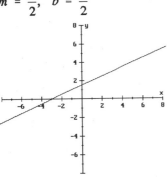

37. $y = 3x - 2$

$m = 3, \quad b = -2$

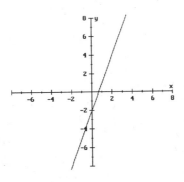

39. $(2, 5), \quad (1, -3)$

$m = \dfrac{5 - (-3)}{2 - 1} = \dfrac{8}{1} = 8$

Parallel lines have equal slopes.

41. $(2, -3), \quad (7, -3)$

$m = \dfrac{-3 - (-3)}{2 - 7} = \dfrac{0}{-5} = 0$

Parallel lines have equal slopes.

43. $4x + 3y = 24$

$3y = -4x + 24$

$y = -\dfrac{4}{3}x + 8$

$m = -\dfrac{4}{3}$

$m_\perp = -\dfrac{1}{-\dfrac{4}{3}} = \dfrac{3}{4}$

45. $(4, 4), \quad (5, 2)$

$m = \dfrac{4 - 2}{4 - 5} = \dfrac{2}{-1} = -2$

47. $(3, 2), \quad (6, -3)$

$m = \dfrac{2 - (-3)}{3 - 6} = \dfrac{5}{-3} = -\dfrac{5}{3}$

49. $(0, 3), \quad (6, -1)$

$m = \dfrac{3 - (-1)}{0 - 6} = \dfrac{4}{-6} = -\dfrac{2}{3}$

51. $(-2, 2), \quad (-5, 6)$

$m = \dfrac{6 - 2}{-5 - (-2)} = \dfrac{4}{-3} = -\dfrac{4}{3}$

53. $(-6, 5), \quad (6, -4)$

$m = \dfrac{5 - (-4)}{-6 - 6} = \dfrac{9}{-12} = -\dfrac{3}{4}$

55. $(1, 3), \quad (5, 3)$

$m = \dfrac{3 - 3}{5 - 1} = \dfrac{0}{4} = 0$

57. $(-4, -3),\quad (-8, -1)$

$$m = \frac{-3 - (-1)}{-4 - (-8)} = \frac{-2}{4} = -\frac{1}{2}$$

59. $(0, 3),\quad (-4, 0)$

$$m = \frac{3 - 0}{0 - (-4)} = \frac{3}{4}$$

61. $\left(-\dfrac{1}{2}, \dfrac{2}{3}\right),\quad \left(\dfrac{1}{2}, \dfrac{5}{3}\right)$

$$m = \frac{\dfrac{5}{3} - \dfrac{2}{3}}{\dfrac{1}{2} - \left(-\dfrac{1}{2}\right)} = \frac{1}{1} = 1$$

63. $\left(\dfrac{2}{5}, -\dfrac{7}{2}\right),\quad \left(\dfrac{4}{10}, -\dfrac{3}{2}\right)$

$$m = \frac{-\dfrac{7}{2} - \left(-\dfrac{3}{2}\right)}{\dfrac{2}{5} - \dfrac{4}{10}} = -\frac{2}{0} \text{ undefined}$$

65. $\left(\dfrac{1}{5}, -\dfrac{2}{5}\right),\quad \left(\dfrac{1}{3}, -\dfrac{2}{5}\right)$

$$m = \frac{-\dfrac{2}{5} - \left(-\dfrac{2}{5}\right)}{\dfrac{1}{5} - \dfrac{1}{3}} = \frac{0}{-\dfrac{2}{15}} = 0$$

67. $3x + 4y = 12$

(a) $(0, 3),\quad (4, 0)$

$$m = \frac{3 - 0}{0 - 4} = -\frac{3}{4}$$

(b) $3x + 4y = 12$

$$4y = -3x + 12$$
$$y = -\frac{3}{4}x + 3$$
$$m = -\frac{3}{4}$$

69. $x + 5y = 10$

(a) $(0, 2),\quad (10, 0)$

$$m = \frac{2 - 0}{0 - 10} = \frac{2}{-10} = -\frac{1}{5}$$

(b) $x + 5y = 10$

$$5y = -x + 10$$
$$y = -\frac{1}{5}x + 2$$
$$m = -\frac{1}{5}$$

71. $4x - y = 7$

(a) $(0, -7),\quad \left(\dfrac{7}{4}, 0\right)$

$$m = \frac{-7 - 0}{0 - \dfrac{7}{4}} = \frac{-7}{-\dfrac{7}{4}} = 4$$

(b) $4x - y = 7$

$$-y = -4x + 7$$
$$y = 4x - 7$$
$$m = 4$$

73. $3x + 7y = 0$

(a) $(0, 0),\quad \left(1, -\dfrac{3}{7}\right)$

$$m = \frac{-\dfrac{3}{7} - 0}{1 - 0} = \frac{-\dfrac{3}{7}}{1} = -\frac{3}{7}$$

(b) $3x + 7y = 0$

$$7y = -3x$$
$$y = -\frac{3}{7}x$$
$$m = -\frac{3}{7}$$

75. (1, 6), (−1, 3)

$$m = \frac{6 - 3}{1 - (-1)} = \frac{3}{2}$$

Parallel lines have equal slopes.

77. (5, −9), (5, −3)

$$m = \frac{-9 - (-3)}{5 - 5} = \frac{-6}{0} \text{ undefined}$$

Parallel lines have equal slopes.

79. $2x - 3y = 24$

$$-3y = -2x + 24$$

$$y = \frac{2}{3}x - 8$$

$$m = \frac{2}{3}$$

$$m_{\perp} = -\frac{1}{\frac{2}{3}} = -\frac{3}{2}$$

81. $y = x - 2$

$m = 1,\ b = -2$

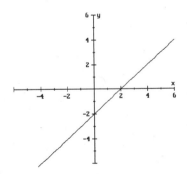

83. $y = \frac{1}{3}x + \frac{2}{3}$

$m = \frac{1}{3},\ b = \frac{2}{3}$

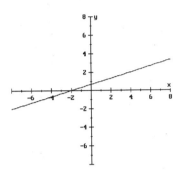

85. $y = -3x + 2$

$m = -3,\ b = 2$

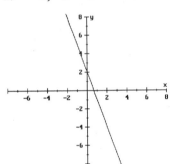

87. $y = 2x + 2$

$m = 2$

$$m_{l_2} = -\frac{1}{2}$$

Using (−4, 3),

$$y - 3 = -\frac{1}{2}(x + 4)$$

$$y - 3 = -\frac{1}{2}x - 2$$

$$y = -\frac{1}{2}x + 1$$

Problem Set 6.4

1. $(1, 2)$; $m = 3$

$$3 = \frac{y - 2}{x - 1}$$

$$3(x - 1) = y - 2$$

$$3x - 3 = y - 2$$

$$3x - y = 1$$

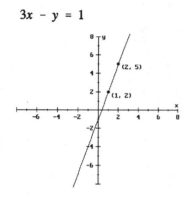

3. $(4, 3)$; $m = -2$

$$-2 = \frac{y - 3}{x - 4}$$

$$-2(x - 4) = y - 3$$

$$-2x + 8 = y - 3$$

$$11 = 2x + y$$

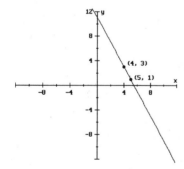

5. $(4, -6)$; $m = \frac{1}{2}$

$$\frac{1}{2} = \frac{y - (-6)}{x - 4}$$

$$x - 4 = 2(y + 6)$$

$$x - 4 = 2y + 12$$

$$x - 2y = 16$$

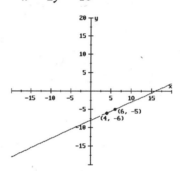

7. $(2, 3)$, $(4, 5)$

$$m = \frac{5 - 3}{4 - 2} = \frac{2}{2} = 1$$

$$y - 3 = 1(x - 2)$$

$$y - 3 = x - 2$$

$$-1 = x - y$$

9. $(2, -3)$, $(5, 0)$

$$m = \frac{-3 - 0}{2 - 5} = \frac{-3}{-3} = 1$$

$$y - 0 = 1(x - 5)$$

$$y = x - 5$$

$$5 = x - y$$

11. $(0, 0)$, $(-1, -5)$

$$m = \frac{-5 - 0}{-1 - 0} = 5$$

$$y - 0 = 5(x - 0)$$

$$y = 5x$$

$$0 = 5x - y$$

268

13. $4x - 2y = 13$

$\quad 4x - 13 = 2y$

$\quad 2x - \dfrac{13}{2} = y$

$\quad m = 2$

Using $(-7, 0)$,

$\quad y - 0 = 2[x - (-7)]$

$\qquad y = 2(x + 7)$

$\qquad y = 2x + 14$

$\quad -14 = 2x - y$

15. $2x + 3y = 4$

$\quad 3y = -2x + 4$

$\quad y = -\dfrac{2}{3}x + \dfrac{4}{3}$

$\quad m = -\dfrac{2}{3}$

$\quad m_\perp = -\dfrac{1}{-\dfrac{2}{3}} = \dfrac{3}{2}$

Using $(1, 5)$,

$\quad y - 5 = \dfrac{3}{2}(x - 1)$

$\quad 2y - 10 = 3(x - 1)$

$\quad 2y - 10 = 3x - 3$

$\quad -7 = 3x - 2y$

17. $b = 2, \quad m = \dfrac{1}{2}$

$\quad y = mx + b$

$\quad y = \dfrac{1}{2}x + 2$

19. Since it is parallel to the x-axis, it is a horizontal line through $(-4, -1)$.

$\quad y = -1$

21. $(32, 0), \quad (212, 100)$

$\quad m = \dfrac{100 - 0}{212 - 32} = \dfrac{100}{180} = \dfrac{5}{9}$

$\quad C - 0 = \dfrac{5}{9}(F - 32)$

$\qquad C = \dfrac{5}{9}(F - 32)$

23. $(2, 1); \quad m = 2$

$\quad 2 = \dfrac{y - 1}{x - 2}$

$\quad 2(x - 2) = y - 1$

$\quad 2x - 4 = y - 1$

$\quad 2x - y = 3$

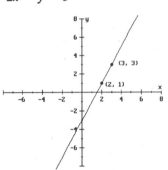

25. $(3, 4); \quad m = -1$

$\quad -1 = \dfrac{y - 4}{x - 3}$

$\quad -1(x - 3) = y - 4$

$\quad -x + 3 = y - 4$

$\quad 7 = x + y$

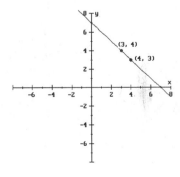

27. $(2, -3);\ m = -\dfrac{1}{2}$

$$-\dfrac{1}{2} = \dfrac{y - (-3)}{x - 2}$$

$$-1(x - 2) = 2(y + 3)$$

$$-x + 2 = 2y + 6$$

$$-4 = x + 2y$$

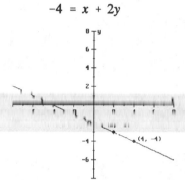

29. $(1, 2),\ (3, 4)$

$$m = \dfrac{4 - 2}{3 - 1} = \dfrac{2}{2} = 1$$

$$y - 2 = 1(x - 1)$$

$$y - 2 = x - 1$$

$$-1 = x - y$$

31. $(3, 0),\ (2, -3)$

$$m = \dfrac{0 - (-3)}{3 - 2} = \dfrac{3}{1} = 3$$

$$y - 0 = 3(x - 3)$$

$$y = 3x - 9$$

$$9 = 3x - y$$

33. $(-2, -4),\ (0, 0)$

$$m = \dfrac{-4 - 0}{-2 - 0} = 2$$

$$y - 0 = 2(x - 0)$$

$$y = 2x$$

$$0 = 2x - y$$

35. $b = 4,\ m = \dfrac{1}{3}$

$$y = mx + b$$

$$y = \dfrac{1}{3}x + 4$$

$$x - 3y = -12$$

37. $\left(-\dfrac{1}{2}, 0\right);\ m = 1$

$$y - 0 = 1\left[x - \left(-\dfrac{1}{2}\right)\right]$$

$$y - x = \dfrac{1}{2}$$

$$2x - 2y = -1$$

39. $(-2, 0),\ (0, 3)$

$$m = \dfrac{3 - 0}{0 - (-2)} = \dfrac{3}{2}$$

$$y = mx + b$$

$$y = \dfrac{3}{2}x + 3$$

$$3x - 2y = -3$$

41. $y = 4x - 3$

$m = 4$

Using $(-1, 4)$,

$$y - 4 = 4[x - (-1)]$$

$$y - 4 = 4(x + 1)$$

$$y - 4 = 4x + 4$$

$$-8 = 4x - y$$

43. $x + 2y = 3$

$$2y = -x + 3$$

$$y = -\dfrac{1}{2}x + \dfrac{3}{2}$$

$$m = -\dfrac{1}{2} \qquad m_{\perp} = -\dfrac{1}{-\dfrac{1}{2}} = 2$$

270

Using (-1, 4),

$$y - 4 = 2[x - (-1)]$$

$$y - 4 = 2(x + 1)$$

$$y - 4 = 2x + 2$$

$$-6 = 2x - y$$

45. Since it is parallel to the y-axis, it is a vertical line through $(2, -3)$.

$$x = 2$$

47. $x = -y$

$$y = -x$$

$$m = -1$$

$$m_\perp = -\frac{1}{-1} = 1$$

Using (0, 0),

$$y = mx + b$$

$$y = 1x + 0$$

$$0 = x - y$$

49. $(-5, 3)$, $(3, 3)$

$$m = \frac{3 - 3}{-5 - 3} = \frac{0}{-8} = 0$$

Horizontal line through $(2, 0)$: $y = 0$

51. $b = 11$

$$4x - y = 0$$

$$4x = y$$

$$m = 4$$

$$m_\perp = -\frac{1}{4}$$

$$y = mx + b$$

$$y = -\frac{1}{4}x + 11$$

$$4y = -x + 44$$

$$x + 4y = 44$$

53. $(1, 2.9)$, $(5, 4.5)$

$$m = \frac{4.5 - 2.9}{5 - 1} = \frac{1.6}{4} = 0.4$$

$$s - 2.9 = 0.4(t - 1)$$

$$s - 2.9 = 0.4t - 0.4$$

$$-2.5 = 0.4t - s$$

Let $t = 10$: $-2.5 = 0.4(10) - s$

$$-2.5 = 4 - s$$

$$-6.5 = -s$$

$$s = 6.5$$

The sales were 6.5 billion dollars.

55. (a) $(10, 1.75)$, $(70, 4.15)$

$$m = \frac{4.15 - 1.75}{70 - 10} = \frac{2.4}{60} = 0.04$$

$$C - 1.75 = 0.04(p - 10)$$

$$C - 1.75 = 0.04p - 0.4$$

$$-1.35 = 0.04p - C$$

(b) Let $p = 40$: $-1.35 = 0.04(40) - C$

$$-1.35 = 1.6 - C$$

$$-2.95 = -C$$

$$C = 2.95$$

The charge would be $2.95.

(c) Let $C = 3.35$: $-1.35 = 0.04p - 3.35$

$$2 = 0.04p$$

$$50 = p$$

It was 50 lbs.

(d) As the weight increases, the charge increases.

57. $px + qy = C$

$$qy = -px + C$$

$$y = -\frac{p}{q}x + \frac{C}{q}$$

$$m = -\frac{p}{q}$$

$$m_{\perp} = -\frac{1}{-\dfrac{p}{q}} = \frac{q}{p}$$

Using (p, q),

$$y - q = \frac{q}{p}(x - p)$$

$$y - q = \frac{q}{p}x - q$$

$$py - pq = qx - pq$$

$$0 = qx - py$$

59. $2x + 3y = 6$

$$\frac{x}{3} + \frac{y}{2} = 1$$

The x-intercept is 3 and the y-intercept is 2.

61. $2x + 3y = 1$

$$\frac{x}{\dfrac{1}{2}} + \frac{y}{\dfrac{1}{3}} = 1$$

The x-intercept is $\dfrac{1}{2}$ and the y-intercept is $\dfrac{1}{3}$.

63. $3x = 5$

$$\frac{3x}{5} = 1$$

$$\frac{x}{\dfrac{5}{3}} = 1$$

The x-intercept is $\dfrac{5}{3}$, no y-intercept.

Problem Set 6.5

1. Yes; $(10, 100)$, $(20, 400)$

3. Yes; $(\sqrt{5}, 1)$, $\left(\dfrac{1}{2}, 1\right)$, $(10, \sqrt{7})$, $(12, \sqrt{7})$

5. $f(x) = 2x - 5$

$\quad f(2) = 2(2) - 5 = -1$

7. $f(x) = 2x - 5$

$\quad f(0) = 2(0) - 5 = -5$

9. $g(x) = \sqrt{x + 1}$

$\quad g(3) = \sqrt{3 + 1} = \sqrt{4} = 2$

11. $g(x) = \sqrt{x + 1}$

$\quad g(-1) = \sqrt{-1 + 1} = \sqrt{0} = 0$

13. $f(x) = x^2 \qquad g(x) = 2x - 3$

$\quad f(3) - g(3) = 3^2 - [2(3) - 3]$

$$= 9 - 3$$
$$= 6$$

(15.)

13. $f(x) = x^2 \qquad g(x) = 2x - 3$

$$\frac{f(3)}{g(3)} = \frac{3^2}{2(3) - 3}$$

$$= \frac{9}{3}$$

$$= 3$$

17. $f(x) = 2x^2 + x - 1$

$\quad f(t) = 2t^2 + t - 1$

19. $\qquad f(x) = 2x^2 + x - 1$

$\quad f(a) + f(b) = (2a^2 + a - 1) + (2b^2 + b - 1)$

$$= 2a^2 + a + 2b^2 + b - 2$$

21. $f(x) = 2x^2 + x - 1$

$\quad f(\sqrt{a}) = \left[2(\sqrt{a})^2 + \sqrt{a} - 1\right]$

$$= 2a + \sqrt{a} - 1$$

23. $h(x) = x^2 - 4$

Natural domain: $\{x \,|\, x \text{ is a real number}\}$

272

25. $f(x) = |x|$

Natural domain: $\{x \mid x \text{ is a real number}\}$

27. $f(x) = \dfrac{3}{x} + 2$

Natural domain: $\{x \mid x \neq 0\}$

29. $g(x) = \dfrac{6}{(x + 2)(x - 7)}$

Natural domain: $\{x \mid x \neq -2, \ x \neq 7\}$

31. $g(x) = \sqrt{x - 1}$

$x - 1 \geq 0$

$x \geq 1$

Natural domain: $\{x \mid x \geq 1\}$

33. $p(x) = \sqrt{x^2 + x - 2}$

$x^2 + x - 2 \geq 0$

Boundary Numbers:

$x^2 + x - 2 = 0$

$(x + 2)(x - 1) = 0$

$x = -2, \quad x = 1$

$$\dfrac{\quad A \quad}{\qquad} \underset{-2}{\Big|} \dfrac{\quad B \quad}{\qquad} \underset{1}{\Big|} \dfrac{\quad C \quad}{\qquad}$$

A: $x = -3;$ $\quad (-3)^2 - 3 - 2 \geq 0$

$4 \geq 0 \quad$ T

B: $x = 0;$ $\quad 0^2 + 0 - 2 \geq 0$

$-2 \geq 0 \quad$ F

C: $x = 2;$ $\quad 2^2 + 2 - 2 \geq 0$

$4 \geq 0 \quad$ T

$(-\infty, -2] \ \cup \ [1, \infty)$

Natural domain: $\{x \mid x \leq -2 \text{ or } x \geq 1\}$

35. $f(x) = x - 7$

$f(1) = 1 - 7 = -6$

$f(4) = 4 - 7 = -3$

$f(7) = 7 - 7 = 0$

$f(11) = 11 - 7 = 4$

$f(14) = 14 - 7 = 7$

Range: $\ -6, -3, 0, 4, 7$

37. $f(x) = 2x - 1$

Let $2n$ represent an even natural number.

$f(2n) = 2(2n) - 1$

$2(2n) - 1$ is an odd natural number.

Range: $\{y \mid y \text{ is an odd natural number}\}$

39. $f(x) = 11$

Range: $\{11\}$

41. $g(x) = x - 1$

Natural domain: $\{x \mid x \text{ is a real number}\}$

43. $h(x) = \dfrac{x}{3} - \dfrac{1}{4}$

Natural domain: $\{x \mid x \text{ is a real number}\}$

45. $r(x) = \dfrac{3}{x - 1}$

Natural domain: $\{x \mid x \neq 1\}$

47. $h(x) = \dfrac{x}{x - 3}$

Natural domain: $\{x \mid x \neq 3\}$

49. $h(x) = \dfrac{x}{x + 3}$

Natural domain: $\{x \mid x \neq -3\}$

273

51. $f(x) = \sqrt{x - 2}$

$x - 2 \geq 0$

$x \geq 2$

Natural domain: $\{x \mid x \geq 2\}$

53. $h(x) = x^2 - 3$

Natural domain: $\{x \mid x \text{ is a real number}\}$

55. $d(x) = \sqrt[3]{x} + 1$

Natural domain: $\{x \mid x \text{ is a real number}\}$

57. $g(x) = \sqrt{x^2 + 1}$

$x^2 + 1 \geq 0$ for real values of x.

Natural domain: $\{x \mid x \text{ is a real number}\}$

59. $f(x) = |x - 5|$

Natural domain: $\{x \mid x \text{ is a real number}\}$

61. $f(x) = 3x + 1$

$f(2) = 3(2) + 1$

$= 7$

$f(x) = 3x + 1$

$f(0) = 3(0) + 1$

$= 1$

65. $f(x) = 3x + 1$

$f(\sqrt{3}) = 3(\sqrt{3}) + 1$

67. $f(x) = 3x + 1$

$f(a - b) = 3(a - b) + 1$

$= 3a - 3b + 1$

69. $f(x) = 3x + 1$

$f(2t) = 3(2t) + 1$

$= 6t + 1$

71. $f(x) = 3x + 1$

$f(a) = 3a + 1$

$\sqrt{f(a)} = \sqrt{3a + 1}$

73. $g(x) = \sqrt{x - 1}$

$g(6) = \sqrt{6 - 1}$

$= \sqrt{5}$

75. $g(x) = \sqrt{x - 1}$

$g(2) = \sqrt{2 - 1}$

$= \sqrt{1}$

$= 1$

77. $g(x) = \sqrt{x - 1}$

$g(5) = \sqrt{5 - 1}$

$= \sqrt{4}$

$= 2$

$2g(5) - 1 = 2(2) - 1$

$= 4 - 1$

$= 3$

79. $g(x) = \sqrt{x - 1}$

$g(2k + 3) = \sqrt{2k + 3 - 1}$

$= \sqrt{2k + 2}$

81. $g(x) = \sqrt{x - 1}$

$g(a) = \sqrt{a - 1}$

$[g(a)]^2 = (\sqrt{a - 1})^2$

$= a - 1$

83. $f(x) = 1 - 2x^2$

$f(t + h) = 1 - 2(t + h)^2$

$= 1 - 2(t^2 + 2th + h^2)$

$= 1 - 2t^2 - 4th - 2h^2$

85. $f(x) = 1 - 2x^2$ (See #83)

$$f(t + h) = 1 - 2t^2 - 4th - 2h^2$$

$$f(t) = 1 - 2t^2$$

$$f(t + h) - f(t) = 1 - 2t^2 - 4th - 2h^2 - (1 - 2t^2)$$

$$= 1 - 2t^2 - 4th - 2h^2 - 1 + 2t^2$$

$$= -4th - 2h^2$$

87. $f(x) = 1 - 2x^2$

$$f(t) = 1 - 2t^2$$

$$2f(t) = 2(1 - 2t^2)$$

$$= 2 - 4t^2$$

89. $f(x) = 1 - 2x^2$

$$f(t) = 1 - 2t^2$$

$$kf(t) = k(1 - 2t^2)$$

$$= k - 2kt^2$$

91. $f(x) = 1 - 2x^2$

$$f(3t) = 1 - 2(3t)^2$$

$$= 1 - 2(9t^2)$$

$$= 1 - 18t^2$$

$$f(2t) = 1 - 2(2t)^2$$

$$= 1 - 2(4t^2)$$

$$= 1 - 8t^2$$

$$f(3t) - f(2t) = (1 - 18t^2) - (1 - 8t^2)$$

$$= 1 - 18t^2 - 1 + 8t^2$$

$$= -10t^2$$

93. $g(x) = \dfrac{1}{x - 1}$

$$g(0) = \dfrac{1}{0 - 1} = -1$$

95. $g(x) = \dfrac{1}{x - 1}$

$$g(1) = \dfrac{1}{1 - 1} = \dfrac{1}{0} \quad \text{undefined}$$

97. $g(x) = \dfrac{1}{x - 1}$

$$g(7) = \dfrac{1}{7 - 1} = \dfrac{1}{6}$$

$$\dfrac{3}{g(7)} = \dfrac{3}{\dfrac{1}{6}} = 18$$

99. $g(x) = \dfrac{1}{x - 1}$

$$g(1) = \dfrac{1}{1 - 1} = \dfrac{1}{0} \quad \text{undefined}$$

Thus $\dfrac{g(1)}{g(2)}$ is undefined.

101. $f(x) = 2x^2 \qquad g(x) = x + 5$

$\qquad f(3) = 2(3)^2 \qquad g(3) = 3 + 5$

$\qquad\qquad = 2(9) \qquad\qquad = 8$

$\qquad\qquad = 18$

$\qquad f(3) + g(3) = 18 + 8 = 26$

103. $f(x) = 2x^2 \qquad g(x) = x + 5$

$\qquad f(3) = 18 \qquad g(3) = 8 \qquad$ (See #101)

$\qquad f(3)g(3) = (18)(8) = 144$

105. $f(x) = 2x^2 \qquad g(x) = x + 5$

$\qquad g(3) = 3 + 5 = 8$

$\qquad f(g(3)) = f(8)$

$\qquad\qquad = 2(8)^2$

$\qquad\qquad = 2(64)$

$\qquad\qquad = 128$

107. $F(x) = 0.25x + 1.5$

    (a) $F(10) = 0.25(10) + 1.5$

                  $= 2.5 + 1.5$

                  $= 4$

The fare is $4.

    (b) $F(3)$ represents the fare for a 3-mile ride.

    (c) $3 = 0.25x + 1.5$

        $1.5 = 0.25x$

          $6 = x$

It would take you 6 miles.

109. $N(t) = t^3 - 2t^2 + t$

    (a) $N(1) = 1^3 - 2(1)^2 + 1$

              $= 1 - 2 + 1$

              $= 0$

There are 0 ants.

    (b) $N(2) = 2^3 - 2(2)^2 + 2$

              $= 8 - 8 + 2$

              $= 2$

There are 2 ants.

    (c) $N(10)$ represents the number of ants after 10 minutes.

111. $f(x) = \dfrac{\sqrt{x + 1}}{\sqrt{x - 2}}$

    $x - 2 > 0$

        $x > 2$

    Natural domain: $\{x \mid x > 2\}$

Problem Set 6.6

1. $f(x) = 2x + 1$

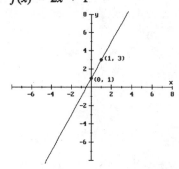

3. $h(x) = \dfrac{1}{2}x - 1$

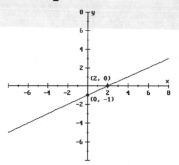

5. $G(x) = 20x + 50$

    (a)

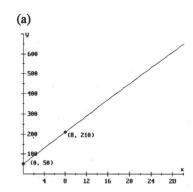

    (b) The y-intercept indicates the tank started with 50 gal of oil.

    (c) $210 = 20x + 50$

          $160 = 20x$

            $8 = x$

It will take 8 minutes.

7. D: $\{x \mid x$ is a real number$\}$

 R: $\{y \mid y \geq -1\}$

9. D: $\{x \mid x$ is a real number$\}$

 R: $\{y \mid y \geq -2\}$

11. D: $\{x \mid x \geq 0\}$

 R: $\{y \mid y \geq 0\}$

13. Yes, passes vertical line test.

15. No, fails vertical line test.

17. Yes, passes vertical line test.

19. $y = x^2 + 3$

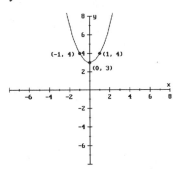

Axis of symmetry: y-axis

21. $y = 3x^2$

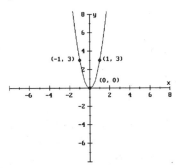

Axis of symmetry: y-axis

23. $y = (x - 3)^2$

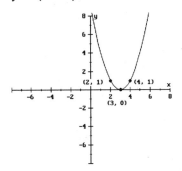

Axis of symmetry: $x = 3$

25. $y = (x + 2)^2 + 1$

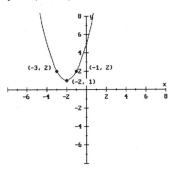

Axis of symmetry: $x = -2$

27. $y = -2x^2 + 4x$

 $y = -2(x^2 - 2x)$

 $y = -2(x^2 - 2x + 1) + 2$

 $y = -2(x - 1)^2 + 2$

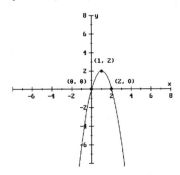

29. $g(x) = x^2 - 2x + 4$

$g(x) = (x^2 - 2x + 1) + 4 - 1$

$g(x) = (x - 1)^2 + 3$

Vertex: (1, 3)

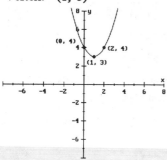

D: $\{x \mid x \text{ is a real number}\}$

R: $\{y \mid y \text{ is a real number}\}$

37. $h(x) = \dfrac{1}{4}x + 2$

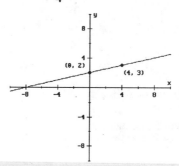

D: $\{x \mid x \text{ is a real number}\}$

R: $\{y \mid y \text{ is a real number}\}$

31. $f(x) = x^2 - 6x$

$f(x) = x^2 - 6x + 9 - 9$

$f(x) = (x - 3)^2 - 9$

Vertex: (3, -9)

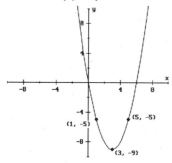

39. $f(x) = 4x^2$

Vertex: (0, 0)

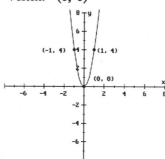

D: $\{x \mid x \text{ is a real number}\}$

R: $\{y \mid y \geq 0\}$

33. $f(x) = 0$

$x = -2$ or $x = 4$

35. $f(x) = 3x + 2$

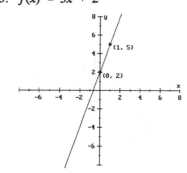

41. $h(x) = \dfrac{1}{3}x^2 - 1$

Vertex: (0, -1)

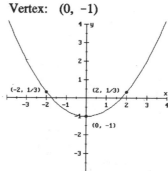

278

41. (con't.)
    D: $\{x \mid x$ is a real number$\}$

    R: $\{y \mid y \geq -1\}$

43. $f(x) = x^2 - 4$

    Vertex: $(0, -4)$

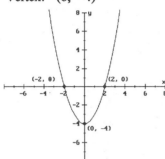

    D: $\{x \mid x$ is a real number$\}$

    R: $\{y \mid y \geq -4\}$

45. $g(x) = (x - 3)^2$

    Vertex: $(3, 0)$

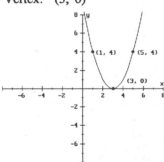

    D: $\{x \mid x$ is a real number$\}$

    R: $\{y \mid y \geq 0\}$

47. $h(x) = \frac{1}{2}(x - 2)^2 + 1$

    Vertex: $(2, 1)$

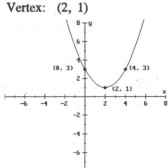

D: $\{x \mid x$ is a real number$\}$

R: $\{y \mid y \geq 1\}$

49. $f(x) = -(x - 1)^2 + 4$

    Vertex: $(1, 4)$

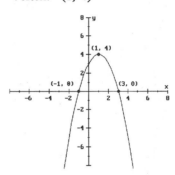

    D: $\{x \mid x$ is a real number$\}$

    R: $\{y \mid y \leq 4\}$

51. $g(x) = 3x^2 + 6x + 5$

    $g(x) = 3(x^2 + 2x) + 5$

    $g(x) = 3(x^2 + 2x + 1) + 5 - 3$

    $g(x) = 3(x + 1)^2 + 2$

    Vertex: $(-1, 2)$

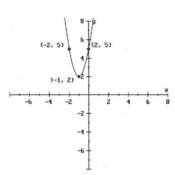

    D: $\{x \mid x$ is a real number$\}$

    R: $\{y \mid y \geq 2\}$

53. $f(x) = -x^2 - 2x - 3$

$f(x) = -(x^2 + 2x) - 3$

$f(x) = -(x^2 + 2x + 1) - 3 + 1$

$f(x) = -(x + 1)^2 - 2$

Vertex: $(-1, -2)$

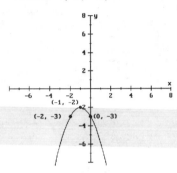

D: $\{x \mid x \text{ is a real number}\}$

R: $\{y \mid y \leq -2\}$

55. $f(x) = x^2 + 1$

Vertex: $(0, 1)$

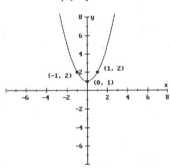

Minimum Value: 1

57. $y = 0.16x^2 - 0.76x + 16.8$

a.) Let $x = 7$:

$y = 0.16(7)^2 - 0.76(7) + 16.8$

$= 19.32$

The baby would weigh approximately 19.32 lbs.

b.) 20 years = 240 months

$y = 0.16(240)^2 - 0.76(240) + 168$

$= 9201.6$

No, it is not reasonable. A twenty year old would weigh 9201.6 lbs.

59. $p = -3x + 600$

a.) $R(x) = x(-3x + 600)$

$R(x) = -3x^2 + 600x$

b.) $R(x) = -3x^2 + 600x$

$= -3(x^2 - 200x)$

$= -3(x^2 - 200x + 10000) + 30000$

$= -3(x - 100)^2 + 30000$

Vertex: $(100, 30000)$

Maximum Revenue: $30,000

61. $s(t) = -16t^2 + 96t + 100$

$s(t) = -16(t^2 - 6t) + 100$

$s(t) = -16(t^2 - 6t + 9) + 100 + 144$

$s(t) = -16(t - 3)^2 + 244$

Vertex: $(3, 244)$

Maximum height: 244 ft

63. (a) $h(x) = -0.005x^2 + x + 6$

$h(20) = -0.005(20)^2 + 20 + 6 = 24$

$h(60) = -0.005(60)^2 + 60 + 6 = 48$

$h(100) = -0.005(100)^2 + 100 + 6 = 56$

$h(140) = -0.005(140)^2 + 140 + 6 = 48$

$h(180) = -0.005(180)^2 + 180 + 6 = 24$

$h(200) = -0.005(200)^2 + 200 + 6 = 6$

(b) $h(60) = 48$ ft

(c) When the javelin is 20 ft from the foul line, the javelin is 24 ft high.

(d) As the javelin goes up and then comes back down, at two different distances from the foul line it will be 24 ft high.

65. $f(x) = \sqrt{x - 1}$

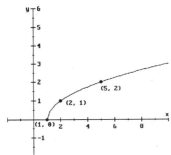

D: $\{x \mid x \geq 1\}$

R: $\{y \mid y \geq 0\}$

67. $f(x) = -\sqrt{x}$

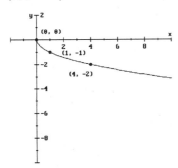

D: $\{x \mid x \geq 0\}$

R: $\{y \mid y \leq 0\}$

69. $f(x) = \sqrt{1 - x^2}$

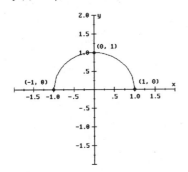

$1 - x^2 \geq 0$

Boundary Numbers:

$1 - x^2 = 0$

$\quad 1 = x^2$

$\quad x = \pm 1$

$$\begin{array}{c|c|c} A & B & C \\ \hline & -1 \quad\; 1 \end{array}$$

A: $x = -2$; $1 - (-2)^2 \geq 0$

$\qquad\qquad\qquad\qquad -3 \geq 0$ F

B: $x = 0$; $1 - 0^2 \geq 0$

$\qquad\qquad\qquad\qquad 1 \geq 0$ T

C: $x = 2$; $1 - 2^2 \geq 0$

$\qquad\qquad\qquad\qquad -3 \geq 0$ F

$[-1, 1]$

D: $\{x \mid -1 \leq x \leq 1\}$

R: $\{y \mid 0 \leq y \leq 1\}$

71. $f(x) = x^3 + 1$

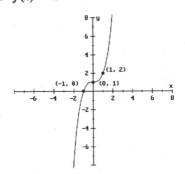

D: $\{x \mid x$ is a real number$\}$

R: $\{y \mid y$ is a real number$\}$

73. $f(x) = x^4$

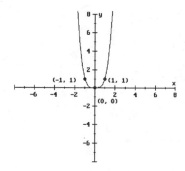

D: $\{x \mid x \text{ is a real number}\}$

R: $\{y \mid y \geq 0\}$

75. (a) $2w + l \leq 300$

$l \leq 300 - 2w$

(b) $A = wl$

$\leq w(300 - 2w)$

$\leq 300w - 2w^2$

(c) $A = 300w - 2w^2$

Vertex: $(75, 11250)$

The largest area is 11,250 ft^2.

Problem Set 6.7

1. $A = kr^2$

3. $V = kr^3$

5. $u = \dfrac{kv}{w}$

7. $x = ky^2$

$12 = k(2)^2$

$3 = k$

$x = 3y^2$

$x = 3(5)^2$

$x = 75$

9. $s = kt^2g$

$36 = k(2)^2(3)$

$36 = 12k$

$3 = k$

$s = 3t^2g$

$s = 3(\sqrt{2})^2(32.2)$

$s = 3(2)(32.2)$

$s = 193.2$

11. $F = kd$

$10 = k(5)$

$2 = k$

$F = 2d$

$F = 2(10) = 20$

20 lb of force would be required to stretch the spring 10 in.

13. $d = kt^2$

$64.4 = k(2)^2$

$16.1 = k$

$d = 16.1t^2$

$d = 16.1(3)^2 = 144.9$

The stone will have fallen 144.9 ft.

15. $p = k\sqrt{l}$

$\dfrac{1}{2} = k\sqrt{16}$

$\dfrac{1}{2} = 4k$

$\dfrac{1}{8} = k$

$p = \dfrac{1}{8}\sqrt{l}$

$\dfrac{1}{4} = \dfrac{1}{8}\sqrt{l}$

$2 = \sqrt{l}$

$4 = l$

The length is 4 in.

17. $s = kv$

19. $V = klwh$

21. $h = \dfrac{k\sqrt{g}}{d}$

23. $x = \dfrac{k}{y^2}$

$2 = \dfrac{k}{3^2}$

$18 = k$

$x = \dfrac{18}{y^2}$

$x = \dfrac{18}{5^2} = \dfrac{18}{25}$

25. $N = \dfrac{kL^2}{M^3}$

$9 = \dfrac{k(\sqrt{3})^2}{2^3}$

$9 = \dfrac{3k}{8}$

$72 = 3k$

$24 = k$

$N = \dfrac{24L^2}{M^3}$

$N = \dfrac{24\left(\dfrac{\sqrt{2}}{2}\right)^2}{\left(\dfrac{3}{2}\right)^3}$

$= \dfrac{24\left(\dfrac{2}{4}\right)}{\dfrac{27}{8}}$

$= \dfrac{12}{\dfrac{27}{8}}$

$= \dfrac{32}{9}$

27. $m_1 = \dfrac{k}{m_2}$

$\dfrac{2}{3} = \dfrac{k}{-\dfrac{3}{2}}$

$\left(\dfrac{2}{3}\right)\left(-\dfrac{3}{2}\right) = k$

$-1 = k$

$m_1 = \dfrac{-1}{m_2}$

$m_1 = \dfrac{-1}{-\dfrac{3}{4}} = \dfrac{4}{3}$

29. $F = kd$

$5 = k(2)$

$\dfrac{5}{2} = k$

$F = \dfrac{5}{2}d$

$F = \dfrac{5}{2}(1) = \dfrac{5}{2}$

A force of $2\dfrac{1}{2}$ lb is required to compress it 1 in.

Chapter 6 Review Problems

1. $(1, 1), \ (5, 4)$

$d = \sqrt{(1 - 5)^2 + (1 - 4)^2}$

$ = \sqrt{(-4)^2 + (-3)^2}$

$ = \sqrt{16 + 9}$

$ = \sqrt{25}$

$ = 5$

$x = \dfrac{1 + 5}{2} = \dfrac{6}{2} = 3$

$y = \dfrac{1 + 4}{2} = \dfrac{5}{2}$

midpoint: $\left(3, \dfrac{5}{2}\right)$

3. $(2, 1), \ (1, 3)$

$d = \sqrt{(2 - 1)^2 + (1 - 3)^2}$

$ = \sqrt{1^2 + (-2)^2}$

$ = \sqrt{1 + 4}$

$ = \sqrt{5}$

$x = \dfrac{2 + 1}{2} = \dfrac{3}{2}$

$y = \dfrac{1 + 3}{2} = \dfrac{4}{2} = 2$

midpoint: $\left(\dfrac{3}{2}, 2\right)$

5. $(8, -3), \ (9, -2)$

$d = \sqrt{(8 - 9)^2 + [-3 - (-2)]^2}$

$ = \sqrt{(-1)^2 + (-1)^2}$

$ = \sqrt{1 + 1}$

$ = \sqrt{2}$

$x = \dfrac{8 + 9}{2} = \dfrac{17}{2}$

$y = \dfrac{-3 + (-2)}{2} = -\dfrac{5}{2}$

midpoint: $\left(\dfrac{17}{2}, -\dfrac{5}{2}\right)$

7. $2x + 3y = 12$

$(0, 4), \ (6, 0)$

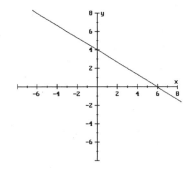

9. $y = \dfrac{1}{2}x - 2$

$(0, -2), \ (4, 0)$

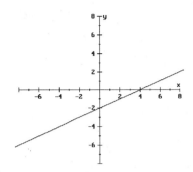

11. $6x + y = 6$

$(0, 6), \quad (1, 0)$

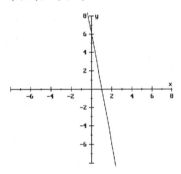

13. $x = -3$

$(-3, 0), \quad (-3, 1)$

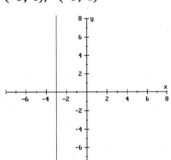

15. $4y = 7$

$y = \dfrac{7}{4}$

$(0, \dfrac{7}{4}), \quad (1, \dfrac{7}{4})$

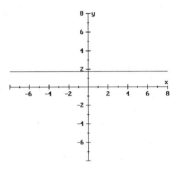

17. $(5, 9), \quad (3, 11)$

$m = \dfrac{11 - 9}{3 - 5} = \dfrac{2}{-2} = -1$

19. $(3, -2), \quad (-2, 8)$

$m = \dfrac{-2 - 8}{3 - (-2)} = \dfrac{-10}{5} = -2$

21. $(-5, 0), \quad (0, 3)$

$m = \dfrac{0 - 3}{-5 - 0} = \dfrac{3}{5}$

23. $\left(-\dfrac{4}{3}, \dfrac{1}{2}\right), \quad \left(\dfrac{1}{3}, -\dfrac{5}{2}\right)$

$m = \dfrac{\dfrac{1}{2} - \left(-\dfrac{5}{2}\right)}{-\dfrac{4}{3} - \dfrac{1}{3}} = \dfrac{3}{-\dfrac{5}{3}} = -\dfrac{9}{5}$

25. $2x - 3y = 6$

$-3y = -2x + 6$

$y = \dfrac{2}{3}x - 2$

$m = \dfrac{2}{3}, \quad b = -2$

27. $2y + 3 = 0$

$2y = -3$

$y = -\dfrac{3}{2}$

Horizontal line: $m = 0, \quad b = -\dfrac{3}{2}$

29. $(4, -2)$, $m = -1$

$$y - (-2) = -1(x - 4)$$
$$y + 2 = -x + 4$$
$$x + y = 2$$

31. $(-9, -6)$, $(-2, -1)$

$$m = \frac{-6 - (-1)}{-9 - (-2)} = \frac{-5}{-7} = \frac{5}{7}$$

$$y - (-6) = \frac{5}{7}[x - (-9)]$$

$$y + 6 = \frac{5}{7}(x + 9)$$

$$7y + 42 = 5x + 45$$

$$-3 = 5x - 7y$$

33. $2x + 3y = 4$

$$3y = -2x + 4$$

$$y = -\frac{2}{3}x + \frac{4}{3}$$

$$m = -\frac{2}{3}$$

Using $(1, -7)$,

$$y - (-7) = -\frac{2}{3}(x - 1)$$

$$y + 7 = -\frac{2}{3}(x - 1)$$

$$3y + 21 = -2x + 2$$

$$2x + 3y = -19$$

35. $3x = 4y$

$$\frac{3}{4}x = y$$

$$m = \frac{3}{4}$$

$$m_1 = -\frac{1}{\frac{3}{4}} = -\frac{4}{3}$$

Using $(3, 1)$,

$$y - 1 = -\frac{4}{3}(x - 3)$$

$$3y - 3 = -4x + 12$$

$$4x + 3y = 15$$

37. $f(x) = -x$

If x is a positive integer,

then $-x$ is a negative integer.

R: $\{y \,|\, y$ is a negative integer$\}$

39. $f(x) = |x - 7|$

Natural domain: $\{x \,|\, x$ is a real number$\}$

41. $f(x) = \sqrt{x^2 - x}$

$$x^2 - x \geq 0$$

Boundary Numbers:

$$x^2 - x = 0$$

$$x(x - 1) = 0$$

$$x = 0, \quad x = 1$$

$$\underbrace{\quad A \quad}_{0} \Big| \underbrace{\quad B \quad}_{1} \Big| \quad C \quad$$

A: $x = -1$; $(-1)^2 - (-1) \geq 0$

$$2 \geq 0 \quad T$$

B: $x = \frac{1}{2}$; $\left(\frac{1}{2}\right)^2 - \frac{1}{2} \geq 0$

$$-\frac{1}{4} \geq 0 \quad F$$

C: $x = 2$; $2^2 - 2 \geq 0$

$$2 \geq 0 \quad T$$

$(-\infty, 0] \cup [1, \infty)$

Natural domain: $\{x \,|\, x \leq 0$ or $x \geq 1\}$

43. $f(x) = \dfrac{x + 3}{(x + 1)(x - 2)}$

   Natural domain: $\{x \mid x \neq -1,\ x \neq 2\}$

45. $f(x) = \dfrac{3}{x + 5}$

   $f(-5) = \dfrac{3}{-5 + 5}$

   $= \dfrac{3}{0}$ undefined

47. $f(x) = \dfrac{3}{x + 5}$

   $f(1) = \dfrac{3}{1 + 5}$

   $= \dfrac{3}{6}$

   $= \dfrac{1}{2}$

   $f(f(1)) = f\left(\dfrac{1}{2}\right)$

   $= \dfrac{3}{\dfrac{1}{2} + 5}$

   $= \dfrac{3}{\dfrac{11}{2}}$

   $= \dfrac{6}{11}$

49. $f(x) = \dfrac{3}{x + 5}$

   $f(-3) = \dfrac{3}{-3 + 5}$

   $= \dfrac{3}{2}$

51. $f(x) = \dfrac{3}{x + 5}$

   $f(a^2) = \dfrac{3}{a^2 + 5}$

53. $f(x) = \dfrac{1}{4}x^2 + 3$

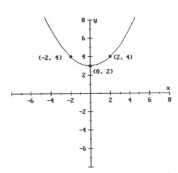

D: $\{x \mid x \text{ is a real number}\}$

R: $\{y \mid y \geq 3\}$

55. $f(x) = (x - 2)^2 + 3$

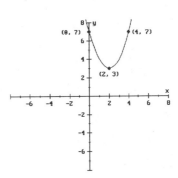

D: $\{x \mid x \text{ is a real number}\}$

R: $\{y \mid y \geq 3\}$

57. $f(x) = 2(x + 1)^2 - 5$

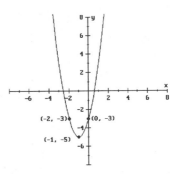

D: $\{x \mid x \text{ is a real number}\}$

R: $\{y \mid y \geq -5\}$

287

59. $f(x) = x^2 - 4x + 3$

$f(x) = (x^2 - 4x + 4) + 3 - 4$

$f(x) = (x - 2)^2 - 1$

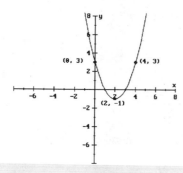

D: $\{x \mid x \text{ is a real number}\}$

R: $\{y \mid y \geq -1\}$

61. $(5, 30)$, $(10, 60)$

(a) $m = \dfrac{60 - 30}{10 - 5} = \dfrac{30}{5} = 6$

$y - 30 = 6(x - 5)$

$y - 30 = 6x - 30$

$0 = 6x - y$

(b) Let $x = 1$: $0 = 6(1) - y$

$y = 6$

$(1, 6)$

(c) $(1, 6)$ indicates that 1 fathom equals 6 ft.

63. $u = kv$

$8 = k(4)$

$2 = k$

$u = 2v$

$u = 2(12)$

$= 24$

65. $F = \dfrac{k}{d^2}$

$50 = \dfrac{k}{12^2}$

$50 = \dfrac{k}{144}$

$(50)(144) = k$

$k = 7200$

$F = \dfrac{7200}{d^2}$

$F = \dfrac{7200}{20^2}$

$= 18$

The result is 18 force units.

67. $27 + a^3 = 3^3 + a^3$

$= (3 + a)(9 - 3a + a^2)$

69. Factored.

71. $(f - g)^2 - 9 = (f - g)^2 - 3^2$

$= [(f - g) - 3][(f - g) + 3]$

$= (f - g - 3)(f - g + 3)$

73. 2 terms;

$y - 7$ is not a factor

75. 3 terms;

$2y^2 - 11y - 21 = (y - 7)(2y + 3)$

$y - 7$ is a factor

77. $9^{-2} = \dfrac{1}{9^2} = \dfrac{1}{81}$

79. $(-9)^2 = 81$

81. Cannot simplify.

83. $\left(a^{2/3}b^{3/2}\right)^6 = \left(a^{2/3}\right)^6\left(b^{3/2}\right)^6$

$\qquad\qquad = a^4 b^9$

85. Will not reduce.

87. $\dfrac{x}{x-1} - \dfrac{3}{1-x} = \dfrac{x}{x-1} + \dfrac{3}{x-1}$

$\qquad\qquad\qquad = \dfrac{x+3}{x-1}$

89. $\dfrac{\dfrac{3}{x^2-1}}{\dfrac{1}{x+1} - \dfrac{1}{x-1}}$

$= \dfrac{\dfrac{3}{(x+1)(x-1)}}{\dfrac{1}{x+1} - \dfrac{1}{x-1}}$

$= \dfrac{\left(\dfrac{3}{(x+1)(x-1)}\right)(x+1)(x-1)}{\left(\dfrac{1}{x+1} - \dfrac{1}{x-1}\right)(x+1)(x-1)}$

$= \dfrac{3}{x-1-(x+1)}$

$= \dfrac{3}{x-1-x-1}$

$= \dfrac{3}{-2}$

$= -\dfrac{3}{2}$

91. Expression;

$\dfrac{s-5}{2s-12} + \dfrac{s-4}{12-2s}$

$= \dfrac{s-5}{2(s-6)} + \dfrac{s-4}{2(6-s)}$

$= \dfrac{s-5}{2(s-6)} - \dfrac{s-4}{2(s-6)}$

$= \dfrac{s-5-s+4}{2(s-6)}$

$= \dfrac{-1}{2(s-6)}$

$= -\dfrac{1}{2(s-6)}$

93. Inequality;

$6x^2 - 5x < 6$

Boundary Numbers:

$6x^2 - 5x = 6$

$6x^2 - 5x - 6 = 0$

$(3x+2)(2x-3) = 0$

$x = -\dfrac{2}{3}, \quad x = \dfrac{3}{2}$

$$\underbrace{\quad A \quad}_{} \Big|_{-\frac{2}{3}} \underbrace{\quad B \quad}_{} \Big|_{\frac{3}{2}} \underbrace{\quad C \quad}_{}$$

$A:\ x = -1; \qquad 6(-1)^2 - 5(-1) < 6$

$\qquad\qquad\qquad\qquad\qquad 11 < 6 \quad F$

$B:\ x = 0; \qquad 6(0)^2 - 5(0) < 6$

$\qquad\qquad\qquad\qquad\qquad 0 < 6 \quad T$

$C:\ x = 2; \qquad 6(2)^2 - 5(2) < 6$

$\qquad\qquad\qquad\qquad\qquad 14 < 6 \quad F$

$\left(-\dfrac{2}{3}, \dfrac{3}{2}\right)$

95. Equation;

$$x^4 - 3x^2 - 4 = 0$$

$$(x^2 - 4)(x^2 + 1) = 0$$

$$x^2 = 4 \quad \text{or} \quad x^2 + 1 = 0$$

$$x^2 = 4 \quad \text{or} \quad x^2 = -1$$

$$x = \pm 2 \quad \text{or} \quad x = \pm i$$

$$\{\pm 2, \pm i\}$$

Chapter 6 Test

1. $(2, -3)$, $(-1, 5)$

$$d = \sqrt{(-3 - 5)^2 + [2 - (-1)]^2}$$

$$= \sqrt{(-8)^2 + 3^2}$$

$$= \sqrt{64 + 9}$$

$$= \sqrt{73}$$

B

2. $(5, -1)$, $(2, 4)$

$$m = \frac{-1 - 4}{5 - 2} = \frac{-5}{3} = -\frac{5}{3}$$

C

3. $(3, 7)$; $m = -\frac{2}{3}$

$$y - 7 = -\frac{2}{3}(x - 3)$$

$$3y - 21 = -2x + 6$$

$$2x + 3y = 27$$

A

4. $3x - 4y = -1$

$$-4y = -3x - 1$$

$$y = \frac{3}{4}x + \frac{1}{4}$$

$$m = \frac{3}{4}$$

$$m_1 = -\frac{1}{\frac{3}{4}} = -\frac{4}{3}$$

Using $(0, 0)$,

$$y = mx + b$$

$$y = -\frac{4}{3}x + 0$$

$$3y = -4x$$

$$4x + 3y = 0$$

B

5. Horizontal lines have a slope of 0.

C

6. C

7. $g(x) = x^2$

$$g(-2) = (-2)^2 = 4$$

$$g(-1) = (-1)^2 = 1$$

$$g(0) = 0^2 = 0$$

$$g(1) = 1^2 = 1$$

$$g(2) = 2^2 = 4$$

R: $\{0, 1, 4\}$

C

8. $f(x) = 1 - 2x + x^2$

$$f(-3) = 1 - 2(-3) + (-3)^2$$

$$= 1 + 6 + 9$$

$$= 16$$

B

9. $f(x) = x^2 + 2x + 2$

$f(x) = (x^2 + 2x + 1) + 2 - 1$

$f(x) = (x + 1)^2 + 1$

Vertex: $(-1, 1)$

A

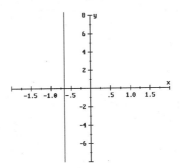

10. A; it passes the vertical line test.

11. $x + 2y = 2$

$(0, 1), \quad (2, 0)$

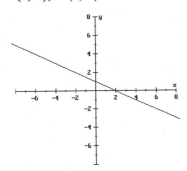

12. $2x - y = 0$

$(0, 0), \quad (1, 2)$

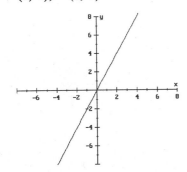

13. $x = -\dfrac{2}{3}$

$\left(-\dfrac{2}{3}, 0\right), \quad \left(-\dfrac{2}{3}, 1\right)$

14. $y = \dfrac{3}{2}x + \dfrac{1}{2}$

$\left(0, \dfrac{1}{2}\right), \quad (1, 2)$

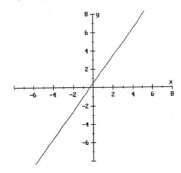

15. (a) $f(x) = x^2 - 1$

Vertex: $(0, -1)$

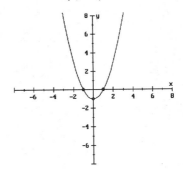

(b) $f(x) = -2(x - 1)^2 + 3$

Vertex: $(1, 3)$

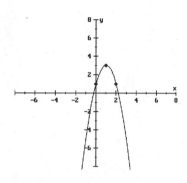

16.

(a) $f(x) = \sqrt{(x + 1)(x - 2)}$

$(x + 1)(x - 2) \geq 0$

Boundary Numbers:

$(x + 1)(x - 2) = 0$

$x = -1, \quad x = 2$

$$\frac{\quad A \quad|\quad B \quad|\quad C \quad}{\qquad -1 \qquad 2 \qquad}$$

A: $x = -2$; $\qquad (-2 + 1)(-2 - 2) \geq 0$

$\qquad\qquad\qquad\qquad\qquad\qquad 4 \geq 0 \quad$ T

B: $x = 0$; $\qquad (0 + 1)(0 - 2) \geq 0$

$\qquad\qquad\qquad\qquad\qquad\qquad -2 \geq 0 \quad$ F

C: $x = 3$; $\qquad (3 + 1)(3 - 2) \geq 0$

$\qquad\qquad\qquad\qquad\qquad\qquad 4 \geq 0 \quad$ T

$(-\infty, -1] \ \cup \ [2, \infty)$

Natural domain: $\{x \mid x \leq -1 \text{ or } x \geq 2\}$

(b) $g(x) = \dfrac{1}{x - 7}$

Natural domain: $\{x \mid x \neq 7\}$

17. $(1, 2), \quad (3, 4)$

$$m = \frac{4 - 2}{3 - 1} = \frac{2}{2} = 1$$

Using $(-1, 0)$,

$$y - 0 = 1[x - (-1)]$$

$$y = x + 1$$

$$-1 = x - y$$

18. A line perpendicular to the x-axis is a vertical line.

Vertical line passing through $(-2, -3)$: $x = -2$

19. $(5, -6), \ (-7, -1)$

$$d = \sqrt{[-6 - (-1)]^2 + [5 - (-7)]^2}$$

$$= \sqrt{(-5)^2 + (12)^2}$$

$$= \sqrt{25 + 144}$$

$$= \sqrt{169}$$

$$= 13$$

20. $(5, -6), \ (-7, -1)$

$$x = \frac{5 + (-7)}{2} = \frac{-2}{2} = -1$$

$$y = \frac{-6 + (-1)}{2} = \frac{-7}{2} = -\frac{7}{2}$$

midpoint: $\left(-1, \ -\dfrac{7}{2}\right)$

21. $(3, 0), \quad (0, 2)$

$$m = \frac{0 - 2}{3 - 0} = -\frac{2}{3}$$

22. $5x + 3y = 7$

$$3y = -5x + 7$$

$$y = -\frac{5}{3}x + \frac{7}{3}$$

$$m = -\frac{5}{3}, \quad b = \frac{7}{3}$$

23.　$x = \dfrac{k\sqrt{y}}{z^2}$

$6 = \dfrac{k\sqrt{9}}{2^2}$

$6(4) = 3k$

$8 = k$

$x = \dfrac{8\sqrt{y}}{z^2}$

$x = \dfrac{8\sqrt{16}}{4^2}$

$= \dfrac{8(4)}{16}$

$= 2$

CHAPTER 7

<u>Problem Set 7.1</u>

1. $\begin{cases} x - 2y = 0 \\ 2x + y = 5 \end{cases}$

$x = 2y$

So, $2(2y) + y = 5$

$\qquad 4y + y = 5$

$\qquad 5y = 5$

$\qquad y = 1$

$x = 2y = 2(1) = 2$

$\{(2, 1)\}$

3. $\begin{cases} 2x - y = 0 \\ x + 4y = 9 \end{cases}$

$2x = y$

So, $x + 4(2x) = 9$

$\qquad x + 8x = 9$

$\qquad 9x = 9$

$\qquad x = 1$

$y = 2x = 2(1) = 2$

$\{(1, 2)\}$

5. $\begin{cases} 2x - 5y = 4 \\ 3x - 2y = -5 \end{cases}$

$2x = 4 + 5y$

$x = 2 + \dfrac{5}{2}y$

So, $3\left(2 + \dfrac{5}{2}y\right) - 2y = -5$

$\qquad 6 + \dfrac{15}{2}y - 2y = -5$

$6 + \dfrac{11}{2}y = -5$

$\dfrac{11}{2}y = -11$

$y = -2$

$x = 2 + \dfrac{5}{2}y = 2 + \dfrac{5}{2}(-2) = 2 - 5 = -3$

$\{(-3, -2)\}$

7. $\begin{cases} x + y = 4 \\ 3x + 3y = 12 \end{cases}$

$x = 4 - y$

So, $3(4 - y) + 3y = 12$

$\qquad 12 - 3y + 3y = 12$

$\qquad 12 = 12$

$\{(x, y) \mid x + y = 4\}$

9. $\begin{cases} \dfrac{1}{3}x - \dfrac{3}{4}y = 1 \\ 4x - 9y = 6 \end{cases}$

$\begin{cases} 4x - 9y = 12 \\ 4x - 9y = 6 \end{cases}$

$4x = 9y + 12$

$x = \dfrac{9}{4}y + 3$

So, $4\left(\dfrac{9}{4}y + 3\right) - 9y = 6$

$\qquad 9y + 12 - 9y = 6$

$\qquad 12 = 6$

$\emptyset$

294

11. $\begin{cases} x^2 - y = -1 \\ x - y = -1 \end{cases}$

$x = y - 1$

So, $(y - 1)^2 - y = -1$

$y^2 - 2y + 1 - y = -1$

$y^2 - 3y + 2 = 0$

$(y - 2)(y - 1) = 0$

$y = 2$ or $y = 1$

$x = y - 1 = 2 - 1 = 1$ or $x = 1 - 1 = 0$

$\{(1, 2), (0, 1)\}$

13. $\begin{cases} x + y = 1 \\ x - y = 3 \end{cases}$

$\begin{aligned} x + y &= 1 \\ \underline{x - y} &= \underline{3} \\ 2x &= 4 \end{aligned}$

$x = 2$

$2 + y = 1$

$y = -1$

$\{(2, -1)\}$

15. $\begin{cases} 2x - y = -5 \\ 3x + 2y = 3 \end{cases}$

$\begin{aligned} 4x - 2y &= -10 \\ \underline{3x + 2y} &= \underline{3} \\ 7x &= -7 \end{aligned}$

$x = -1$

$2(-1) - y = -5$

$-2 - y = -5$

$-y = -3$

$y = 3$

$\{(-1, 3)\}$

17. $\begin{cases} 3x + 2y = 5 \\ 6x - y = 0 \end{cases}$

$\begin{aligned} 3x + 2y &= 5 \\ \underline{12x - 2y} &= \underline{0} \\ 15x &= 5 \end{aligned}$

$x = \dfrac{1}{3}$

$3\left(\dfrac{1}{3}\right) + 2y = 5$

$1 + 2y = 5$

$2y = 4$

$y = 2$

$\left\{\left(\dfrac{1}{3}, 2\right)\right\}$

19. $\begin{cases} 2x + 3y = 5 \\ 3x - 2y = 1 \end{cases}$

$\begin{aligned} 4x + 6y &= 10 \\ \underline{9x - 6y} &= \underline{3} \\ 13x &= 13 \end{aligned}$

$x = 1$

$2(1) + 3y = 5$

$2 + 3y = 5$

$3y = 3$

$y = 1$

$\{(1, 1)\}$

21. $\begin{cases} 5x - 2y = 0 \\ 7x - 3y = -1 \end{cases}$

$-15x + 6y = 0$
$\underline{14x - 6y = -2}$
$-x \quad\quad = -2$

$x \quad = 2$

$5(2) - 2y = 0$

$10 - 2y = 0$

$-2y = -10$

$y = 5$

$\{(2, 5)\}$

23. $\begin{cases} \dfrac{4}{7}x - \dfrac{3}{5}y = 0 \\ -\dfrac{2}{5}x + \dfrac{1}{7}y = 0 \end{cases}$

$\begin{cases} 20x - 21y = 0 \\ -14x + 5y = 0 \end{cases}$

$100x - 105y = 0$
$\underline{-294x + 105y = 0}$
$-194x \quad\quad = 0$

$x \quad = 0$

$\dfrac{4}{7}(0) - \dfrac{3}{5}y = 0$

$0 - \dfrac{3}{5}y = 0$

$-\dfrac{3}{5}y = 0$

$y = 0$

$\{(0, 0)\}$

25. $\begin{cases} 6x - 4y = 23 \\ -9x + 6y = 12 \end{cases}$

$18x - 12y = 69$
$\underline{-18x + 12y = 24}$
$0 = 93$

$\emptyset$

27. $\begin{cases} \dfrac{2}{3}x + 8y = \dfrac{6}{7} \\ \dfrac{3}{4}x + 9y = 3 \end{cases}$

$\begin{cases} 14x + 168y = 18 \\ 3x + 36y = 12 \end{cases}$

$-42x - 504y = 54$
$\underline{42x + 504y = 168}$
$0 = 2?$

$\emptyset$

29. 1st numb : x
2nd number: y

$\begin{cases} y \quad y = 18 \\ x + y = 0 \end{cases}$

$x - y = 18$
$\underline{x + y = 0}$
$2x \quad = 18$

$x \quad = 9$

$9 + y = 0$

$y = -9$

The numbers are 9 and −9.

31. $\begin{cases} x + 2y = 5 \\ 2x - y = 0 \end{cases}$

$x + 2y = 5$

$x = 5 - 2y$

296

So, $2(5 - 2y) - y = 0$

$10 - 4y - y = 0$

$-5y = -10$

$y = 2$

$x = 5 - 2y = 5 - 2(2) = 1$

$\{(1, 2)\}$

33. $\begin{cases} 2x + y = 5 \\ x - 4y = -2 \end{cases}$

$x - 4y = -2$

$x = 4y - 2$

So, $2(4y - 2) + y = 5$

$8y - 4 + y = 5$

$9y = 9$

$y = 1$

$x = 4y - 2 = 4(1) - 2 = 2$

$\{(2, 1)\}$

35. $\begin{cases} 2x + 5y = -10 \\ 3x - 2y = 4 \end{cases}$

$2x + 5y = -10$

$2x = -5y - 10$

$x = -\dfrac{5}{2}y - 5$

So, $3\left(-\dfrac{5}{2}y - 5\right) - 2y = 4$

$-\dfrac{15}{2}y - 15 - 2y = 4$

$-\dfrac{19}{2}y - 15 = 4$

$-\dfrac{19}{2}y = 19$

$y = -2$

$x = -\dfrac{5}{2}y - 5 = -\dfrac{5}{2}(-2) - 5 = 0$

$\{(0, -2)\}$

37. $\begin{cases} 4x - 5y = -8 \\ 2x + 3y = 7 \end{cases}$

$2x + 3y = 7$

$2x = 7 - 3y$

$x = \dfrac{7}{2} - \dfrac{3}{2}y$

So, $4\left(\dfrac{7}{2} - \dfrac{3}{2}y\right) - 5y = -8$

$14 - 6y - 5y = -8$

$14 - 11y = -8$

$-11y = -22$

$y = 2$

$x = \dfrac{7}{2} - \dfrac{3}{2}y = \dfrac{7}{2} - \dfrac{3}{2}(2) = \dfrac{7}{2} - 3 = \dfrac{1}{2}$

$\left\{\left(\dfrac{1}{2}, 2\right)\right\}$

39. $\begin{cases} 4x + 6y = 1 \\ 2x + 3y = 2 \end{cases}$

$2x = 2 - 3y$

$x = 1 - \dfrac{3}{2}y$

So, $4\left(1 - \dfrac{3}{2}y\right) + 6y = 1$

$$4 - 6y + 6y = 1$$

$$4 = 1$$

$$\varnothing$$

41. $\begin{cases} 2x + 6y = 9 \\ 4x + y = 7 \end{cases}$

$$2x = 9 - 6y$$

$$x = \frac{9}{2} - 3y$$

So, $4\left(\dfrac{9}{2} - 3y\right) + y = 7$

$$18 - 12y + y = 7$$

$$18 - 11y = 7$$

$$-11y = -11$$

$$y = 1$$

$$x = \frac{9}{2} - 3y = \frac{9}{2} - 3(1) = \frac{9}{2} - 3 = \frac{3}{2}$$

$$\left\{\left(\frac{3}{2},\ 1\right)\right\}$$

43. $\begin{cases} 2x - y = 4 \\ 6x - 3y = 8 \end{cases}$

$$-y = 4 - 2x$$

$$y = -4 + 2x$$

So, $6x - 3(-4 + 2x) = 8$

$$6x + 12 - 6x = 8$$

$$12 = 8$$

$$\varnothing$$

45. $\begin{cases} x - 2y = 0 \\ 3x - 4y = 0 \end{cases}$

$$x = 2y$$

So, $3(2y) - 4y = 0$

$$6y - 4y = 0$$

$$2y = 0$$

$$y = 0$$

$$x = 2y = 2(0) = 0$$

$$\{(0,\ 0)\}$$

47.

$\begin{cases} x + y = 3 \\ xy = 2 \end{cases}$

$$x = 3 - y$$

So, $(3 - y)y = 2$

$$3y - y^2 = 2$$

$$0 = y^2 - 3y + 2$$

$$0 = (y - 2)(y - 1)$$

$$\begin{array}{ll} y = 2 & \text{or} \quad y = 1 \\ x = 3 - y = 3 - 2 = 1 & \quad x = 3 - 1 = 2 \end{array}$$

$$\{(1,\ 2),\ (2,\ 1)\}$$

49. $\begin{cases} 2x - y = 7 \\ 3x + y = 8 \end{cases}$

$$\begin{array}{r} 2x - y = 7 \\ \underline{3x + y = 8} \\ 5x = 15 \end{array}$$

$$x = 3$$

$$2(3) - y = 7$$

$$6 - y = 7$$

$$-y = 1$$

$$y = -1$$

$$\{(3, -1)\}$$

51. $\begin{cases} 3x + y = 5 \\ 4x - 2y = 10 \end{cases}$

$$\begin{array}{r} 6x + 2y = 10 \\ \underline{4x - 2y = 10} \\ 10x = 20 \end{array}$$

$$x = 2$$

$$3(2) + y = 5$$

$$6 + y = 5$$

$$y = -1$$

$$\{(2, -1)\}$$

53. $\begin{cases} 2x - 3y = -5 \\ 3x + y = -2 \end{cases}$

$$\begin{array}{r} 2x - 3y = -5 \\ \underline{9x + 3y = -6} \\ 11x = -11 \end{array}$$

$$x = -1$$

$$3(-1) + y = -2$$

$$-3 + y = -2$$

$$y = 1$$

$$\{(-1, 1)\}$$

55. $\begin{cases} 2x - 3y = -3 \\ 3x + 2y = 15 \end{cases}$

$$\begin{array}{r} 4x - 6y = -6 \\ \underline{9x + 6y = 45} \\ 13x = 39 \end{array}$$

$$x = 3$$

$$3(3) + 2y = 15$$

$$9 + 12y = 15$$

$$2y = 6$$

$$y = 3$$

$$\{(3, 3)\}$$

57. $\begin{cases} 7x + 4y = -1 \\ 9x - 6y = -5 \end{cases}$

$$\begin{array}{r} 21x + 12y = -3 \\ \underline{18x - 12y = -10} \\ 39x = -13 \end{array}$$

$$x = -\frac{1}{3}$$

$$9\left(-\frac{1}{3}\right) - 6y = -5$$

$$-3 - 6y = -5$$

$$-6y = -2$$

$$y = \frac{1}{3}$$

$$\left\{\left(-\frac{1}{3}, \frac{1}{3}\right)\right\}$$

59. $\begin{cases} 5x - 2y = 0 \\ 3x - 7y = 0 \end{cases}$

$$\begin{array}{r} -15x + 6y = 0 \\ \underline{15x - 35y = 0} \\ -29y = 0 \end{array}$$

$$y = 0$$

$$5x - 2(0) = 0$$

$$5x - 0 = 0$$

$$5x = 0$$

$$x = 0$$

$$\{(0, 0)\}$$

61. $\begin{cases} 3x + 4y = 0 \\ 4x + 3y = 0 \end{cases}$

$$\begin{array}{r} -12x - 16y = 0 \\ \underline{12x + 9y = 0} \\ -7y = 0 \end{array}$$

$$y = 0$$

$$3x + 4(0) = 0$$

$$3x + 0 = 0$$

$$3x = 0$$

$$x = 0$$

$$\{(0, 0)\}$$

63. $\begin{cases} 6x - 5y = 6 \\ 9x + 7y = -20 \end{cases}$

$$\begin{array}{r} 42x - 35y = 42 \\ \underline{45x + 35y = -100} \\ 87x = -58 \end{array}$$

$$x = -\frac{2}{3}$$

$$6\left(-\frac{2}{3}\right) - 5y = 6$$

$$-4 - 5y = 6$$

$$-5y = 10$$

$$y = -2$$

$$\left\{\left(-\frac{2}{3}, -2\right)\right\}$$

65. $\begin{cases} -4x + 15y = -2 \\ 12x + 5y = -4 \end{cases}$

$$\begin{array}{r} -12x + 45y = -6 \\ \underline{12x + 5y = -4} \\ 50y = -10 \end{array}$$

$$y = -\frac{1}{5}$$

$$12x + 5\left(-\frac{1}{5}\right) = -4$$

$$12x - 1 = -4$$

$$12x = -3$$

$$x = -\frac{1}{4}$$

$$\left\{\left(-\frac{1}{4}, -\frac{1}{5}\right)\right\}$$

67. $\begin{cases} 3x - 4y = 5 \\ 4x - 7y = 9 \end{cases}$

$$\begin{array}{r} -12x + 16y = -20 \\ \underline{12x - 21y = 27} \\ -5y = 7 \end{array}$$

$$y = -\frac{7}{5}$$

$$3x - 4\left(-\frac{7}{5}\right) = 5$$

$$3x + \frac{28}{5} = 5$$

$$3x = -\frac{3}{5}$$

$$x = -\frac{1}{5}$$

$$\left\{\left(-\frac{1}{5}, -\frac{7}{5}\right)\right\}$$

69.
$$\begin{cases} 17x - 11y = -1 \\ -34x + 22y = 2 \end{cases}$$

$$\begin{array}{r} 34x - 22y = -2 \\ \underline{-34x + 22y = 2} \\ 0 = 0 \end{array}$$

$$\{(x,\ y)\ |\ 17x - 11y = -1\}$$

71.
$$\begin{cases} \dfrac{1}{5}x - \dfrac{2}{3}y = \dfrac{1}{15} \\[2mm] \dfrac{3}{4}x - \dfrac{5}{6}y = \dfrac{1}{3} \end{cases}$$

$$\begin{cases} 3x - 10y = 1 \\ 9x - 10y = 4 \end{cases}$$

$$\begin{array}{r} -3x + 10y = -1 \\ \underline{9x - 10y = 4} \\ 6x = 3 \end{array}$$

$$x = \dfrac{1}{2}$$

$$3\left(\dfrac{1}{2}\right) - 10y = 1$$

$$\dfrac{3}{2} - 10y = 1$$

$$-10y = -\dfrac{1}{2}$$

$$y = \dfrac{1}{20}$$

$$\left\{\left(\dfrac{1}{2},\ \dfrac{1}{20}\right)\right\}$$

73.
$$\begin{cases} \dfrac{2}{7}x + \dfrac{2}{5}y = \dfrac{3}{7} \\[2mm] \dfrac{3}{5}x + \dfrac{7}{3}y = \dfrac{5}{2} \end{cases}$$

$$\begin{cases} 10x + 14y = 15 \\ 18x + 70y = 75 \end{cases}$$

$$\begin{array}{r} -50x - 70y = -75 \\ \underline{18x + 70y = 75} \\ -32x = 0 \end{array}$$

$$x = 0$$

$$10\,(0) + 14y = 15$$

$$14y = 15$$

$$y = \dfrac{15}{14}$$

$$\left\{\left(0,\ \dfrac{15}{14}\right)\right\}$$

75.
$$\begin{cases} \dfrac{3}{8}x - \dfrac{1}{9}y = 0 \\[2mm] \dfrac{9}{2}x - \dfrac{5}{3}y = 0 \end{cases}$$

$$\begin{cases} 27x - 8y = 0 \\ 27x - 10y = 0 \end{cases}$$

$$\begin{array}{r} 27x - 8y = 0 \\ \underline{-27x + 10y = 0} \\ 2y = 0 \end{array}$$

$$y = 0$$

$$\dfrac{3}{8}x - \dfrac{1}{9}(0) = 0$$

$$\dfrac{3}{8}x = 0$$

$$x = 0$$

$$\{(0,\ 0)\}$$

77.
$$\begin{cases} 3x^2 - 2y^2 = 4 \\ 2x - y^2 = 0 \end{cases}$$

$$3x^2 - 2y^2 = 4$$
$$\underline{-4x + 2y^2 = 0}$$
$$3x^2 - 4x = 4$$

$$3x^2 - 4x - 4 = 0$$

$$(3x + 2)(x - 2) = 0$$

$$x = -\frac{2}{3} \qquad \text{or} \qquad x = 2$$

$$2\left(-\frac{2}{3}\right) - y^2 = 0 \qquad 2(2) - y^2 = 0$$

$$-\frac{4}{3} - y^2 = 0 \qquad\qquad 4 = y^2$$

$$-\frac{4}{3} = y^2 \qquad\qquad \pm 2 = y$$

No Solution.

$\{(2, 2), (2, -2)\}$

79.
$$\begin{cases} x^2 + y^2 = 1 \\ x^2 - y = 1 \end{cases}$$

$$x^2 + y^2 = 1$$
$$\underline{-x^2 + y = -1}$$
$$y^2 + y = 0$$

$$y(y + 1) = 0$$

$$y = 0 \qquad \text{or} \qquad y = -1$$

$$x^2 - 0 = 1 \qquad x^2 - (-1) = 1$$

$$x^2 = 1 \qquad\quad x^2 + 1 = 1$$

$$x = \pm 1 \qquad\qquad x^2 = 0$$

$$x = 0$$

$\{(1, 0), (-1, 0), (0, -1)\}$

81. 1st number: x
2nd number: y

$$\begin{cases} x - y = 2 \\ 3x + 5y = 94 \end{cases}$$

$$5x - 5y = 10$$
$$\underline{3x + 5y = 94}$$
$$8x = 104$$

$$x = 13$$

$$13 - y = 2$$

$$-y = -11$$

$$y = 11$$

The numbers are 13 and 11.

83. 1st number: x
2nd number: y
other number: $2\left(\dfrac{x + y}{2}\right) = x + y$

$$\begin{cases} x + y + x + y = 68 \\ x - y = 20 \end{cases}$$

$$\begin{cases} 2x + 2y = 68 \\ x - y = 20 \end{cases}$$

$$x + y = 34$$
$$\underline{x - y = 20}$$
$$2x = 54$$

$$x = 27$$

$$27 - y = 20$$

$$-y = -7$$

$$y = 7$$

$$x + y = 27 + 7 = 34$$

The numbers are 27, 7 and 34.

11. 1st number: x
2nd number: y

$x + y = 9$

$x^2 + y^2 = 185$

$x = 19 - y$

$(19 - y)^2 + y^2 = 185$

$361 - 38y + y^2 + y^2 = 185$

$2y^2 - 38y + 176 = 0$

$y^2 - 19y + 88 = 0$

$(y - 8)(y - 11) = 0$

$y = 8$ or $y = 11$

$x = 19 - 8 = 11$ $x = 19 - 11 = 8$

The numbers are 8 and 11.

13. price per lemon: x
price per pound of grapes: y

$\begin{cases} 3y + 6x = 357 \\ 2y + 5x = 248 \end{cases}$

$\begin{array}{r} -6y - 12x = -714 \\ \underline{6y + 15x = 744} \\ 3x = 30 \end{array}$

$x = 10$

$3y + 6(10) = 357$

$3y + 60 = 357$

$3y = 297$

$y = 99$

Each lemon is 10 cents and the grapes are 99 cents per pound.

15.

	time for whole job	part done in 1 hr
Tom	x	$1/x$
Randi	y	$1/y$
together	$1\ 1/5 = 6/5$	$1/(6/5) = 5/6$

$\begin{cases} \dfrac{1}{x} + \dfrac{1}{y} = \dfrac{5}{6} \\[2mm] \left(\dfrac{18}{60}\right)\left(\dfrac{5}{6}\right) + \left(\dfrac{3}{2}\right)\left(\dfrac{1}{x}\right) = 1 \end{cases}$

$\begin{cases} \dfrac{1}{x} + \dfrac{1}{y} = \dfrac{5}{6} \\[2mm] \dfrac{1}{4} + \dfrac{3}{2x} = 1 \end{cases}$

Let $u = \dfrac{1}{x}$ and $v = \dfrac{1}{y}$.

$\begin{cases} u + v = \dfrac{5}{6} \\[2mm] \dfrac{1}{4} + \dfrac{3}{2}u = 1 \end{cases}$

$\begin{cases} 6u + 6v = 5 \\ 1 + 6u = 4 \end{cases}$

$6u = 3$

$u = \dfrac{1}{2}$

$6\left(\dfrac{1}{2}\right) + 6v = 5$

$3 + 6v = 5$

$6v = 2$

$v = \dfrac{1}{3}$

305

$$u = \frac{1}{x} \qquad v = \frac{1}{y}$$

$$\frac{1}{2} = \frac{1}{x} \qquad \frac{1}{3} = \frac{1}{y}$$

$$x = 2 \qquad y = 3$$

It would take Tom 2 hours and Randi 3 hours.

17. width: w
    length: l

$$\begin{cases} 2w + 2l = 1200 \\ l = 4w \end{cases}$$

$$2w + 2(4w) = 1200$$

$$2w + 8w = 1200$$

$$10w = 1200$$

$$w = 120$$

$$l = 4(120) = 480$$

The pasture is 120 ft by 480 ft.

19.

	D	R	T = D/R
Carolyn	10	x	10/x
Bill	6	y	6/y

$$\begin{cases} \dfrac{10}{x} = \dfrac{6}{y} \\ x = 2y - 1 \end{cases}$$

$$\begin{cases} 10y = 6x \\ x = 2y - 1 \end{cases}$$

$$10y = 6(2y - 1)$$

$$10y = 12y - 6$$

$$-2y = -6$$

$$y = 3$$

$$x = 2(3) - 1 = 5$$

Carolyn walks 5 mph and Bill walks 3 mph.

21. shortest side: x
    length of other sides: y

$$\begin{cases} x + 2y = 48 \\ x = y - 3 \end{cases}$$

$$x = y - 3$$

So, $y - 3 + 2y = 48$

$$3y - 3 = 48$$

$$3y = 51$$

$$y = 17$$

$$x = 17 - 3 = 14$$

The triangle is 14 in by 17 in by 17 in.

23. width: x
    length: y

$$\begin{cases} 2x + 2y = 60 \\ xy = 216 \end{cases}$$

$$xy = 216$$

$$x = \frac{216}{y}$$

$$2\left(\frac{216}{y}\right) + 2y = 60$$

$$\frac{216}{y} + y = 30$$

$$216 + y^2 = 30y$$

306

$$y^2 - 30y + 216 = 0$$

$$(y - 18)(y - 12) = 0$$

$$y = 18 \qquad \text{or} \qquad y = 12$$

$$x = \frac{216}{18} = 12 \qquad\qquad x = \frac{216}{12} = 18$$

The dimensions are 12 in by 18 in.

25. price per leotard: x
price per pair of tights: y

$$\begin{cases} 3x + 4y = 185 \\ 2x + 3y = 127.50 \end{cases}$$

$$-6x - 8y = -370$$

$$6x + 9y = 382.50$$

$$y = 12.50$$

$$3x + 4(12.50) = 185$$

$$3x + 50 = 185$$

$$3x = 135$$

$$x = 45$$

A leotard is $45 and a pair of tights is $12.50.

27. number of hours walking: x
number of hours running: y

$$\begin{cases} 200y + 150x = 150 \\ 250y + 180x = 185 \end{cases}$$

$$\begin{cases} 4y + 3x = 3 \\ 50y + 36x = 37 \end{cases}$$

$$-48y - 36x = -36$$
$$\underline{50y + 36x = 37}$$
$$2y \qquad\quad = 1$$

$$y \qquad\quad = \frac{1}{2}$$

$$4\left(\frac{1}{2}\right) + 3x = 3$$

$$2 + 3x = 3$$

$$3x = 1$$

$$x = \frac{1}{3}$$

They walked $\frac{1}{3}$ hr and ran $\frac{1}{2}$ hr.

29.

x: hours for $\frac{1}{4}$ inch pipe to fill tank, working alone, drain closed.

y: hours for $\frac{1}{2}$ inch pipe to fill tank, working alone, drain closed.

z: hours for drain to empty the tank, working alone.

$$40 \text{ min} = \frac{40}{60} = \frac{2}{3} \text{ hr}$$

$$12 \text{ min} = \frac{12}{60} = \frac{1}{5} \text{ hr}$$

$$\begin{cases} \dfrac{1}{x} + \dfrac{1}{y} = \dfrac{1}{\frac{1}{5}} \\[2mm] \dfrac{2}{3}\left(\dfrac{1}{y}\right) - \dfrac{2}{3}\left(\dfrac{1}{z}\right) = 1 \\[2mm] 2\left(\dfrac{1}{x}\right) - 2\left(\dfrac{1}{z}\right) = 1 \end{cases}$$

Let $u = \dfrac{1}{x}$, $v = \dfrac{1}{y}$, $w = \dfrac{1}{z}$

$$\begin{cases} u + v = 5 \\[2mm] \dfrac{2}{3}v - \dfrac{2}{3}w = 1 \\[2mm] 2u - 2w = 1 \end{cases}$$

$u = 5 - v$

$$\frac{2}{3}v - \frac{2}{3}w = 1$$

$2(5 - v) - 2w = 1$

$2v - 2w = 3$

$10 - 2v - 2w = 1$

$2v - 2w = 3$
$\underline{-2v - 2w = -9}$
$-4w = -6$

$$w = \frac{3}{2}$$

$2v - 2\left(\dfrac{3}{2}\right) = 3$

$2v - 3 = 3$

$2v = 6$

$v = 3$

$u = 5 - 3 = 2$

$u = \dfrac{1}{x} \qquad v = \dfrac{1}{y} \qquad w = \dfrac{1}{z}$

$2 = \dfrac{1}{x} \qquad 3 = \dfrac{1}{y} \qquad \dfrac{3}{2} = \dfrac{1}{z}$

$x = \dfrac{1}{2} \qquad y = \dfrac{1}{3} \qquad z = \dfrac{2}{3}$

It takes the drain $\dfrac{2}{3}$ hr or 40 minutes
to empty a full tank with the pipes closed.

Problem Set 7.3

1. $\begin{cases} x + y + z = 6 \\ x + 2y - z = 2 \\ 2x - 2y + z = 1 \end{cases}$

$\begin{array}{l} x + y + z = 6 \\ \underline{x + 2y - z = 2} \\ 2x + 3y = 8 \end{array} \qquad \begin{array}{l} x + 2y - z = 2 \\ \underline{2x - 2y + z = 1} \\ 3x = 3 \\ x = 1 \end{array}$

$2(1) + 3y = 8$

$2 + 3y = 8$

$3y = 6$

$y = 2$

$1 + 2 + z = 6$

$3 + z = 6$

$z = 3$

$\{(1, 2, 3)\}$

3. $\begin{cases} 2x - y + 2z = 2 \\ -2x + 3y + 4z = 10 \\ -2x + y - z = 0 \end{cases}$

$\begin{array}{l} 2x - y + 2z = 2 \\ \underline{-2x + 3y + 4z = 10} \\ 2y + 6z = 8 \end{array} \qquad \begin{array}{l} 2x - y + 2z = 2 \\ \underline{-2x + y - z = 0} \\ z = 2 \end{array}$

$2y + 6(2) = 12$

$2y + 12 = 12$

$2y = 0$

$y = 0$

$2x - 0 + 2(2) = 2$

$2x + 4 = 2$

$2x = -2$

$x = -1$

$\{(-1, 0, 2)\}$

5. $\begin{cases} x + 2y + 3z = 1 \\ 2x - 2y - 3z = -1 \\ 3x + y + 6z = 2 \end{cases}$

$\begin{array}{l} x + 2y + 3z = 1 \\ \underline{2x - 2y - 3z = -1} \\ 3x \qquad\qquad = 0 \end{array}$ $\qquad$ $\begin{array}{l} x + 2y + 3z = 1 \\ \underline{-6x - 2y - 12z = -4} \\ -5x \qquad - 9z = -3 \end{array}$

$\begin{array}{l} x \qquad\qquad = 0 \\ -5(0) - 9z = -3 \end{array}$

$-9z = -3$

$z = \dfrac{1}{3}$

$0 + 2y + 3\left(\dfrac{1}{3}\right) = 1$

$2y + 1 = 1$

$2y = 0$

$y = 0$

$\left\{\left(0,\ 0,\ \dfrac{1}{3}\right)\right\}$

7. $\begin{cases} 3x + 3y - z = 5 \\ x + y + z = 5 \\ -2x - 2y + z = -3 \end{cases}$

$\begin{array}{l} 3x + 3y - z = 5 \\ \underline{x + y + z = 5} \\ 4x + 4y \qquad = 10 \end{array}$ $\qquad$ $\begin{array}{l} 3x + 3y - z = 5 \\ \underline{-2x - 2y + z = -3} \\ x + y \qquad = 2 \end{array}$

$4x + 4y = 10$

$\underline{-4x - 4y = -8}$

$0 = 2$

$\emptyset$

9. $\begin{cases} 3x - 5y - 2z = 9 \\ x - y + 2z = 1 \\ 2x - 3y = 1 \end{cases}$

$\begin{array}{l} 3x - 5y - 2z = 9 \\ \underline{x - y + 2z = 1} \\ 4x - 6y \qquad = 10 \end{array}$

$\begin{array}{l} 4x - 6y = 10 \\ \underline{-4x + 6y = -2} \\ 0 = 8 \end{array}$

$\emptyset$

11. $\begin{cases} x + y + z = 6 \\ x + 2y + z = 9 \\ 2x + 2y - z = 6 \end{cases}$

$\begin{array}{l} x + y + z = 6 \\ \underline{2x + 2y - z = 6} \\ 3x + 3y \qquad = 12 \end{array}$ $\qquad$ $\begin{array}{l} x + 2y + z = 9 \\ \underline{2x + 2y - z = 6} \\ 3x + 4y \qquad = 15 \end{array}$

$\begin{array}{l} 3x + 3y = 12 \\ \underline{-3x - 4y = -15} \\ -y = -3 \end{array}$

$y = 3$

$3x + 3(3) = 12$

$3x + 9 = 12$

$3x = 3$

$x = 1$

$1 + 3 + z = 6$

$4 + z = 6$

$z = 2$

$\{(1,\ 3,\ 2)\}$

13.
$$\begin{cases} 3x_1 + x_2 + 2x_3 = 10 \\ -3x_1 - 2x_2 + 4x_3 = -11 \\ 2x_2 + x_3 = 2 \end{cases}$$

$$\begin{array}{l} 3x_1 + x_2 + 2x_3 = 10 \\ \underline{-3x_1 - 2x_2 + 4x_3 = -11} \\ -x_2 + 6x_3 = -1 \end{array}$$

$$\begin{array}{l} -2x_2 + 12x_3 = -2 \\ \underline{2x_2 + x_3 = 2} \\ 13x_3 = 0 \end{array}$$

$$x_3 = 0$$

$$2x_2 + 0 = 2$$

$$2x_2 = 2$$

$$x_2 = 1$$

$$3x_1 + 1 + 2(0) = 10$$

$$3x_1 + 1 = 10$$

$$3x_1 = 9$$

$$x_1 = 3$$

$$\{(3, 1, 0)\}$$

15.
$$\begin{cases} 3x + 3y - z = 3 \\ 5x + y + 3z = 1 \\ 2x + 4y - 3z = 4 \end{cases}$$

$$\begin{array}{ll} \begin{array}{l} 9x + 9y - 3z = 9 \\ \underline{5x + y + 3z = 1} \\ 14x + 10y = 10 \end{array} & \begin{array}{l} 5x + y + 3z = 1 \\ \underline{2x + 4y - 3z = 4} \\ 7x + 5y = 5 \end{array} \end{array}$$

$$\begin{array}{l} 14x + 10y = 10 \\ \underline{-14x - 10y = -10} \\ 0 = 0 \end{array}$$

Dependent

17.
$$\begin{cases} 2x + 2y - z = 1 \\ x + 2y - 3z = 4 \\ 5x + 6y - 5z = 3 \end{cases}$$

$$\begin{array}{ll} \begin{array}{l} 2x + 2y - z = 1 \\ \underline{-x - 2y + 3z = -4} \\ x + 2z = -3 \end{array} & \begin{array}{l} -3x - 6y + 9z = -12 \\ \underline{5x + 6y - 5z = 3} \\ 2x + 4z = -9 \end{array} \end{array}$$

$$\begin{array}{l} -2x - 4z = 6 \\ \underline{2x + 4z = -9} \\ 0 = -3 \end{array}$$

$\emptyset$

19.
$$\begin{cases} 3x + y - z = 8 \\ 2x - y + 2z = 3 \\ x + 2y - 3z = 5 \end{cases}$$

$$\begin{array}{ll} \begin{array}{l} 3x + y - z = 8 \\ \underline{2x - y + 2z = 3} \\ 5x + z = 11 \end{array} & \begin{array}{l} 4x - 2y + 4z = 6 \\ \underline{x + 2y - 3z = 5} \\ 5x + z = 11 \end{array} \end{array}$$

$$\begin{array}{l} 5x + z = 11 \\ \underline{-5x - z = -11} \\ 0 = 0 \end{array}$$

Dependent

21.
$$\begin{cases} 2x - 3y - 10z = 4 \\ 4x + - 5z = 3 \\ 6y + 5z = -3 \end{cases}$$

$$\begin{array}{l} -4x + 6y + 20z = -8 \\ \underline{4x - 5z = 3} \\ 6y + 15z = -5 \end{array}$$

$$\begin{array}{l} 6y + 15z = -5 \\ \underline{-6y - 5z = 3} \\ 10z = -2 \end{array}$$

$$z = -\frac{1}{5}$$

$$6y + 5\left(-\frac{1}{5}\right) = -3$$

$$6y - 1 = -3$$

$$6y = -2$$

$$y = -\frac{1}{3}$$

$$4x - 5\left(-\frac{1}{5}\right) = 3$$

$$4x + 1 = 3$$

$$4x = 2$$

$$x = \frac{1}{2}$$

$$\left\{\left(\frac{1}{2},\ -\frac{1}{3},\ -\frac{1}{5}\right)\right\}$$

23. $\begin{cases} 2x + 6y = 5 \\ 3y - z = 2 \\ -3x - 3z = 1 \end{cases}$

$$\begin{array}{r} -9y + 3z = -6 \\ \underline{-3x \qquad - 3z = 1} \\ -3x - 9y \qquad = -5 \end{array}$$

$\begin{cases} -3x - 9y = -5 \\ 2x + 6y = 5 \end{cases}$

$$\begin{array}{r} -6x - 18y = -10 \\ \underline{6x + 18y = 15} \\ 0 = 5 \end{array}$$

$$\emptyset$$

25. $\begin{cases} 2x - 3y = 6 \\ -2y - 3z = 4 \\ 3x - 2z = 0 \end{cases}$

$$\begin{array}{r} -4x + 6y \qquad = -12 \\ \underline{\qquad - 6y - 9z = 12} \\ -4x \qquad - 9z = 0 \end{array}$$

$\begin{cases} 3x - 2z = 0 \\ -4x - 9z = 0 \end{cases}$

$$\begin{array}{r} 12x - 8z = 0 \\ \underline{-12x - 27z = 0} \\ - 35z = 0 \end{array}$$

$$z = 0$$

$$3x - 2(0) = 0$$

$$3x - 0 = 0$$

$$3x = 0$$

$$x = 0$$

$$2(0) - 3y = 6$$

$$0 - 3y = 6$$

$$-3y = 6$$

$$y = -2$$

$$\{(0,\ -2,\ 0)\}$$

27. $\begin{cases} x + y - z = -8 \\ 4x + 5y - 6z = -2 \\ 2x + 3y - 4z = 14 \end{cases}$

$$\begin{array}{r} -2x - 2y + 2z = 16 \\ \underline{2x + 3y - 4z = 14} \\ y - 2z = 30 \end{array} \qquad \begin{array}{r} -4x - 4y + 4z = 32 \\ \underline{4x + 5y - 6z = -2} \\ y - 2z = 30 \end{array}$$

$$\begin{array}{r} y - 2z = 30 \\ \underline{-y + 2z = -30} \\ 0 = 0 \end{array}$$

Dependent

311

29. $\begin{cases} 2x + 4y - z = 3 \\ x + y = 1 \\ 2x + 3y + z = 2 \end{cases}$

$2x + 4y - z = 3$
$\underline{2x + 3y + z = 2}$
$4x + 7y = 5$

$4x + 7y = 5$
$\underline{-4x - 4y = -4}$
$3y = 1$

$y = \dfrac{1}{3}$

$x + \dfrac{1}{3} = 1$

$x = \dfrac{2}{3}$

$2\left(\dfrac{2}{3}\right) + 4\left(\dfrac{1}{3}\right) - z = 3$

$\dfrac{4}{3} + \dfrac{4}{3} - z = 3$

$\dfrac{8}{3} - z = 3$

$-z = \dfrac{1}{3}$

$z = -\dfrac{1}{3}$

$\left\{\left(\dfrac{2}{3}, \dfrac{1}{3}, -\dfrac{1}{3}\right)\right\}$

31. $\begin{cases} 3x - 2y - z = 4 \\ x + 4y + 2z = -1 \\ 2x - 4y - 3z = 6 \end{cases}$

$6x - 4y - 2z = 8 \qquad x + 4y + 2z = -1$
$\underline{x + 4y + 2z = -1} \qquad \underline{2x - 4y - 3z = 6}$
$7x = 7 \qquad 3x - z = 5$

$x = 1$

$3(1) - z = 5$

$3 - z = 5$

$-z = 2$

$z = -2$

$1 + 4y + 2(-2) = -1$

$4y - 3 = -1$

$4y = 2$

$y = \dfrac{1}{2}$

$\left\{\left(1, \dfrac{1}{2}, -2\right)\right\}$

33. 1st number: x
2nd number: y
3rd number: z

$\begin{cases} x + y = 6 \\ x + z = 7 \\ x + y + z = 12 \end{cases}$

$x + z = 7$
$\underline{-x - y - z = -12}$
$ - y = -5$

$y = 5$

$x + 5 = 6$

$x = 1$

$1 + z = 7$

$z = 6$

The numbers are 1, 5, and 6.

35. 1st number: x
2nd number: y
3rd number: z

$$\begin{cases} x + y \phantom{{}+z} = 1 \\ x \phantom{{}+y} + z = 8 \\ x + y + z = 6 \end{cases}$$

$$\begin{aligned} -x \phantom{{}-y} - z &= -8 \\ \underline{x + y + z} &= \underline{6} \\ y \phantom{{}+x+z} &= -2 \end{aligned}$$

$x - 2 = 1$

$x = 3$

$3 + z = 8$

$z = 5$

$\{(3, -2, 5)\}$

37. price of small drink: x
price of medium drink: y
price of popcorn: z

$$\begin{cases} 2x + 3y + z = 525 \\ x \phantom{{}+3y} + z = 160 \\ \phantom{x+{}} y + 2z = 290 \end{cases}$$

$$\begin{aligned} 2x + 3y + z &= 525 \\ \underline{-2x \phantom{{}+3y} - 2z} &= \underline{-320} \\ 3y - z &= 205 \end{aligned}$$

$$\begin{aligned} y + 2z &= 290 \\ \underline{6y - 2z} &= \underline{410} \\ 7y \phantom{{}- 2z} &= 700 \\ y \phantom{{}- 2z} &= 100 \end{aligned}$$

$3(100) - z = 205$

$300 - z = 205$

$-z = -95$

$z = 95$

$x + 95 = 160$

$x = 65$

A small drink is 65 cents, a medium drink is \$1.00 and popcorn is 95 cents.

39. 1st number: x
2nd number: y
3rd number: z
4th number: w

$$\begin{cases} x + y + z + w = 11 \\ 2x \phantom{{}+y} + z \phantom{{}+w} = 2 \\ 3x \phantom{{}+y} + 2z \phantom{{}+w} = 5 \\ \phantom{3x+{}} 3y \phantom{{}+2z} + 2w = 17 \end{cases}$$

$$\begin{aligned} -4x - 2z &= -4 \\ \underline{3x + 2z} &= \underline{5} \\ -x \phantom{{}- 2z} &= 1 \\ x \phantom{{}- 2z} &= -1 \end{aligned}$$

$2(-1) + z = 2$

$-2 + z = 2$

$z = 4$

$-1 + y + 4 + w = 11$

$y + w = 8$

$$\begin{aligned} 3y + 2w &= 17 \\ \underline{-2y - 2w} &= \underline{-16} \\ y \phantom{{}- 2w} &= 1 \end{aligned}$$

$1 + w = 8$

$w = 7$

The numbers are -1, 1, 4, and 7.

41.
$$\begin{cases} x - 2y - z + 2w = -2 \\ 2x - y + z - w = 0 \\ -x + 3y - w = 4 \\ 5x - 3y = -4 \end{cases}$$

$$\begin{array}{r} x - 2y - z + 2w = -2 \\ \underline{2x - y + z - w = 0} \\ 3x - 3y + w = -2 \end{array}$$

$$\begin{array}{r} -x + 3y - w = 4 \\ \underline{3x - 3y + w = -2} \\ 2x = 2 \end{array}$$

$$x = 1$$

$$5(1) - 3y = -4$$
$$5 - 3y = -4$$
$$-3y = -9$$
$$y = 3$$

$$-1 + 3(3) - w = 4$$
$$-1 + 9 - w = 4$$
$$8 - w = 4$$
$$-w = -4$$
$$w = 4$$

$$1 - 2(3) - z + 2(4) = -2$$
$$1 - 6 - z + 8 = -2$$
$$3 - z = -2$$
$$-z = -5$$
$$z = 5$$

$$\{(1, 3, 5, 4)\}$$

43.
$$\begin{cases} x_1 + x_2 = 1 \\ x_2 - x_3 = -2 \\ 2x_3 - 2x_4 = 5 \\ 2x_4 + 2x_5 = 1 \\ - 2x_5 = 0 \end{cases}$$

$$-2x_5 = 0$$
$$x_5 = 0$$

$$2x_4 + 2x_5 = 0$$
$$2x_4 + 0 = 1$$
$$x_4 = \frac{1}{2}$$

$$2x_3 - 2x_4 = 5$$
$$2x_3 - 2\left(\frac{1}{2}\right) = 5$$
$$2x_3 - 1 = 5$$
$$2x_3 = 6$$
$$x_3 = 3$$

$$x_2 - x_3 = -2$$
$$x_2 - 3 = -2$$
$$x_2 = 1$$

$$x_1 + x_2 = 1$$
$$x_1 + 1 = 1$$
$$x_1 = 0$$

$$\left\{\left(0, 1, 3, \frac{1}{2}, 0\right)\right\}$$

Problem Set 7.4

1. $x - y < 2$

   B.L.: $x - y = 2$ (not included)

   Test $(0, 0)$

   $0 - 0 < 2$

   $\quad 0 < 2 \quad$ T

   Shade region.

   Test $(3, 0)$

   $3 - 0 < 2$

   $\quad 3 < 2 \quad$ F

   Do not shade region.

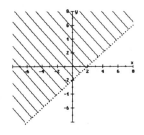

3. $y - \dfrac{1}{2}x > 4$

   B.L.: $y - \dfrac{1}{2}x = 4$ (not included)

   Test $(0, 0)$

   $0 - \dfrac{1}{2}(0) > 4$

   $\quad 0 > 4 \quad$ F

   Do not shade region.

Test $(0, 5)$

$5 - \dfrac{1}{2}(0) > 4$

$\quad 5 > 4 \quad$ T

Shade region.

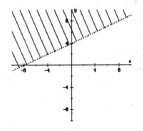

5. $x - y \leq 0$

   B.L.: $x - y = 0$ (included)

   Test $(1, 0)$

   $1 - 0 \leq 0$

   $\quad 1 \leq 0 \quad$ F

   Do not shade region.

   Test $(0, 1)$

   $0 - 1 \leq 0$

   $\quad -1 \leq 0 \quad$ T

   Shade region.

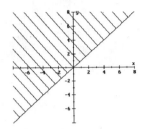

7. $x \geq 3$

   B.L.: $x = 3$ (included)

   Test $(0, 0)$

   $0 \geq 3$ F

   Do not shade region.

   Test $(4, 0)$

   $4 \geq 3$ T

   Shade region.

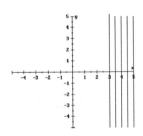

9. $y \leq -1$

   B.L.: $y = -1$ (included)

   Test $(0, 0)$

   $0 \leq -1$ F

   Do not shade region.

   Test $(0, -2)$

   $-2 \leq -1$ T

   Shade region.

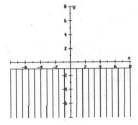

11. $x + y < 3$

   B.L.: $x + y = 3$ (not included)

   Test $(0, 0)$

   $0 + 0 < 3$

       $0 < 3$ T

   Shade region.

   Test $(4, 0)$

   $4 + 0 < 3$

       $4 < 3$ F

   Do not shade region.

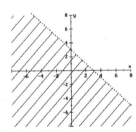

13. $3x + y \leq 5$

   B.L.: $3x + y = 5$ (included)

   Test $(0, 0)$

   $3(0) + 0 \leq 5$

       $0 \leq 5$ T

   Shade region.

316

Test (2, 0)

$3(2) + 0 \leq 5$

$\qquad 6 \leq 5 \qquad$ F

Do not shade region.

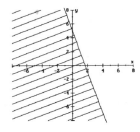

15. $x + y \geq 0$

B.L.: $x + y = 0$ (included)

Test (0, 1)

$0 + 1 \geq 0$

$\qquad 1 \geq 0 \qquad$ T

Shade region.

Test (−1, 0)

$-1 + 0 \geq 0$

$\qquad -1 \geq 0 \qquad$ F

Do not shade region.

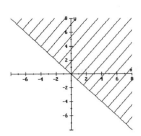

17. $y > 2x - 3$

B.L.: $y = 2x - 3$ (not included)

Test (0, 0)

$0 > 2(0) - 3$

$0 > -3 \qquad$ T

Shade region.

Test (2, 0)

$0 > 2(2) - 3$

$0 > 1 \qquad$ F

Do not shade region.

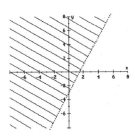

19. $x \leq 1$

B.L.: $x = 1$ (included)

Test (0, 0)

$0 \leq 1 \qquad$ T

Shade region.

Test (2, 0)

$2 \leq 1 \qquad$ F

Do not shade region.

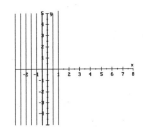

317

21. $y < 3x + 4$

B.L.: $y = 3x + 4$ (not included)

Test $(0, 0)$

$0 < 3(0) + 4$

$0 < 4$ T

Shade region.

Test $(-2, 0)$

$0 < 3(-2) + 4$

$0 < -2$ F

Do not shade region.

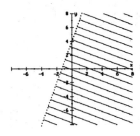

23. $x > 0$

B.L.: $x = 0$ (not included)

Test $(1, 0)$

$1 > 0$ T

Shade region.

Test $(-1, 0)$

$-1 > 0$ F

Do not shade region.

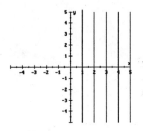

25. $2x - 3y \le 4$

B.L.: $2x - 3y = 4$ (included)

Test $(0, 0)$

$2(0) - 3(0) \le 4$

$0 \le 4$ T

Shade region.

Test $(3, 0)$

$2(3) - 3(0) \le 4$

$6 \le 4$ F

Do not shade region.

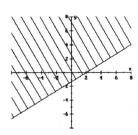

27. $3x - 2y \le 5$

B.L.: $3x - 2y = 5$ (included)

Test $(0, 0)$

$3(0) - 2(0) \le 5$

$0 \le 5$ T

Shade region.

Test (2, 0)

$3(2) - 2(0) \le 5$

$\qquad 6 \le 5 \quad$ F

Do not shade region.

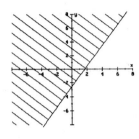

29. $3x - y < 6$

B.L.: $3x - y = 6$ (not included)

Test (0, 0)

$3(0) - 0 < 6$

$\qquad 0 < 6 \quad$ T

Shade region.

Test (3, 0)

$3(3) - 0 < 6$

$\qquad 9 < 6 \quad$ F

Do not shade region.

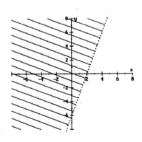

31. $\dfrac{1}{2}x - \dfrac{1}{6}y \le 1$

B.L.: $\dfrac{1}{2}x - \dfrac{1}{6}y = 1$ (included)

$\qquad 3x - y = 6$

Test (0, 0)

$\dfrac{1}{2}(0) - \dfrac{1}{6}(0) \le 6$

$\qquad 0 \le 6 \quad$ T

Shade region.

Test (4, 0)

$\dfrac{1}{2}(4) - \dfrac{1}{6}(0) \le 1$

$\qquad 2 \le 1 \quad$ F

Do not shade region.

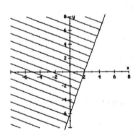

33. $2.4x - 3.2y \ge 0$

B.L.: $2.4x - 3.2y = 0$ (included)

$\qquad 24x - 32y = 0$

$\qquad 3x - 4y = 0$

Test (1, 0)

$2.4(1) - 3.2(0) \ge 0$

$\qquad 2.4 \ge 0 \quad$ T

Shade region.

Test $(0, 1)$

$2.4(0) - 3.2(1) \geq 0$

$\qquad -3.2 \geq 0 \quad$ F

Do not shade region.

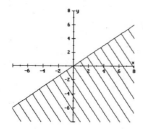

35. $3y - 6 \geq 0$

B.L.: $3y - 6 = 0$ (included)

$\qquad y = 2$

Test $(0, 0)$

$3(0) - 6 \geq 0$

$\qquad -6 \geq 0 \quad$ F

Do not shade region.

Test $(0, 3)$

$3(3) - 6 \geq 0$

$\qquad 3 \geq 0 \quad$ T

Shade region.

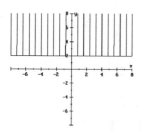

37. $y - \dfrac{1}{3}x < \dfrac{4}{3}$

B.L.: $y - \dfrac{1}{3}x = \dfrac{4}{3}$ (not included)

$\qquad 3y - x = 4$

Test $(0, 0)$

$0 - \dfrac{1}{3}(0) < \dfrac{4}{3}$

$\qquad 0 < \dfrac{4}{3} \quad$ T

Shade region.

Test $(0, 2)$

$2 - \dfrac{1}{3}(0) < \dfrac{4}{3}$

$\qquad 2 < \dfrac{4}{3} \quad$ F

Do not shade region.

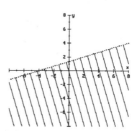

39. $\dfrac{1}{4}x - \dfrac{1}{2}y < 2$

B.L.: $\dfrac{1}{4}x - \dfrac{1}{2}y = 2$ (not included)

$\qquad x - 2y = 8$

Test $(0, 0)$

$\dfrac{1}{4}(0) - \dfrac{1}{2}(0) < 2$

$$0 < 2 \quad \text{T}$$

Shade region.

Test $(0, -6)$

$$\frac{1}{4}(0) - \frac{1}{2}(-6) < 2$$

$$3 < 2 \quad \text{F}$$

Do not shade region.

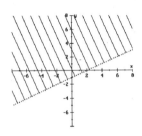

41. $2.1x - 1.2y < 1.1$

B.L.: $2.1x - 1.2y = 1.1$ (not included)

$$21x - 12y = 11$$

Test $(0, 0)$

$$2.1(0) - 1.2(0) < 1.1$$

$$0 < 1.1 \quad \text{T}$$

Shade region.

Test $(1, 0)$

$$2.1(1) - 1.2(0) < 1.1$$

$$2.1 < 1.1 \quad \text{F}$$

Do not shade region.

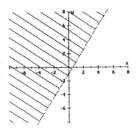

43. $3x \le 6$

B.L.: $3x = 6$ (included)

$$x = 2$$

Test $(0, 0)$

$$3(0) \le 6$$

$$0 \le 6 \quad \text{T}$$

Shade region.

Test $(3, 0)$

$$3(3) \le 6$$

$$9 \le 6 \quad \text{F}$$

Do not shade region.

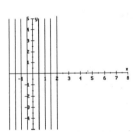

45. $6x - y > 2$

B.L.: $6x - y = 2$ (not included)

Test $(0, 0)$

$$6(0) - 0 > 2$$

$$0 > 2 \quad \text{F}$$

Do not shade region.

Test (1, 0)

6(1) − 0 > 2

$\qquad$ 6 > 2   T

Shade region.

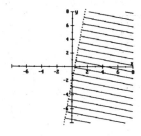

47. number of guitars : x
number of violins : y

$350x + 600y \le 20000$

B.L.: $350x + 600y = 20000$ (included)

$\qquad$ $7x + 12y = 400$

Test (0, 0)

$350(0) + 600(0) \le 20000$

$\qquad$ $0 \le 20000$ T

Shade region.

Test (60, 0)

$350(60) + 600(0) \le 20000$

$\qquad$ $21000 \le 20000$ F

Do not shade region.

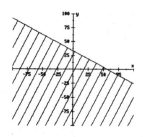

49. $y \le mx$; $m > 0$

B.L.: $y = mx$ (included)

Test (0, 1)

$1 \le m(0)$

$1 \le 0$ F

Do not shade region.

Test (1, 0)

$0 \le m(1)$

$0 \le m$ T

Shade region.

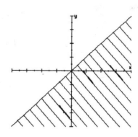

51. $y \le mx + b$; $m > 0$, $b > 0$

B.L.: $y = mx + b$ (included)

Test (0, 0)

$0 \le m(0) + b$

$0 \le b$ T

Shade region.

Test (0, $b + 1$)

$b + 1 \le m(0) + b$

$b + 1 \le b$ F

Do not shade region.

322

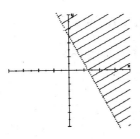

53. $y > mx + b$; $m < 0$, $b > 0$

   B.L.: $y = mx + b$ (not included)

   Test (0, 0)

   $0 > m(0) + b$

   $0 > b$ **F**

   Do not shade region.

   Test (0, $b + 1$)

   $b + 1 > m(0) + b$

   $b + 1 > b$ **T**

   Shade region.

55. $y \le a$; $a > 0$

   B.L.: $y = a$ (included)

   Test (0, 0)

   $0 \le a$ **T**

   Shade region.

Test (0, $a + 1$)

$a + 1 \le a$ **F**

Do not shade region.

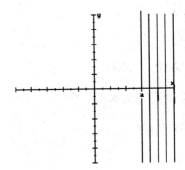

57. $x \ge a$; $a > 0$

   B.L.: $x = a$ (included)

   Test (0, 0)

   $0 \ge a$ **F**

   Do not shade region.

   Test ($a + 1$, 0)

   $a + 1 \ge a$ **T**

   Shade region.

Problem Set 7.5

1. $\begin{cases} x - y \le 4 \\ x + y \ge 2 \end{cases}$

   B.L.: $x - y = 4$ (included)

   B.L.: $x + y = 2$ (included)

   Test $(0, 3)$

   $0 - 3 \le 4$

   $\quad -3 \le 4 \quad$ T

   $0 + 3 \ge 2$

   $\quad 3 \ge 2 \quad$ T

   Region in solution set.

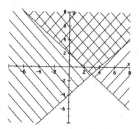

3. $\begin{cases} x + y \le 0 \\ x - y > 0 \end{cases}$

   B.L.: $x + y = 0$ (included)

   B.L.: $x - y = 0$ (not included)

   Test $(0, -1)$

   $0 - 1 \le 0$

   $\quad -1 \le 0 \quad$ T

   $0 - (-1) > 0$

   $\quad 1 > 0 \quad$ T

   Region in solution set.

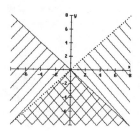

5. $\begin{cases} 3x - y < 1 \\ x + 2y > 5 \end{cases}$

   B.L.: $3x - y = 1$ (not included)

   B.L.: $x + 2y = 5$ (not included)

   Test $(1, 3)$

   $3(1) - 3 < 1$

   $\quad 0 < 1 \quad$ T

   $1 + 2(3) > 5$

   $\quad 7 > 5 \quad$ T

   Region in solution set.

7. $\begin{cases} 2x + y \ge 3 \\ x - 2y > 4 \end{cases}$

   B.L.: $2x + y = 3$ (included)

   B.L.: $x - 2y = 4$ (not included)

   Test $(5, 0)$

   $2(5) + 0 \ge 3$

   $\quad 10 \ge 3 \quad$ T

$$5 - 2(0) > 4$$

$$5 > 4 \quad T$$

Region in solution set.

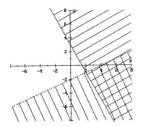

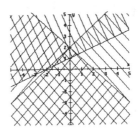

9. $\begin{cases} x - y \geq -2 \\ x + y \leq 2 \\ x - 2y \leq -2 \end{cases}$

B.L.: $x - y = -2$ (included)

B.L.: $x + y = 2$ (included)

B.L.: $x - 2y = -2$ (included)

Test $\left(0, \dfrac{3}{2}\right)$

$$0 - \frac{3}{2} \geq -2$$

$$-\frac{3}{2} \geq -2 \quad T$$

$$0 + \frac{3}{2} \leq 2$$

$$\frac{3}{2} \leq 2 \quad T$$

$$0 - 2\left(\frac{3}{2}\right) \leq -2$$

$$-3 \leq -2 \quad T$$

Region in solution set.

11. $\begin{cases} x - \dfrac{1}{3}y \leq -1 \\ 3x - y \geq -9 \end{cases}$

B.L.: $x - \dfrac{1}{3}y = -1$

$$3x - y = -3 \text{ (included)}$$

B.L.: $3x - y = -9$ (included)

Test $(-2, 0)$

$$-2 - \frac{1}{3}(0) \leq -1$$

$$-2 \leq -1 \quad T$$

$$3(-2) + 0 \geq -9$$

$$-6 \geq -9 \quad T$$

Region in solution set.

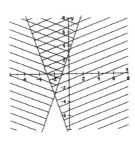

13. $\begin{cases} x \geq 0 \\ y \geq 0 \end{cases}$

B.L.: $x = 0$ (included)

B.L.: $y = 0$ (included)

Test $(1, 1)$

$1 \geq 0$ T

$1 \geq 0$ T

Region in solution set.

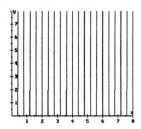

15. $\begin{cases} x < 4 \\ y > 2 \end{cases}$

B.L.: $x = 4$ (not included)

B.L.: $y = 2$ (not included)

Test $(0, 3)$

$0 < 4$ T

$3 > 2$ T

Region in solution set.

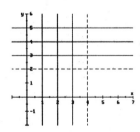

17. $\begin{cases} 2x - y \leq 4 \\ 2x - y \geq 5 \end{cases}$

B.L.: $2x - y = 4$ (included)

B.L.: $2x - y = 5$ (included)

No points in common.

$\emptyset$

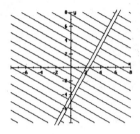

19. $\begin{cases} x + 3y \geq 6 \\ 2x + 6y \leq 4 \end{cases}$

B.L.: $x + 3y = 6$ (included)

B.L.: $2x + 6y = 4$ (included)

No points in common.

$\emptyset$

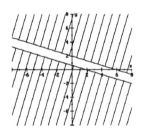

21. $\begin{cases} 2x + 5y > -9 \\ 7x + 3y < 12 \end{cases}$

B.L.: $2x + 5y = -9$ (not included)

B.L.: $7x + 3y = 12$ (not included)

Test $(0, 0)$

$2(0) + 5(0) > -9$

$0 > -9$ T

$$7(0) + 3(0) < 12$$

$$0 < 12 \quad T$$

Region in solution set.

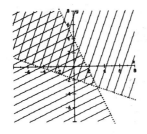

25. $\begin{cases} x + 1 \le 0 \\ y + 3 > 0 \end{cases}$

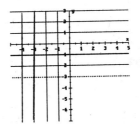

23. $\begin{cases} x - 3y > 0 \\ x - 3y > 2 \end{cases}$

B.L.: $x - 3y = 0$ (not included)

B.L.: $x - 3y = 2$ (not included)

Test $(3, 0)$

$$3 - 3(0) > 0$$

$$3 > 0 \quad T$$

$$3 - 3(0) > 2$$

$$3 > 2 \quad T$$

Region in solution set.

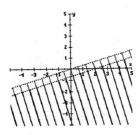

27. $\begin{cases} x - y \le 1 \\ x + 2y \le 1 \end{cases}$

B.L.: $x - y = 1$ (included)

B.L.: $x + 2y = 1$ (included)

Test $(0, 0)$

$$0 - 0 \le 1$$

$$0 \le 1 \quad T$$

$$0 + 2(0) \le 1$$

$$0 \le 1 \quad T$$

Region in solution set.

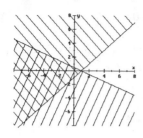

29. $\begin{cases} x - 2y \le 3 \\ 2x - 4y > 1 \end{cases}$

B.L.: $x - 2y = 3$ (included)

B.L.: $2x - 4y = 1$ (not included)

Test $(2, 0)$

$2 - 2(0) \le 3$

$2 \le 3$ T

$2(2) - 4(0) > 1$

$4 > 1$ T

Region in solution set.

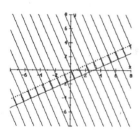

31. $\begin{cases} 2x + y < 2 \\ x + \frac{1}{2}y < 0 \end{cases}$

B.L.: $2x + y = 2$ (not included)

B.L.: $x + \frac{1}{2}y = 0$

$2x + y = 0$ (not included)

Test $(-1, 0)$

$2(-1) + 0 < 2$

$-2 < 2$ T

$-1 + \frac{1}{2}(0) < 0$

$-1 < 0$ T

Region in solution set.

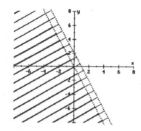

33. $\begin{cases} x < 0 \\ y \le 0 \end{cases}$

B.L.: $x = 0$ (not included)

B.L.: $y = 0$ (included)

Test $(-1, -1)$

$-1 < 0$ T

$-1 \le 0$ T

Region in solution set.

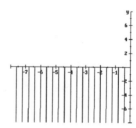

35. $\begin{cases} x < -3 \\ y \geq 1 \end{cases}$

B.L.: $x = -3$ (not included)

B.L.: $y = 1$ (included)

Test $(-4, 2)$

$-4 < -3 \quad$ T

$2 \geq 1 \quad$ T

Region in solution set.

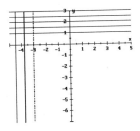

37. $\begin{cases} \frac{1}{2}x - y > 5 \\ x + 2y < 14 \end{cases}$

B.L.: $\frac{1}{2}x - y = 5$

$\qquad x - 2y = 10$ (not included)

B.L.: $x + 2y = 14$ (not included)

Test $(10, -1)$

$\frac{1}{2}(10) - (-1) > 5$

$\qquad\qquad 6 > 5 \quad$ T

$10 + 2(-1) < 14$

$\qquad\qquad 8 < 14 \quad$ T

Region in solution set.

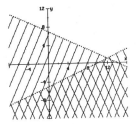

39. $\begin{cases} 3x + 2y < 6 \\ 3x + 2y \geq 1 \end{cases}$

B.L.: $3x + 2y = 6$ (not included)

B.L.: $3x + 2y = 1$ (included)

Test $(0, 1)$

$3(0) + 2(1) < 6$

$\qquad\qquad 2 < 6 \quad$ T

$3(0) + 2(1) \geq 1$

$\qquad\qquad 2 \geq 1 \quad$ T

Region in solution set.

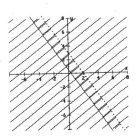

41. $\begin{cases} x - y > -3 \\ x - 2y \leq 0 \\ 3x + y \leq 3 \end{cases}$

B.L.: $x - y = -3$ (not included)

B.L.: $x - 2y = 0$ (included)

B.L.: $3x + y = 3$ (included)

Test (-1, 0)

-1 - 0 > -3

    -1 > -3   T

-1 - 2(0) ≤ 0

    -1 ≤ 0   T

3(-1) + 0 ≤ 3

    -3 ≤ 3   T

Region in solution set.

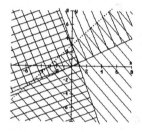

2(1) + 1 ≤ 40

    3 ≤ 40   T

1 ≥ 0   T

1 ≥ 0   T

Region in solution set.

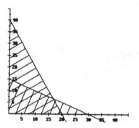

43. number of Deluxe cans: d
     number of Premium cans: p

$$\begin{cases} d + 2p \le 30 \\ 2d + p \le 40 \\ p \ge 0 \\ d \ge 0 \end{cases}$$

B.L.: $d + 2p = 30$ (included)

B.L.: $2d + p = 40$ (included)

B.L.: $p = 0$ (included)

B.L.: $d = 0$ (included)

Test (1, 1)

1 + 2(1) ≤ 30

    3 ≤ 30   T

45. $\begin{cases} y \le mx \; ; \; m > 0 \\ y < nx \; ; \; n < 0 \end{cases}$

B.L.: $y = mx$ (included)

B.L.: $y = nx$ (not included)

Test (0, -1)

-1 ≤ m(0)

-1 ≤ 0   T

-1 < n(0)

-1 < 0   T

Region in solution set.

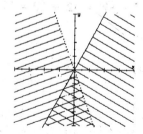

47.
$$\begin{cases} y < a \;\; ; \;\; a > 0 \\ x < b \;\; ; \;\; b > 0 \end{cases}$$

B.L.: $y = a$ (not included)

B.L.: $x = b$ (not included)

Test $(-1, -1)$

$-1 < a$ T

$-1 < b$ T

Region in solution set.

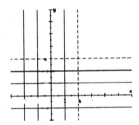

Problem Set 7.6

1. $\begin{vmatrix} 1 & 2 \\ 2 & 5 \end{vmatrix} = 1 \cdot 5 - 2 \cdot 2 = 5 - 4 = 1$

3. $\begin{vmatrix} 3 & 2 \\ 6 & 4 \end{vmatrix} = 3 \cdot 4 - 2 \cdot 6 = 12 - 12 = 0$

5. $\begin{vmatrix} 3 & 5 \\ -2 & 4 \end{vmatrix} = 3 \cdot 4 - 5 \cdot (-2) = 12 + 10 = 22$

7. Expand about 1st row:

$\begin{vmatrix} 1 & 1 & 2 \\ 2 & 3 & 1 \\ 1 & -1 & 1 \end{vmatrix} = 1 \begin{vmatrix} 3 & 1 \\ -1 & 1 \end{vmatrix} - 1 \begin{vmatrix} 2 & 1 \\ 1 & 1 \end{vmatrix} + 2 \begin{vmatrix} 2 & 3 \\ 1 & -1 \end{vmatrix}$

$= 3 - (-1) - (2 - 1) + 2(-2 - 3)$

$= 4 - 1 + 2(-5)$

$= 3 - 10$

$= -7$

9. Expand about 2nd row:

$\begin{vmatrix} 4 & 2 & 1 \\ 0 & 0 & 3 \\ 3 & 5 & 3 \end{vmatrix} = -3 \begin{vmatrix} 4 & 2 \\ 3 & 5 \end{vmatrix}$

$= -3(20 - 6)$

$= -3(14)$

$= -42$

11. Expand about 2nd row:

$\begin{vmatrix} 0 & 1 & 0 \\ 0 & 0 & 1 \\ 2 & 0 & 0 \end{vmatrix} = - \begin{vmatrix} 0 & 1 \\ 2 & 0 \end{vmatrix} = -(0 - 2) = 2$

13.
$$\begin{cases} x + y = 1 \\ x - y = 3 \end{cases}$$

$x = \dfrac{\begin{vmatrix} 1 & 1 \\ 3 & -1 \end{vmatrix}}{\begin{vmatrix} 1 & 1 \\ 1 & -1 \end{vmatrix}} = \dfrac{1(-1) - 1 \cdot 3}{1(-1) - 1 \cdot 1}$

$= \dfrac{-1 - 3}{-1 - 1}$

$= \dfrac{-4}{-2}$

$= 2$

$$y = \frac{\begin{vmatrix} 1 & 1 \\ 1 & 3 \end{vmatrix}}{\begin{vmatrix} 1 & 1 \\ 1 & -1 \end{vmatrix}} = \frac{1 \cdot 3 - 1 \cdot 1}{-2}$$

$$= \frac{3 - 1}{-2}$$

$$= \frac{2}{-2}$$

$$= -1$$

$\{(2, -1)\}$

15. $\begin{cases} 2x - y = -5 \\ 3x + 2y = 3 \end{cases}$

$$x = \frac{\begin{vmatrix} -5 & -1 \\ 3 & 2 \end{vmatrix}}{\begin{vmatrix} 2 & -1 \\ 3 & 2 \end{vmatrix}} = \frac{-5 \cdot 2 - (-1) \cdot 3}{2 \cdot 2 - (-1) \cdot 3}$$

$$= \frac{-10 + 3}{4 + 3}$$

$$= \frac{-7}{7}$$

$$= -1$$

$$y = \frac{\begin{vmatrix} 2 & -5 \\ 3 & 3 \end{vmatrix}}{\begin{vmatrix} 2 & -1 \\ 3 & 2 \end{vmatrix}} = \frac{2 \cdot 3 - 5 \cdot 3}{7}$$

$$= \frac{6 + 15}{7}$$

$$= \frac{21}{7}$$

$$= 3$$

$\{(-1, 3)\}$

17. $\begin{cases} 3x + 2y = 5 \\ 6x - y = 0 \end{cases}$

$$x = \frac{\begin{vmatrix} 5 & 2 \\ 0 & -1 \end{vmatrix}}{\begin{vmatrix} 3 & 2 \\ 6 & -1 \end{vmatrix}} = \frac{5 \cdot (-1) - 2 \cdot 0}{3 \cdot (-1) - 2 \cdot 6}$$

$$= \frac{-5 - 0}{-3 - 12}$$

$$= \frac{-5}{-15}$$

$$= \frac{1}{3}$$

$$y = \frac{\begin{vmatrix} 3 & 5 \\ 6 & 0 \end{vmatrix}}{\begin{vmatrix} 3 & 2 \\ 6 & -1 \end{vmatrix}} = \frac{3 \cdot 0 - 5 \cdot 6}{-15}$$

$$= \frac{0 - 30}{-15}$$

$$= \frac{-30}{-15}$$

$$= 2$$

$\left\{\left(\frac{1}{3}, 2\right)\right\}$

19. $\begin{cases} x + y + z = 6 \\ x + 2y - z = 2 \\ 2x - 2y + z = 1 \end{cases}$

$$x = \frac{\begin{vmatrix} 6 & 1 & 1 \\ 2 & 2 & -1 \\ 1 & -2 & 1 \end{vmatrix}}{\begin{vmatrix} 1 & 1 & 1 \\ 1 & 2 & -1 \\ 2 & -2 & 1 \end{vmatrix}}$$

$$= \frac{6\begin{vmatrix} 2 & -1 \\ -2 & 1 \end{vmatrix} - 1\begin{vmatrix} 2 & -1 \\ 1 & 1 \end{vmatrix} + 1\begin{vmatrix} 2 & 2 \\ 1 & -2 \end{vmatrix}}{1\begin{vmatrix} 2 & -1 \\ -2 & 1 \end{vmatrix} - 1\begin{vmatrix} 1 & -1 \\ 2 & 1 \end{vmatrix} + 1\begin{vmatrix} 1 & 2 \\ 2 & -2 \end{vmatrix}}$$

$$= \frac{6(2 - 2) - [2 - (-1)] + (-4 - 2)}{(2 - 2) - [1 - (-2)] + (-2 - 4)}$$

$$= \frac{6(0) - 3 - 6}{0 - 3 - 6}$$

$$= \frac{-9}{-9}$$

$$= 1$$

$$y = \frac{\begin{vmatrix} 1 & 6 & 1 \\ 1 & 2 & -1 \\ 2 & 1 & 1 \end{vmatrix}}{-9}$$

$$= \frac{1\begin{vmatrix} 2 & -1 \\ 1 & 1 \end{vmatrix} - 6\begin{vmatrix} 1 & -1 \\ 2 & 1 \end{vmatrix} + 1\begin{vmatrix} 1 & 2 \\ 2 & 1 \end{vmatrix}}{-9}$$

$$= \frac{[2 - (-1)] - 6(1 + 2) + (1 - 4)}{-9}$$

$$= \frac{3 - 6(3) - 3}{-9}$$

$$= \frac{3 - 18 - 3}{-9}$$

$$= \frac{-18}{-9}$$

$$= 2$$

$$z = \frac{\begin{vmatrix} 1 & 1 & 6 \\ 1 & 2 & 2 \\ 2 & -2 & 1 \end{vmatrix}}{-9}$$

$$= \frac{1\begin{vmatrix} 2 & 2 \\ -2 & 1 \end{vmatrix} - 1\begin{vmatrix} 1 & 2 \\ 2 & 1 \end{vmatrix} + 6\begin{vmatrix} 1 & 2 \\ 2 & -2 \end{vmatrix}}{-9}$$

$$= \frac{2 - (-4) - (1 - 4) + 6(-2 - 4)}{-9}$$

$$= \frac{2 + 4 + 3 + 6(-6)}{-9}$$

$$= \frac{9 - 36}{-9}$$

$$= \frac{-27}{-9}$$

$$= 3$$

$$\{(1, 2, 3)\}$$

21. $\begin{cases} 2x - y + 2z = 2 \\ -2x + 3y + 4z = 10 \\ -2x + y - z = 0 \end{cases}$

$$x = \frac{\begin{vmatrix} 2 & -1 & 2 \\ 10 & 3 & 4 \\ 0 & 1 & -1 \end{vmatrix}}{\begin{vmatrix} 2 & -1 & 2 \\ -2 & 3 & 4 \\ -2 & 1 & -1 \end{vmatrix}}$$

$$= \frac{-1\begin{vmatrix} 2 & 2 \\ 10 & 4 \end{vmatrix} + (-1)\begin{vmatrix} 2 & -1 \\ 10 & 3 \end{vmatrix}}{2\begin{vmatrix} 3 & 4 \\ 1 & -1 \end{vmatrix} - (-1)\begin{vmatrix} -2 & 4 \\ -2 & -1 \end{vmatrix} + 2\begin{vmatrix} -2 & 3 \\ -2 & 1 \end{vmatrix}}$$

$$= \frac{-(8 - 20) - [6 - (-10)]}{2(-3 - 4) + 1[2 - (-8)] + 2[-2 - (-6)]}$$

$$= \frac{12 - 16}{-14 + 10 + 8}$$

$$= \frac{-4}{4}$$

$$= -1$$

$$y = \frac{\begin{vmatrix} 2 & 2 & 2 \\ -2 & 10 & 4 \\ -2 & 0 & -1 \end{vmatrix}}{4}$$

$$= \frac{-2 \begin{vmatrix} 2 & 2 \\ 10 & 4 \end{vmatrix} - 1 \begin{vmatrix} 2 & 2 \\ -2 & 10 \end{vmatrix}}{4}$$

$$= \frac{-2 (8 - 20) - [20 - (-4)]}{4}$$

$$= \frac{24 - 24}{4}$$

$$= \frac{0}{4}$$

$$= 0$$

$$z = \frac{\begin{vmatrix} 2 & -1 & 2 \\ -2 & 3 & 10 \\ -2 & 1 & 0 \end{vmatrix}}{4}$$

$$= \frac{-2 \begin{vmatrix} -1 & 2 \\ 3 & 10 \end{vmatrix} - 1 \begin{vmatrix} 2 & 2 \\ -2 & 10 \end{vmatrix}}{4}$$

$$= \frac{-2 (-10 - 6) - [20 - (-4)]}{4}$$

$$= \frac{32 - 24}{4}$$

$$= \frac{8}{4}$$

$$= 2$$

$$\{(-1, 0, 2)\}$$

23. $\begin{vmatrix} 2 & 1 \\ 5 & 3 \end{vmatrix} = 2 \cdot 3 - 1 \cdot 5 = 6 - 5 = 1$

25. $\begin{vmatrix} 2 & 6 \\ 3 & 9 \end{vmatrix} = 2 \cdot 9 - 6 \cdot 3 = 18 - 18 = 0$

27. $\begin{vmatrix} 4 & 5 \\ -3 & 3 \end{vmatrix} = 4 \cdot 3 - 5 \cdot (-3) = 12 + 15 = 27$

29. Expand about 1st row:

$$\begin{vmatrix} 2 & 1 & 1 \\ 1 & 3 & -1 \\ 1 & -2 & 4 \end{vmatrix}$$

$$= 2 \begin{vmatrix} 3 & -1 \\ -2 & 4 \end{vmatrix} - 1 \begin{vmatrix} 1 & -1 \\ 1 & 4 \end{vmatrix} + 1 \begin{vmatrix} 1 & 3 \\ 1 & -2 \end{vmatrix}$$

$$= 2 (12 - 2) - [4 - (-1)] + (-2 - 3)$$

$$= 2 (10) - 5 - 5$$

$$= 10$$

31. Expand about 3rd column:

$$\begin{vmatrix} 1 & 2 & 0 \\ -1 & -1 & 0 \\ 3 & 4 & 2 \end{vmatrix} = 2 \begin{vmatrix} 1 & 2 \\ -1 & -1 \end{vmatrix}$$

$$= 2 [-1 - (-2)]$$

$$= 2 (1)$$

$$= 2$$

33. $\begin{vmatrix} 2 & 0 \\ 0 & -3 \end{vmatrix} = 2 \cdot (-3) - 0 \cdot 0 = -6$

35. Expand about 3rd row:

$$\begin{vmatrix} 3 & 3 & 4 \\ 2 & 2 & 1 \\ 1 & 1 & 3 \end{vmatrix} = 1 \begin{vmatrix} 3 & 4 \\ 2 & 1 \end{vmatrix} - 1 \begin{vmatrix} 3 & 4 \\ 2 & 1 \end{vmatrix} + 3 \begin{vmatrix} 3 & 3 \\ 2 & 2 \end{vmatrix}$$

$$= 3 - 8 - (3 - 8) + 3(6 - 6)$$

$$= -5 + 5 + 3(0)$$

$$= 0$$

37. Expand about 1st row:

$$\begin{vmatrix} -2 & 3 & 5 \\ 4 & 2 & 3 \\ 7 & -3 & 2 \end{vmatrix}$$

$$= -2 \begin{vmatrix} 2 & 3 \\ -3 & 2 \end{vmatrix} - 3 \begin{vmatrix} 4 & 3 \\ 7 & 2 \end{vmatrix} + 5 \begin{vmatrix} 4 & 2 \\ 7 & -3 \end{vmatrix}$$

$$= -2 [4 - (-9)] - 3 (8 - 21) + 5 (-12 - 14)$$

$$= -2 (13) - 3(-13) + 5 (-26)$$

$$= -26 + 39 - 130$$

$$= -117$$

39.

$$\begin{vmatrix} 4 & 0 \\ 2 & -6 \end{vmatrix} = 4 \cdot (-6) - 0 \cdot 2 = -24 - 0 = -24$$

41. $\begin{cases} 3x + 4y = 0 \\ 4x + 3y = 0 \end{cases}$

$$x = \frac{\begin{vmatrix} 0 & 4 \\ 0 & 3 \end{vmatrix}}{\begin{vmatrix} 3 & 4 \\ 4 & 3 \end{vmatrix}} = \frac{0 - 0}{9 - 16} = \frac{0}{-7} = 0$$

$$y = \frac{\begin{vmatrix} 3 & 0 \\ 4 & 0 \end{vmatrix}}{-7} = \frac{0 - 0}{-7} = \frac{0}{-7} = 0$$

$$\{(0, 0)\}$$

43. $\begin{cases} 6x - 5y = 6 \\ 9x + 7y = -20 \end{cases}$

$$x = \frac{\begin{vmatrix} 6 & -5 \\ -20 & 7 \end{vmatrix}}{\begin{vmatrix} 6 & -5 \\ 9 & 7 \end{vmatrix}} = \frac{42 - 100}{42 - (-45)} = \frac{-58}{87} = \frac{-2}{3}$$

$$y = \frac{\begin{vmatrix} 6 & 6 \\ 9 & -20 \end{vmatrix}}{87} = \frac{-120 - 54}{87} = \frac{-174}{87} = -2$$

$$\left\{ \left(-\frac{2}{3}, -2 \right) \right\}$$

45. $\begin{cases} -4x + 15y = -2 \\ 12x + 5y = -4 \end{cases}$

$$x = \frac{\begin{vmatrix} -2 & 15 \\ -4 & 5 \end{vmatrix}}{\begin{vmatrix} -4 & 15 \\ 12 & 5 \end{vmatrix}} = \frac{-10 - (-60)}{-20 - 180} = \frac{50}{-200} = -\frac{1}{4}$$

$$y = \frac{\begin{vmatrix} -4 & -2 \\ 12 & -4 \end{vmatrix}}{-200} = \frac{16 - (-24)}{-200} = \frac{40}{-200} = -\frac{1}{5}$$

$$\left\{ \left(-\frac{1}{4}, -\frac{1}{5} \right) \right\}$$

47. $\begin{cases} 3x - 4y = 5 \\ 4x - 7y = 9 \end{cases}$

$$x = \frac{\begin{vmatrix} 5 & -4 \\ 9 & -7 \end{vmatrix}}{\begin{vmatrix} 3 & -4 \\ 4 & -7 \end{vmatrix}} = \frac{-35 - (-36)}{-21 - (-16)} = \frac{1}{-5}$$

$$y = \frac{\begin{vmatrix} 3 & 5 \\ 4 & 9 \end{vmatrix}}{-5} = \frac{27 - 20}{-5} = \frac{7}{-5}$$

$$\left\{ \left(-\frac{1}{5}, \ -\frac{7}{5} \right) \right\}$$

49. $\begin{cases} 17x - 11y = -1 \\ -34x + 22y = 2 \end{cases}$

Since $\begin{vmatrix} 17 & -11 \\ -34 & 22 \end{vmatrix} = 374 - 374 = 0$

Cramer's Rule does not apply.

51. $\begin{cases} x - \dfrac{2}{3}y = \dfrac{1}{15} \\ \dfrac{3}{4}x - \dfrac{5}{6}y = \dfrac{1}{3} \end{cases}$

$$x = \frac{\begin{vmatrix} \dfrac{1}{15} & -\dfrac{2}{3} \\ \dfrac{1}{3} & -\dfrac{5}{6} \end{vmatrix}}{\begin{vmatrix} \dfrac{1}{5} & -\dfrac{2}{3} \\ \dfrac{3}{4} & -\dfrac{5}{6} \end{vmatrix}} = \frac{\dfrac{1}{15}\left(-\dfrac{5}{6}\right) - \left(-\dfrac{2}{3}\right)\left(\dfrac{1}{3}\right)}{\left(\dfrac{1}{5}\right)\left(-\dfrac{5}{6}\right) - \left(-\dfrac{2}{3}\right)\left(\dfrac{3}{4}\right)}$$

$$= \frac{-\dfrac{1}{18} + \dfrac{2}{9}}{-\dfrac{1}{6} + \dfrac{1}{2}}$$

$$= \frac{\dfrac{1}{6}}{\dfrac{1}{3}}$$

$$= \frac{1}{2}$$

$$y = \frac{\begin{vmatrix} \dfrac{1}{5} & \dfrac{1}{15} \\ \dfrac{3}{4} & \dfrac{1}{3} \end{vmatrix}}{\dfrac{1}{3}}$$

$$= \frac{\dfrac{1}{5} \cdot \dfrac{1}{3} - \dfrac{1}{15} \cdot \dfrac{3}{4}}{\dfrac{1}{3}}$$

$$= \frac{\dfrac{1}{60}}{\dfrac{1}{3}}$$

$$= \frac{1}{20}$$

$$\left\{ \left(\dfrac{1}{2}, \ \dfrac{1}{20} \right) \right\}$$

53. $\begin{cases} x + y + z = 6 \\ x + 2y + z = 9 \\ 2x + 2y - z = 6 \end{cases}$

$$x = \frac{\begin{vmatrix} 6 & 1 & 1 \\ 9 & 2 & 1 \\ 6 & 2 & -1 \end{vmatrix}}{\begin{vmatrix} 1 & 1 & 1 \\ 1 & 2 & 1 \\ 2 & 2 & -1 \end{vmatrix}}$$

$$= \frac{6\begin{vmatrix} 2 & 1 \\ 2 & -1 \end{vmatrix} - 1\begin{vmatrix} 9 & 1 \\ 6 & -1 \end{vmatrix} + 1\begin{vmatrix} 9 & 2 \\ 6 & 2 \end{vmatrix}}{1\begin{vmatrix} 2 & 1 \\ 2 & -1 \end{vmatrix} - 1\begin{vmatrix} 1 & 1 \\ 2 & -1 \end{vmatrix} + 1\begin{vmatrix} 1 & 2 \\ 2 & 2 \end{vmatrix}}$$

$$= \frac{6(-2 - 2) - (-9 - 6) + (18 - 12)}{(-2 - 2) - (-1 - 2) + (2 - 4)}$$

$$= \frac{-24 + 15 + 6}{-4 - (-3) - 2}$$

$$= \frac{-3}{-3}$$

$$= 1$$

$$y = \frac{\begin{vmatrix} 1 & 6 & 1 \\ 1 & 9 & 1 \\ 2 & 6 & -1 \end{vmatrix}}{-3}$$

$$= \frac{1\begin{vmatrix} 9 & 1 \\ 6 & -1 \end{vmatrix} - 6\begin{vmatrix} 1 & 1 \\ 2 & -1 \end{vmatrix} + 1\begin{vmatrix} 1 & 9 \\ 2 & 6 \end{vmatrix}}{-3}$$

$$= \frac{(-9 - 6) - 6(-1 - 2) + (6 - 18)}{-3}$$

$$= \frac{-15 + 18 - 12}{-3}$$

$$= \frac{-9}{-3}$$

$$= 3$$

$$z = \frac{\begin{vmatrix} 1 & 1 & 6 \\ 1 & 2 & 9 \\ 2 & 2 & 6 \end{vmatrix}}{-3}$$

$$= \frac{1\begin{vmatrix} 2 & 9 \\ 2 & 6 \end{vmatrix} - 1\begin{vmatrix} 1 & 9 \\ 2 & 6 \end{vmatrix} + 6\begin{vmatrix} 1 & 2 \\ 2 & 2 \end{vmatrix}}{-3}$$

$$= \frac{(12 - 18) - (6 - 18) + 6(2 - 4)}{-3}$$

$$= \frac{-6 + 12 - 12}{-3}$$

$$= \frac{-6}{-3}$$

$$= 2$$

$$\{(1, 3, 2)\}$$

55. $\begin{cases} 3x_1 + x_2 + 2x_3 = 10 \\ -3x_1 - 2x_2 + 4x_3 = -11 \\ + 2x_2 + x_3 = 2 \end{cases}$

$$x = \frac{\begin{vmatrix} 10 & 1 & 2 \\ -11 & -2 & 4 \\ 2 & 2 & 1 \end{vmatrix}}{\begin{vmatrix} 3 & 1 & 2 \\ -3 & -2 & 4 \\ 0 & 2 & 1 \end{vmatrix}}$$

$$= \frac{10\begin{vmatrix} -2 & 4 \\ 2 & 1 \end{vmatrix} - 1\begin{vmatrix} -11 & 4 \\ 2 & 1 \end{vmatrix} + 2\begin{vmatrix} -11 & -2 \\ 2 & 2 \end{vmatrix}}{-2\begin{vmatrix} 3 & 2 \\ -3 & 4 \end{vmatrix} + 1\begin{vmatrix} 3 & 1 \\ -3 & -2 \end{vmatrix}}$$

$$= \frac{10(-2 - 8) - (-11 - 8) + 2[-22 - (-4)]}{-2[12 - (-6)] + [-6 - (-3)]}$$

$$= \frac{-100 + 19 - 36}{-36 - 3}$$

$$= \frac{-117}{-39}$$

$$= 3$$

$$y = \frac{\begin{vmatrix} 3 & 10 & 2 \\ -3 & -11 & 4 \\ 0 & 2 & 1 \end{vmatrix}}{-39}$$

$$= \frac{-2\begin{vmatrix} 3 & 2 \\ -3 & 4 \end{vmatrix} + 1\begin{vmatrix} 3 & 10 \\ -3 & -11 \end{vmatrix}}{-39}$$

$$= \frac{-2[12 - (-6)] + [-33 - (-30)]}{-39}$$

$$= \frac{-36 - 3}{-39}$$

$$= \frac{-39}{-39}$$

$= 1$

$$z = \frac{\begin{vmatrix} 3 & 1 & 10 \\ -3 & -2 & -11 \\ 0 & 2 & 2 \end{vmatrix}}{-39}$$

$$= \frac{-2\begin{vmatrix} 3 & 10 \\ -3 & -11 \end{vmatrix} + 2\begin{vmatrix} 3 & 1 \\ -3 & -2 \end{vmatrix}}{-39}$$

$$= \frac{-2[-33 - (-30)] + 2[-6 - (-3)]}{-39}$$

$$= \frac{6 - 6}{-39}$$

$$= \frac{0}{-39}$$

$$= 0$$

$\{(3, 1, 0)\}$

57. $\begin{cases} 3x + 3y - z = 3 \\ 5x + y + 3z = 1 \\ 2x + 4y - 3z = 4 \end{cases}$

Since $\begin{vmatrix} 3 & 3 & -1 \\ 5 & 1 & 3 \\ 2 & 4 & -3 \end{vmatrix}$

$$= 3\begin{vmatrix} 1 & 3 \\ 4 & -3 \end{vmatrix} - 3\begin{vmatrix} 5 & 3 \\ 2 & -3 \end{vmatrix} - 1\begin{vmatrix} 5 & 1 \\ 2 & 4 \end{vmatrix}$$

$$= 3(-3 - 12) - 3(-15 - 6) - (20 - 2)$$

$$= -45 + 63 - 18$$

$$= 0$$

Cramer's Rule does not apply.

59. $\begin{cases} 2x + 2y - z = 1 \\ x + 2y - 3z = 4 \\ 5x + 6y - 5z = 3 \end{cases}$

Since $\begin{vmatrix} 2 & 2 & -1 \\ 1 & 2 & -3 \\ 5 & 6 & -5 \end{vmatrix}$

$$= 2\begin{vmatrix} 2 & -3 \\ 6 & -5 \end{vmatrix} - 2\begin{vmatrix} 1 & -3 \\ 5 & -5 \end{vmatrix} - 1\begin{vmatrix} 1 & 2 \\ 5 & 6 \end{vmatrix}$$

$$= 2[-10 - (-18)] - 2[-5 - (-15)] - (6 - 10)$$

$$= 16 - 20 + 4$$

$$= 0$$

Cramer's Rule does not apply.

61. **Expand about 2nd column:**

$$\begin{vmatrix} 1 & 1 & 2 & 1 \\ 2 & 1 & 0 & -3 \\ -2 & 0 & 1 & 2 \\ 4 & 0 & 5 & 1 \end{vmatrix}$$

$$= -1 \begin{vmatrix} 2 & 0 & -3 \\ -2 & 1 & 2 \\ 4 & 5 & 1 \end{vmatrix} + 1 \begin{vmatrix} 1 & 2 & 1 \\ -2 & 1 & 2 \\ 4 & 5 & 1 \end{vmatrix}$$

$$= - \left[2 \begin{vmatrix} 1 & 2 \\ 5 & 1 \end{vmatrix} - 3 \begin{vmatrix} -2 & 1 \\ 4 & 5 \end{vmatrix} \right] + \left[1 \begin{vmatrix} 1 & 2 \\ 5 & 1 \end{vmatrix} - 2 \begin{vmatrix} -2 & 2 \\ 4 & 1 \end{vmatrix} + 1 \begin{vmatrix} -2 & 1 \\ 4 & 5 \end{vmatrix} \right]$$

$$= - [2(1 - 10) - 3(-10 - 4)] + [(1 - 10) - 2(-2 - 8) + (-10 - 4)]$$

$$= - (-18 + 42) + (-9 + 20 - 14)$$

$$= -27$$

63. **Expand about 1st row:**

$$\begin{vmatrix} 1 & 1 & 0 & 0 & 0 \\ 0 & 2 & 2 & 0 & 0 \\ 0 & 0 & 3 & 3 & 0 \\ 0 & 0 & 0 & 4 & 4 \\ 5 & 0 & 0 & 0 & 5 \end{vmatrix}$$

$$= 1 \begin{vmatrix} 2 & 2 & 0 & 0 \\ 0 & 3 & 3 & 0 \\ 0 & 0 & 4 & 4 \\ 0 & 0 & 0 & 5 \end{vmatrix} - 1 \begin{vmatrix} 0 & 2 & 0 & 0 \\ 0 & 3 & 3 & 0 \\ 0 & 0 & 4 & 4 \\ 5 & 0 & 0 & 5 \end{vmatrix}$$

$$= 2 \begin{vmatrix} 3 & 3 & 0 \\ 0 & 4 & 4 \\ 0 & 0 & 5 \end{vmatrix} - (-2) \begin{vmatrix} 0 & 3 & 0 \\ 0 & 4 & 4 \\ 5 & 0 & 5 \end{vmatrix}$$

$$= 2(3) \begin{vmatrix} 4 & 4 \\ 0 & 5 \end{vmatrix} + 2(5) \begin{vmatrix} 3 & 0 \\ 4 & 4 \end{vmatrix}$$

$$= 6(20 - 0) + 10(12 - 0)$$

$$= 120 + 120$$

$$= 240$$

65. $\begin{cases} x - 2y - z + 2w = -2 \\ 2x - y + z - w = 0 \\ -x + 3y \quad - w = 4 \\ 5x - 3y \qquad = -4 \end{cases}$

$$x = \dfrac{\begin{vmatrix} -2 & -2 & -1 & 2 \\ 0 & -1 & 1 & -1 \\ 4 & 3 & 0 & -1 \\ -4 & -3 & 0 & 0 \end{vmatrix}}{\begin{vmatrix} 1 & -2 & -1 & 2 \\ 2 & -1 & 1 & -1 \\ -1 & 3 & 0 & -1 \\ 5 & -3 & 0 & 0 \end{vmatrix}} = \dfrac{-1\begin{vmatrix} 0 & -1 & -1 \\ 4 & 3 & -1 \\ -4 & -3 & 0 \end{vmatrix} - 1\begin{vmatrix} -2 & -2 & 2 \\ 4 & 3 & -1 \\ -4 & -3 & 0 \end{vmatrix}}{-1\begin{vmatrix} 2 & -1 & -1 \\ -1 & 3 & -1 \\ 5 & -3 & 0 \end{vmatrix} - 1\begin{vmatrix} 1 & -2 & 2 \\ -1 & 3 & -1 \\ 5 & -3 & 0 \end{vmatrix}}$$

$$= \dfrac{-\left[-(-1)\begin{vmatrix} 4 & -1 \\ -4 & 0 \end{vmatrix} - 1\begin{vmatrix} 4 & 3 \\ -4 & -3 \end{vmatrix}\right] - \left[2\begin{vmatrix} 4 & 3 \\ -4 & -3 \end{vmatrix} - (-1)\begin{vmatrix} -2 & -2 \\ -4 & -3 \end{vmatrix}\right]}{-\left[-1\begin{vmatrix} -1 & 3 \\ 5 & -3 \end{vmatrix} - (-1)\begin{vmatrix} 2 & -1 \\ 5 & -3 \end{vmatrix}\right] - \left[2\begin{vmatrix} -1 & 3 \\ 5 & -3 \end{vmatrix} - (-1)\begin{vmatrix} 1 & -2 \\ 5 & -3 \end{vmatrix}\right]}$$

$$= \dfrac{-\left[(0 - 4) - (-12 + 12)\right] - \left[2(-12 + 12) + (6 - 8)\right]}{-\left[-(3 - 15) + (-6 + 5)\right] - \left[2(3 - 15) + (-3 + 10)\right]}$$

$$= \dfrac{-(-4) - [2(0) - 2]}{-(12 - 1) - [2(-12) + 7]}$$

$$= \dfrac{6}{6}$$

$$= 1$$

$$y = \dfrac{\begin{vmatrix} 1 & -2 & -1 & 2 \\ 2 & 0 & 1 & -1 \\ -1 & 4 & 0 & -1 \\ 5 & -4 & 0 & 0 \end{vmatrix}}{6}$$

$$= \dfrac{-1\begin{vmatrix} 2 & 0 & -1 \\ -1 & 4 & -1 \\ 5 & -4 & 0 \end{vmatrix} - 1\begin{vmatrix} 1 & -2 & 2 \\ -1 & 4 & -1 \\ 5 & -4 & 0 \end{vmatrix}}{6}$$

$$= \frac{-\left[2\begin{vmatrix} 4 & -1 \\ -4 & 0 \end{vmatrix} - 1\begin{vmatrix} -1 & 4 \\ 5 & -4 \end{vmatrix}\right] - \left[2\begin{vmatrix} -1 & 4 \\ 5 & -4 \end{vmatrix} - (-1)\begin{vmatrix} 1 & -2 \\ 5 & -4 \end{vmatrix}\right]}{6}$$

$$= \frac{-[2(0 - 4) - 1(4 - 20)] - [2(4 - 20) + (-4 + 10)]}{6}$$

$$= \frac{-(-8 + 16) - (-32 + 6)}{6}$$

$$= \frac{-8 + 26}{6}$$

$$= \frac{18}{6}$$

$$= 3$$

$$z = \frac{\begin{vmatrix} 1 & -2 & -2 & 2 \\ 2 & -1 & 0 & -1 \\ -1 & 3 & 4 & -1 \\ 5 & -3 & -4 & 0 \end{vmatrix}}{6}$$

$$= \frac{-2\begin{vmatrix} 2 & -1 & 0 \\ -1 & 3 & 4 \\ 5 & -3 & -4 \end{vmatrix} - 1\begin{vmatrix} 1 & -2 & -2 \\ -1 & 3 & 4 \\ 5 & -3 & -4 \end{vmatrix} - (-1)\begin{vmatrix} 1 & -2 & -2 \\ 2 & -1 & 0 \\ 5 & -3 & -4 \end{vmatrix}}{6}$$

$$= \frac{-2\left[2\begin{vmatrix} 3 & 4 \\ -3 & -4 \end{vmatrix} - (-1)\begin{vmatrix} -1 & 4 \\ 5 & -4 \end{vmatrix}\right] - \left[1\begin{vmatrix} 3 & 4 \\ -3 & -4 \end{vmatrix} - (-2)\begin{vmatrix} -1 & 4 \\ 5 & -4 \end{vmatrix} - 2\begin{vmatrix} -1 & 3 \\ 5 & -3 \end{vmatrix}\right] + \left[-2\begin{vmatrix} -2 & -2 \\ -3 & -4 \end{vmatrix} - 1\begin{vmatrix} 1 & -2 \\ 5 & -4 \end{vmatrix}\right]}{6}$$

$$= \frac{-2[2(-12 + 12) + (4 - 20)] - [(-12 + 12) + 2(4 - 20) - 2(3 - 15)] + [-2(8 - 6) - (-4 + 10)]}{6}$$

$$= \frac{-2(-16) - (-32 + 24) + (-4 - 6)}{6}$$

$$= \frac{32 + 8 - 10}{6}$$

$$= \frac{30}{6}$$

$$= 5$$

$$w = \frac{\begin{vmatrix} 1 & -2 & -1 & -2 \\ 2 & -1 & 1 & 0 \\ -1 & 3 & 0 & 4 \\ 5 & -3 & 0 & -4 \end{vmatrix}}{6}$$

$$= \frac{-1 \begin{vmatrix} 2 & -1 & 0 \\ -1 & 3 & 4 \\ 5 & -3 & -4 \end{vmatrix} - 1 \begin{vmatrix} 1 & -2 & -2 \\ -1 & 3 & 4 \\ 5 & -3 & -4 \end{vmatrix}}{6}$$

$$= \frac{-\left[2 \begin{vmatrix} 3 & 4 \\ -3 & -4 \end{vmatrix} - (-1) \begin{vmatrix} -1 & 4 \\ 5 & -4 \end{vmatrix} \right] - \left[1 \begin{vmatrix} 3 & 4 \\ -3 & -4 \end{vmatrix} - (-2) \begin{vmatrix} -1 & 4 \\ 5 & -4 \end{vmatrix} - 2 \begin{vmatrix} -1 & 3 \\ 5 & -3 \end{vmatrix} \right]}{6}$$

$$= \frac{-[2(-12 + 12) + (4 - 20)] - [(-12 + 12) + 2(4 - 20) - 2(3 - 15)]}{6}$$

$$= \frac{-(-16) - (-32 + 24)}{6}$$

$$= \frac{16 - (-8)}{6}$$

$$= \frac{24}{6}$$

$$= 4$$

$$\{(1, 3, 5, 4)\}$$

Problem Set 7.7

1. $\begin{cases} 2x + y = -4 \\ 5x - 6y = 1 \end{cases}$

$\begin{bmatrix} 2 & 3 & | & -4 \\ 5 & -6 & | & 1 \end{bmatrix}$

3. $\begin{cases} 2x - 3y + 4z = 1 \\ 5x - y + z = 0 \\ 7x + 6y - 2z = 5 \end{cases}$

$\begin{bmatrix} 2 & -3 & 4 & | & 1 \\ 5 & -1 & 1 & | & 0 \\ 7 & 6 & -2 & | & 5 \end{bmatrix}$

5. $\begin{cases} x + 2y = z \\ -x - 11 = y \\ y - 13 = 4z \end{cases}$

$\begin{cases} x + 2y - z = 0 \\ -x - y = 11 \\ y - 4z = 13 \end{cases}$

$\begin{bmatrix} 1 & 2 & -1 & | & 0 \\ -1 & -1 & 0 & | & 11 \\ 0 & 1 & -4 & | & 13 \end{bmatrix}$

7. $\begin{bmatrix} 1 & 2 & | & 4 \\ 9 & -8 & | & -1 \end{bmatrix}$

$x + 2y = 4$

$9x - 8y = -1$

9. $\begin{bmatrix} 1 & 0 & 0 & | & 5 \\ 0 & 1 & 0 & | & -3 \\ 0 & 0 & 1 & | & -2 \end{bmatrix}$

$\begin{cases} x = 5 \\ y = -3 \\ z = -2 \end{cases}$

11. $\begin{cases} x + 3y = 7 \\ -2x + y = 0 \end{cases}$

$\begin{bmatrix} 1 & 3 & | & 7 \\ -2 & 1 & | & 0 \end{bmatrix}$ $2R_1 + R_2 \rightarrow R_2$

$\begin{bmatrix} 1 & 3 & | & 7 \\ 0 & 7 & | & 14 \end{bmatrix}$ $\frac{1}{7}R_2 \rightarrow R_2$

$\begin{bmatrix} 1 & 3 & | & 7 \\ 0 & 1 & | & 2 \end{bmatrix}$

$y = 2$

$x + 3y = 7$

$x + 3(2) = 7$

$x = 1$

$\{(1, 2)\}$

13. $\begin{cases} 2x + 4y = 2 \\ 3x + 7y = 1 \end{cases}$

$\begin{bmatrix} 2 & 4 & | & 2 \\ 3 & 7 & | & 1 \end{bmatrix}$ $\frac{1}{2}R_1 \rightarrow R_1$

$\begin{bmatrix} 1 & 2 & | & 1 \\ 3 & 7 & | & 1 \end{bmatrix}$ $-3R_1 + R_2 \rightarrow R_2$

$\begin{bmatrix} 1 & 2 & | & 1 \\ 0 & 1 & | & -2 \end{bmatrix}$

$y = -2$

$x + 2y = 1$

343

$$x + 2(-2) = 1$$

$$x = 5$$

$$\{(5, -2)\}$$

15. $\begin{cases} x + 2y + 2z = 3 \\ 2x + 3y + 6z = 2 \\ -x + y + z = 0 \end{cases}$

$\begin{bmatrix} 1 & 2 & 2 & | & 3 \\ ② & 3 & 6 & | & 2 \\ -1 & 1 & 1 & | & 0 \end{bmatrix}$ $-2R_1 + R_2 \rightarrow R_2$

$\begin{bmatrix} 1 & 2 & 2 & | & 3 \\ 0 & -1 & 2 & | & -4 \\ -1 & 1 & 1 & | & 0 \end{bmatrix}$ $R_1 + R_3 \rightarrow R_3$

$\begin{bmatrix} 1 & 2 & 2 & | & 3 \\ 0 & -1 & 2 & | & -4 \\ 0 & 3 & 3 & | & 3 \end{bmatrix}$ $-1R_2 \rightarrow R_2$

$\begin{bmatrix} 1 & 2 & 2 & | & 3 \\ 0 & 1 & -2 & | & 4 \\ 0 & 3 & 3 & | & 3 \end{bmatrix}$ $-3R_2 + R_3 \rightarrow R_3$

$\begin{bmatrix} 1 & 2 & 2 & | & 3 \\ 0 & 1 & -2 & | & 4 \\ 0 & 0 & 9 & | & -9 \end{bmatrix}$ $\frac{1}{9}R_3 \rightarrow R_3$

$\begin{bmatrix} 1 & 2 & 2 & | & 3 \\ 0 & 1 & -2 & | & 4 \\ 0 & 0 & 1 & | & -1 \end{bmatrix}$

$$z = -1$$

$$y - 2z = 4$$

$$y - 2(-1) = 4$$

$$y = 2$$

$$x + 2y + 2z = 3$$

$$x + 2(2) + 2(-1) = 3$$

$$x = 1$$

$$\{(1, 2, -1)\}$$

17. $\begin{cases} 3x + 2y = 5 \\ 6x - 5y = 3 \end{cases}$

$\begin{bmatrix} 3 & 2 & | & 5 \\ 6 & -5 & | & 3 \end{bmatrix}$

19. $\begin{cases} 3x - 2y + 5y = 4 \\ 2x + 3y - z = 0 \\ x - y - 2z = 1 \end{cases}$

$\begin{bmatrix} 3 & -2 & 5 & | & 4 \\ 2 & 3 & -1 & | & 0 \\ 1 & -1 & -2 & | & 1 \end{bmatrix}$

21. $\begin{cases} x + 3y = 2y \\ -x - 10 = z \\ y + 3 = 4x - z \end{cases}$

$\begin{cases} x - 2y + 3z = 0 \\ -x \quad\quad - z = 10 \\ -4x + y + z = -3 \end{cases}$

$\begin{bmatrix} 1 & -2 & 3 & | & 0 \\ -1 & 0 & -1 & | & 10 \\ -4 & 1 & 1 & | & -3 \end{bmatrix}$

23. $\begin{bmatrix} 2 & 1 & | & 3 \\ 6 & -5 & | & 11 \end{bmatrix}$

$\begin{cases} 2x + y = 3 \\ 6x - 5y = 11 \end{cases}$

25. $\begin{bmatrix} 0 & 0 & 1 & | & -3 \\ 0 & 1 & 0 & | & 4 \\ 1 & 0 & 0 & | & 2 \end{bmatrix}$

$\begin{cases} z = -3 \\ y = 4 \\ x = 2 \end{cases}$

27. $\begin{cases} x - 2y = -1 \\ -3x + 8y = 5 \end{cases}$

$\begin{bmatrix} 1 & -2 & | & -1 \\ -3 & 8 & | & 5 \end{bmatrix}$ $3R_1 + R_2 \to R_2$

$\begin{bmatrix} 1 & -2 & | & -1 \\ 0 & 2 & | & 2 \end{bmatrix}$ $\frac{1}{2}R_2 \to R_2$

$\begin{bmatrix} 1 & -2 & | & -1 \\ 0 & 1 & | & 1 \end{bmatrix}$

$y = 1$

$x - 2y = -1$

$x - 2(1) = -1$

$x = 1$

$\{(1, 1)\}$

29. $\begin{cases} 3x - 9y = 3 \\ 2x - 5y = 4 \end{cases}$

$\begin{bmatrix} 3 & -9 & | & 3 \\ 2 & -5 & | & 4 \end{bmatrix}$ $\frac{1}{3}R_1 \to R_2$

$\begin{bmatrix} 1 & -3 & | & 1 \\ 2 & -5 & | & 4 \end{bmatrix}$ $-2R_1 + R_2 \to R_2$

$\begin{bmatrix} 1 & -3 & | & 1 \\ 0 & 1 & | & 2 \end{bmatrix}$

$y = 2$

$x - 3y = 1$

$x - 3(2) = 1$

$x = 7$

$\{(7, 2)\}$

31. $\begin{cases} x + 3y - 2z = 1 \\ 2x + 5y - 2z = 6 \\ -2x - 4y + 3z = -1 \end{cases}$

$\begin{bmatrix} 1 & 3 & -2 & | & 1 \\ 2 & 5 & -2 & | & 6 \\ -2 & -4 & 3 & | & -1 \end{bmatrix}$ $R_2 + R_3 \to R_3$

$\begin{bmatrix} 1 & 3 & -2 & | & 1 \\ 2 & 5 & -2 & | & 6 \\ 0 & 1 & 1 & | & 5 \end{bmatrix}$ $-2R_1 + R_2 \to R_2$

$\begin{bmatrix} 1 & 3 & -2 & | & 1 \\ 0 & -1 & 2 & | & 4 \\ 0 & 1 & 1 & | & 5 \end{bmatrix}$ $R_2 + R_3 \to R_3$

$\begin{bmatrix} 1 & 3 & -2 & | & 1 \\ 0 & -1 & 2 & | & 4 \\ 0 & 0 & 3 & | & 9 \end{bmatrix}$ $-1R_2 \to R_2$

$$\begin{bmatrix} 1 & 3 & -2 & | & 1 \\ 0 & 1 & -2 & | & -4 \\ 0 & 0 & 3 & | & 9 \end{bmatrix} \quad \frac{1}{3}R_3 \to R_3$$

$$\begin{bmatrix} 1 & 3 & -2 & | & 1 \\ 0 & 1 & -2 & | & -4 \\ 0 & 0 & 1 & | & 3 \end{bmatrix}$$

$$z = 3$$

$$y - 2z = -4$$

$$y - 2(3) = -4$$

$$y = 2$$

$$x + 3y - 2z = 1$$

$$x + 3(2) - 2(3) = 1$$

$$x = 1$$

$$\{(1, 2, 3)\}$$

33. $\begin{cases} 2x - 6y + 4z = 0 \\ 3x + y - 4z = 10 \\ 4x + 3y - 9z = 13 \end{cases}$

$$\begin{bmatrix} 2 & -6 & 4 & | & 0 \\ 3 & 1 & -4 & | & 10 \\ 4 & 3 & -9 & | & 13 \end{bmatrix} \quad \frac{1}{2}R_1 \to R_1$$

$$\begin{bmatrix} 1 & -3 & 2 & | & 0 \\ 3 & 1 & -4 & | & 10 \\ 4 & 3 & -9 & | & 13 \end{bmatrix} \quad -3R_1 + R_2 \to R_2$$

$$\begin{bmatrix} 1 & -3 & 2 & | & 0 \\ 0 & 10 & -10 & | & 10 \\ 4 & 3 & -9 & | & 13 \end{bmatrix} \quad -4R_1 + R_3 \to R_3$$

$$\begin{bmatrix} 1 & -3 & 2 & | & 0 \\ 0 & 10 & -10 & | & 10 \\ 0 & 15 & -17 & | & 13 \end{bmatrix} \quad \frac{1}{10}R_2 \to R_2$$

$$\begin{bmatrix} 1 & -3 & 2 & | & 0 \\ 0 & 1 & -1 & | & 1 \\ 0 & 15 & -17 & | & 13 \end{bmatrix} \quad -15R_2 + R_3 \to R_3$$

$$\begin{bmatrix} 1 & -3 & 2 & | & 0 \\ 0 & 1 & -1 & | & 1 \\ 0 & 0 & -2 & | & -2 \end{bmatrix} \quad -\frac{1}{2}R_3 \to R_3$$

$$\begin{bmatrix} 1 & -3 & 2 & | & 0 \\ 0 & 1 & -1 & | & 1 \\ 0 & 0 & 1 & | & 1 \end{bmatrix}$$

$$z = 1$$

$$y - z = 1$$

$$y - 1 = 1$$

$$y = 2$$

$$x - 3y + 2z = 0$$

$$x - 3(2) + 2(1) = 0$$

$$x = 4$$

$$\{(4, 2, 1)\}$$

35. $\begin{cases} 3x + 9y - 6z = -15 \\ 2x + 3y + 2z = 2 \\ 5x + 8y + 14z = 13 \end{cases}$

$$\begin{bmatrix} 3 & 9 & -6 & | & -15 \\ 2 & 3 & 2 & | & 2 \\ 5 & 8 & 14 & | & 13 \end{bmatrix} \quad \frac{1}{3}R_1 \to R_1$$

$$\begin{bmatrix} 1 & 3 & -2 & | & -5 \\ 2 & 3 & 2 & | & 2 \\ 5 & 8 & 14 & | & 13 \end{bmatrix} \quad -2R_1 + R_2 \to R_2$$

$$\begin{bmatrix} 1 & 3 & -2 & | & -5 \\ 0 & -3 & 6 & | & 12 \\ 5 & 8 & 14 & | & 13 \end{bmatrix} \quad -5R_1 + R_3 \to R_3$$

$$\begin{bmatrix} 1 & 3 & -2 & | & -5 \\ 0 & -3 & 6 & | & 12 \\ 0 & -7 & 24 & | & 38 \end{bmatrix} \quad -\frac{1}{3}R_2 \rightarrow R_2$$

$$\begin{bmatrix} 1 & 3 & -2 & | & -5 \\ 0 & 1 & -2 & | & -4 \\ 0 & -7 & 24 & | & 38 \end{bmatrix} \quad 7R_2 + R_3 \rightarrow R_3$$

$$\begin{bmatrix} 1 & 3 & -2 & | & -5 \\ 0 & 1 & -2 & | & -4 \\ 0 & 0 & 10 & | & 10 \end{bmatrix} \quad \frac{1}{10}R_3 \rightarrow R_3$$

$$\begin{bmatrix} 1 & 3 & -2 & | & -5 \\ 0 & 1 & -2 & | & -4 \\ 0 & 0 & 1 & | & 1 \end{bmatrix}$$

$$z = 1$$
$$y - 2z = -4$$
$$y - 2(1) = -4$$
$$y = -2$$
$$x + 3y - 2z = -5$$
$$x + 3(-2) - 2(1) = -5$$
$$x = 3$$

$$\{(3, -2, 1)\}$$

37. $\begin{cases} 3x + 4y = 0 \\ 4x + 3y = 0 \end{cases}$

$$\begin{bmatrix} 3 & 4 & | & 0 \\ 4 & 3 & | & 0 \end{bmatrix} \quad \frac{1}{3}R_1 \rightarrow R_1$$

$$\begin{bmatrix} 1 & \frac{4}{3} & | & 0 \\ 4 & 3 & | & 0 \end{bmatrix} \quad -4R_1 + R_2 \rightarrow R_2$$

$$\begin{bmatrix} 1 & \frac{4}{3} & | & 0 \\ 0 & -\frac{7}{3} & | & 0 \end{bmatrix} \quad -\frac{3}{7}R_2 \rightarrow R_2$$

$$\begin{bmatrix} 1 & \frac{4}{3} & | & 0 \\ 0 & 1 & | & 0 \end{bmatrix}$$

$$y = 0$$
$$x + \frac{4}{3}y = 0$$
$$x + \frac{4}{3}(0) = 0$$
$$x = 0$$

$$\{(0, 0)\}$$

39. $\begin{cases} 6x - 5y = 6 \\ 9x + 7y = -20 \end{cases}$

$$\begin{bmatrix} 6 & -5 & | & 6 \\ 9 & 7 & | & -20 \end{bmatrix} \quad \frac{1}{6}R_1 \rightarrow R_1$$

$$\begin{bmatrix} 1 & -\frac{5}{6} & | & 1 \\ 9 & 7 & | & -20 \end{bmatrix} \quad -9R_1 + R_2 \rightarrow R_2$$

$$\begin{bmatrix} 1 & -\frac{5}{6} & | & 1 \\ 0 & \frac{29}{2} & | & -29 \end{bmatrix} \quad \frac{2}{29}R_2 \rightarrow R_2$$

$$\begin{bmatrix} 1 & -\frac{5}{6} & | & 1 \\ 0 & 1 & | & -2 \end{bmatrix}$$

$$y = -2$$
$$x - \frac{5}{6}y = 1$$

$$x - \frac{5}{6}(-2) = 1$$

$$x = -\frac{2}{3}$$

$$\left\{\left(-\frac{2}{3},\ -2\right)\right\}$$

41. $\begin{cases} -4x + 15y = -2 \\ 12x + 5y = -4 \end{cases}$

$$\begin{bmatrix} -4 & 15 & | & -2 \\ 12 & 5 & | & -4 \end{bmatrix} \quad -\frac{1}{4}R_1 \rightarrow R_1$$

$$\begin{bmatrix} 1 & -\frac{15}{4} & | & \frac{1}{2} \\ 12 & 5 & | & -4 \end{bmatrix} \quad -12R_1 + R_2 \rightarrow R_2$$

$$\begin{bmatrix} 1 & -\frac{15}{4} & | & \frac{1}{2} \\ 0 & 50 & | & -10 \end{bmatrix} \quad \frac{1}{50}R_2 \rightarrow R_2$$

$$\begin{bmatrix} 1 & -\frac{15}{4} & | & \frac{1}{2} \\ 0 & 1 & | & -\frac{1}{5} \end{bmatrix}$$

$$y = -\frac{1}{5}$$

$$x - \frac{15}{4}y = \frac{1}{2}$$

$$x - \frac{15}{4}\left(-\frac{1}{5}\right) = \frac{1}{2}$$

$$x = -\frac{1}{4}$$

$$\left\{\left(-\frac{1}{4},\ -\frac{1}{5}\right)\right\}$$

43. $\begin{cases} 3x - 4y = 5 \\ 4x - 7y = 9 \end{cases}$

$$\begin{bmatrix} 3 & -4 & | & 5 \\ 4 & -7 & | & 9 \end{bmatrix} \quad \frac{1}{3}R_1 \rightarrow R_1$$

$$\begin{bmatrix} 1 & -\frac{4}{3} & | & \frac{5}{3} \\ 4 & -7 & | & 9 \end{bmatrix} \quad -4R_1 + R_2 \rightarrow R_2$$

$$\begin{bmatrix} 1 & -\frac{4}{3} & | & \frac{5}{3} \\ 0 & -\frac{5}{3} & | & \frac{7}{3} \end{bmatrix} \quad -\frac{3}{5}R_2 \rightarrow R_2$$

$$\begin{bmatrix} 1 & -\frac{4}{3} & | & \frac{5}{3} \\ 0 & 1 & | & -\frac{7}{5} \end{bmatrix}$$

$$y = -\frac{7}{5}$$

$$x - \frac{4}{3}y = \frac{5}{3}$$

$$x - \frac{4}{3}\left(-\frac{7}{5}\right) = \frac{5}{3}$$

$$x = -\frac{1}{5}$$

$$\left\{\left(-\frac{1}{5},\ -\frac{7}{5}\right)\right\}$$

45. $\begin{cases} 17x - 11y = -1 \\ -34x + 22y = 2 \end{cases}$

$$\begin{bmatrix} 17 & -11 & | & -1 \\ -34 & 22 & | & 2 \end{bmatrix} \quad 2R_1 + R_2 \rightarrow R_2$$

$$\begin{bmatrix} 17 & -11 & | & -1 \\ 0 & 0 & | & 0 \end{bmatrix}$$

Dependent $\{(x, y) \mid 17x - 11y = -1\}$

47.
$$\begin{cases} \dfrac{1}{5}x - \dfrac{2}{3}y = \dfrac{1}{15} \\ \dfrac{3}{4}x - \dfrac{5}{6}y = \dfrac{1}{3} \end{cases}$$

$$\begin{cases} 3x - 10y = 1 \\ 9x - 10y = 4 \end{cases}$$

$$\begin{bmatrix} 3 & -10 & | & 1 \\ 9 & -10 & | & 4 \end{bmatrix} \quad -3R_1 + R_2 \to R_2$$

$$\begin{bmatrix} 3 & -10 & | & 1 \\ 0 & 20 & | & 1 \end{bmatrix} \quad \dfrac{1}{3}R_1 \to R_1$$

$$\begin{bmatrix} 1 & -\dfrac{10}{3} & | & \dfrac{1}{3} \\ 0 & -20 & | & 1 \end{bmatrix} \quad \dfrac{1}{20}R_2 \to R_2$$

$$\begin{bmatrix} 1 & -\dfrac{10}{3} & | & \dfrac{1}{3} \\ 0 & 1 & | & \dfrac{1}{20} \end{bmatrix}$$

$$y = \dfrac{1}{20}$$

$$x - \dfrac{10}{3}y = \dfrac{1}{3}$$

$$x - \dfrac{10}{3}\left(\dfrac{1}{20}\right) = \dfrac{1}{3}$$

$$x = \dfrac{1}{2}$$

$$\left\{\left(\dfrac{1}{2}, \dfrac{1}{20}\right)\right\}$$

49.
$$\begin{cases} x + y + z = 6 \\ x + 2y + z = 9 \\ 2x + 2y - z = 6 \end{cases}$$

$$\begin{bmatrix} 1 & 1 & 1 & | & 6 \\ 1 & 2 & 1 & | & 9 \\ 2 & 2 & -1 & | & 6 \end{bmatrix} \quad R_1 - R_2 \to R_2$$

$$\begin{bmatrix} 1 & 1 & 1 & | & 6 \\ 0 & -1 & 0 & | & -3 \\ 2 & 2 & -1 & | & 6 \end{bmatrix} \quad -2R_1 + R_3 \to R_3$$

$$\begin{bmatrix} 1 & 1 & 1 & | & 6 \\ 0 & -1 & 0 & | & -3 \\ 0 & 0 & -3 & | & -6 \end{bmatrix} \quad -1R_2 \to R_2$$

$$\begin{bmatrix} 1 & 1 & 1 & | & 6 \\ 0 & 1 & 0 & | & 3 \\ 0 & 0 & -3 & | & -6 \end{bmatrix} \quad -\dfrac{1}{3}R_3 \to R_3$$

$$\begin{bmatrix} 1 & 1 & 1 & | & 6 \\ 0 & 1 & 0 & | & 3 \\ 0 & 0 & 1 & | & 2 \end{bmatrix}$$

$$z = 2$$

$$y = 3$$

$$x + y + z = 6$$

$$x + 3 + 2 = 6$$

$$x = 1$$

$$\{(1, 3, 2)\}$$

51.

$$\begin{cases} 3x_1 + x_2 + 2x_3 = 10 \\ -3x_1 - 2x_2 + 4x_3 = -11 \\ 2x_2 + x_3 = 2 \end{cases}$$

$$\begin{bmatrix} 3 & 1 & 2 & | & 10 \\ -3 & -2 & 4 & | & -11 \\ 0 & 2 & 1 & | & 2 \end{bmatrix} \quad R_1 + R_2 \rightarrow R_2$$

$$\begin{bmatrix} 3 & 1 & 2 & | & 10 \\ 0 & -1 & 6 & | & -1 \\ 0 & 2 & 1 & | & 2 \end{bmatrix} \quad 2R_2 + R_3 \rightarrow R_3$$

$$\begin{bmatrix} 3 & 1 & 2 & | & 10 \\ 0 & -1 & 6 & | & -1 \\ 0 & 0 & 13 & | & 0 \end{bmatrix} \quad -1R_2 \rightarrow R_2$$

$$\begin{bmatrix} 3 & 1 & 2 & | & 10 \\ 0 & 1 & -6 & | & 1 \\ 0 & 0 & 13 & | & 0 \end{bmatrix} \quad \frac{1}{13}R_3 \rightarrow R_3$$

$$\begin{bmatrix} 3 & 1 & 2 & | & 10 \\ 0 & 1 & -6 & | & 1 \\ 0 & 0 & 1 & | & 0 \end{bmatrix}$$

$$z = 0$$

$$y - 6z = 1$$

$$y - 6(0) = 1$$

$$y = 1$$

$$3x + y + 2z = 10$$

$$3x + 1 + 2(0) = 10$$

$$3x = 9$$

$$x = 3$$

$$\{(3, 1, 0)\}$$

53.

$$\begin{cases} 3x + 3y - z = 3 \\ 5x + y + 3z = 1 \\ 2x + 4y - 3z = 4 \end{cases}$$

$$\begin{bmatrix} 3 & 3 & -1 & | & 3 \\ 5 & 1 & 3 & | & 1 \\ 2 & 4 & -3 & | & 4 \end{bmatrix} \quad \frac{1}{3}R_1 \rightarrow R_1$$

$$\begin{bmatrix} 1 & 1 & -\frac{1}{3} & | & 1 \\ 5 & 1 & 3 & | & 1 \\ 2 & 4 & -3 & | & 4 \end{bmatrix} \quad -5R_1 + R_2 \rightarrow R_2$$

$$\begin{bmatrix} 1 & 1 & -\frac{1}{3} & | & 1 \\ 0 & -4 & \frac{14}{3} & | & -4 \\ 2 & 4 & -3 & | & 4 \end{bmatrix} \quad -2R_1 + R_3 \rightarrow R_3$$

$$\begin{bmatrix} 1 & 1 & -\frac{1}{3} & | & 1 \\ 0 & -4 & \frac{14}{3} & | & -4 \\ 0 & 2 & -\frac{7}{3} & | & 2 \end{bmatrix} \quad 2R_3 + R_2 \rightarrow R_3$$

$$\begin{bmatrix} 1 & 1 & -\frac{1}{3} & | & 1 \\ 0 & -4 & \frac{14}{3} & | & -4 \\ 0 & 0 & 0 & | & 0 \end{bmatrix}$$

Dependent: $\{(x, y, z) \mid 3x + 3y - z = 3\}$

55.

$$\begin{cases} 2x + 2y - z = 1 \\ x + 2y - 3z = 4 \\ 5x + 6y - 5z = 3 \end{cases}$$

350

$$\begin{bmatrix} 2 & 2 & -1 & | & 1 \\ 1 & 2 & -3 & | & 4 \\ 5 & 6 & -5 & | & 3 \end{bmatrix} \quad R_1 \leftrightarrow R_2$$

$$\begin{bmatrix} 1 & 2 & -3 & | & 4 \\ 2 & 2 & -1 & | & 1 \\ 5 & 6 & -5 & | & 3 \end{bmatrix} \quad -2R_1 + R_2 \rightarrow R_2$$

$$\begin{bmatrix} 1 & 2 & -3 & | & 4 \\ 0 & -2 & 5 & | & -7 \\ 5 & 6 & -5 & | & 3 \end{bmatrix} \quad -5R_1 + R_3 \rightarrow R_3$$

$$\begin{bmatrix} 1 & 2 & -3 & | & 4 \\ 0 & -2 & 5 & | & -7 \\ 0 & -4 & 10 & | & -17 \end{bmatrix} \quad -2R_2 + R_3 \rightarrow R_3$$

$$\begin{bmatrix} 1 & 2 & -3 & | & 4 \\ 0 & -2 & 5 & | & -7 \\ 0 & 0 & 0 & | & -3 \end{bmatrix}$$

$$0 = -3$$

Inconsistent System

$$\emptyset$$

57.
$$\begin{cases} x + y + z + w = 2 \\ x - y + 2z = -1 \\ 2y + 3z - 3w = -9 \\ 2x - 3y + 2w = 6 \end{cases}$$

$$\begin{bmatrix} 1 & 1 & 1 & 1 & | & 2 \\ 1 & -1 & 2 & 0 & | & -1 \\ 0 & 2 & 3 & -3 & | & -9 \\ 2 & -3 & 0 & 2 & | & 6 \end{bmatrix} \quad \begin{array}{l} R_1 - R_2 \rightarrow R_2 \\ -2R_1 + R_4 \rightarrow R_4 \end{array}$$

$$\begin{bmatrix} 1 & 1 & 1 & 1 & | & 2 \\ 0 & 2 & -1 & 1 & | & 3 \\ 0 & 2 & 3 & -3 & | & -9 \\ 0 & -5 & -2 & 0 & | & 2 \end{bmatrix} \quad R_2 - R_3 \rightarrow R_3$$

$$\begin{bmatrix} 1 & 1 & 1 & 1 & | & 2 \\ 0 & 2 & -1 & 1 & | & 3 \\ 0 & 0 & -4 & 4 & | & 12 \\ 0 & -5 & -2 & 0 & | & 2 \end{bmatrix} \quad \begin{array}{l} \dfrac{1}{2}R_2 \rightarrow R_2 \\[2mm] -\dfrac{1}{4}R_3 \rightarrow R_3 \end{array}$$

$$\begin{bmatrix} 1 & 1 & 1 & 1 & | & 2 \\ 0 & 1 & -\dfrac{1}{2} & \dfrac{1}{2} & | & \dfrac{3}{2} \\ 0 & 0 & 1 & -1 & | & -3 \\ 0 & -5 & -2 & 0 & | & 2 \end{bmatrix} \quad 5R_2 + R_4 \rightarrow R_4$$

$$\begin{bmatrix} 1 & 1 & 1 & 1 & | & 2 \\ 0 & 1 & -\dfrac{1}{2} & \dfrac{1}{2} & | & \dfrac{3}{2} \\ 0 & 0 & 1 & -1 & | & -3 \\ 0 & 0 & -\dfrac{9}{2} & \dfrac{5}{2} & | & \dfrac{19}{2} \end{bmatrix} \quad -\dfrac{2}{9}R_4 \rightarrow R_4$$

$$\begin{bmatrix} 1 & 1 & 1 & 1 & | & 2 \\ 0 & 1 & -\dfrac{1}{2} & \dfrac{1}{2} & | & \dfrac{3}{2} \\ 0 & 0 & 1 & -1 & | & -3 \\ 0 & 0 & 1 & -\dfrac{5}{9} & | & -\dfrac{19}{9} \end{bmatrix} \quad R_3 - R_4 \rightarrow R_4$$

$$\begin{bmatrix} 1 & 1 & 1 & 1 & | & 2 \\ 0 & 1 & -\dfrac{1}{2} & \dfrac{1}{2} & | & \dfrac{3}{2} \\ 0 & 0 & 1 & -1 & | & -3 \\ 0 & 0 & 0 & -\dfrac{4}{9} & | & -\dfrac{8}{9} \end{bmatrix} \quad -\dfrac{9}{4}R_4 \rightarrow R_4$$

$$\begin{bmatrix} 1 & 1 & 1 & 1 & | & 2 \\ 0 & 1 & -\dfrac{1}{2} & \dfrac{1}{2} & | & \dfrac{3}{2} \\ 0 & 0 & 1 & -1 & | & -3 \\ 0 & 0 & 0 & 1 & | & 2 \end{bmatrix}$$

$$w = 2$$

$$z - w = -3$$

$$z - 2 = -3$$

$$z = -1$$

$$y - \frac{1}{2}z + \frac{1}{2}w = \frac{3}{2}$$

$$y - \frac{1}{2}(-1) + \frac{1}{2}(2) = \frac{3}{2}$$

$$y = 0$$

$$x + y + z + w = 2$$

$$x + 0 - 1 + 2 = 2$$

$$x = 1$$

$$\{(1, 0, -1, 2)\}$$

59. $\begin{cases} x_1 + 2x_2 - x_3 = 2 \\ 2x_1 + x_3 + x_4 = 9 \\ x_2 - x_4 = -2 \\ 3x_1 + 4x_2 = 11 \end{cases}$

$\begin{bmatrix} 1 & 2 & -1 & 0 & | & 2 \\ 2 & 0 & 1 & 1 & | & 9 \\ 0 & 1 & 0 & -1 & | & -2 \\ 3 & 4 & 0 & 0 & | & 11 \end{bmatrix}$ $\begin{matrix} -2R_1 + R_2 \rightarrow R_2 \\ -3R_1 + R_4 \rightarrow R_4 \end{matrix}$

$\begin{bmatrix} 1 & 2 & -1 & 0 & | & 2 \\ 0 & -4 & 3 & 1 & | & 5 \\ 0 & 1 & 0 & -1 & | & -2 \\ 0 & -2 & 3 & 0 & | & 5 \end{bmatrix}$ $R_2 \leftrightarrow R_3$

$\begin{bmatrix} 1 & 2 & -1 & 0 & | & 2 \\ 0 & 1 & 0 & -1 & | & -2 \\ 0 & -4 & 3 & 1 & | & 5 \\ 0 & -2 & 3 & 0 & | & 5 \end{bmatrix}$ $\begin{matrix} 4R_2 + R_3 \rightarrow R_3 \\ 2R_2 + R_4 \rightarrow R_4 \end{matrix}$

$\begin{bmatrix} 1 & 2 & -1 & 0 & | & 2 \\ 0 & 1 & 0 & -1 & | & -2 \\ 0 & 0 & 3 & -3 & | & -3 \\ 0 & 0 & 3 & -2 & | & 1 \end{bmatrix}$ $\begin{matrix} R_3 - R_4 \rightarrow R_4 \\ \frac{1}{3}R_3 \rightarrow R_3 \end{matrix}$

$\begin{bmatrix} 1 & 2 & -1 & 0 & | & 2 \\ 0 & 1 & 0 & -1 & | & -2 \\ 0 & 0 & 1 & -1 & | & -1 \\ 0 & 0 & 0 & -1 & | & -4 \end{bmatrix}$ $-1R_4 \rightarrow R_4$

$\begin{bmatrix} 1 & 2 & -1 & 0 & | & 2 \\ 0 & 1 & 0 & -1 & | & -2 \\ 0 & 0 & 1 & -1 & | & -1 \\ 0 & 0 & 0 & 1 & | & 4 \end{bmatrix}$

$$w = 4$$

$$z - w = -1$$

$$z - 4 = -1$$

$$z = 3$$

$$y - w = -2$$

$$y - 4 = -2$$

$$y = 2$$

$$x + 2y - z = 2$$

$$x + 2(2) - 3 = 2$$

$$x = 1$$

$$\{(1, 2, 3, 4)\}$$

Chapter 7 Review Problems

1. $\begin{cases} 2x + y = 4 \\ 3x - 2y = -1 \end{cases}$

$$y = 4 - 2x$$

So, $\quad 3x - 2(4 - 2x) = -1$

$$3x - 8 + 4x = -1$$

$$7x = 7$$

$$x = 1$$

$y = 4 - 2(1) = 2 = 2$

$\{(1,\ 2)\}$

3. $\begin{cases} 5x + 3y = 1 \\ 7x + 2y = 8 \end{cases}$

$2y = 8 - 7x$

$y = 4 - \dfrac{7}{2}x$

So, $\quad 5x + 3\left(4 - \dfrac{7}{2}x\right) = 1$

$\quad\quad 5x + 12 - \dfrac{21}{2}x = 1$

$\quad\quad\quad\quad -\dfrac{11}{2}x = -11$

$\quad\quad\quad\quad\quad\quad x = 2$

$y = 4 - \dfrac{7}{2}(2) = -3$

$\{(2,\ -3)\}$

5. $\begin{cases} 5x^2 - 4y^2 = 9 \\ x - y = 3 \end{cases}$

$x = y + 3$

$\quad\quad (y + 3)^2 - 4y^2 = 9$

$y^2 + 6y + 9 - 4y^2 = 9$

$\quad\quad\quad\quad -3y^2 + 6y = 0$

$\quad\quad\quad\quad -3y\,(y - 2) = 0$

$y = 0 \quad\quad \text{or} \quad y = 2$

$x = 0 + 3 \quad\quad x = 2 + 3 = 5$

$\{(3,\ 0),\ (5,\ 2)\}$

7. $\begin{cases} x + 2y = 2 \\ x - 2y = 6 \end{cases}$

$\begin{aligned} x + 2y &= 2 \\ \underline{x - 2y} &= \underline{6} \\ 2x &= 8 \end{aligned}$

$x \quad\quad = 4$

$4 + 2y = 2$

$\quad\quad 2y = -2$

$\quad\quad\ y = -1$

$\{(4,\ -1)\}$

9. $\begin{cases} 2x + 3y = 15 \\ 2x - 7y = 5 \end{cases}$

$\begin{aligned} 2x + 3y &= 15 \\ \underline{-2x + 7y} &= \underline{-5} \\ 10y &= 10 \end{aligned}$

$\quad\quad\quad y = 1$

$2x + 3(1) = 15$

$\quad\quad 2x = 12$

$\quad\quad\ x = 6$

$\{(6,\ 1)\}$

11. $\begin{cases} 5x + 6y = -3 \\ -4x + 9y = 7 \end{cases}$

$\begin{aligned} 20x + 24y &= -12 \\ \underline{-20x + 45y} &= \underline{35} \\ 69y &= 23 \end{aligned}$

$\quad\quad y = \dfrac{1}{3}$

$$5x + 6\left(\frac{1}{3}\right) = -3$$

$$5x = -5$$

$$x = -1$$

$$\left\{\left(-1,\ \frac{1}{3}\right)\right\}$$

13. $\begin{cases} 24x - 18y = -1 \\ 6x - 10y = -3 \end{cases}$

$$\begin{array}{r} 24x - 18y = -1 \\ \underline{-24x + 40y = 12} \\ 22y = 11 \end{array}$$

$$y = \frac{1}{2}$$

$$6x - 10\left(\frac{1}{2}\right) = -3$$

$$6x = 2$$

$$x = \frac{1}{3}$$

$$\left\{\left(\frac{1}{3},\ \frac{1}{2}\right)\right\}$$

15. $\begin{cases} 5x + 4y = 2 \\ 10x + 12y = 5 \end{cases}$

$$\begin{array}{r} -10x - 8y = -4 \\ \underline{10x + 12y = 5} \\ 4y = 1 \end{array}$$

$$y = \frac{1}{4}$$

$$5x + 4\left(\frac{1}{4}\right) = 2$$

$$5x = 1$$

$$x = \frac{1}{5}$$

$$\left\{\left(\frac{1}{5},\ \frac{1}{4}\right)\right\}$$

17. $\begin{cases} x - 2y + z = 2 \\ x + y + z = 8 \\ x - y - z = 2 \end{cases}$

$$\begin{array}{rr} x - 2y + z = 2 & x + y + z = 8 \\ \underline{-x - y - z = -8} & \underline{x - y - z = 2} \\ -3y = -6 & 2x = 10 \end{array}$$

$$y = 2 \qquad\qquad x = 5$$

$$5 - 2(2) + z = 2$$

$$z = 1$$

$$\{(5,\ 2,\ 1)\}$$

19. $\begin{cases} 2x + 3y + 5z = 0 \\ x - 2y - z = 0 \\ 3x + 2y - 2z = 0 \end{cases}$

$$\begin{array}{rr} x - 2y - z = 0 & 4x + 6y + 10z = 0 \\ \underline{3x + 2y - 2z = 0} & \underline{3x - 6y - 3z = 0} \\ 4x - 3z = 0 & 7x + 7z = 0 \\ & x + z = 0 \end{array}$$

$$\begin{array}{r} 4x - 3z = 0 \\ \underline{3x + 3z = 0} \\ 7x = 0 \end{array}$$

$$x = 0$$

$$0 + z = 0$$

$$z = 0$$

$$0 - 2y - 0 = 0$$

$$2y = 0$$

$$y = 0$$

354

$\{(0,\ 0,\ 0)\}$

$x = -2$

$-2 + 3y + 3 = 1$

$3y = 0$

$y = 0$

$\{(-2,\ 0,\ 3)\}$

21. $\begin{cases} 2x + 5y - z = -6 \\ x - y + 3z = 18 \\ 2x + 3y - 2z = -12 \end{cases}$

$\begin{array}{l} 2x + 5y - z = -6 \\ \underline{-2x - 3y + 2z = 12} \\ 2y + z = 6 \end{array}$ $\begin{array}{l} 2x + 5y - z = -6 \\ \underline{-2x + 2y - 6z = -36} \\ 7y - 7z = -42 \\ y - z = -6 \end{array}$

25. $\begin{cases} x - y + 3z = 6 \\ 2x + y - z = -3 \\ 3x - 2y + z = -4 \end{cases}$

$\begin{array}{l} 2y + z = 6 \\ \underline{y - z = -6} \\ 3y = 0 \\[4pt] y = 0 \\[6pt] 0 - z = -6 \\[4pt] z = 6 \\[6pt] x - 0 + 3(6) = 18 \\[4pt] x = 0 \end{array}$

$\begin{array}{l} x - y + 3z = 6 \\ \underline{2x + y - z = -3} \\ 3x + 2z = 3 \end{array}$ $\begin{array}{l} 4x + 2y - 2z = -6 \\ \underline{3x - 2y + z = -4} \\ 7x - z = -10 \end{array}$

$\begin{array}{l} 3x + 2z = 3 \\ \underline{14x - 2z = -20} \\ 17x = -17 \\[4pt] x = -1 \\[6pt] 3(-1) + 2z = 3 \\[4pt] 2z = 6 \\[4pt] z = 3 \\[6pt] -1 - y + 3(3) = 6 \\[4pt] -y = -2 \\[4pt] y = 2 \end{array}$

$\{(0,\ 0,\ 6)\}$

23. $\begin{cases} x + 3y + z = 1 \\ 2x + y + 2z = 2 \\ 3x - y + 2z = 0 \end{cases}$

$\begin{array}{l} 2x + y + 2z = 2 \\ \underline{3x - y + 2z = 0} \\ 5x + 4z = 2 \end{array}$ $\begin{array}{l} x + 3y + z = 1 \\ \underline{9x - 3y + 6z = 0} \\ 10x + 7z = 1 \end{array}$

$\begin{array}{l} -10y - 8z = -4 \\ \underline{10y + 7z = 1} \\ -z = -3 \\[4pt] z = 3 \end{array}$

$5x + 4(3) = 2$

$5x = -10$

$\{(-2,\ 0,\ 3)\}$

$\{(-1,\ 2,\ 3)\}$

27. number of cantaloupes: x
number of honeydews: y
number of watermelons: z

$\begin{cases} 3x + 4y + z = 1192 \\ x + y + z = 477 \\ x + 2z = 517 \end{cases}$

355

$$3x + 4y + z = 1192$$
$$\underline{-4x - 4y - 4z = -1908}$$
$$- x \qquad - 3z = -716$$

$$-x - 3z = -716$$
$$\underline{x + 2z = 517}$$
$$- z = -199$$

$$z = 199$$

$$x + 2(199) = 517$$

$$x = 119$$

$$119 + y + 199 = 477$$

$$y = 159$$

A cantaloupe costs $1.19,
a honeydew costs $1.59,
a watermelon costs $1.99.

29. 1st number: x
2nd number: y

$$x^2 + y^2 = 58$$

$$x - y = 10$$

$$x = y + 10$$

$$(y + 10)^2 + y^2 = 58$$

$$y^2 + 20y + 100 + y^2 = 58$$

$$2y^2 + 20y + 42 = 0$$

$$y^2 + 10y + 21 = 0$$

$$(y + 7)(y + 3) = 0$$

$$y = -7 \qquad \text{or} \qquad y = -3$$

$$x = -7 + 10 = 3 \qquad x = -3 + 10 = 7$$

The numbers are 3 and −7
or 7 and −3.

31. $x + 2y \geq 3$

B.L.: $x + 2y = 3$ (included)

Test (0, 0)

Test (0, 0)

$$0 + 2(0) \geq 3$$

$$0 \geq 3 \qquad F$$

Do not shade region.

Test (0, 4)

$$0 + 2(4) \geq 3$$

$$0 \geq 8 \qquad T$$

Shade region.

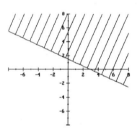

33. $x - y < 0$

B.L.: $x - y = 0$ (not included)

Test (1, 0)

$$1 - 0 < 0$$

$$1 < 0 \qquad F$$

Do not shade region.

Test (0, 1)

$$0 - 1 < 0$$

$$-1 < 0 \qquad T$$

Shade region.

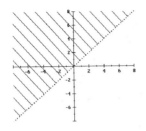

35. $y < 3x - 2$

B.L.: $y = 3x - 2$ (not included)

Test $(0, 0)$

$0 < 3(0) - 2$

$0 < -2$ F

Do not shade region.

Test $(1, 0)$

$0 < 3(1) - 2$

$0 < 1$ T

Shade region.

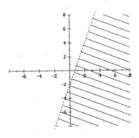

37. $\begin{cases} 2x + y < 7 \\ x - y > 5 \end{cases}$

B.L.: $2x + y = 7$ (not included)

B.L.: $x - y = 5$ (not included)

Test $(4, -2)$

$2(4) - 2 < 7$

     $6 < 7$ T

$4 - (-2) > 5$

     $6 > 5$ T

Region in solution set.

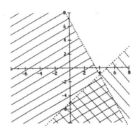

39. $\begin{cases} y \le 2 \\ x \ge 0 \end{cases}$

B.L.: $y = 2$ (included)

B.L.: $x = 0$ (included)

Test $(1, 0)$

$0 \le 2$ T

$1 \ge 0$ T

Region in solution set.

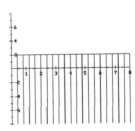

41. $2x \le y + 3$

B.L.: $2x = y + 3$ (included)

Test $(0, 0)$

$2(0) \le 0 + 3$

     $0 \le 3$ T

357

Shade region.

Test (2, 0)

$2(2) \le 0 + 3$

$\quad 4 \le 3 \quad$ F

Do not shade region.

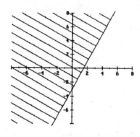

43. $y < 2x$

B.L.: $y = 2x$ (not included)

Test (1, 0)

$0 < 2(1)$

$0 < 2 \quad$ T

Shade region.

Test (0, 1)

$1 < 2(0)$

$1 < 0 \quad$ F

Do not shade region.

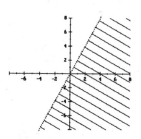

45. $\begin{cases} x - y > 7 \\ x - y < 5 \end{cases}$

B.L.: $x - y = 7$ (not included)

B.L.: $x - y < 5$ (not included)

No points in common.

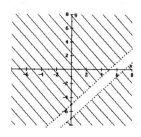

$\emptyset$

47. $2x - 3y \le 4$

B.L.: $2x - 3y = 4$ (included)

Test (0, 0)

$2(0) - 3(0) \le 4$

$\quad\quad 0 \le 4 \quad$ T

Shade region.

Test (3, 0)

$2(3) - 3(0) \le 4$

$\quad\quad 6 \le 4 \quad$ F

Do not shade region.

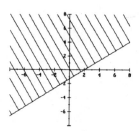

49. $5x - 7y > 0$

B.L.: $5x - 7y = 0$ (included)

Test $(1, 0)$

$5(1) - 7(0) > 0$

$\qquad 5 > 0 \quad$ T

Shade region.

Test $(0, 1)$

$5(0) - 7(1) > 0$

$\qquad -7 > 0 \quad$ F

Do not shade region.

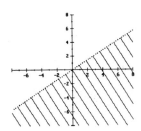

51. $\dfrac{1}{2}x - y < 1$

B.L.: $\dfrac{1}{2}x - y = 1$ (not included)

$\qquad x - 2y = 2$

Test $(0, 0)$

$\dfrac{1}{2}(0) - 0 < 1$

$\qquad 0 < 1 \quad$ T

Shade region.

Test $(0, -2)$

$\dfrac{1}{2}(0) - (-2) < 1$

$\qquad 2 < 1 \quad$ F

Do not shade region.

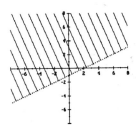

53. $\begin{cases} 2x - 5y \le 5 \\ 4x - 10y \le 0 \end{cases}$

B.L.: $2x - 5y = 5$ (included)

B.L.: $4x - 10y = 0$ (included)

Test $(0, 1)$

$2(0) - 5(1) \le 5$

$\qquad -5 \le 5 \quad$ T

$4(0) - 10(1) \le 0$

$\qquad -10 \le 0 \quad$ T

Region in solution set.

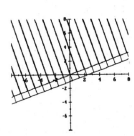

55. $\begin{cases} 3x + y = 1 \\ 4x + y = -1 \end{cases}$

$$x = \frac{\begin{vmatrix} 1 & 1 \\ -1 & 1 \end{vmatrix}}{\begin{vmatrix} 3 & 1 \\ 4 & 1 \end{vmatrix}} = \frac{1 + 1}{3 - 4} = \frac{2}{-1} = -2$$

$$y = \frac{\begin{vmatrix} 3 & 1 \\ 4 & -1 \end{vmatrix}}{-1} = \frac{-3 - 4}{-1} = \frac{-7}{-1} = 7$$

$$\{(-2, 7)\}$$

57. $\begin{cases} 2x - 3y = -5 \\ 3x + 2y = -1 \end{cases}$

$$x = \frac{\begin{vmatrix} -5 & -3 \\ -1 & 2 \end{vmatrix}}{\begin{vmatrix} 2 & -3 \\ 3 & 2 \end{vmatrix}} = \frac{-10 - 3}{4 + 9} = -\frac{13}{13} = -1$$

$$y = \frac{\begin{vmatrix} 2 & -5 \\ 3 & -1 \end{vmatrix}}{13} = \frac{-2 + 15}{13} = \frac{13}{13} = 1$$

$$\{(-1, 1)\}$$

59. $\begin{cases} 5x + 4y = 12 \\ -6x + 8y = -8 \end{cases}$

$$x = \frac{\begin{vmatrix} 12 & 4 \\ -8 & 8 \end{vmatrix}}{\begin{vmatrix} 5 & 4 \\ -6 & 8 \end{vmatrix}} = \frac{96 + 32}{40 + 24} = \frac{128}{64} = 2$$

$$y = \frac{\begin{vmatrix} 5 & 12 \\ -6 & -8 \end{vmatrix}}{64} = \frac{-40 + 72}{64} = \frac{32}{64} = \frac{1}{2}$$

$$\left\{ \left(2, \frac{1}{2} \right) \right\}$$

61. $\begin{cases} -x + 3y = 4 \\ 2x - 6y = -8 \end{cases}$

Since $\begin{vmatrix} -1 & 3 \\ 2 & -6 \end{vmatrix} = 6 - 6 = 0$,

Cramer's Rule does not apply.

63. $\begin{cases} x + y - z = -3 \\ x - y + z = -1 \\ x + y + z = -1 \end{cases}$

$$x = \frac{\begin{vmatrix} -3 & 1 & -1 \\ -1 & -1 & 1 \\ -1 & 1 & 1 \end{vmatrix}}{\begin{vmatrix} 1 & 1 & -1 \\ 1 & -1 & 1 \\ 1 & 1 & 1 \end{vmatrix}}$$

$$= \frac{-3 \begin{vmatrix} -1 & 1 \\ 1 & 1 \end{vmatrix} - 1 \begin{vmatrix} -1 & 1 \\ -1 & 1 \end{vmatrix} - 1 \begin{vmatrix} -1 & -1 \\ -1 & 1 \end{vmatrix}}{1 \begin{vmatrix} -1 & 1 \\ 1 & 1 \end{vmatrix} - 1 \begin{vmatrix} 1 & 1 \\ 1 & 1 \end{vmatrix} - 1 \begin{vmatrix} 1 & -1 \\ 1 & 1 \end{vmatrix}}$$

$$= \frac{-3(-1 - 1) - (-1 + 1) - (-1 - 1)}{(-1 - 1) - (1 - 1) - (1 + 1)}$$

$$= \frac{6 + 2}{-2 - 2}$$

$$= \frac{8}{-4}$$

$$= -2$$

$$y = \frac{\begin{vmatrix} 1 & -3 & -1 \\ 1 & -1 & 1 \\ 1 & -1 & 1 \end{vmatrix}}{-4}$$

$$= \frac{1 \begin{vmatrix} -1 & 1 \\ -1 & 1 \end{vmatrix} - (-3) \begin{vmatrix} 1 & 1 \\ 1 & 1 \end{vmatrix} - 1 \begin{vmatrix} 1 & -1 \\ 1 & -1 \end{vmatrix}}{-4}$$

$$= \frac{(-1 + 1) + 3(1 - 1) - (-1 + 1)}{-4}$$

$$= \frac{0}{-4}$$

$$= 0$$

$$z = \frac{\begin{vmatrix} 1 & 1 & -3 \\ 1 & -1 & -1 \\ 1 & 1 & -1 \end{vmatrix}}{-4}$$

$$= \frac{1\begin{vmatrix} -1 & -1 \\ 1 & -1 \end{vmatrix} - 1\begin{vmatrix} 1 & -1 \\ 1 & -1 \end{vmatrix} - 3\begin{vmatrix} 1 & -1 \\ 1 & 1 \end{vmatrix}}{-4}$$

$$= \frac{(1 + 1) - (-1 + 1) - 3(1 + 1)}{-4}$$

$$= \frac{2 - 6}{-4}$$

$$= \frac{-4}{-4}$$

$$= 1$$

$$\{(-2, 0, 1)\}$$

65. $\begin{cases} x - y - z = 0 \\ x + 2y - z = 3 \\ x + y + 2z = 5 \end{cases}$

$$x = \frac{\begin{vmatrix} 0 & -1 & -1 \\ 3 & 2 & -1 \\ 5 & 1 & 2 \end{vmatrix}}{\begin{vmatrix} 1 & -1 & -1 \\ 1 & 2 & -1 \\ 1 & 1 & 2 \end{vmatrix}}$$

$$= \frac{-(-1)\begin{vmatrix} 3 & -1 \\ 5 & 2 \end{vmatrix} - 1\begin{vmatrix} 3 & 2 \\ 5 & 1 \end{vmatrix}}{1\begin{vmatrix} 2 & -1 \\ 1 & 2 \end{vmatrix} - (-1)\begin{vmatrix} 1 & -1 \\ 1 & 2 \end{vmatrix} - 1\begin{vmatrix} 1 & 2 \\ 1 & 1 \end{vmatrix}}$$

$$= \frac{(6 + 5) - (3 - 10)}{(4 + 1) + (2 + 1) - (1 - 2)}$$

$$= \frac{11 + 7}{5 + 3 + 1}$$

$$= \frac{18}{9}$$

$$= 2$$

$$y = \frac{\begin{vmatrix} 1 & 0 & -1 \\ 1 & 3 & -1 \\ 1 & 5 & 2 \end{vmatrix}}{9}$$

$$= \frac{1\begin{vmatrix} 3 & -1 \\ 5 & 2 \end{vmatrix} - 1\begin{vmatrix} 1 & 3 \\ 1 & 5 \end{vmatrix}}{9}$$

$$= \frac{(6 + 5) - (5 - 3)}{9}$$

$$= \frac{11 - 2}{9}$$

$$= \frac{9}{9}$$

$$= 1$$

$$z = \frac{\begin{vmatrix} 1 & -1 & 0 \\ 1 & 2 & 3 \\ 1 & 1 & 5 \end{vmatrix}}{9}$$

$$= \frac{1\begin{vmatrix} 2 & 3 \\ 1 & 5 \end{vmatrix} - (-1)\begin{vmatrix} 1 & 3 \\ 1 & 5 \end{vmatrix}}{9}$$

$$= \frac{(10 - 3) + (5 - 3)}{9}$$

$$= \frac{7 + 2}{9}$$

$$= \frac{9}{9}$$

$$= 1$$

$$\{(2, 1, 1)\}$$

67.

$$\begin{cases} x - 2y + z = 17 \\ 2x + 3y + 2z = 13 \\ 3x + y - z = 14 \end{cases}$$

$$x = \frac{\begin{vmatrix} 17 & -2 & 1 \\ 13 & 3 & 2 \\ 14 & 1 & -1 \end{vmatrix}}{\begin{vmatrix} 1 & -2 & 1 \\ 2 & 3 & 2 \\ 3 & 1 & -1 \end{vmatrix}}$$

$$= \frac{17 \begin{vmatrix} 3 & 2 \\ 1 & -1 \end{vmatrix} - (-2) \begin{vmatrix} 13 & 2 \\ 14 & -1 \end{vmatrix} + 1 \begin{vmatrix} 13 & 3 \\ 14 & 1 \end{vmatrix}}{1 \begin{vmatrix} 3 & 2 \\ 1 & -1 \end{vmatrix} - (-2) \begin{vmatrix} 2 & 2 \\ 3 & -1 \end{vmatrix} + 1 \begin{vmatrix} 2 & 3 \\ 3 & 1 \end{vmatrix}}$$

$$= \frac{17(-3 - 2) + 2(-13 - 28) + (13 - 42)}{(-3 - 2) + 2(-2 - 6) + (2 - 9)}$$

$$= \frac{-85 - 82 - 29}{-5 - 16 - 7}$$

$$= \frac{-196}{-28}$$

$$= 7$$

$$y = \frac{\begin{vmatrix} 1 & 17 & 1 \\ 2 & 13 & 2 \\ 3 & 14 & -1 \end{vmatrix}}{-28}$$

$$= \frac{1 \begin{vmatrix} 13 & 2 \\ 14 & -1 \end{vmatrix} - 17 \begin{vmatrix} 2 & 2 \\ 3 & -1 \end{vmatrix} + 1 \begin{vmatrix} 2 & 13 \\ 3 & 14 \end{vmatrix}}{-28}$$

$$= \frac{(-13 - 28) - 17(-2 - 6) + (28 - 39)}{-28}$$

$$= \frac{-41 + 136 - 11}{-28}$$

$$= \frac{84}{-28}$$

$$= -3$$

$$z = \frac{\begin{vmatrix} 1 & -2 & 17 \\ 2 & 3 & 13 \\ 3 & 1 & 14 \end{vmatrix}}{-28}$$

$$= \frac{1 \begin{vmatrix} 3 & 13 \\ 1 & 14 \end{vmatrix} - (-2) \begin{vmatrix} 2 & 13 \\ 3 & 14 \end{vmatrix} + 17 \begin{vmatrix} 2 & 3 \\ 3 & 1 \end{vmatrix}}{-28}$$

$$= \frac{(42 - 13) + 2(28 - 39) + 17(2 - 9)}{-28}$$

$$= \frac{29 - 22 - 119}{-28}$$

$$= \frac{-112}{-28}$$

$$= 4$$

$$\{(7, -3, 4)\}$$

69.

$$\begin{cases} 3x - y + 4z = 9 \\ x - 2y + 3z = -2 \\ 2x - y + z = 3 \end{cases}$$

$$x = \frac{\begin{vmatrix} 9 & -1 & 4 \\ -2 & -2 & 3 \\ 3 & -1 & 1 \end{vmatrix}}{\begin{vmatrix} 3 & -1 & 4 \\ 1 & -2 & 3 \\ 2 & -1 & 1 \end{vmatrix}}$$

$$= \frac{9\begin{vmatrix} -2 & 3 \\ -1 & 1 \end{vmatrix} - (-1)\begin{vmatrix} -2 & 3 \\ 3 & 1 \end{vmatrix} + 4\begin{vmatrix} -2 & -2 \\ 3 & -1 \end{vmatrix}}{3\begin{vmatrix} -2 & 3 \\ -1 & 1 \end{vmatrix} - (-1)\begin{vmatrix} 1 & 3 \\ 2 & 1 \end{vmatrix} + 4\begin{vmatrix} 1 & -2 \\ 2 & -1 \end{vmatrix}}$$

$$= \frac{9(-2 + 3) + (-2 - 9) + 4(2 + 6)}{3(-2 + 3) + (1 - 6) + 4(-1 + 4)}$$

$$= \frac{9 - 11 + 32}{3 - 5 + 12}$$

$$= \frac{30}{10}$$

$$= 3$$

$$y = \frac{\begin{vmatrix} 3 & 9 & 4 \\ 1 & -2 & 3 \\ 2 & 3 & 1 \end{vmatrix}}{10}$$

$$= \frac{3\begin{vmatrix} -2 & 3 \\ 3 & 1 \end{vmatrix} - 9\begin{vmatrix} 1 & 3 \\ 2 & 1 \end{vmatrix} + 4\begin{vmatrix} 1 & -2 \\ 2 & 3 \end{vmatrix}}{10}$$

$$= \frac{3(-2 - 9) - 9(1 - 6) + 4(3 + 4)}{10}$$

$$= \frac{-33 + 45 + 28}{10}$$

$$= \frac{40}{10}$$

$$= 4$$

$$z = \frac{\begin{vmatrix} 3 & -1 & 9 \\ 1 & -2 & -2 \\ 2 & -1 & 3 \end{vmatrix}}{10}$$

$$= \frac{3\begin{vmatrix} -2 & -2 \\ -1 & 3 \end{vmatrix} - (-1)\begin{vmatrix} 1 & -2 \\ 2 & 3 \end{vmatrix} + 9\begin{vmatrix} 1 & -2 \\ 2 & -1 \end{vmatrix}}{10}$$

$$= \frac{3(-6 - 2) + (3 + 4) + 9(-1 + 4)}{10}$$

$$= \frac{-24 + 7 + 27}{10}$$

$$= \frac{10}{10}$$

$$= 1$$

$$\{(3, 4, 1)\}$$

71.

$$\begin{cases} 2x - 3y - 4z = 4 \\ 3x - y + z = -2 \\ x + 2y - 3z = -6 \end{cases}$$

Since $\begin{vmatrix} 2 & -3 & 4 \\ 3 & -1 & 1 \\ 1 & 2 & -3 \end{vmatrix}$

$$= 2\begin{vmatrix} -1 & 1 \\ 2 & -3 \end{vmatrix} - (-3)\begin{vmatrix} 3 & 1 \\ 1 & -3 \end{vmatrix} + 4\begin{vmatrix} 3 & -1 \\ 1 & 2 \end{vmatrix}$$

$$= 2(3 - 2) + 3(-9 - 1) + 4(6 + 1)$$

$$= 2 - 30 + 28$$

$$= 0$$

Cramer's Rule does not apply.

363

73. $\begin{cases} x + 2y = 2 \\ x - 2y = 6 \end{cases}$

$\begin{bmatrix} 1 & 2 & | & 2 \\ 1 & -2 & | & 6 \end{bmatrix}$ $R_1 - R_2 \to R_2$

$\begin{bmatrix} 1 & 2 & | & 2 \\ 0 & 4 & | & -4 \end{bmatrix}$ $\frac{1}{4}R_2 \to R_2$

$\begin{bmatrix} 1 & 2 & | & 2 \\ 0 & 1 & | & -1 \end{bmatrix}$

$$y = -1$$

$$x + 2y = 2$$

$$x + 2(-1) = 2$$

$$x = 4$$

$$\{(4, -1)\}$$

75. $\begin{cases} 2x + 3y = 15 \\ 2x - 7y = 5 \end{cases}$

$\begin{bmatrix} 2 & 3 & | & 15 \\ 2 & -7 & | & 5 \end{bmatrix}$ $R_1 - R_2 \to R_2$

$\begin{bmatrix} 2 & 3 & | & 15 \\ 0 & 10 & | & 10 \end{bmatrix}$ $\begin{array}{c} \frac{1}{2}R_1 \to R_1 \\ \frac{1}{10}R_2 \to R_2 \end{array}$

$\begin{bmatrix} 1 & \frac{3}{2} & | & \frac{15}{2} \\ 0 & 1 & | & 1 \end{bmatrix}$

$$y = 1$$

$$x + \frac{3}{2}y = \frac{15}{2}$$

$$x + \frac{3}{2}(1) = \frac{15}{2}$$

$$x = 6$$

$$\{(6, 1)\}$$

77. $\begin{cases} 5x + 6y = -3 \\ -4x + 9y = 7 \end{cases}$

$\begin{bmatrix} 5 & 6 & | & -3 \\ -4 & 9 & | & 7 \end{bmatrix}$ $R_1 + R_2 \to R_1$

$\begin{bmatrix} 1 & 15 & | & 4 \\ -4 & 9 & | & 7 \end{bmatrix}$ $4R_1 + R_2 \to R_2$

$\begin{bmatrix} 1 & 15 & | & 4 \\ 0 & 69 & | & 23 \end{bmatrix}$ $\frac{1}{69}R_2 \to R_2$

$\begin{bmatrix} 1 & 15 & | & 4 \\ 0 & 1 & | & \frac{1}{3} \end{bmatrix}$

$$y = \frac{1}{3}$$

$$x + 15y = 4$$

$$x + 15\left(\frac{1}{3}\right) = 4$$

$$x = -1$$

$$\left\{\left(-1, \frac{1}{3}\right)\right\}$$

79. $\begin{cases} 24x - 18y = -1 \\ 6x - 10y = -3 \end{cases}$

$\begin{bmatrix} 24 & -18 & | & -1 \\ 6 & -10 & | & -3 \end{bmatrix}$ $\begin{array}{c} -4R_2 + R_1 \to R_2 \\ \frac{1}{24}R_1 \to R_1 \end{array}$

$\begin{bmatrix} 1 & -\frac{3}{4} & | & -\frac{1}{24} \\ 0 & 22 & | & 11 \end{bmatrix}$ $\frac{1}{22}R_2 \to R_2$

$\begin{bmatrix} 1 & -\frac{3}{4} & | & -\frac{1}{24} \\ 0 & 1 & | & \frac{1}{2} \end{bmatrix}$

$$y = \frac{1}{2}$$

$$x - \frac{3}{4}y = -\frac{1}{24}$$

$$x - \frac{3}{4}\left(\frac{1}{2}\right) = -\frac{1}{24}$$

$$x = \frac{1}{3}$$

$$\left\{\left(\frac{1}{3}, \frac{1}{2}\right)\right\}$$

81. $\begin{cases} 5x + 4y = 2 \\ 10x + 12y = 5 \end{cases}$

$$\begin{bmatrix} 5 & 4 & | & 2 \\ 10 & 12 & | & 5 \end{bmatrix} \quad \begin{array}{c} -2R_1 + R_2 \to R_2 \\ \frac{1}{5}R_1 \to R_1 \end{array}$$

$$\begin{bmatrix} 1 & \frac{4}{5} & | & \frac{2}{5} \\ 0 & 4 & | & 1 \end{bmatrix} \quad \frac{1}{4}R_2 \to R_2$$

$$\begin{bmatrix} 1 & \frac{4}{5} & | & \frac{2}{5} \\ 0 & 1 & | & \frac{1}{4} \end{bmatrix}$$

$$y = \frac{1}{4}$$

$$x + \frac{4}{5}y = \frac{2}{5}$$

$$x + \frac{4}{5}\left(\frac{1}{4}\right) = \frac{2}{5}$$

$$x = \frac{1}{5}$$

$$\left\{\left(\frac{1}{5}, \frac{1}{4}\right)\right\}$$

83. $\begin{cases} x - 2y + z = 2 \\ x + y + z = 8 \\ x - y - z = 2 \end{cases}$

$$\begin{bmatrix} 1 & -2 & 1 & | & 2 \\ 1 & 1 & 1 & | & 8 \\ 1 & -1 & -1 & | & 2 \end{bmatrix} \quad \begin{array}{c} R_1 - R_2 \to R_2 \\ R_1 - R_3 \to R_3 \end{array}$$

$$\begin{bmatrix} 1 & -2 & 1 & | & 2 \\ 0 & -3 & 0 & | & -6 \\ 0 & -1 & 2 & | & 0 \end{bmatrix} \quad \begin{array}{c} -3R_3 + R_2 \to R_3 \\ -\frac{1}{3}R_2 \to R_2 \end{array}$$

$$\begin{bmatrix} 1 & -2 & 1 & | & 2 \\ 0 & 1 & 0 & | & 2 \\ 0 & 0 & -6 & | & -6 \end{bmatrix} \quad -\frac{1}{6}R_3 \to R_3$$

$$\begin{bmatrix} 1 & -2 & 1 & | & 2 \\ 0 & 1 & 0 & | & 2 \\ 0 & 0 & 1 & | & 1 \end{bmatrix}$$

$$z = 1$$

$$y = 2$$

$$x - 2y + z = 2$$

$$x - 2(2) + 1 = 2$$

$$x = 5$$

$$\{(5, 2, 1)\}$$

85. $\begin{cases} 2x + 3y + 5z = 0 \\ x - 2y - z = 0 \\ 3x + 2y - 2z = 0 \end{cases}$

$$\begin{bmatrix} 2 & 3 & 5 & | & 0 \\ 1 & -2 & -1 & | & 0 \\ 3 & 2 & -2 & | & 0 \end{bmatrix} \quad R_1 \leftrightarrow R_2$$

$$\begin{bmatrix} 1 & -2 & -1 & | & 0 \\ 2 & 3 & 5 & | & 0 \\ 3 & 2 & -2 & | & 0 \end{bmatrix} \quad \begin{matrix} -2R_1 + R_2 \to R_2 \\ -3R_1 + R_3 \to R_3 \end{matrix}$$

$$\begin{bmatrix} 1 & -2 & -1 & | & 0 \\ 0 & 7 & 7 & | & 0 \\ 0 & 8 & 1 & | & 0 \end{bmatrix} \quad \tfrac{1}{7}R_2 \to R_2$$

$$\begin{bmatrix} 1 & -2 & -1 & | & 0 \\ 0 & 1 & 1 & | & 0 \\ 0 & 8 & 1 & | & 0 \end{bmatrix} \quad -8R_2 + R_3 \to R_3$$

$$\begin{bmatrix} 1 & -2 & -1 & | & 0 \\ 0 & 1 & 1 & | & 0 \\ 0 & 0 & -7 & | & 0 \end{bmatrix} \quad -\tfrac{1}{7}R_3 \to R_3$$

$$\begin{bmatrix} 1 & -2 & -1 & | & 0 \\ 0 & 1 & 1 & | & 0 \\ 0 & 0 & 1 & | & 0 \end{bmatrix}$$

$$z = 0$$
$$y + z = 0$$
$$y + 0 = 0$$
$$y = 0$$
$$x - 2y - z = 0$$
$$x - 2(0) - 0 = 0$$
$$x = 0$$
$$\{(0, 0, 0)\}$$

$$\begin{bmatrix} 1 & -1 & 3 & | & 18 \\ 2 & 5 & -1 & | & -6 \\ 2 & 3 & -2 & | & -12 \end{bmatrix} \quad \begin{matrix} -2R_1 + R_2 \to R_2 \\ R_2 - R_3 \to R_3 \end{matrix}$$

$$\begin{bmatrix} 1 & -1 & 3 & | & 18 \\ 0 & 7 & -7 & | & -42 \\ 0 & 2 & 1 & | & 6 \end{bmatrix} \quad \tfrac{1}{7}R_2 \to R_2$$

$$\begin{bmatrix} 1 & -1 & 3 & | & 18 \\ 0 & 1 & -1 & | & -6 \\ 0 & 2 & 1 & | & 6 \end{bmatrix} \quad -2R_2 + R_3 \to R_3$$

$$\begin{bmatrix} 1 & -1 & 3 & | & 18 \\ 0 & 1 & -1 & | & -6 \\ 0 & 0 & 3 & | & 18 \end{bmatrix} \quad \tfrac{1}{3}R_3 \to R_3$$

$$\begin{bmatrix} 1 & -1 & 3 & | & 18 \\ 0 & 1 & -1 & | & -6 \\ 0 & 0 & 1 & | & 6 \end{bmatrix}$$

$$z = 6$$
$$y - z = -6$$
$$y - 6 = -6$$
$$y = 0$$
$$x - y + 3z = 18$$
$$x - 0 + 3(6) = 18$$
$$x = 0$$
$$\{(0, 0, 6)\}$$

87. $$\begin{cases} 2x + 5y - z = -6 \\ x - y + 3z = 18 \\ 2x + 3y - 2z = -12 \end{cases}$$

$$\begin{bmatrix} 2 & 5 & -1 & | & -6 \\ 1 & -1 & 3 & | & 18 \\ 2 & 3 & -2 & | & -12 \end{bmatrix} \quad R_1 \leftrightarrow R_2$$

89. $$\begin{cases} x + 3y + z = 1 \\ 2x + y + 2z = 2 \\ 3x - y + 2z = 0 \end{cases}$$

$$\begin{bmatrix} 1 & 3 & 1 & | & 1 \\ 2 & 1 & 2 & | & 2 \\ 3 & -1 & 2 & | & 0 \end{bmatrix} \quad \begin{matrix} -2R_1 + R_2 \to R_2 \\ -3R_1 + R_3 \to R_3 \end{matrix}$$

$$\begin{bmatrix} 1 & 3 & 1 & | & 1 \\ 0 & -5 & 0 & | & 0 \\ 0 & -10 & -1 & | & -3 \end{bmatrix} \quad \begin{array}{l} -2R_2 + R_3 \to R_3 \\ -\dfrac{1}{5}R_2 \to R_2 \end{array}$$

$$\begin{bmatrix} 1 & 3 & 1 & | & 1 \\ 0 & 1 & 0 & | & 0 \\ 0 & 0 & -1 & | & -3 \end{bmatrix} \quad -1R_3 \to R_3$$

$$\begin{bmatrix} 1 & 3 & 1 & | & 1 \\ 0 & 1 & 0 & | & 0 \\ 0 & 0 & 1 & | & 3 \end{bmatrix}$$

$$z = 3$$
$$y = 0$$
$$x + 3y + z = 1$$
$$x + 3(0) + 3 = 1$$
$$x = -2$$
$$\{(-2, 0, 3)\}$$

91.
$$\begin{cases} x - y + 3z = 6 \\ 2x + y - z = -3 \\ 3x - 2y + z = -4 \end{cases}$$

$$\begin{bmatrix} 1 & -1 & 3 & | & 6 \\ 2 & 1 & -1 & | & -3 \\ 3 & -2 & 1 & | & -4 \end{bmatrix} \quad \begin{array}{l} -2R_1 + R_2 \to R_2 \\ -3R_1 + R_3 \to R_3 \end{array}$$

$$\begin{bmatrix} 1 & -1 & 3 & | & 6 \\ 0 & 3 & -7 & | & -15 \\ 0 & 1 & -8 & | & -22 \end{bmatrix} \quad R_2 \leftrightarrow R_3$$

$$\begin{bmatrix} 1 & -1 & 3 & | & 6 \\ 0 & 1 & -8 & | & -22 \\ 0 & 3 & -7 & | & -15 \end{bmatrix} \quad -3R_2 + R_3 \to R_3$$

$$\begin{bmatrix} 1 & -1 & 3 & | & 6 \\ 0 & 1 & -8 & | & -22 \\ 0 & 0 & 17 & | & 51 \end{bmatrix} \quad \dfrac{1}{17}R_3 \to R_3$$

$$\begin{bmatrix} 1 & -1 & 3 & | & 6 \\ 0 & 1 & -8 & | & -22 \\ 0 & 0 & 1 & | & 3 \end{bmatrix}$$

$$z = 3$$
$$y - 8z = -22$$
$$y - 8(3) = -22$$
$$y = 2$$
$$x - y + 3z = 6$$
$$x - 2 + 3(3) = 6$$
$$x = -1$$
$$\{(-1, 2, 3)\}$$

93. Factored

95. $x^3 + 8 = x^3 + 2^3$
$$= (x + 2)(x^2 - 2x + 4)$$

97. 2 terms;

$x + 2$ is not a factor.

99. 4 terms;

$$ax - bx + 2a - 2b = x(a - b) + 2(a - b)$$
$$= (a - b)(x + 2)$$

$x + 2$ is a factor.

101. $-2^{-4} = -\dfrac{1}{2^4} = -\dfrac{1}{16}$

103. $(-2)^{-3} = -\dfrac{1}{(-2)^3} = -\dfrac{1}{8}$

105.

$$\left(\frac{8x^{-6}y^9}{27y^3}\right)^{-\frac{1}{3}} = \frac{8^{-\frac{1}{3}}(x^{-6})^{-\frac{1}{3}}(y^9)^{-\frac{1}{3}}}{27^{-\frac{1}{3}}(y^3)^{-\frac{1}{3}}}$$

$$= \frac{27^{\frac{1}{3}}x^2y^{-3}}{8^{\frac{1}{3}}y^{-1}}$$

$$= \frac{3x^2y}{2y^3}$$

$$= \frac{3x^2}{2y^2}$$

107. will not reduce.

109.

$$\frac{1}{1-x} + \frac{2}{x^2-1} = 1$$

$$\frac{-1}{x-1} + \frac{2}{(x+1)(x-1)} = 1$$

$$(x+1)(x-1)\left[\frac{-1}{x-1} + \frac{2}{(x+1)(x-1)}\right] = (x+1)(x-1)(1)$$

$$-x - 1 + 2 = x^2 - 1$$

$$x^2 + x - 2 = 0$$

$$(x+2)(x-1) = 0$$

$$x = -2 \quad \text{or} \quad x = 1$$

$$\{-2, 1\}$$

111. Inequality;

$$|5 - x| < 3$$

$$-3 < 5 - x < 3$$

$$-8 < \quad -x < -2$$

$$8 > \quad x > 2$$
$$\{x \mid 2 < x < 8\}$$

113. Expression;

$$\sqrt[3]{8x^3 + 64x^6} = \sqrt[3]{8x^3(1 + 8x^3)}$$

$$= 2x\sqrt[3]{1 + 8x^3}$$

Chapter 7 Test

1. $\begin{cases} 3x + y = 1 \\ y = x - 1 \end{cases}$

$$3x + (x - 1) = 1$$

$$4x = 2$$

$$x = \frac{1}{2}$$

$$y = \frac{1}{2} - 1 = -\frac{1}{2}$$

$$\left\{\left(\frac{1}{2}, -\frac{1}{2}\right)\right\}$$

D

2. $\begin{cases} 3x + 2y = -1 \\ 3x - 4y = 11 \end{cases}$

$$\begin{array}{r} 3x + 2y = -1 \\ \underline{-3x + 4y = -11} \\ 6y = -12 \end{array}$$

$$y = -2$$

$$3x + 2(-2) = -1$$

$$3x = 3$$

$$x = 1$$

$$\{(1, -2)\}$$

B

3. $\begin{cases} x + y = -1 \\ y + z = 1 \\ x + z = 4 \end{cases}$

$$\begin{array}{r} x + y \quad = -1 \\ \underline{\quad - y - z = -1} \\ x \quad - z = -2 \end{array}$$

3. $x - z = -2$
   $\underline{x + z = 4}$
   $2x \quad\;\; = 2$

   $x \quad\;\; = 1$

   $1 - z = -2$

   $\quad -z = -3$

   $\quad\;\; z = 3$

   $y + 3 = 1$

   $\quad\;\; y = -2$

   $\{(1, \; -2, \; 3)\}$

   D

4. $\begin{cases} x + y \quad\quad = -1 \\ 2x + 4y + z = 1 \\ \quad\quad 2y + z = 1 \end{cases}$

   $2x + 4y + z = 1$
   $\underline{\quad\; - 2y - z = -1}$
   $2x + 2y \quad\quad = 0$

   $\quad 2x + 2y = 0$
   $\quad \underline{-2x - 2y = 2}$
   $\quad\quad\quad\quad 0 = 2$

   $\emptyset$

   B

5. $\begin{cases} 2x + 3y = 1 \\ 4x + 6y = 0 \end{cases}$

   $-4x - 6y = -2$
   $\underline{\;\; 4x + 6y = 0}$
   $\quad\quad\quad 0 = -2$

   $\emptyset$

   C

6. $x + 2y \le 2$

   B.L.: $x + 2y = 2$ (included)

   Test $(0, 0)$

   $0 + 2(0) \le 2$

   $\quad\quad 0 \le 2 \quad$ T

   Shade region.

   Test $(0, 4)$

   $0 + 2(4) \le 2$

   $\quad\quad 8 \le 2 \quad$ F

   Do not shade region.

   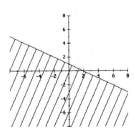

7. $2x - y > 0$

   B.L.: $2x - y = 0$ (not included)

   Test $(-1, 0)$

   $2(-1) - 0 > 0$

   $\quad\quad -2 > 0 \quad$ F

   Do not shade region.

   Test $(1, 0)$

   $2(1) - 0 > 0$

   $\quad\quad 2 > 0 \quad$ T

   Shade region.

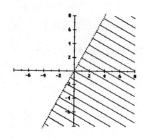

8. $\begin{cases} x - 2y \geq -2 \\ 2x - y \geq 0 \end{cases}$

B.L.: $x - 2y = -2$ (included)

B.L.: $2x - y = 0$ (included)

Test (1, 0)

$1 - 2(0) \geq -2$

$1 \geq -2$ T

$2(1) - 0 \geq 0$

$2 \geq 0$ T

Region in solution set.

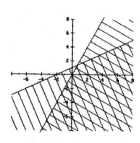

9. $\begin{cases} x + y \geq 1 \\ x < 1 \end{cases}$

B.L.: $x + y = 1$ (included)

B.L.: $x = 1$ (not included)

Test (0, 2)

$0 + 2 \geq 1$

$2 \geq 1$ T

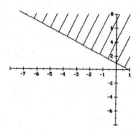

10. $\begin{cases} x + y = 3 \\ x^2 + y^2 = 17 \end{cases}$

$x = 3 - y$

So, $(3 - y)^2 + y^2 = 17$

$9 - 6y + y^2 + y^2 = 17$

$2y^2 - 6y - 8 = 0$

$y^2 - 3y - 4 = 0$

$(y - 4)(y + 1) = 0$

$y = 4$ or $y = -1$

$x = 3 - 4 = -1$ $x = 3 - (-1) = 4$

$\{(-1, 4), (4, -1)\}$

11. $\begin{cases} 3x - 5y = 6 \\ 6x + y = 1 \end{cases}$

$y = 1 - 6x$

$3x - 5(1 - 6x) = 6$

$3x - 5 + 30x = 6$

$33x = 11$

$x = \dfrac{1}{3}$

370

$$y = 1 - 6\left(\frac{1}{3}\right) = -1$$

$$\left\{\left(\frac{1}{3},\ -1\right)\right\}$$

12. $\begin{cases} 7x + 2y = 3 \\ 9x + 3y = 3 \end{cases}$

$$\begin{array}{l} -21x - 6y = -9 \\ \underline{18x + 6y = 6} \\ -3x \quad\quad = -3 \end{array}$$

$$x \quad = 1$$

$$7(1) + 2y = 3$$

$$2y = -4$$

$$y = -2$$

$$\{(1,\ -2)\}$$

13. $\begin{cases} y = 3x - 6 \\ 9x - 3y = 18 \end{cases}$

$$9x - 3(3x - 6) = 18$$

$$9x - 9x + 18 = 18$$

$$18 = 18$$

Dependent: $\{(x,\ y) \mid y = 3x - 6\}$

14. $\begin{cases} 3x + 2y \quad\quad = 1 \\ \quad\quad y - 3z = -10 \\ x + y + z = 3 \end{cases}$

$$\begin{array}{l} y - 3z = -10 \\ \underline{3x + 3y + 3z = 9} \\ 3x + 4y \quad\quad = -1 \end{array}$$

$$\begin{array}{l} 3x + 4y = -1 \\ \underline{-3x - 2y = -1} \\ \quad\quad 2y = -2 \end{array}$$

$$y = -1$$

$$-1 - 3z = -10$$

$$-3z = -9$$

$$z = 3$$

$$x - 1 + 3 = 3$$

$$x = 1$$

$$\{(1,\ -1,\ 3)\}$$

15. cost of lb of bananas: b
cost of lb of apples: a

$$3b + 2a = 297$$

$$1a + 5b = 264$$

$$a = 264 - 5b$$

$$3b + 2(264 - 5b) = 297$$

$$3b + 528 - 10b = 297$$

$$-7b = -231$$

$$b = 33$$

Bananas cost 33 cents per lb.

16. $\begin{vmatrix} 1 & 2 & 1 \\ 2 & 0 & -1 \\ 3 & 1 & 0 \end{vmatrix} = -2\begin{vmatrix} 2 & -1 \\ 3 & 0 \end{vmatrix} - 1\begin{vmatrix} 1 & 1 \\ 2 & -1 \end{vmatrix}$

$$= -2(0 + 3) - (-1 - 2)$$

$$= -6 + 3$$

$$= -3$$

17. $\begin{cases} 2x + 3y = 1 \\ 3x + 2y = 2 \end{cases}$

$$x = \frac{\begin{vmatrix} 1 & 3 \\ 2 & 2 \end{vmatrix}}{\begin{vmatrix} 2 & 3 \\ 3 & 2 \end{vmatrix}} = \frac{2-6}{4-9} = \frac{-4}{-5} = \frac{4}{5}$$

$$y = \frac{\begin{vmatrix} 2 & 1 \\ 3 & 2 \end{vmatrix}}{-5} = \frac{4-3}{-5} = \frac{1}{-5}$$

$$\left\{ \left(\frac{4}{5}, -\frac{1}{5} \right) \right\}$$

18. $\begin{cases} x + 2y + z = 0 \\ \quad\;\; -y + z = 1 \\ 2x + 3y \quad\;\; = 2 \end{cases}$

$$x = \frac{\begin{vmatrix} 0 & 2 & 1 \\ 1 & -1 & 1 \\ 2 & 3 & 0 \end{vmatrix}}{\begin{vmatrix} 1 & 2 & 1 \\ 0 & -1 & 1 \\ 2 & 3 & 0 \end{vmatrix}}$$

$$= \frac{-2\begin{vmatrix} 1 & 1 \\ 2 & 0 \end{vmatrix} + 1\begin{vmatrix} 1 & -1 \\ 2 & 3 \end{vmatrix}}{1\begin{vmatrix} -1 & 1 \\ 3 & 0 \end{vmatrix} + 2\begin{vmatrix} 2 & 1 \\ -1 & 1 \end{vmatrix}}$$

$$= \frac{-2(0-2) + (3+2)}{(0-3) + 2(2+1)}$$

$$= \frac{4+5}{-3+6}$$

$$= \frac{9}{3}$$

$$= 3$$

$$y = \frac{\begin{vmatrix} 1 & 0 & 1 \\ 0 & 1 & 1 \\ 2 & 2 & 0 \end{vmatrix}}{3}$$

$$= \frac{1\begin{vmatrix} 1 & 1 \\ 2 & 0 \end{vmatrix} + 1\begin{vmatrix} 0 & 1 \\ 2 & 2 \end{vmatrix}}{3}$$

$$= \frac{(0-2) + (0-2)}{3}$$

$$= -\frac{4}{3}$$

$$z = \frac{\begin{vmatrix} 1 & 2 & 0 \\ 0 & -1 & 1 \\ 2 & 3 & 2 \end{vmatrix}}{3}$$

$$= \frac{1\begin{vmatrix} -1 & 1 \\ 3 & 2 \end{vmatrix} - 2\begin{vmatrix} 0 & 1 \\ 2 & 2 \end{vmatrix}}{3}$$

$$= \frac{(-2-3) - 2(0-2)}{3}$$

$$= \frac{-5+4}{3}$$

$$= -\frac{1}{3}$$

$$\left\{ \left(3, -\frac{4}{3}, -\frac{1}{3} \right) \right\}$$

19. $\begin{cases} 2x + y = 1 \\ x - 7y = -2 \end{cases}$

$$\begin{bmatrix} 2 & 1 & | & 1 \\ 1 & -7 & | & -2 \end{bmatrix} \quad R_1 \leftrightarrow R_2$$

$$\begin{bmatrix} 1 & -7 & | & -2 \\ 2 & 1 & | & 1 \end{bmatrix} \quad -2R_1 + R_2 \rightarrow R_2$$

$$\begin{bmatrix} 1 & -7 & | & -2 \\ 0 & 15 & | & 5 \end{bmatrix} \quad \frac{1}{15}R_2 \rightarrow R_2$$

$$\begin{bmatrix} 1 & -7 & | & -2 \\ 0 & 1 & | & \frac{1}{3} \end{bmatrix}$$

$$y = \frac{1}{3}$$

$$x - 7y = -2$$

$$x - 7\left(\frac{1}{3}\right) = -2$$

$$x = \frac{1}{3}$$

$$\left\{ \left(\frac{1}{3}, \frac{1}{3} \right) \right\}$$

20. $\begin{cases} 2x \quad - z = 1 \\ x + y \quad = 2 \\ x + 3y - 4z = -2 \end{cases}$

$$\begin{bmatrix} 2 & 0 & -1 & | & 1 \\ 1 & 1 & 0 & | & 2 \\ 1 & 3 & -4 & | & -2 \end{bmatrix} \quad R_1 \leftrightarrow R_2$$

$$\begin{bmatrix} 1 & 1 & 0 & | & 2 \\ 2 & 0 & -1 & | & 1 \\ 1 & 3 & -4 & | & -2 \end{bmatrix} \quad \begin{array}{l} -2R_1 + R_2 \rightarrow R_2 \\ R_1 - R_3 \rightarrow R_3 \end{array}$$

$$\begin{bmatrix} 1 & 1 & 0 & | & 2 \\ 0 & -2 & -1 & | & -3 \\ 0 & -2 & 4 & | & 4 \end{bmatrix} \quad \begin{array}{l} R_3 - R_2 \rightarrow R_3 \\ -\frac{1}{2}R_2 \rightarrow R_2 \end{array}$$

$$\begin{bmatrix} 1 & 1 & 0 & | & 2 \\ 0 & 1 & \frac{1}{2} & | & \frac{3}{2} \\ 0 & 0 & 5 & | & 7 \end{bmatrix} \quad \frac{1}{5}R_3 \rightarrow R_3$$

$$\begin{bmatrix} 1 & 1 & 0 & | & 2 \\ 0 & 1 & \frac{1}{2} & | & \frac{3}{2} \\ 0 & 0 & 1 & | & \frac{7}{5} \end{bmatrix}$$

$$z = \frac{7}{5}$$

$$y + \frac{1}{2}z = \frac{3}{2}$$

$$y + \frac{1}{2}\left(\frac{7}{5}\right) = \frac{3}{2}$$

$$y = \frac{4}{5}$$

$$x + y = 2$$

$$x + \frac{4}{5} = 2$$

$$x = \frac{6}{5}$$

$$\left\{ \left(\frac{6}{5}, \frac{4}{5}, \frac{7}{5} \right) \right\}$$

CHAPTER 8

Problem Set 8.1

1. $f(x) = 2x;$ $g(x) = x - 5$

$(f + g)(x) = 2x + x - 5 = 3x - 5$

$(f - g)(x) = 2x - (x - 5)$

$\qquad = 2x - x + 5 = x + 5$

$(fg)(x) = 2x(x - 5) = 2x^2 - 10x$

$\left(\dfrac{f}{g}\right)(x) = \dfrac{2x}{x - 5}$

3. $f(x) = 3 - x^2;$ $g(x) = x + 1$

$(f + g)(x) = 3 - x^2 + x + 1$

$\qquad = -x^2 + x + 4$

$(f - g)(x) = 3 - x^2 - (x + 1)$

$\qquad = 3 - x^2 - x - 1$

$\qquad = -x^2 - x + 2$

$(fg)(x) = (3 - x^2)(x + 1)$

$\qquad = -x^3 - x^2 + 3x + 3$

$\left(\dfrac{f}{g}\right)(x) = \dfrac{3 - x^2}{x + 1}$

5. $f(x) = x - 1;$ $g(x) = 1 - x^2$

$(g - f)(0) = (1 - 0^2) - (0 - 1)$

$\qquad = 1 + 1$

$\qquad = 2$

7. $f(x) = x - 1;$ $g(x) = 1 - x^2$

$\left(\dfrac{f}{g}\right)(2) = \dfrac{2 - 1}{1 - 2^2}$

$\qquad = \dfrac{1}{-3}$

$\qquad = -\dfrac{1}{3}$

9. $f(x) = 5x + 8;$ $g(x) = 7x - 4$

$(f \circ g)(x) = f(g(x))$

$\qquad = f(7x - 4)$

$\qquad = 5(7x - 4) + 8$

$\qquad = 35x - 20 + 8$

$\qquad = 35x - 12$

$(g \circ f)(x) = g(f(x))$

$\qquad = g(5x + 8)$

$\qquad = 7(5x + 8) - 4$

$\qquad = 35x + 56 - 4$

$\qquad = 35x + 52$

11. $f(x) = x^2 + 2;$ $g(x) = x - 3$

$(f \circ g)(x) = f(g(x))$

$\qquad = f(x - 3)$

$\qquad = (x - 3)^2 + 2$

$\qquad = x^2 - 6x + 9 + 2$

$\qquad = x^2 - 6x + 11$

$(g \circ f)(x) = g(f(x))$

$\qquad = g(x^2 + 2)$

$\qquad = x^2 + 2 - 3$

$\qquad = x^2 - 1$

13. $f(x) = x - 1;$ $g(x) = 1 - x^2$

$(f \circ g)(-1) = f(g(-1))$

$\qquad = f[1 - (-1)^2]$

$\qquad = f(0)$

$\qquad = 0 - 1$

$\qquad = -1$

15. $f(x) = x - 1; \quad g(x) = 1 - x^2$

$(f \circ f)(1) = f(f(1))$

$\qquad\qquad = f(1 - 1)$

$\qquad\qquad = f(0)$

$\qquad\qquad = 0 - 1$

$\qquad\qquad = -1$

17. $f(x) = 3x; \quad g(x) = x + 8$

$(f + g)(x) = 3x + x + 8 = 4x + 8$

$(f - g)(x) = 3x - (x + 8)$

$\qquad\qquad = 3x - x - 8 = 2x - 8$

$(fg)(x) = 3x(x + 8) = 3x^2 + 24x$

$\left(\dfrac{f}{g}\right)(x) = \dfrac{3x}{x + 8}$

$(f \circ g)(x) = f(g(x))$

$\qquad\qquad = f(x + 8)$

$\qquad\qquad = 3(x + 8)$

$\qquad\qquad = 3x + 24$

19. $f(x) = 4x + 2; \quad g(x) = 3x - 9$

$(f + g)(x) = 4x + 2 + 3x - 9 = 7x - 7$

$(f - g)(x) = 4x + 2 - (3x - 9)$

$\qquad\qquad = 4x + 2 - 3x + 9$

$\qquad\qquad = x + 11$

$(fg)(x) = (4x + 2)(3x - 9) = 12x^2 - 30x - 18$

$\left(\dfrac{f}{g}\right)(x) = \dfrac{4x + 2}{3x - 9}$

$(f \circ g)(x) = f(g(x))$

$\qquad\qquad = f(3x - 9)$

$\qquad\qquad = 4(3x - 9) + 2$

$\qquad\qquad = 12x - 36 + 2$

$\qquad\qquad = 12x - 34$

21. $f(x) = x - 3; \quad g(x) = \dfrac{1}{3}x + 3$

$(f + g)(x) = x - 3 + \dfrac{1}{3}x + 3 = \dfrac{4}{3}x$

$(f - g)(x) = x - 3 - \left(\dfrac{1}{3}x + 3\right)$

$\qquad\qquad = x - 3 - \dfrac{1}{3}x - 3$

$\qquad\qquad = \dfrac{2}{3}x - 6$

$(fg)(x) = (x - 3)\left(\dfrac{1}{2}x + 3\right) = \dfrac{1}{3}x^2 + 2x - 9$

$\left(\dfrac{f}{g}\right)(x) = \dfrac{x - 3}{\dfrac{1}{3}x + 3} = \dfrac{3(x - 3)}{x + 9}$

$(f \circ g)(x) = f(g(x))$

$\qquad\qquad = f\left(\dfrac{1}{3}x + 3\right)$

$\qquad\qquad = \dfrac{1}{3}x + 3 - 3)$

$\qquad\qquad = \dfrac{1}{3}x$

23. $f(x) = 2x^2 + x + 3; \quad g(x) = 2x - 5$

$(f + g)(x) = 2x^2 + x + 3 + 2x - 5$

$\qquad\qquad = 2x^2 + 3x - 2$

$(f - g)(x) = 2x^2 + x + 3 - (2x - 5)$

$\qquad\qquad = 2x^2 + x + 3 - 2x + 5$

$\qquad\qquad = 2x^2 - x + 8$

$(fg)(x) = (2x^2 + x + 3)(2x - 5)$

$\qquad\qquad = 4x^3 + 2x^2 + 6x - 10x^2 - 5x - 15$

$\qquad\qquad = 4x^3 - 8x^2 + x - 15$

$\left(\dfrac{f}{g}\right)(x) = \dfrac{2x^2 + x + 3}{2x - 5}$

23. (con't.)

$(f \circ g)(x) = f(g(x))$

$\quad = f(2x - 5)$

$\quad = 2(2x - 5)^2 + 2x - 5 + 3$

$\quad = 2(4x^2 - 20x + 25) + 2x - 2$

$\quad = 8x^2 - 40x + 50 + 2x - 2)$

$\quad = 8x^2 - 38x + 48$

25. $f(x) = 1 - x^2; \qquad g(x) = x + 3$

$(f + g)(x) = 1 - x^2 + x + 3$

$\quad = -x^2 + x + 4$

$(f - g)(x) = 1 - x^2 - (x + 3)$

$\quad = 1 - x^2 - x - 3$

$\quad = -x^2 - x - 2$

$(fg)(x) = (1 - x^2)(x + 3)$

$\quad = -x^3 - 3x^2 + x + 3$

$\left(\dfrac{f}{g}\right)(x) = \dfrac{1 - x^2}{x + 3}$

$(f \circ g)(x) = f(g(x))$

$\quad = f(x + 3)$

$\quad = 1 - (x + 3)^2$

$\quad = 1 - (x^2 + 6x + 9)$

$\quad = 1 - x^2 - 6x - 9$

$\quad = -x^2 - 6x - 8$

27. $f(x) = x^2 - x - 2; \qquad g(x) = x^2 - 2x$

$(f + g)(x) = x^2 - x - 2 + x^2 - 2x$

$\quad = 2x^2 - 3x - 2$

$(f - g)(x) = x^2 - x - 2 - (x^2 - 2x)$

$\quad = x^2 - x - 2 - x^2 + 2x$

$\quad = x - 2$

$(fg)(x) = (x^2 - x - 2)(x^2 - 2x)$

$\quad = x^4 - x^3 - 2x^2 - 2x^3 + 2x^2 + 4x$

$\quad = x^4 - 3x^3 + 4x$

$\left(\dfrac{f}{g}\right)(x) = \dfrac{x^2 - x - 2}{x^2 - 2x}$

$\quad = \dfrac{(x - 2)(x + 1)}{x(x - 2)}$

$\quad = \dfrac{x + 1}{x}, \quad x \neq 2$

$(f \circ g)(x) = f(g(x))$

$\quad = f(x^2 - 2x)$

$\quad = (x^2 - 2x)^2 - (x^2 - 2x) - 2$

$\quad = x^4 - 4x^3 + 4x^2 - x^2 + 2x - 2$

$\quad = x^4 - 4x^3 + 3x^2 + 2x - 2$

29. $f(x) = 2x - 5; \qquad g(x) = 4 - x^2$

$(g - f)(0) = 4 - 0^2 - [2(0) - 5]$

$\quad = 4 + 5$

$\quad = 9$

31. $f(x) = 2x - 5; \qquad g(x) = 4 - x^2$

$\left(\dfrac{f}{g}\right)(1) = \dfrac{2(1) - 5}{4 - 1^2}$

$\quad = \dfrac{-3}{3}$

$\quad = -1$

33. $f(x) = 2x - 5; \qquad g(x) = 4 - x^2$

$(f \circ g)(-1) = f(g(-1))$

$\quad = f[4 - (-1)^2]$

$\quad = f(3)$

$\quad = 2(3) - 5$

$\quad = 1$

35. $f(x) = 2x - 5$

$$(f \circ f)(1) = f(f(1))$$
$$= f[2(1) - 5]$$
$$= f(-3)$$
$$= 2(-3) - 5$$
$$= -11$$

37. $f(x) = x + 3; \quad g(x) = x - 3$

$$(f \circ g)(x) = f(g(x))$$
$$= f(x - 3)$$
$$= x - 3 + 3$$
$$= x$$

$$(g \circ f)(x) = g(f(x))$$
$$= g(x + 3)$$
$$= x + 3 - 3$$
$$= x$$

39. $f(x) = x - 1; \quad g(x) = 1 - x^2$

$$\left(\frac{f}{g}\right)(x) = \frac{x - 1}{1 - x^2}$$

$$= \frac{x - 1}{(1 - x)(1 + x)}$$

$$= \frac{-(1 - x)}{(1 - x)(1 + x)}$$

$$= -\frac{1}{1 + x}, \quad x \neq 1$$

No

Problem Set 8.2

1. Yes; passes horizontal line test.

3. Yes; passes horizontal line test.

5. No; fails horizontal line test.

7. Yes; passes horizontal line test.

9. $f(x) = 3x - 1$

$$y = 3x - 1$$
$$x = 3y - 1$$
$$x + 1 = 3y$$
$$\frac{x + 1}{3} = y$$
$$f^{-1}(x) = \frac{x + 1}{3}$$

11. $f(x) = \frac{1}{2}x + 3$

$$y = \frac{1}{2}x + 3$$
$$x = \frac{1}{2}y + 3$$
$$x - 3 = \frac{1}{2}y$$
$$2(x - 3) = y$$
$$f^{-1}(x) = 2(x - 3)$$

13. $f(x) = x^3$

$$y = x^3$$
$$x = y^3$$
$$\sqrt[3]{x} = y$$
$$f^{-1}(x) = \sqrt[3]{x}$$

15. $f(x) = 3x + 1$

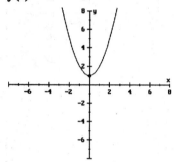

One-to-one

17. $f(x) = x^2 + 1$

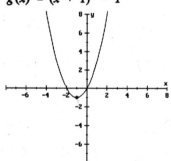

Not one-to-one

19. $g(x) = (x + 1)^2 - 1$

Not one-to-one

21. $f(x) = 5x - 1$

$y = 5x - 1$

$x = 5y - 1$

$x + 1 = 5y$

$\dfrac{x + 1}{5} = y$

$f^{-1}(x) = \dfrac{x + 1}{5}$

23. $f(x) = \dfrac{1}{3}x - 1$

$y = \dfrac{1}{3}x - 1$

$x = \dfrac{1}{3}y - 1$

$x + 1 = \dfrac{1}{3}y$

$3(x + 1) = y$

$f^{-1}(x) = 3(x + 1)$

25. $f(x) = x^3 + 2$

$y = x^3 + 2$

$x = y^3 + 2$

$x - 2 = y^3$

$\sqrt[3]{x - 2} = y$

$f^{-1} = \sqrt[3]{x - 2}$

27. $f(x) = 2x + 1; \quad f^{-1}(x) = \dfrac{x - 1}{2}$

$f(f^{-1}(x)) = f\left(\dfrac{x - 1}{2}\right)$

$\qquad = 2\left(\dfrac{x - 1}{2}\right) + 1$

$\qquad = x - 1 + 1$

$\qquad = x$

$f^{-1}(f(x)) = f^{-1}(2x + 1)$

$\qquad = \dfrac{2x + 1 - 1}{2}$

$\qquad = \dfrac{2x}{2}$

$\qquad = x$

378

29. $f(x) = x + 1;$ $f^{-1}(x) = x - 1$

$$f(f^{-1}(x)) = f(x - 1)$$
$$= x - 1 + 1$$
$$= x$$

$$f^{-1}(f(x)) = f^{-1}(x + 1)$$
$$= x + 1 - 1$$
$$= x$$

31. $f(x) = x + 2;$ $f^{-1}(x) = x - 2$

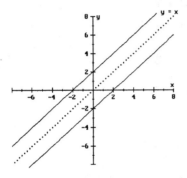

They are symmetric about the line $y = x$.

Problem Set 8.3

1. $3^{1.5} \approx 5.196$

3. $10^{5/3} \approx 46.416$

5. $f(x) = 3^x$

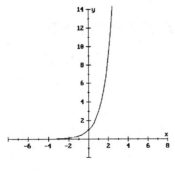

x	y
-2	1/9
-1	1/3
0	1
1	3
2	9

7. $h(x) = \left(\dfrac{1}{3}\right)^x$

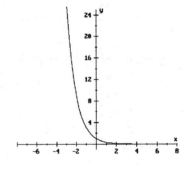

x	y
-2	9
-1	3
0	1
1	1/3
2	1/9

9. $5^x = 625$

$$5^x = 5^4$$
$$x = 4$$
$$\{4\}$$

11. $6^{2x + 1} = 36$

$$6^{2x + 1} = 6^2$$
$$2x + 1 = 2$$
$$2x = 1$$
$$x = \frac{1}{2}$$
$$\left\{\frac{1}{2}\right\}$$

13. $e^3 \approx 20.0855$

15. $e^{-4} \approx 0.0183$

17. $P = 14.7e^{-0.00004x}$
$$= 14.7e^{-0.00004(5280)}$$
$$= 11.9$$
$$11.9 \text{ lb/in}^2$$

19. $A = (10000)3^{0.3t}$

$= (10000)3^{0.3(2.5)}$

$= 22795.071$

22,795 bacteria

21. $3^{1/3} \approx 1.4422$

23. $2^{3.51} \approx 11.3924$

25. $5^{7/3} \approx 42.7494$

27. $e^2 \approx 7.3891$

29. $6^{\sqrt{3}} \approx 22.2740$

31. $e^{-2} \approx 0.1353$

33. $e^{\pi - 2} \approx 3.1318$

35. $f(x) = 4^x$

x	y
-2	1/16
-1	1/4
0	1
1	4
2	16

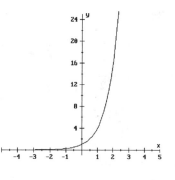

37. $h(x) = \left(\dfrac{1}{5}\right)^x$

x	y
-2	25
-1	5
0	1
1	1/5
2	1/25

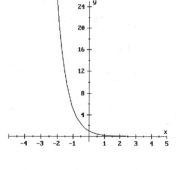

39. $p(x) = 5^{-x}$

x	y
-2	25
-1	5
0	1
1	1/5
2	1/25

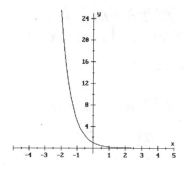

41. $2^x = 64$

$2^x = 2^6$

$x = 6$

$\{6\}$

43. $6^{x + 1} = 36$

$6^{x + 1} = 6^2$

$x + 1 = 2$

$x = 1$

$x = 1$

$\{1\}$

45. $3^{2x} = 27$

$3^{2x} = 3^3$

$2x = 3$

$x = \dfrac{3}{2}$

$\left\{\dfrac{3}{2}\right\}$

47. $2^x = \dfrac{1}{8}$

$2^x = 2^{-3}$

$x = -3$

$\{-3\}$

380

49. $27^x = 9$

$(3^3)^x = 3^2$

$3^{3x} = 3^2$

$3x = 2$

$x = \dfrac{2}{3}$

$\left\{\dfrac{2}{3}\right\}$

51. $4^{x-1} = 8^{x+1}$

$(2^2)^{x-1} = (2^3)^{x+1}$

$2^{2x-2} = 2^{3x+3}$

$2x - 2 = 3x + 3$

$-5 = x$

$\{-5\}$

53. $25^{2x-1} = 125^{3x+2}$

$(5^2)^{2x-1} = (5^3)^{3x+2}$

$5^{4x-2} = 5^{9x+6}$

$4x - 2 = 9x + 6$

$-8 = 5x$

$-\dfrac{8}{5} = x$

$\left\{-\dfrac{8}{5}\right\}$

55. $3^5 = 243$

$5^3 = 125$

3^5 is larger.

57. $3^{0.5} = 1.732$

$5^{0.3} = 1.621$

$3^{0.5}$ is larger.

59. $3^{\sqrt{5}} = 11.665$

$5^{\sqrt{3}} = 16.242$

$5^{\sqrt{3}}$ is larger.

61. $P = P_0 e^{0.05t}$

$= 50000 e^{0.05(30)}$

≈ 224084

224,084 people

63. $A(t) = A_0 \cdot 2^{-\frac{t}{5.2}}$

$= 100 \cdot 2^{-\frac{10}{5.2}}$

≈ 26.369

26.369 gm

65. $A(t) = A_0 \cdot 2^{-\frac{t}{5.2}}$

$= 2.5 \cdot 2^{-\frac{5}{5.2}}$

≈ 1.28

1.28 million curies remained in 1967.

67. $f(x) = 5^x$

a.)

x	y
1	5
2	25
3	125
4	625
5	3125

b.) $5^8 = 390,625$

By the end of the eighth week, over a quarter of a million people are involved.

69. $40,000

71. $162,000

73. $f(t)$ is 22%; $g(t)$ is 10.7%.

75. $3\frac{1}{2}$ yrs;   7yrs

77. $f(5) = 10000(1.22)^5 \approx \$27{,}000$

79. $f(10) = 10000(1.22)^{10} \approx \$73{,}000$

81. $f(x) = 2^x + 1$

x	y
-2	5/4
-1	3/2
0	2
1	3
2	5

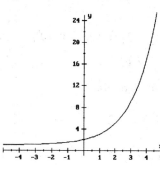

83. $f(x) = 2^{x+1}$

x	y
-2	1/2
-1	1
0	2
1	4
2	8

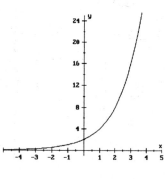

85. $f(x) = e^{x+1} + 2$

x	y
-2	2.4
-1	3
0	4.7
1	9.4
2	22.1

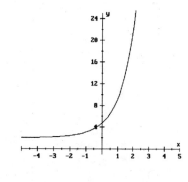

Problem Set 8.4

1. $\log_2 8 = 3$

    $2^3 = 8$

3. $\log_8 64 = 2$

    $8^2 = 64$

5. $2^3 = 8$

    $\log_2 8 = 3$

7. $\left(\frac{1}{3}\right)^2 = \frac{1}{9}$

    $\log_{1/3} \frac{1}{9} = 2$

9. $\log_4 x = 2$

    $4^2 = x$

    $16 = x$

    $\{16\}$

11. $\log_5 x = 3$

    $5^3 = x$

    $125 = x$

    $\{125\}$

13. $\log_5 125 = x$

    $5^x = 125$

    $5^x = 5^3$

    $x = 3$

    $\{3\}$

15. $\log_3(-9)$ undefined

17. $\log_3 3^2 = x$

    $3^x = 3^2$

    $x = 2$

    $\{2\}$

19. $\log_{100} 10000 = x$

$\qquad 100^x = 10000$

$\qquad 100^x = 100^2$

$\qquad x = 2$

$\qquad \{2\}$

21. $\log 0.01 = \log_{10} 0.01 = x$

$\qquad 10^x = 0.01$

$\qquad 10^x = 10^{-2}$

$\qquad x = -2$

$\qquad \{-2\}$

23. $\ln 1 = \log_e 1 = x$

$\qquad e^x = 1$

$\qquad e^x = e^0$

$\qquad x = 0$

$\qquad \{0\}$

25. $\log x = 3$

$\qquad \log_{10} x = 3$

$\qquad 10^3 = x$

$\qquad 1000 = x$

$\qquad \{1000\}$

27. $f(x) = \log_6 x$

$\qquad 6^y = x$

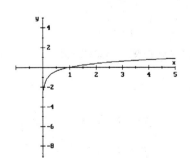

x	y
0.3	-0.6
1	0
3	0.6
6	1
18	1.6

29. $\beta = 10 \log \dfrac{I}{I_0}$

$\qquad = 10 \log \dfrac{10^{-11}}{10^{-12}}$

$\qquad = 10 \log 10$

$\qquad = 10(1)$

$\qquad = 10$

10 decibels

31. $\beta = 10 \log \dfrac{3.1 \times 10^{-6}}{10^{-12}}$

$\qquad = 10 \log (3.1 \times 10^6)$

$\qquad \approx 65$

65 decibels

33. $\log_5 x = 2$

$\qquad 5^2 = x$

$\qquad 25 = x$

$\qquad \{25\}$

35. $\log_3 27 = x$

$\qquad 3^x = 27$

$\qquad 3^x = 3^3$

$\qquad x = 3$

$\qquad \{3\}$

37. $\log_2 x = 4$

$\qquad 2^4 = x$

$\qquad 16 = x$

$\qquad \{16\}$

383

39. $\log_{1/2} x = -1$

$$\left(\frac{1}{2}\right)^{-1} = x$$

$$2 = x$$

$$\{2\}$$

41. $\log_5 5 = x$

$$5^x = 5^1$$

$$x = 1$$

$$\{1\}$$

43. $\log_3 9 = x$

$$3^x = 9$$

$$3^x = 3^2$$

$$x = 2$$

$$\{2\}$$

45. $\log_3 1 = x - 1$

$$3^{x-1} = 1$$

$$3^{x-1} = 3^0$$

$$x - 1 = 0$$

$$x = 1$$

$$\{1\}$$

47. $\log_{10}(x + 1) = 1$

$$10^1 = x + 1$$

$$9 = x$$

$$\{9\}$$

49. $\log_3 81 = x$

$$3^x = 81$$

$$3^x = 3^4$$

$$x = 4$$

$$\{4\}$$

51. $\log_{1/2} 4 = x$

$$\left(\frac{1}{2}\right)^x = 4$$

$$(2^{-1})^x = 2^2$$

$$2^{-x} = 2^2$$

$$-x = 2$$

$$x = -2$$

$$\{-2\}$$

53. $\log_5 5^2 = x$

$$5^x = 5^2$$

$$x = 2$$

$$\{2\}$$

55. $\log_{100} 10 = x$

$$100^x = 10$$

$$(10^2)^x = 10^1$$

$$10^{2x} = 10^1$$

$$2x = 1$$

$$x = \frac{1}{2}$$

$$\left\{\frac{1}{2}\right\}$$

57. $\log_2 x = 1$

$$2^1 = x$$

$$2 = x$$

$$\{2\}$$

59. $\log_{1/3} x = -1$

$$\left(\frac{1}{3}\right)^{-1} = x$$

$$3 = x$$

$$\{3\}$$

61. $\log_7 7 = x$

$\qquad 7^x = 7^1$

$\qquad x = 1$

$\qquad \{1\}$

63. $\log_{10}(x + 2) = 2$

$\qquad 10^2 = x + 2$

$\qquad 100 = x + 2$

$\qquad 98 = x$

$\qquad \{98\}$

65. $f(x) = \log_3 x$

$\qquad 3^y = x$

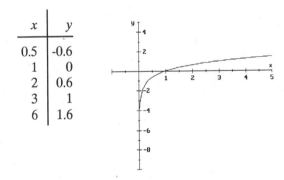

x	y
0.5	-0.6
1	0
2	0.6
3	1
6	1.6

67. $f(x) = \log_5 x$

$\qquad 5^y = x$

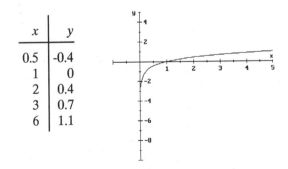

x	y
0.5	-0.4
1	0
2	0.4
3	0.7
6	1.1

69. $\log 15 \approx 1.1761$

71. $\log (3 + 5) = \log 8 \approx 0.9031$

73. $\log 32 - \log 2 \approx 1.5052 - 0.3010$

$\qquad\qquad\qquad \approx 1.2042$

75. $\log (0.8)^2 \approx -0.1938$

77. $2 \log 0.8 \approx -0.1938$

79. $\dfrac{1}{2} \log 2 \approx 0.1505$

81. $G = 10 \log \dfrac{P_{out}}{P_{in}}$

$\qquad = 10 \log \dfrac{50}{0.1}$

$\qquad = 10 \log 500$

$\qquad \approx 26.99$

$\qquad$ 26.99 decibels

83. $f(x) = \log_2 x$

$\qquad 2^y = x$

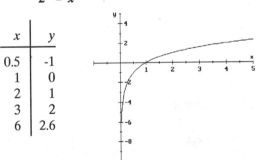

x	y
0.5	-1
1	0
2	1
3	2
6	2.6

85. $f(x) = \log_2(x - 1)$

$\qquad 2^y = x - 1$

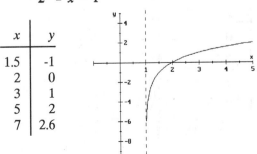

x	y
1.5	-1
2	0
3	1
5	2
7	2.6

87. $f(x) = \log_2 x - 1$

$y + 1 = \log_2 x$

$2^{y+1} = x$

x	y
0.5	-2
1	-1
2	0
3	0.6
6	1.6

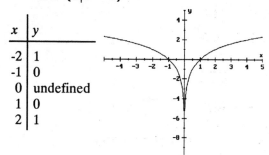

89. $f(x) = \log_2 |x|$

$2^y = |x|$

D: $\{x \mid x \neq 0\}$

x	y
-2	1
-1	0
0	undefined
1	0
2	1

91. $f(x) = 2 \log_2 x$

$\dfrac{y}{2} = \log_2 x$

$2^{y/2} = x$

D: $\{x \mid x > 0\}$

x	y
1	0
2	2
3	3.2
4	4
5	4.6

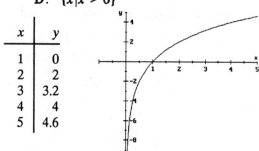

Problem Set 8.5

1. $\log_b 6 = \log_b(2 \cdot 3)$

$\quad\quad = \log_b 2 + \log_b 3$

$\quad\quad \approx 1.32 + 1.62$

$\quad\quad \approx 2.94$

3. $\log_b 2^2 = 2 \log_b 2$

$\quad\quad \approx 2(1.32)$

$\quad\quad \approx 2.64$

5. $\log_b \dfrac{1}{2} = \log_b 1 - \log_b 2$

$\quad\quad \approx 0 - 1.32$

$\quad\quad \approx -1.32$

7. $\log_b \dfrac{4}{3} = \log_b 4 - \log_b 3$

$\quad\quad = \log_b 2^2 - \log_b 3$

$\quad\quad = 2 \log_2 - \log_b 3$

$\quad\quad \approx 2(1.32) - 1.62$

$\quad\quad \approx 1.02$

9. $\log_3 x + \log_3 2 = 2$

$\quad\quad \log_3 2x = 2$

$\quad\quad\quad 3^2 = 2x$

$\quad\quad\quad 9 = 2x$

$\quad\quad \dfrac{9}{2} = x$

$\quad\quad \left\{\dfrac{9}{2}\right\}$

11. $\log x + \log 2 = 0$

$$\log_{10} 2x = 0$$

$$10^0 = 2x$$

$$1 = 2x$$

$$\frac{1}{2} = x$$

$$\left\{\frac{1}{2}\right\}$$

13. $\log_{16} x + \log_{16} 4 = 1$

$$\log_{16} 4x = 1$$

$$16^1 = 4x$$

$$4 = x$$

$$\{4\}$$

15. $\log_7(x + 1) - \log_7 49 = 1$

$$\log_7 \frac{x + 1}{49} = 1$$

$$7^1 = \frac{x + 1}{49}$$

$$343 = x + 1$$

$$342 = x$$

$$\{342\}$$

17. $\log_8(2x + 2) = \log_8 8$

$$2x + 2 = 8$$

$$2x = 6$$

$$x = 3$$

$$\{3\}$$

19. $\log x + \log (x - 3) = 1$

$$\log_{10}[x(x - 3)] = 1$$

$$10^1 = x(x - 3)$$

$$10 = x^2 - 3x$$

$$0 = x^2 - 3x - 10$$

$$0 = (x - 5)(x + 2)$$

$$x = 5 \quad \text{or} \quad x = -2$$

Since log (-2) is undefined, $x = 5$ is the only solution.

$$\{5\}$$

21. $\log_2 x^2 = 2$

$$2^2 = x^2$$

$$4 = x^2$$

$$\pm 2 = x$$

$$\{\pm 2\}$$

23. $7^x = 5$

$$\log 7^x = \log 5$$

$$x \log 7 = \log 5$$

$$x = \frac{\log 5}{\log 7}$$

$$x \approx 0.827$$

25.
$$2^x \cdot 3^{2x - 5} = 588$$

$$\log (2^x \cdot 3^{2x - 5}) = \log 588$$

$$\log 2^x + \log 3^{2x - 5} = \log 588$$

$$x \log 2 + (2x - 5) \log 3 = \log 588$$

$$x \log 2 + 2x \log 3 - 5 \log 3 = \log 588$$

$$x \log 2 + 2x \log 3 = 5 \log 3 + \log 588$$

$$x(\log 2 + 2 \log 3) = 5 \log 3 + \log 588$$

$$x = \frac{5 \log 3 + \log 588}{\log 2 + 2 \log 3}$$

$$x \approx 4.107$$

27. $A = P\left(1 + \dfrac{r}{n}\right)^{nt}$

    $20000 = 10000\left(1 + \dfrac{0.05}{1}\right)^{t}$

      $2 = 1.05^{t}$

    $\log 2 = \log 1.05^{t}$

    $\log 2 = t \log 1.05$

    $\dfrac{\log 2}{\log 1.05} = t$

      $t \approx 14.2$

    14.2 years

29. $\log_b 15 = \log_b(3 \cdot 5)$

        $= \log_b 3 + \log_b 5$

        $\approx 1.62 + 2.52$

        ≈ 4.14

31. $\log_b 5^2 = 2 \log_b 5$

        $\approx 2(2.52)$

        ≈ 5.04

33. $\log_b \dfrac{1}{3} = \log_b 1 - \log_b 3$

        $\approx 0 - 1.62$

        ≈ -1.62

35. $\log_b \dfrac{9}{5} = \log_b 9 - \log_b 5$

        $= \log_b 3^2 - \log_b 5$

        $= 2 \log_b 3 - \log_b 5$

        $\approx 2(1.62) - 2.52$

        ≈ 0.72

37. $\log_b 14 = \log_b(7 \cdot 2) = \log_b 7 + \log_b 2$

    T

39. F

41. F

43. F

45. $\log_2 5x = \log_2(x + 1)$

      $5x = x + 1$

      $4x = 1$

       $x = \dfrac{1}{4}$

      $\left\{\dfrac{1}{4}\right\}$

47. $\log_2 x - \log_2 3 = 4$

      $\log_2 \dfrac{x}{3} = 4$

       $2^4 = \dfrac{x}{3}$

    $3 \cdot 2^4 = x$

      $48 = x$

      $\{48\}$

49. $\log_{16} 2x + \log_{16} 2 = 2$

    $\log_{16}(2x \cdot 2) = 2$

      $\log_{16} 4x = 2$

       $16^2 = 4x$

       $\dfrac{16^2}{4} = x$

      $64 = x$

      $\{64\}$

51. $\log_2(x - 1) - \log_2 16 = 1$

$$\log_2 \frac{x - 1}{16} = 1$$

$$2^1 = \frac{x - 1}{16}$$

$$32 = x - 1$$

$$33 = x$$

$$\{33\}$$

53. $\ln (2x + 1) = \ln e$

$$2x + 1 = e$$

$$2x = e - 1$$

$$x = \frac{e - 1}{2}$$

$$\left\{\frac{e - 1}{2}\right\}$$

55. $\log_3 |x| = 1$

$$3 = |x|$$

$$x = 3 \quad \text{or} \quad x = -3$$

$$\{3, -3\}$$

57. $\log_2 x(x - 3) = 2$

$$2^2 = x(x - 3)$$

$$4 = x^2 - 3x$$

$$0 = x^2 - 3x - 4$$

$$0 = (x - 4)(x + 1)$$

$$x = 4 \quad \text{or} \quad x = -1$$

$$\{4, -1\}$$

59. $\log_5 x + \log_5 (x - 4) = 1$

$$\log_5 [x(x - 4)] = 1$$

$$5^1 = x(x - 4)$$

$$5 = x^2 - 4x$$

$$0 = x^2 - 4x - 5$$

$$0 = (x - 5)(x + 1)$$

$$x = 5 \quad \text{or} \quad x = -1$$

Since $\log_5 -1$ is undefined, $x = 5$ is the only solution.

$$\{5\}$$

61. $\qquad 5^{x + 1} = 6^x$

$$\log 5^{x + 1} = \log 6^x$$

$$(x + 1)\log 5 = x \log 6$$

$$x \log 5 + \log 5 = x \log 6$$

$$\log 5 = x \log 6 - x \log 5$$

$$\log 5 = x(\log 6 - \log 5)$$

$$\frac{\log 5}{\log 6 - \log 5} = x$$

$$x \approx 8.827$$

$$\{8.827\}$$

63. $\qquad 13^{z - 2} = 17^{z + 3}$

$$\log 13^{z - 2} = \log 17^{z + 3}$$

$$(z - 2)\log 13 = (z + 3)\log 17$$

$$z \log 13 - 2 \log 13 = z \log 17 + 3 \log 17$$

$$z \log 13 - z \log 17 = 2 \log 13 + 3 \log 17$$

$$z(\log 13 - \log 17) = 2 \log 13 + 3 \log 17$$

$$z = \frac{2 \log 13 + 3 \log 17}{\log 13 - \log 17}$$

$$z \approx -50.806$$

$$\{-50.806\}$$

65. $3e^{2s} = 1$

$$e^{2s} = \frac{1}{3}$$

$$\ln e^{2s} = \ln \frac{1}{3}$$

389

65. (con't.)

$$2s \ln e = \ln \frac{1}{3}$$

$$s = \frac{\ln \frac{1}{3}}{2}$$

$$s \approx -0.549$$

$$\{-0.549\}$$

67.

$$7^{12x} \cdot 2^{7x} = 6.3314 \times 10^{11}$$

$$\log (7^{12x} \cdot 2^{7x}) = \log (6.3314 \times 10^{11})$$

$$\log 7^{12x} + \log 2^{7x} = \log 6.3314 + \log 10^{11}$$

$$12x \log 7 + 7x \log 2 = \log 6.3314 + 11$$

$$x(12 \log 7 + 7 \log 2) = \log 6.3314 + 11$$

$$x = \frac{\log 6.3314 + 11}{12 \log 7 + 7 \log 2}$$

$$x \approx 0.964$$

$$\{0.964\}$$

69.

$$A = P\left(1 + \frac{r}{n}\right)^{nt}$$

$$2000 = 1000\left(1 + \frac{0.06}{4}\right)^{4t}$$

$$2 = 1.015^{4t}$$

$$\log 2 = \log 1.015^{4t}$$

$$\log 2 = 4t \log 1.015$$

$$\frac{\log 2}{4 \log 1.015} = t$$

$$t \approx 11.6$$

11.6 years

71.

$$A = P\left(1 + \frac{r}{n}\right)^{nt}$$

$$3000 = 1000\left(1 + \frac{0.075}{12}\right)^{12t}$$

$$3 = 1.00625^{12t}$$

$$\log 3 = \log 1.00625^{12t}$$

$$\log 3 = 12t \log 1.00625$$

$$\frac{\log 3}{12 \log 1.00625} = t$$

$$t \approx 14.7$$

14.7 years

73.

$$R(t) = M(t)$$

$$2.81\left(\frac{7}{8}\right)^t = 1.52\left(\frac{11}{12}\right)^t$$

$$\frac{2.81}{1.52} = \frac{\left(\frac{11}{12}\right)^t}{\left(\frac{7}{8}\right)^t}$$

$$1.85 = \left(\frac{\frac{11}{12}}{\frac{7}{8}}\right)^t$$

$$1.85 = \left(\frac{22}{21}\right)^t$$

$$\log 1.85 = \log \left(\frac{22}{21}\right)^t$$

$$\log 1.85 = t \log \frac{22}{21}$$

$$\frac{\log 1.85}{\log \frac{22}{21}} = t$$

$$t \approx 13.2$$

They will be equal in approximately 13.2 years.

$$R(13.2) = 2.81\left(\frac{7}{8}\right)^{13.2} \approx 0.4821838 \text{ million}$$

$$\approx 480000$$

Each will number 480,000.
Neither will ever equal zero using the given models.

1. $f(x) = 3x - 5$; $g(x) = x^2$

  $(f + g)(x) = 3x - 5 + x^2$

  $\qquad = x^2 + 3x - 5$

3. $f(x) = 3x - 5$; $g(x) = x^2$

  $(f - g)(x) = 3x - 5 - x^2$

  $\qquad = -x^2 + 3x - 5$

5. $f(x) = 3x - 5$; $g(x) = x^2$

  $(fg)(x) = (3x - 5)x^2$

  $\qquad = 3x^3 - 5x^2$

7. $f(x) = 3x - 5$; $g(x) = x^2$

  $(f + g)(3) = 3(3) - 5 + 3^2$

  $\qquad = 9 - 5 + 9$

  $\qquad = 13$

9. $f(x) = 3x - 5$; $g(x) = x^2$

  $\left(\dfrac{f}{g}\right)(-1) = \dfrac{3(-1) - 5}{(-1)^2}$

  $\qquad = -8$

11. $f(x) = 3x - 5$; $g(x) = x^2$

  $(f \circ g)\left(\dfrac{1}{3}\right) = f\left(g\left(\dfrac{1}{3}\right)\right)$

  $\qquad = f\left[\left(\dfrac{1}{3}\right)^2\right]$

  $\qquad = f\left(\dfrac{1}{9}\right)$

  $\qquad = 3\left(\dfrac{1}{9}\right) - 5$

$\qquad = \dfrac{1}{3} - 5$

$\qquad = -\dfrac{14}{3}$

13. $f(x) = 3x - 5$; $g(x) = x^2$

  $f(0) - g(0) = 3(0) - 5 - 0^2$

  $\qquad = -5$

15. $f(x) = 3x - 5$; $g(x) = x^2$

  $f\left(\dfrac{1}{3}\right)g\left(\dfrac{1}{3}\right) = \left[3\left(\dfrac{1}{3}\right) - 5\right]\left(\dfrac{1}{3}\right)^2$

  $\qquad = (-4)\left(\dfrac{1}{9}\right)$

  $\qquad = -\dfrac{4}{9}$

17. $f(x) = 1 - 2x$

  $y = 1 - 2x$

  $x = 1 - 2y$

  $x - 1 = -2y$

  $\dfrac{x - 1}{-2} = y$

  $\dfrac{1 - x}{2} = y$

  $f^{-1}(x) = \dfrac{1 - x}{2}$

19. $f(x) = x^3 - 7$

  $y = x^3 - 7$

  $x = y^3 - 7$

  $x + 7 = y^3$

  $\sqrt[3]{x + 7} = y$

  $f^{-1}(x) = \sqrt[3]{x + 7}$

21. $f(x) = 7^x$; $g(x) = \log_7 x$

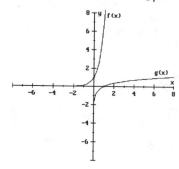

23. $f(x) = 10^x$; $g(x) = \log_{10} x$

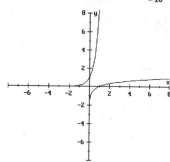

25. $2^x = 128$

$2^x = 2^7$

$x = 7$

{7}

27. $\log_{10} 10^x = 1$

$10^1 = 10^x$

$1 = x$

{1}

29. $\ln (x + e) - \ln e = 1$

$\ln (x + e) - 1 = 1$

$\ln (x + e) = 2$

$e^2 = x + e$

$e^2 - e = x$

$\{e^2 - e\}$

31. $3^{x+1} = 9$

$3^{x+1} = 3^2$

$x + 1 = 2$

$x = 1$

{1}

33. $(25)^x = 625$

$25^x = 25^2$

$x = 2$

{2}

35. $\log x + \log (x + 3) = 1$

$\log x(x + 3) = 1$

$10^1 = x(x + 3)$

$0 = x^2 + 3x - 10$

$0 = (x + 5)(x - 2)$

$x = -5$ or $x = 2$

Since log (−5) is undefined, the only solution is $x = 2$.

{2}

37. $\log (x + 14) = \log (2x + 7)$

$x + 14 = 2x + 7$

$7 = x$

{7}

39. $\log_2 x + \log_2(x - 1) = 1$

$\log_2 x(x - 1) = 1$

$2^1 = x(x - 1)$

$0 = x^2 - x - 2$

$0 = (x + 1)(x - 2)$

$x = -1$ or $x = 2$

Since $\log_2(-1)$ is undefined, the only solution is $x = 2$.

{2}

41. $\log(x + 3) - \log x = 1$

$$\log \frac{x + 3}{x} = 1$$

$$10^1 = \frac{x + 3}{x}$$

$$10x = x + 3$$

$$9x = 3$$

$$x = \frac{1}{3}$$

$$\left\{ \frac{1}{3} \right\}$$

43. $11^x = 5$

$$\log 11^x = \log 5$$

$$x \log 11 = \log 5$$

$$x = \frac{\log 5}{\log 11}$$

$$x \approx 0.671$$

$$\{0.671\}$$

45. $2^{x + 3} = 3^x$

$$\log 2^{x + 3} = \log 3^x$$

$$(x + 3)\log 2 = x \log 3$$

$$x \log 2 + 3 \log 2 = x \log 3$$

$$3 \log 2 = x \log 3 - x \log 2$$

$$3 \log 2 = x(\log 3 - \log 2)$$

$$\frac{3 \log 2}{\log 3 - \log 2} = x$$

$$x \approx 5.129$$

$$\{5.129\}$$

47. $e^{2x} = 5$

$$\ln e^{2x} = \ln 5$$

$$2x = \ln 5$$

$$x = \frac{\ln 5}{2}$$

$$x \approx 0.805$$

$$\{0.805\}$$

49. $9^{2t} = 2^{t - 3}$

$$\log 9^{2t} = \log 2^{t - 3}$$

$$2t \log 9 = (t - 3)\log 2$$

$$2t \log 9 = t \log 2 - 3 \log 2$$

$$2t \log 9 - t \log 2 = -3 \log 2$$

$$t(2 \log 9 - \log 2) = -3 \log 2$$

$$t = \frac{-3 \log 2}{2 \log 9 - \log 2}$$

$$t \approx -0.562$$

$$\{-0.562\}$$

51. $A = P\left(1 + \frac{r}{n}\right)^{nt}$

$$10000 = 5000\left(1 + \frac{0.065}{2}\right)^{2t}$$

$$2 = 1.0325^{2t}$$

$$\log 2 = \log 1.0325^{2t}$$

$$\log 2 = 2t \log 1.0325$$

$$\frac{\log 2}{2 \log 1.0325} = t$$

$$t \approx 10.8$$

10.8 years

393

53.

$$A = P\left(1 + \frac{r}{n}\right)^{nt}$$

$$20000 = 10000\left(1 + \frac{0.07}{12}\right)^{12t}$$

$$2 = 1.00583^{12t}$$

$$\log 2 = \log 1.00583^{12t}$$

$$\log 2 = 12t \log 1.00583$$

$$\frac{\log 2}{12 \log 1.00583} = t$$

$$t \approx 9.9$$

9.9 years

55. Factored

57. $4 - (a + b)^2 = 2^2 - (a + b)^2$

$$= [2 - (a + b)][2 + (a + b)]$$

$$= (2 - a - b)(2 + a + b)$$

59. $a^3b + c^2b = b(a^3 + c^2)$

61. 2 terms;

$$a^3 - b^3 = (a - b)(a^2 + ab + b^2)$$

$a - b$ is a factor.

63. 2 terms;

$$w(a - b) + y(b - a)$$

$$= w(a - b) - y(a - b)$$

$$= (a - b)(w - y)$$

$a - b$ is a factor.

65. $\sqrt[5]{64x^5 - 128y^5} = \sqrt[5]{32(2x^5 - 4y^5)}$

$$= 2\sqrt[5]{2x^5 - 4y^5}$$

67. $\sqrt[3]{(x - 3)^3} = x - 3$

69. $(2^{-1} + 4^{-1})^{-1} = \left(\frac{1}{2} + \frac{1}{4}\right)^{-1}$

$$= \left(\frac{3}{4}\right)^{-1}$$

$$= \frac{4}{3}$$

71. $\dfrac{-5x^{-3}}{y} = -\dfrac{5}{x^3y}$

73. $(3x - 1)^3 = (3x)^3 - 3(3x)^2 \cdot 1 + 3(3x) \cdot 1^2 - 1^3$

$$= 27x^3 - 27x^2 + 9x - 1$$

75. Will not reduce.

77. $\dfrac{3x}{x^2 + 6x + 9} + \dfrac{1}{x + 3}$

$$= \frac{3x}{(x + 3)^2} + \frac{1}{x + 3}$$

$$= \frac{3x}{(x + 3)^2} + \frac{1(x + 3)}{(x + 3)(x + 3)}$$

$$= \frac{3x + x + 3}{(x + 3)^2}$$

$$= \frac{4x + 3}{(x + 3)^2}$$

79. $\dfrac{\dfrac{1}{x + 3} - 1}{\dfrac{x + 2}{x^2 + 6x + 9}}$

$$= \frac{\dfrac{1}{x + 3} - 1}{\dfrac{x + 2}{(x + 3)^2}}$$

$$= \left(\frac{1}{x + 3} - 1\right) \cdot \frac{(x + 3)^2}{x + 2}$$

$$= \left(\frac{1}{x + 3}\right) \cdot \frac{(x + 3)^2}{x + 2} - \frac{(x + 3)^2}{x + 2}$$

79. (con't.)

$$= \frac{x + 3}{x + 2} - \frac{(x + 3)^2}{x + 2}$$

$$= \frac{x + 3}{x + 2} - \frac{x^2 + 6x + 9}{x + 2}$$

$$= \frac{x + 3 - (x^2 + 6x + 9)}{x + 2}$$

$$= \frac{x + 3 - x^2 - 6x - 9}{x + 2}$$

$$= \frac{-x^2 - 5x - 6}{x + 2}$$

$$= \frac{-(x^2 + 5x + 6)}{x + 2}$$

$$= \frac{-(x + 2)(x + 3)}{x + 2}$$

$$= -(x + 3)$$

81. Equation;

$$\frac{1}{x - 2} = \frac{4}{2 - x}$$

$$2 - x = 4(x - 2)$$

$$2 - x = 4x - 8$$

$$10 = 5x$$

$$2 = x$$

Since $x = 2$ causes a denominator to equal 0, there is no solution.

$$\emptyset$$

83. Expression;

$$-2^2 = -4$$

85. Inequality;

$$2x + 3(x - 7) \le 5$$

$$2x + 3x - 21 \le 5$$

$$5x \le 26$$

$$x \le \frac{26}{5}$$

$$\left(-\infty, \frac{26}{5}\right]$$

87. Equation;

$$|x + 11| = 0$$

$$x + 11 = 0$$

$$x = -11$$

$$\{-11\}$$

89. Expression;

$$\sqrt{48} - \sqrt{12} = 4\sqrt{3} - 2\sqrt{3}$$

$$= 2\sqrt{3}$$

91. Expression;

$$a^2(a^{-3}b)^3 = a^2 \cdot a^{-9}b^3$$

$$= a^{-7}b^3$$

$$= \frac{b^3}{a^7}$$

93. Expression;

$$ab^{-3}(ab^4 - b^5) = ab^{-3} \cdot ab^4 - ab^{-3} \cdot b^5$$

$$= a^2b - ab^2$$

95. Inequality;

$$x + 7 \ge 11 \quad \text{or} \quad x + 7 \le -11$$

$$x \ge 4 \quad \text{or} \quad x \le -18$$

$$(-\infty, -18] \cup [4, \infty)$$

1. $\log_a P = M$

   $a^M = P$

   B

2. $\log 43.4 \approx 1.63748973$

   B

3. $\log x = 0.8174$

   $10^{0.8174} = x$

   $x \approx 6.5674987$

   D

4. $e^{\sqrt{3}} \approx 5.65223367$

   A

5. **C**

6. $\log_5 125 = x$

   $5^x = 125$

   $5^x = 5^3$

   $x = 3$

   $\{3\}$

7. $\log_2 \dfrac{1}{64} = x$

   $2^x = \dfrac{1}{64}$

   $2^x = 2^{-6}$

   $x = -6$

   $\{-6\}$

8. $f(x) = \dfrac{2}{x + 3}; \quad g(x) = 2x - 1$

   $(f \circ g)(x) = f(g(x))$

   $\qquad = f(2x - 1)$

$= \dfrac{2}{2x - 1 + 3}$

$= \dfrac{2}{2x + 2}$

$= \dfrac{1}{x + 1}$

9. $\qquad f(x) = \dfrac{2}{x + 3}$

   $\qquad y = \dfrac{2}{x + 3}$

   $\qquad x = \dfrac{2}{y + 3}$

   $(y + 3)x = 2$

   $y + 3 = \dfrac{2}{x}$

   $y = \dfrac{2}{x} - 3$

   $f^{-1}(x) = \dfrac{2}{x} - 3$

10. $f(x) = \dfrac{2}{x + 3}; \quad g(x) = 2x - 1$

   $(g \circ f)(-2) = g(f(-2))$

   $\qquad = g\left(\dfrac{2}{-2 + 3}\right)$

   $\qquad = g(2)$

   $\qquad = 2(2) - 1$

   $\qquad = 3$

11. $\log_3 27 = x$

   $3^x = 27$

   $3^x = 3^3$

   $x = 3$

   $\{3\}$

12. $\log_2 x + \log_2(x + 3) = 2$

$$\log_2 x(x + 3) = 2$$

$$2^2 = x(x + 3)$$

$$0 = x^2 + 3x - 4$$

$$0 = (x + 4)(x - 1)$$

$$x = -4 \quad \text{or} \quad x = 1$$

Since $\log_2(-4)$ is undefined, the only solution is $x = 1$.

$\{1\}$

13. $2^{x-6} = 64$

$$2^{x-6} = 2^6$$

$$x - 6 = 6$$

$$x = 12$$

$\{12\}$

14.
$$2^{x-1} = 3^x$$

$$\log 2^{x-1} = \log 3^x$$

$$(x - 1)\log 2 = x \log 3$$

$$x \log 2 - \log 2 = x \log 3$$

$$-\log 2 = x \log 3 - x \log 2$$

$$-\log 2 = x(\log 3 - \log 2)$$

$$\frac{-\log 2}{\log 3 - \log 2} = x$$

$$x \approx -1.710$$

$\{-1.710\}$

15.
$$A = P\left(1 + \frac{r}{n}\right)^{nt}$$

$$70000 = 35000\left(1 + \frac{0.065}{1}\right)^t$$

$$2 = 1.065^t$$

$$\log 2 = \log 1.065^t$$

$$\log 2 = t \log 1.065$$

$$\frac{\log 2}{\log 1.065} = t$$

$$t \approx 11$$

11 years

CHAPTER 9

Problem Set 9.1

1. $x^2 + y^2 = 4$

   Center: $(0, 0)$
   radius: 2

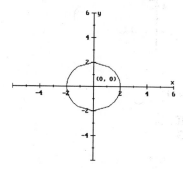

3. $x^2 + (y - 1)^2 = 9$

   Center: $(0, 1)$
   radius: 3

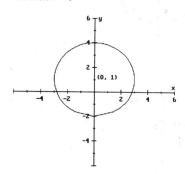

5. $(x + 1)^2 + (y - 2)^2 = 10$

   Center: $(-1, 2)$
   radius: $\sqrt{10}$

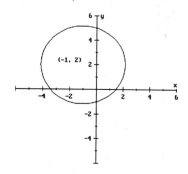

7. $x^2 + y^2 - 2x = 8$

   $x^2 - 2x + y^2 = 8$

   $x^2 - 2x + 1 + y^2 = 8 + 1$

   $(x - 1)^2 + y^2 = 9$

   Center: $(1, 0)$
   radius: 3

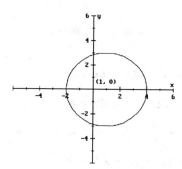

9. $x^2 + y^2 + 8y = 4$

   $x^2 + y^2 + 8y + 16 = 4 + 16$

   $x^2 + (y + 4)^2 = 20$

   Center: $(0, -4)$
   radius: $2\sqrt{5}$

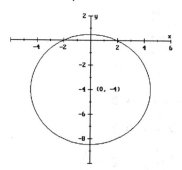

11. $x^2 + y^2 - 10x + 16 = 0$

   $x^2 - 10x + y^2 = -16$

   $x^2 - 10x + 25 + y^2 = -16 + 25$

   $(x - 5)^2 + y^2 = 9$

   Center: $(5, 0)$
   radius: 3

11. (con't.)

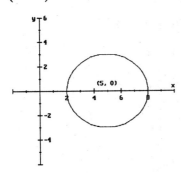

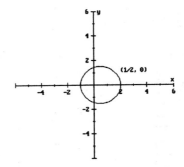

13. $$x^2 + y^2 - 2x - 4y = 4$$

$$x^2 - 2x + y^2 - 4y = 4$$

$$x^2 - 2x + 1 + y^2 - 4y + 4 = 4 + 1 + 4$$

$$(x - 1)^2 + (y - 2)^2 = 9$$

Center: (1, 2)
radius: 3

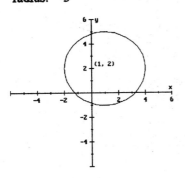

17. $x^2 + y^2 = 9$

Center: (0, 0)
radius: 3

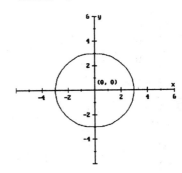

19. $x^2 + (y - 2)^2 = 16$

Center: (0, 2)
radius: 4

15. $x^2 + y^2 - x - 2 = 0$

$$x^2 - x + y^2 = 2$$

$$x^2 - x + \frac{1}{4} + y^2 = 2 + \frac{1}{4}$$

$$\left(x - \frac{1}{2}\right)^2 + y^2 = \frac{9}{4}$$

Center: $\left(\frac{1}{2}, 0\right)$

radius: $\frac{3}{2}$

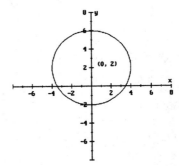

21. $(x + 3)^2 + (y + 2)^2 = 49$

Center: $(-3, -2)$
radius: 7

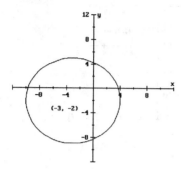

23. $x^2 + y^2 - 4x = 5$

$x^2 - 4x + y^2 = 5$

$x^2 - 4x + 4 + y^2 = 5 + 4$

$(x - 2)^2 + y^2 = 9$

Center: $(2, 0)$
radius: 3

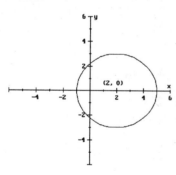

25. $x^2 + y^2 - 8y = 0$

$x^2 + y^2 - 8y + 16 = 0 + 16$

$x^2 + (y - 4)^2 = 16$

Center: $(0, 4)$
radius: 4

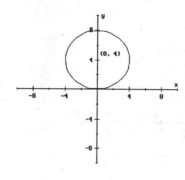

27. $x^2 + y^2 - 6y = 0$

$x^2 + y^2 - 6y + 9 = 0 + 9$

$x^2 + (y - 3)^2 = 9$

Center: $(0, 3)$
radius: 3

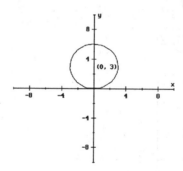

29. $x^2 + y^2 + 8x = -7$

$x^2 + 8x + y^2 = -7$

$x^2 + 8x + 16 + y^2 = -7 + 16$

$(x + 4)^2 + y^2 = 9$

Center: $(-4, 0)$
radius: 3

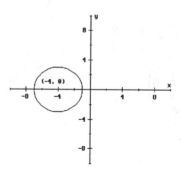

400

31. $x^2 + y^2 + 2y = 7$

$x^2 + y^2 + 2y + 1 = 7 + 1$

$x^2 + (y + 1)^2 = 8$

Center: $(0, -1)$
radius: $2\sqrt{2}$

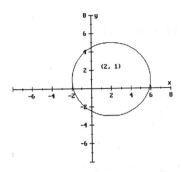

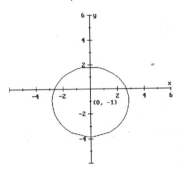

33. $x^2 + y^2 - 12x - 13 = 0$

$x^2 - 12x + y^2 = 13$

$x^2 - 12x + 36 + y^2 = 13 + 36$

$(x - 6)^2 + y^2 = 49$

Center: $(6, 0)$
radius: 7

37. $x^2 + y^2 - 4x + 4y - 1 = 0$

$x^2 - 4x + y^2 + 4y = 1$

$x^2 - 4x + 4 + y^2 + 4y + 4 = 1 + 4 + 4$

$(x - 2)^2 + (y + 2)^2 = 9$

Center: $(2, -2)$
radius: 3

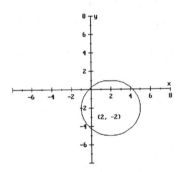

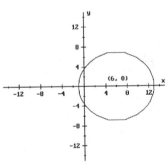

35. $x^2 + y^2 - 4x - 2y = 11$

$x^2 - 4x + y^2 - 2y = 11$

$x^2 - 4x + 4 + y^2 - 2y + 1 = 11 + 4 + 1$

$(x - 2)^2 + (y - 1)^2 = 16$

Center: $(2, 1)$
radius: 4

39. $x^2 + y^2 + 6x - 6y + 9 = 0$

$x^2 + 6x + y^2 - 6y = -9$

$x^2 + 6x + 9 + y^2 - 6y + 9 = -9 + 9 + 9$

$(x + 3)^2 + (y - 3)^2 = 9$

Center: $(-3, 3)$
radius: 3

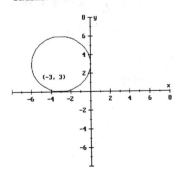

41. $(x + 2)^2 + (y + 3)^2 = 25$

Center: $(-2, -3)$
radius: 5

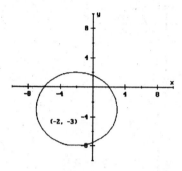

43. $x^2 + y^2 + 4x = -3$

$$x^2 + 4x + y^2 = -3$$

$$x^2 + 4x + 4 + y^2 = -3 + 4$$

$$(x + 2)^2 + y^2 = 1$$

Center: $(-2, 0)$
radius: 1

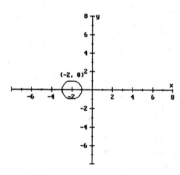

45. $x^2 + y^2 + 12x + 4y = 0$

$$x^2 + 12x + y^2 + 4y = 0$$

$$x^2 + 12x + 36 + y^2 + 4y + 4 = 0 + 36 + 4$$

$$(x + 6)^2 + (y + 2)^2 = 40$$

Center: $(-6, -2)$
radius: $\sqrt{40} = 2\sqrt{10}$

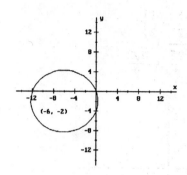

47. $x^2 + y^2 - 5x - y + 4 = 0$

$$x^2 - 5x + y^2 - y = -4$$

$$x^2 - 5x + \frac{25}{4} + y^2 - y + \frac{1}{4} = -4 + \frac{25}{4} + \frac{1}{4}$$

$$\left(x - \frac{5}{2}\right)^2 + \left(y - \frac{1}{2}\right)^2 = \frac{5}{2}$$

Center: $\left(\frac{5}{2}, \frac{1}{2}\right)$

radius: $\sqrt{\frac{5}{2}} = \frac{\sqrt{10}}{2}$

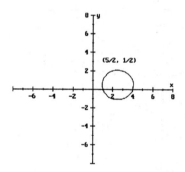

49. $x^2 + y^2 + 7x + y = 0$

$$x^2 + 7x + y^2 + y = 0$$

$$x^2 + 7x + \frac{49}{4} + y^2 + y + \frac{1}{4} = 0 + \frac{49}{4} + \frac{1}{4}$$

$$\left(x + \frac{7}{2}\right)^2 + \left(y + \frac{1}{2}\right)^2 = \frac{25}{2}$$

Center: $\left(-\dfrac{7}{2}, -\dfrac{1}{2}\right)$

radius: $\sqrt{\dfrac{25}{2}} = \dfrac{5\sqrt{2}}{2}$

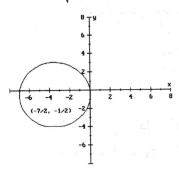

51. $x^2 + y^2 + 6x + 8y + 5 = 0$

$x^2 + 6x + y^2 + 8y = -5$

$x^2 + 6x + 9 + y^2 + 8y + 16 = -5 + 9 + 16$

$(x + 3)^2 + (y + 4)^2 = 20$

Center: $(-3, -4)$
radius: $\sqrt{20} = 2\sqrt{5}$

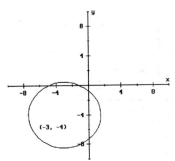

53. $x^2 + y^2 - y - 1 = 0$

$x^2 + y^2 - y = 1$

$x^2 + y^2 - y + \dfrac{1}{4} = 1 + \dfrac{1}{4}$

$x^2 + \left(y - \dfrac{1}{2}\right)^2 = \dfrac{5}{4}$

Center: $\left(0, \dfrac{1}{2}\right)$

radius: $\dfrac{\sqrt{5}}{2}$

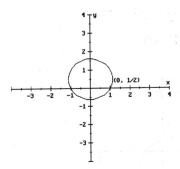

55. $x^2 + y^2 - x - 3y - 6 = 0$

$x^2 - x + y^2 - 3y = 6$

$x^2 - x + \dfrac{1}{4} + y^2 - 3y + \dfrac{9}{4} = 6 + \dfrac{1}{4} + \dfrac{9}{4}$

$\left(x - \dfrac{1}{2}\right)^2 + \left(y - \dfrac{3}{2}\right)^2 = \dfrac{17}{2}$

Center: $\left(\dfrac{1}{2}, \dfrac{3}{2}\right)$

radius: $\sqrt{\dfrac{17}{2}} = \dfrac{\sqrt{34}}{2}$

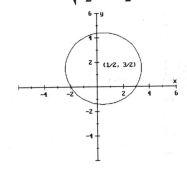

57. $x^2 + y^2 + x + 5y = 0$

$x^2 + x + y^2 + 5y = 0$

$x^2 + x + \dfrac{1}{4} + y^2 + 5y + \dfrac{25}{4} = 0 + \dfrac{1}{4} + \dfrac{25}{4}$

$\left(x + \dfrac{1}{2}\right)^2 + \left(y + \dfrac{5}{2}\right)^2 = \dfrac{13}{2}$

Center: $\left(-\dfrac{1}{2}, -\dfrac{5}{2}\right)$

radius: $\sqrt{\dfrac{13}{2}} = \dfrac{\sqrt{26}}{2}$

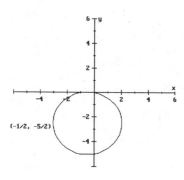

(-1/2, -5/2)

59. Consider a circle with center $C(h, k)$ and radius r. Let $P(x, y)$ be an arbitrary point on the circle.

$$d(C, P) = r$$

$$\sqrt{(x - h)^2 + (y - k)^2} = r$$

$$(x - h)^2 + (y - k)^2 = r^2$$

61. $2x^2 + 2y^2 - 4x - 2y + 1 = 0$

$$2x^2 - 4x + 2y^2 - 2y = -1$$

$$x^2 - 2x + y^2 - y = -\frac{1}{2}$$

$$x^2 - 2x + 1 + y^2 - y + \frac{1}{4} = -\frac{1}{2} + 1 + \frac{1}{4}$$

$$(x - 1)^2 + \left(y - \frac{1}{2}\right)^2 = \frac{3}{4}$$

Center: $\left(1, \dfrac{1}{2}\right)$

radius: $\dfrac{\sqrt{3}}{2}$

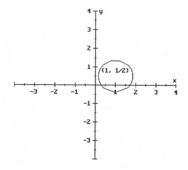

(1, 1/2)

63. $\quad 4 + 2x + 4y - x^2 - y^2 = 0$

$$x^2 - 2x + y^2 - 4y = 4$$

$$x^2 - 2x + 1 + y^2 - 4y + 4 = 4 + 1 + 4$$

$$(x - 1)^2 + (y - 2)^2 = 9$$

Center: (1, 2)
radius: 3

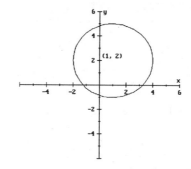

(1, 2)

Problem Set 9.2

1. $x = y^2 - 1$

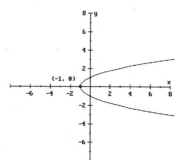

(-1, 0)

3. $x = -\dfrac{1}{2}y^2$

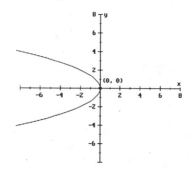

(0, 0)

404

5. $x = (y + 1)^2 + 2$

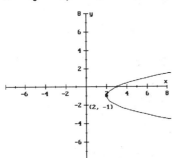

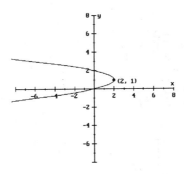

7. $x = -\dfrac{1}{2}(y + 1)^2$

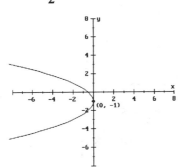

13. $x = (y - 1)^2$

Vertex: $(0, 1)$

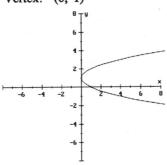

9. $x = y^2 - 3y + 2$

$x = \left(y^2 - 3y + \dfrac{9}{4}\right) + 2 - \dfrac{9}{4}$

$x = \left(y - \dfrac{3}{2}\right)^2 - \dfrac{1}{4}$

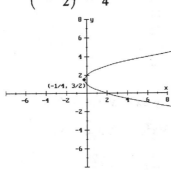

15. $y = (x - 3)^2 + 2$

Vertex: $(3, 2)$

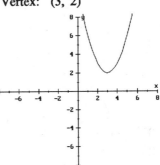

11. $x = -2y^2 + 4y$

$x = -2(y^2 - 2y)$

$x = -2(y^2 - 2y + 1) + 2$

$x = -2(y - 1)^2 + 2$

17. $x = (y + 3)^2 + 3$

Vertex: $(3, -3)$

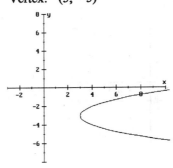

405

19. $y = \dfrac{1}{2}\left(x + \dfrac{2}{3}\right)^2 + \dfrac{1}{3}$

Vertex: $\left(-\dfrac{2}{3}, \dfrac{1}{3}\right)$

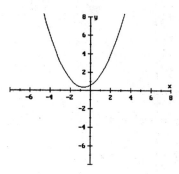

21. $y = -(x + 3)^2 + 3$

Vertex: $(-3, 3)$

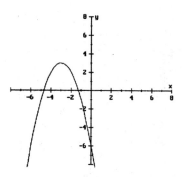

23. $x = \dfrac{1}{3}\left(y + \dfrac{3}{2}\right)^2 - \dfrac{1}{2}$

Vertex: $\left(-\dfrac{1}{2}, -\dfrac{3}{2}\right)$

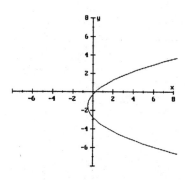

25. $y^2 - 2x - 4 = 0$

$y^2 - 4 = 2x$

$\dfrac{1}{2}y^2 - 2 = x$

Vertex: $(-2, 0)$

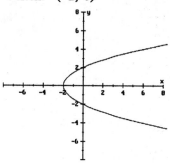

27. $x = -2y^2 - 8y - 8$

$x = -2(y^2 + 4y) - 8$

$x = -2(y^2 + 4y + 4) - 8 + 8$

$x = -2(y + 2)^2$

Vertex: $(0, -2)$

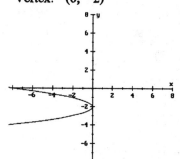

29. $\qquad 3x^2 - 9x - y + 6 = 0$

$3(x^2 - 3x) + 6 = y$

$3\left(x^2 - 3x + \dfrac{9}{4}\right) + 6 - \dfrac{27}{4} = y$

$3\left(x - \dfrac{3}{2}\right)^2 - \dfrac{3}{4} = y$

Vertex: $\left(\dfrac{3}{2}, -\dfrac{3}{4}\right)$

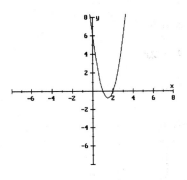

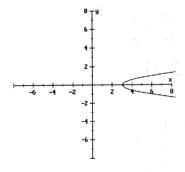

31. $y^2 - x = 0$

$y^2 = x$

Vertex: $(0, 0)$

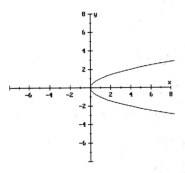

33. $x^2 - 2 = y$

Vertex: $(0, -2)$

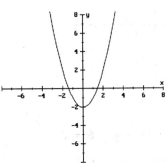

35. $3y^2 + 3 = x$

Vertex: $(3, 0)$

37. $x^2 + 2y + 4 = 0$

$2y = -x^2 - 4$

$y = -\frac{1}{2}x^2 - 2$

Vertex: $(0, -2)$

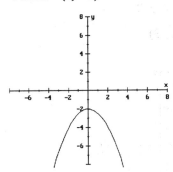

39. $y^2 - 8y = x - 10$

$(y^2 - 8y + 16) + 10 - 16 = x$

$(y - 4)^2 - 6 = x$

Vertex: $(-6, 4)$

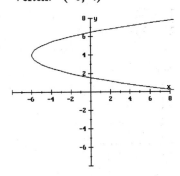

41.
$$x^2 - x + 2 = y$$

$$\left(x^2 - x + \frac{1}{4}\right) + 2 - \frac{1}{4} = y$$

$$\left(x - \frac{1}{2}\right)^2 + \frac{7}{4} = y$$

Vertex: $\left(\dfrac{1}{2}, \dfrac{7}{4}\right)$

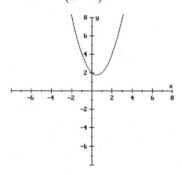

43. $x = -2y^2 + 8y$

$x = -2(y^2 + 4y)$

$x = -2(y^2 + 4y + 4) + 8$

$x = -2(y + 2)^2 + 8$

Vertex: $(8, -2)$

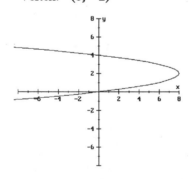

45. $y = -2x^2 + 4x - 2$

$y = -2(x^2 - 2x) - 2$

$y = -2(x^2 - 2x + 1) - 2 + 2$

$y = -2(x - 1)^2$

Vertex: $(1, 0)$

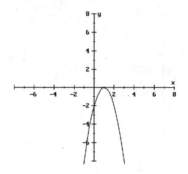

47.
$$3y^2 + 9y - x + 6 = 0$$

$$3(y^2 + 3y) + 6 = x$$

$$3\left(y^2 + 3y + \frac{9}{4}\right) + 6 - \frac{27}{4} = x$$

$$3\left(y + \frac{3}{2}\right)^2 - \frac{3}{4} = x$$

Vertex: $\left(-\dfrac{3}{4}, -\dfrac{3}{2}\right)$

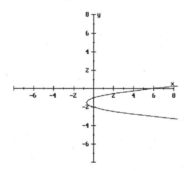

49.
$$2y^2 - x - 3y + 1 = 0$$

$$(2y^2 - 3y) + 1 = x$$

$$2\left(y^2 - \frac{3}{2}y\right) + 1 = x$$

$$2\left(y^2 - \frac{3}{2}y + \frac{9}{16}\right) + 1 - \frac{9}{8} = x$$

$$2\left(y - \frac{3}{4}\right)^2 - \frac{1}{8} = x$$

Vertex: $\left(-\dfrac{1}{8}, \dfrac{3}{4}\right)$

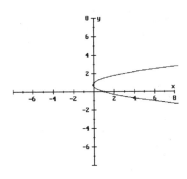

$$y = -2\left(x + \frac{3}{4}\right)^2 - \frac{7}{8}$$

Vertex: $\left(-\frac{3}{4}, -\frac{7}{8}\right)$

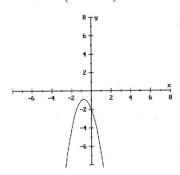

51. $3x^2 - 4x - 2y = 0$

$$3x^2 - 4x = 2y$$

$$3\left(x^2 - \frac{4}{3}x\right) = 2y$$

$$\frac{3}{2}\left(x^2 - \frac{4}{3}x\right) = y$$

$$\frac{3}{2}\left(x^2 - \frac{4}{3}x + \frac{4}{9}\right) - \frac{2}{3} = y$$

$$\frac{3}{2}\left(x - \frac{2}{3}\right)^2 - \frac{2}{3} = y$$

Vertex: $\left(\frac{2}{3}, -\frac{2}{3}\right)$

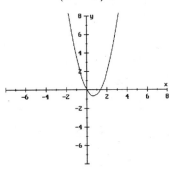

55. $4y^2 - 5y - 2x = 0$

$$4y^2 - 5y = 2x$$

$$4\left(y^2 - \frac{5}{4}y\right) = 2x$$

$$2\left(y^2 - \frac{5}{4}y\right) = x$$

$$2\left(y^2 - \frac{5}{4}y + \frac{25}{64}\right) - \frac{25}{32} = x$$

$$2\left(y - \frac{5}{8}\right)^2 - \frac{25}{32} = x$$

Vertex: $\left(-\frac{25}{32}, \frac{5}{8}\right)$

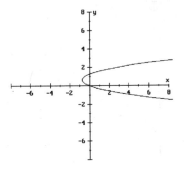

53. $2x^2 + y + 3x + 2 = 0$

$$y = -2x^2 - 3x - 2$$

$$y = -2\left(x^2 + \frac{3}{2}x\right) - 2$$

$$y = -2\left(x^2 + \frac{3}{2}x + \frac{9}{16}\right) - 2 + \frac{9}{8}$$

Problem Set 9.3

1. $\dfrac{x^2}{36} + \dfrac{y^2}{16} = 1$

   $\dfrac{x^2}{6^2} + \dfrac{y^2}{4^2} = 1$

   $a = 6, \quad b = 4$

   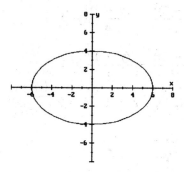

3. $x^2 + \dfrac{y^2}{9} = 1$

   $\dfrac{x^2}{1^2} + \dfrac{y^2}{3^2} = 1$

   $a = 1, \quad b = 3$

   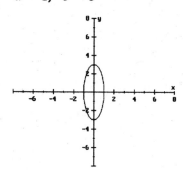

5. $\dfrac{x^2}{16} - \dfrac{y^2}{4} = 1$

   $\dfrac{x^2}{4^2} - \dfrac{y^2}{2^2} = 1$

   $a = 4, \quad b = 2$

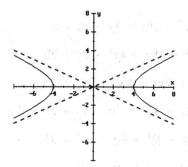

7. $\dfrac{y^2}{4} - x^2 = 1$

   $\dfrac{y^2}{2^2} - \dfrac{x^2}{1^2} = 1$

   $a = 1, \quad b = 2$

   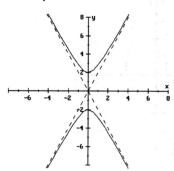

9. $\dfrac{(x - 1)^2}{36} + \dfrac{(y - 2)^2}{49} = 1$

   $\dfrac{(x - 1)^2}{6^2} + \dfrac{(y - 2)^2}{7^2} = 1$

   Center: $(1, 2); \quad a = 6, \quad b = 7$

   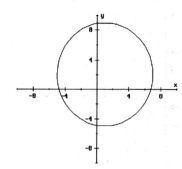

410

11. $9x^2 + 4y^2 - 18x - 8y - 23 = 0$

$$9x^2 - 18x + 4y^2 - 8y = 23$$

$$9(x^2 - 2x) + 4(y^2 - 2y) = 23$$

$$9(x^2 - 2x + 1) + 4(y^2 - 2y + 1) = 23 + 9 + 4$$

$$9(x - 1)^2 + 4(y - 1)^2 = 36$$

$$\frac{(x - 1)^2}{4} + \frac{(y - 1)^2}{9} = 1$$

$$\frac{(x - 1)^2}{2^2} + \frac{(y - 1)^2}{3^2} = 1$$

Center: $(1, 1)$; $a = 2$, $b = 3$

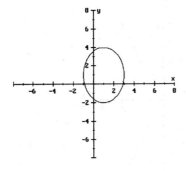

15. $4x^2 - y^2 - 16x + 2y + 11 = 0$

$$(4x^2 - 16x) + (-y^2 + 2y) = -11$$

$$4(x^2 - 4x) - (y^2 - 2y) = -11$$

$$4(x^2 - 4x + 4) - (y^2 - 2y + 1) = -11 + 16 - 1$$

$$4(x - 2)^2 - (y - 1)^2 = 4$$

$$\frac{(x - 2)^2}{1} - \frac{(y - 1)^2}{4} = 1$$

$$\frac{(x - 2)^2}{1^2} - \frac{(y - 1)^2}{2^2} = 1$$

Center: $(2, 1)$; $a = 1$, $b = 2$

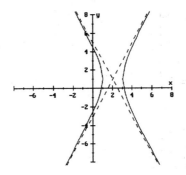

13. $\dfrac{(y - 1)^2}{9} - \dfrac{(x + 1)^2}{4} = 1$

$$\frac{(y - 1)^2}{3^2} + \frac{(x + 1)^2}{2^2} = 1$$

Center: $(-1, 1)$; $a = 2$, $b = 3$

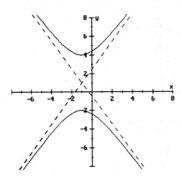

17. $\dfrac{x^2}{36} - \dfrac{y^2}{16} = 1$

$$\frac{x^2}{6^2} - \frac{y^2}{4^2} = 1$$

$a = 6$, $b = 4$

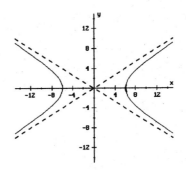

411

19. $\dfrac{y^2}{9} + \dfrac{x^2}{16} = 1$

$\dfrac{y^2}{3^2} + \dfrac{x^2}{4^2} = 1$

$a = 4, \quad b = 3$

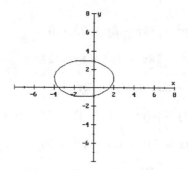

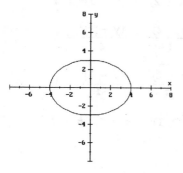

21. $\dfrac{x^2}{4} + y^2 = 1$

$\dfrac{x^2}{2^2} + \dfrac{y^2}{1^2} = 1$

$a = 2, \quad b = 1$

25. $\dfrac{16(x-1)^2}{25} - \dfrac{9(y-1)^2}{4} = 1$

$\dfrac{(x-1)^2}{\dfrac{25}{16}} - \dfrac{(y-1)^2}{\dfrac{4}{9}} = 1$

$\dfrac{(x-1)^2}{\left(\dfrac{5}{4}\right)^2} - \dfrac{(y-1)^2}{\left(\dfrac{2}{3}\right)^2} = 1$

Center: $(1, 1); \quad a = \dfrac{5}{4}, \quad b = \dfrac{2}{3}$

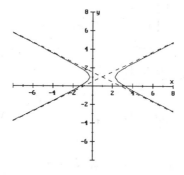

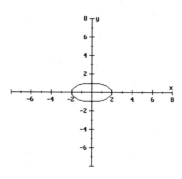

23. $\dfrac{(x+1)^2}{9} + \dfrac{(y-1)^2}{4} = 1$

$\dfrac{(x+1)^2}{3^2} + \dfrac{(y-1)^2}{2^2} = 1$

Center: $(-1, 1); \quad a = 3, \quad b = 2$

27. $2x^2 + y^2 = 8$

$\dfrac{x^2}{4} + \dfrac{y^2}{8} = 1$

$\dfrac{x^2}{2^2} + \dfrac{y^2}{(2\sqrt{2})^2} = 1$

$a = 2, \quad b = 2\sqrt{2}$

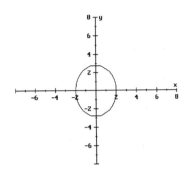

$$\frac{(x+4)^2}{3^2} + \frac{(y-2)^2}{2^2} = 1$$

Center: $(-4, 2)$; $a = 3$, $b = 2$

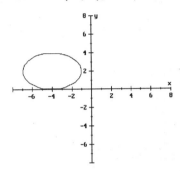

29. $x^2 - 4y^2 + 16y = 0$

$x^2 - 4(y^2 - 4y) = 0$

$x^2 - 4(y^2 - 4y + 4) = 0 - 16$

$x^2 - 4(y - 2)^2 = -16$

$-\frac{x^2}{16} + \frac{(y-2)^2}{4} = 1$

$\frac{(y-2)^2}{2^2} - \frac{x^2}{4^2} = 1$

Center: $(0, 2)$; $a = 4$, $b = 2$

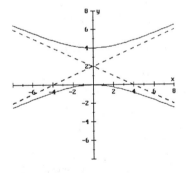

33. $16x^2 + y^2 + 64x - 8y + 64 = 0$

$16x^2 + 64x + y^2 - 8y = -64$

$16(x^2 + 4x) + (y^2 - 8y) = -64$

$16(x^2 + 4x + 4) + (y^2 - 8y + 16) = -64 + 64 + 16$

$16(x + 2)^2 + (y - 4)^2 = 16$

$\frac{(x+2)^2}{1} + \frac{(y-4)^2}{16} = 1$

$\frac{(x+2)^2}{1^2} + \frac{(y-4)^2}{4^2} = 1$

Center: $(-2, 4)$; $a = 1$, $b = 4$

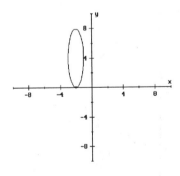

31. $4x^2 + 9y^2 + 32x - 36y + 64 = 0$

$4x^2 + 32x + 9y^2 - 36y = -64$

$4(x^2 + 8x) + 9(y^2 - 4y) = -64$

$4(x^2 + 8x + 16) + 9(y^2 - 4y + 4) = -64 + 64 + 36$

$4(x + 4)^2 + 9(y - 2)^2 = 36$

$\frac{(x+4)^2}{9} + \frac{(y-2)^2}{4} = 1$

35. $9x^2 - 36y^2 + 72y = 0$

$9x^2 - 36(y^2 - 2y) = 0$

$9x^2 - 36(y^2 - 2y + 1) = 0 - 36$

413

$$9x^2 - 36(y - 1)^2 = -36$$

$$-\frac{x^2}{4} + \frac{(y - 1)^2}{1} = 1$$

$$\frac{(y - 1)^2}{1^2} - \frac{x^2}{2^2} = 1$$

Center: $(0, 1)$;　$a = 2$,　$b = 1$

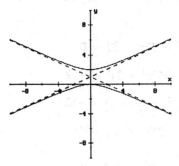

$$x^2 - 4(y - 2)^2 = 16$$

$$\frac{x^2}{16} - \frac{(y - 2)^2}{4} = 1$$

$$\frac{x^2}{4^2} - \frac{(y - 2)^2}{2^2} = 1$$

Center: $(0, 2)$;　$a = 4$,　$b = 2$

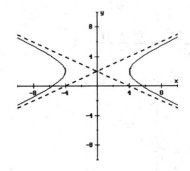

37.　$x^2 - y^2 - 2x + 2y + 4 = 0$

$$(x^2 - 2x) - (y^2 - 2y) = -4$$

$$(x^2 - 2x + 1) - (y^2 - 2y + 1) = -4 + 1 - 1$$

$$(x - 1)^2 - (y - 1)^2 = -4$$

$$\frac{-(x - 1)^2}{4} + \frac{(y - 1)^2}{4} = 1$$

$$\frac{(y - 1)^2}{2^2} - \frac{(x - 1)^2}{2^2} = 1$$

Center: $(1, 1)$;　$a = 2$,　$b = 2$

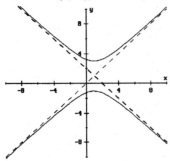

39.　$x^2 - 4y^2 + 16y - 32 = 0$

$$x^2 - 4(y^2 - 4y) = 32$$

$$x^2 - 4(y^2 - 4y + 4) = 32 - 16$$

41.　$25x^2 - 9y^2 + 50x - 200 = 0$

$$25x^2 + 50x - 9y^2 = 200$$

$$25(x^2 + 2x) - 9y^2 = 200$$

$$25(x^2 + 2x + 1) - 9y^2 = 200 + 25$$

$$25(x + 1)^2 - 9y^2 = 225$$

$$\frac{(x + 1)^2}{9} - \frac{y^2}{25} = 1$$

$$\frac{(x + 1)^2}{3^2} - \frac{y^2}{5^2} = 1$$

Center: $(-1, 0)$;　$a = 3$,　$b = 5$

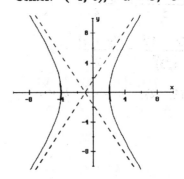

414

43. $x^2 - y^2 + 6x + 4y + 14 = 0$

$$x^2 + 6x - (y^2 - 4y) = -14$$

$$(x^2 + 6x + 9) - (y^2 - 4y + 4) = -14 + 9 - 4$$

$$(x + 3)^2 - (y - 2)^2 = -9$$

$$\frac{-(x + 3)^2}{9} + \frac{(y - 2)^2}{9} = 1$$

$$\frac{(y - 2)^2}{3^2} - \frac{(x + 3)^2}{3^2} = 1$$

Center: $(-3, 2)$; $a = 3$, $b = 3$

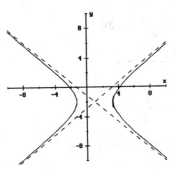

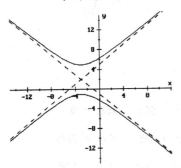

45. $4x^2 - 5y^2 - 8x - 20y - 36 = 0$

$$(4x^2 - 8x) + (-5y^2 - 20y) = 36$$

$$4(x^2 - 2x) - 5(y^2 + 4y) = 36$$

$$4(x^2 - 2x + 1) - 5(y^2 + 4y + 4) = 36 + 4 - 20$$

$$4(x - 1)^2 - 5(y + 2)^2 = 20$$

$$\frac{(x - 1)^2}{5} - \frac{(y + 2)^2}{4} = 1$$

$$\frac{(x - 1)^2}{\left(\sqrt{5}\right)^2} - \frac{(y + 2)^2}{2^2} = 1$$

Center: $(1, -2)$; $a = \sqrt{5}$, $b = 2$

47. $9x^2 + 36y^2 + 72y + 32 = 0$

$$9x^2 + 36(y^2 + 2y) = -32$$

$$9x^2 + 36(y^2 + 2y + 1) = -32 + 36$$

$$9x^2 + 36(y + 1)^2 = 4$$

$$\frac{9x^2}{4} + 9(y + 1)^2 = 1$$

$$\frac{x^2}{\frac{4}{9}} + \frac{(y + 1)^2}{\frac{1}{9}} = 1$$

$$\frac{x^2}{\left(\frac{2}{3}\right)^2} + \frac{(y + 1)^2}{\left(\frac{1}{3}\right)^2} = 1$$

Center: $(0, -1)$; $a = \dfrac{2}{3}$, $b = \dfrac{1}{3}$

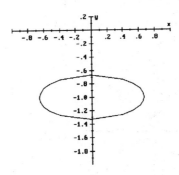

49. $4x^2 + 9y^2 - 4x - 6y - 34 = 0$

$$4x^2 - 4x + 9y^2 - 6y = 34$$

$$4(x^2 - x) + 9\left(y^2 - \frac{2}{3}y\right) = 34$$

$$4\left(x^2 - x + \frac{1}{4}\right) + 9\left(y^2 - \frac{2}{3}y + \frac{1}{9}\right) = 34 + 1 + 1$$

$$4\left(x - \frac{1}{2}\right)^2 + 9\left(y - \frac{1}{3}\right)^2 = 36$$

$$\frac{\left(x - \frac{1}{2}\right)^2}{9} + \frac{\left(y - \frac{1}{3}\right)^2}{4} = 1$$

$$\frac{\left(x - \frac{1}{2}\right)^2}{3^2} + \frac{\left(y - \frac{1}{3}\right)^2}{2^2} = 1$$

Center: $\left(\frac{1}{2}, \ \frac{1}{3}\right);$ $\quad a = 3, \quad b = 2$

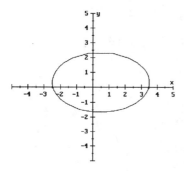

51. $9x^2 - 4y^2 - 3x - 2y + 144 = 0$

$$9x^2 - 3x - 4y^2 - 2y = -144$$

$$9\left(x^2 - \frac{1}{3}x\right) - 4\left(y^2 + \frac{1}{2}y\right) = -144$$

$$9\left(x^2 - \frac{1}{3}x + \frac{1}{36}\right) - 4\left(y^2 + \frac{1}{2}y + \frac{1}{16}\right) = -144 + \frac{1}{4} - \frac{1}{4}$$

$$9\left(x - \frac{1}{6}\right)^2 - 4\left(y + \frac{1}{4}\right)^2 = -144$$

$$\frac{-\left(x - \frac{1}{6}\right)^2}{16} + \frac{\left(y + \frac{1}{4}\right)^2}{36} = 1$$

$$\frac{\left(y + \frac{1}{4}\right)^2}{6^2} - \frac{\left(x - \frac{1}{6}\right)^2}{4^2} = 1$$

Center: $\left(\frac{1}{6}, \ -\frac{1}{4}\right);$ $\quad a = 4, \quad b = 6$

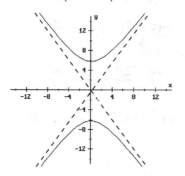

Problem Set 9.4

1. $x^2 - y^2 - 3x + y - 4 = 0$

$A = 1 > 0, \quad C = -1 < 0$

Hyperbola

3. $8x^2 + 5y^2 - 19x - 7y - 25 = 0$

$A = 8 > 0, \quad C = 5 > 0$

Ellipse

5. $3x^2 + 3y^2 + 13y = 0$

$A = C = 3$

Circle

7. $x^2 - 4x + 36y + 31 = 0$

$C = 0$

Parabola

9. $13 - 2y^2 + 3x - 7y = 0$

$A = 0$

Parabola

11. $3x^2 + 2y^2 - 2x - 2y - 14 = 0$

$A = 3 > 0, \quad C = 2 > 0$

Ellipse

13. $11x^2 + 11y - 13x - 5 = 0$

$C = 0$

Parabola

15. $5x^2 + 6x - 6y^2 - 5y = 0$

$A = 5 > 0, \quad C = -6 < 0$

Hyperbola

17. $2x - 3y + 4x^2 + 5y^2 - 6 = 0$

$A = 4 > 0, \quad C = 5 > 0$

Ellipse

19. $7x - 8y - 4x^2 - 4y^2 = 13$

$A = C = -4$

Circle

21. $\qquad 4x^2 + 3y^2 - 8x + 6y + 7 = 0$

$4x^2 - 8x + 3y^2 + 6y = -7$

$4(x^2 - 2x) + 3(y^2 + 2y) = -7$

$4(x^2 - 2x + 1) + 3(y^2 + 2y + 1) = -7 + 4 + 3$

$4(x - 1)^2 + 3(y + 1)^2 = 0$

Point: $(1, -1)$

Chapter 9 Review Problems

1. $x^2 - y^2 - 2x - 6y = 0$

$A = 1 > 0, \quad C = -1 < 0$

Hyperbola

3. $x^2 + 5y^2 - 10x - 13 = 0$

$A = 1 > 0, \quad C = 5 > 0$

Ellipse

5. $2x^2 + 3y^2 - 4x + 6y = 21$

$A = 2 > 0, \quad C = 3 > 0$

Ellipse

7. $3x^2 - 5y^2 - 18x - 20y - 100 = 0$

$A = 3 > 0, \quad C = -5 < 0$

Hyperbola

9. $2x^2 + x - y + 8 = 0$

$C = 0$

Parabola

11. $\dfrac{x^2}{4} + \dfrac{y^2}{25} = 1$

$\dfrac{x^2}{2^2} + \dfrac{y^2}{5^2} = 1$

$a = 2, \quad b = 5$

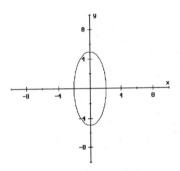

13. $x + 2 = 2(y - 1)^2$

$x = 2(y - 1)^2 - 2$

Vertex: $(-2, 1)$

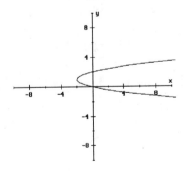

417

15. $x^2 - y^2 - 4x - 4y - 16 = 0$

$(x^2 - 4x) - (y^2 + 4y) = 16$

$(x^2 - 4x + 4) - (y^2 + 4y + 4) = 16 + 4 - 4$

$(x - 2)^2 - (y + 2)^2 = 16$

$\dfrac{(x - 2)^2}{16} - \dfrac{(y + 2)^2}{16} = 1$

$\dfrac{(x - 2)^2}{4^2} - \dfrac{(y + 2)^2}{4^2} = 1$

Center: $(2, -2)$; $a = 4$, $b = 4$

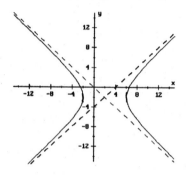

17. $16x^2 + 9y^2 - 32x + 18y - 119 = 0$

$16(x^2 - 2x) + 9(y^2 + 2y) = 119$

$16(x^2 - 2x + 1) + 9(y^2 + 2y + 1) = 119 + 16 + 9$

$16(x - 1)^2 + 9(y + 1)^2 = 144$

$\dfrac{(x - 1)^2}{9} + \dfrac{(y + 1)^2}{16} = 1$

$\dfrac{(x - 1)^2}{3^2} + \dfrac{(y + 1)^2}{4^2} = 1$

Center: $(1, -1)$; $a = 3$, $b = 4$

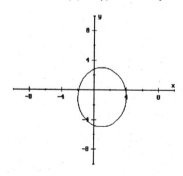

19. $x^2 + y^2 - 4x + 6y + 12 = 0$

$(x^2 - 4x) + (y^2 + 6y) = -12$

$(x^2 - 4x + 4) + (y^2 + 6y + 9) = -12 + 4 + 9$

$(x - 2)^2 + (y + 3)^2 = 1$

Center: $(2, -3)$
radius: 1

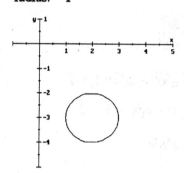

21. Factored

23. $8 + z^3 = 2^3 + z^3$

$= (2 + z)(4 - 2z + z^2)$

25. $63x^2 - 42x + 7 = 7(9x^2 - 6x + 1)$

$= 7(3x - 1)^2$

27. 2 terms

$(x - 3)^2 - (x - 3) = (x - 3)[(x - 3) - 1]$

$= (x - 3)(x - 4)$

$x - 3$ is a factor.

29. 3 terms

$x^2 - 6x + 9 = (x - 3)^2$

$x - 3$ is a factor.

31. $-4^{-4} = -\dfrac{1}{4^4} = -\dfrac{1}{256}$

418

33. $\dfrac{4x^2 - 9}{27 - 8x^3} = -\dfrac{4x^2 - 9}{8x^3 - 27}$

$= -\dfrac{(2x + 3)(2x - 3)}{(2x - 3)(4x^2 + 6x + 9)}$

$= -\dfrac{2x + 3}{4x^2 + 6x + 9}$

35. $\dfrac{-3x^{-2}}{y^{-3}z^2} = -\dfrac{3y^3}{x^2z^2}$

37. $\left(\sqrt{x} + 2\right)^2 = \left(\sqrt{x}\right)^2 + 2\sqrt{x} \cdot 2 + 2^2$

$= x + 4\sqrt{x} + 4$

39. $a^2b(abc) = a^3b^2c$

41. Will not reduce.

43. $\dfrac{x^2}{x^2 - 25} + \dfrac{1}{5 - x} = \dfrac{x}{5 + x}$

$\dfrac{x^2}{(x + 5)(x - 5)} - \dfrac{1}{x - 5} = \dfrac{x}{x + 5}$

$(x + 5)(x - 5)\left[\dfrac{x^2}{(x + 5)(x - 5)} - \dfrac{1}{x - 5}\right] = (x + 5)(x - 5)\left(\dfrac{x}{x + 5}\right)$

$x^2 - (x + 5) = x(x - 5)$

$x^2 - x - 5 = x^2 - 5x$

$-5 = -4x$

$\dfrac{5}{4} = x$

$\left\{\dfrac{5}{4}\right\}$

45. Expression;

$(x - 7)^2 = x^2 - 2x \cdot 7 + 7^2$

$= x^2 - 14x + 49$

47. Equation;

$$\left|\frac{7 - 5x}{11}\right| = 3$$

$$\frac{7 - 5x}{11} = 3 \quad \text{or} \quad \frac{7 - 5x}{11} = -3$$

$$7 - 5x = 33 \qquad 7 - 5x = -33$$

$$-5x = 26 \qquad\qquad -5x = -40$$

$$x = -\frac{26}{5} \qquad\qquad x = 8$$

$$\left\{-\frac{26}{5}, 8\right\}$$

49. Expression;

$$\frac{3 - 2i}{1 + i} = \frac{3 - 2i}{1 + i} \cdot \frac{1 - i}{1 - i}$$

$$= \frac{3 - 3i - 2i + 2i^2}{1 - i^2}$$

$$= \frac{3 - 5i + 2(-1)}{1 - (-1)}$$

$$= \frac{1 - 5i}{2}$$

$$= \frac{1}{2} - \frac{5}{2}i$$

Chapter 9 Test

1.
$$x^2 + y^2 - 2x + 4y = 4$$

$$(x^2 - 2x + 1) + (y^2 + 4y + 4) = 4 + 1 + 4$$

$$(x - 1)^2 + (y + 2)^2 = 9$$

$$r = 3$$

B

2. $x = y^2 - 4y + 3$

$$x = (y^2 - 4y + 4) + 3 - 4$$

$$x = (y - 2)^2 - 1$$

Vertex: $(-1, 2)$

A

3. $5x^2 + 6y^2 + 50x - 24y + 119 = 0$

$$5(x^2 + 10x) + 6(y^2 - 4y) = -119$$

$$5(x^2 + 10x + 25) + 6(y^2 - 4y + 4) = -119 + 125 + 24$$

$$5(x + 5)^2 + 6(y - 2)^2 = 30$$

$$\frac{(x + 5)^2}{6} + \frac{(y - 2)^2}{5} = 1$$

Center: $(-5, 2)$

B

4. $x^2 + y + 2x - 3 = 0$

$$y = -x^2 - 2x + 3$$

$$C = 0, \quad A = -1 < 0$$

Parabola opening downward

D

5. $2x^2 + 2y^2 + 2x - 3 = 0$

$$A = C = 2$$

Circle

B

6. $4x^2 + 8x - 4y^2 + 8y - 4 = 0$

$$A = 4 > 0, \quad C = -4 < 0$$

Hyperbola

D

7. $2x^2 + 2y + y^2 = 0$

   $A = 2 > 0, \quad C = 1 > 0$

   Ellipse

   A

8. $2x^2 + 3y^2 - 8x + 18y + 29 = 0$

   $2(x^2 - 4x) + 3(y^2 + 6y) = -29$

   $2(x^2 - 4x + 4) + 3(y^2 + 6y + 9) = -29 + 8 + 27$

   $2(x - 2)^2 + 3(y + 3)^2 = 6$

   $\dfrac{(x - 2)^2}{3} + \dfrac{(y + 3)^2}{2} = 1$

   Center: $(2, -3)$

   A

9. $\quad y^2 - x - 4y + 5 = 0$

   $(y^2 - 4y) + 5 = x$

   $(y^2 - 4y + 4) + 5 - 4 = x$

   $(y - 2)^2 + 1 = x$

   Vertex: $(1, 2)$

   Opens to the right.

   C

10. $\quad x^2 + y^2 - 4y = 1$

    $x^2 + (y^2 - 4y + 4) = 1 + 4$

    $x^2 + (y - 2)^2 = 5$

    Center: $(0, 2)$
    radius: $\sqrt{5}$

11. $y = -2x^2 - 4x - 3$

    $y = -2(x^2 + 2x) - 3$

    $y = -2(x^2 + 2x + 1) - 3 + 2$

    $y = -2(x + 1)^2 - 1$

    Vertex: $(-1, -1)$

    Opens downward

12. $\quad x^2 - y^2 + 4x + 2y + 4 = 0$

    $(x^2 + 4x) - (y^2 - 2y) = -4$

    $(x^2 + 4x + 4) - (y^2 - 2y + 1) = -4 + 4 - 1$

    $(x + 2)^2 - (y - 1)^2 = -1$

    $\dfrac{(y - 1)^2}{1} - \dfrac{(x + 2)^2}{1} = 1$

    Center: $(-2, 1); \quad a = 1, \quad b = 1$

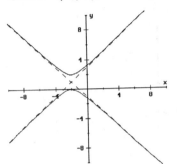

13. $4x^2 + 9y^2 - 16x - 18y - 11 = 0$

    $4(x^2 - 4x) + 9(y^2 - 2y) = 11$

    $4(x^2 - 4x + 4) + 9(y^2 - 2y + 1) = 11 + 16 + 9$

    $4(x - 2)^2 + 9(y - 1)^2 = 36$

    $\dfrac{(x - 2)^2}{9} + \dfrac{(y - 1)^2}{4} = 1$

    $\dfrac{(x - 2)^2}{3^2} + \dfrac{(y - 1)^2}{2^2} = 1$

    Center: $(2, 1); \quad a = 3, \quad b = 2$

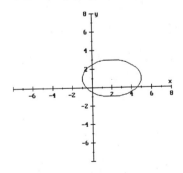

421

14. $x^2 + y^2 - 6x - 6y + 9 = 0$

$(x^2 - 6x) + (y^2 - 6y) = -9$

$(x^2 - 6x + 9) + (y^2 - 6y + 9) = -9 + 9 + 9$

$(x - 3)^2 + (y - 3)^2 = 9$

Center: (3, 3)
radius: 3

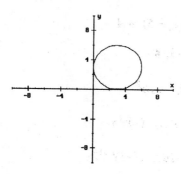

15. $x = y^2 - 6y + 9$

$x = (y - 3)^2$

Vertex: (0, 3)

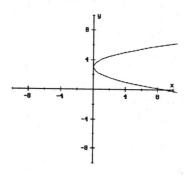

Problem Set 10.1

1. $a_n = n$

   $a_1 = 1$

   $a_2 = 2$

   $a_3 = 3$

   $a_4 = 4$

   $a_5 = 5$

   1, 2, 3, 4, 5

3. $a_n = 2(n + 1)$

   $a_1 = 2(1 + 1) = 4$

   $a_2 = 2(2 + 1) = 6$

   $a_3 = 2(3 + 1) = 8$

   $a_4 = 2(4 + 1) = 10$

   $a_5 = 2(5 + 1) = 12$

   4, 6, 8, 10, 12

5. $c_n = n^2$

   $c_1 = 1^2 = 1$

   $c_2 = 2^2 = 4$

   $c_3 = 3^2 = 9$

   $c_4 = 4^2 = 16$

   $c_5 = 5^2 = 25$

   1, 4, 9, 16, 25

7. $a_n = 3$

   $a_1 = a_2 = a_3 = a_4 = a_5 = 3$

   3, 3, 3, 3, 3

9. $k_n = (-1)^n(1 - n)$

   $k_1 = (-1)^1(1 - 1) = 0$

   $k_2 = (-1)^2(1 - 2) = -1$

   $k_3 = (-1)^3(1 - 3) = 2$

   $k_4 = (-1)^4(1 - 4) = -3$

   $k_5 = (-1)^5(1 - 5) = 4$

   0, -1, 2, -3, 4

11. $1, 4, 9, 16, \ldots, n^2$

13. $-1, 2, -3, 4, \ldots (-1)^n n$

15. $2, -4, 8, -16, \ldots (-1)^{n+1} 2n$

17. $a_n = 3 - 2n$

    1, -1, -3, -5, …

    Decreasing

19. $a_n = (-n)^{-1}$

    $a_n = \dfrac{1}{-n}$

    $-1, -\dfrac{1}{2}, -\dfrac{1}{3}, -\dfrac{1}{4}, \ldots$

    Increasing

21. $k_n = 1 - n^{-1}$

    $k_n = 1 - \dfrac{1}{n}$

    $0, \dfrac{1}{2}, \dfrac{2}{3}, \dfrac{3}{4}, \ldots$

    Increasing

23. 2, 3, 4, ...

$d = 3 - 2 = 1$

$a_n = 2 + (n - 1)1$

25. 5, 15, 25, ...

$d = 15 - 5 = 10$

$a_n = 5 + (n - 1)10$

27. $\frac{1}{2}, 0, -\frac{1}{2}, ...$

$d = 0 - \frac{1}{2} = -\frac{1}{2}$

$a_n = \frac{1}{2} + (n - 1)\left(-\frac{1}{2}\right)$

29. 1, 5, 25, ...

$r = \frac{5}{1} = 5$

$a_n = 1(5)^{n-1}$

31. 2, -4, 8, ...

$r = -\frac{4}{2} = -2$

$a_n = 2(-2)^{n-1}$

33. $1, -\frac{1}{4}, \frac{1}{16}, ...$

$r = \frac{\frac{-1}{4}}{1} = -\frac{1}{4}$

$a_n = 1\left(-\frac{1}{4}\right)^{n-1}$

35. $a_n = 3n - 1$

$a_1 = 3(1) - 1 = 2$

$a_2 = 3(2) - 1 = 5$

$a_3 = 3(3) - 1 = 8$

$a_4 = 3(4) - 1 = 11$

$a_5 = 3(5) - 1 = 14$

2, 5, 8, 11, 14

37. $A_n = 8 - 2n$

$A_1 = 8 - 2(1) = 6$

$A_2 = 8 - 2(2) = 4$

$A_3 = 8 - 2(3) = 2$

$A_4 = 8 - 2(4) = 0$

$A_5 = 8 - 2(5) = -2$

6, 4, 2, 0, -2

39. $h_n = n^2 - n + 1$

$h_1 = 1^2 - 1 + 1 = 1$

$h_2 = 2^2 - 2 + 1 = 3$

$h_3 = 3^2 - 3 + 1 = 7$

$h_4 = 4^2 - 4 + 1 = 13$

$h_5 = 5^2 - 5 + 1 = 21$

1, 3, 7, 13, 21

41. $j_n = (-1)^{n-1}n^2$

$j_1 = (-1)^{1-1}(1)^2 = 1$

$j_2 = (-1)^{2-1}(2)^2 = -4$

$j_3 = (-1)^{3-1}(3)^2 = 9$

$j_4 = (-1)^{4-1}(4)^2 = -16$

$j_5 = (-1)^{5-1}(5)^2 = 25$

1, -4, 9, -16, 25

43. $4, 6, 8, 10, \ldots, 2n + 2$

45. $-1, 3, 7, 11, \ldots, 4n - 5$

47. $0, 1, 8, 27, \ldots, (n - 1)^3$

49. $a_n = -4 + n$

   $-3, -2, -1, 0, \ldots$

   Increasing

51. $b_n = n^{-2} - 2 = \dfrac{1}{n^2} - 2$

   $-1, -\dfrac{7}{4}, -\dfrac{17}{9}, -\dfrac{31}{16}, \ldots$

   $-1, -1\dfrac{3}{4}, -1\dfrac{8}{9}, -1\dfrac{15}{16}, \ldots$

   Decreasing

53. $d_n = (-2)^{-2n} = \dfrac{1}{(-2)^{2n}}$

   $\dfrac{1}{4}, \dfrac{1}{16}, \dfrac{1}{64}, \dfrac{1}{256}, \ldots$

   Decreasing

55. $1, 4, 7, \ldots$

   $d = 4 - 1 = 3$

   $a_n = 1 + (n - 1)3$

57. $10, 8, 6, \ldots$

   $d = 8 - 10 = -2$

   $a_n = 10 + (n - 1)(-2)$

59. $-\dfrac{1}{2}, 0, \dfrac{1}{2}, \ldots$

   $d = 0 - \left(-\dfrac{1}{2}\right) = \dfrac{1}{2}$

   $a_n = -\dfrac{1}{2} + (7 - 1)\left(\dfrac{1}{2}\right)$

61. $1, 3, 9, \ldots$

   $r = \dfrac{3}{1} = 3$

   $a_n = 1(3)^{n-1}$

63. $2, -6, 18, \ldots$

   $r = \dfrac{-6}{2} = -3$

   $a_n = 2(-3)^{n-1}$

65. $27, 9, 3, \ldots$

   $r = \dfrac{9}{27} = \dfrac{1}{3}$

   $a_n = 27\left(\dfrac{1}{3}\right)^{n-1}$

Problem Set 10.2

1. $\displaystyle\sum_{j=1}^{5} 2j = (2 \cdot 1) + (2 \cdot 2) + (2 \cdot 3) + (2 \cdot 4)$

   $+ (2 \cdot 5)$

   $= 2 + 4 + 6 + 8 + 10$

3. $\displaystyle\sum_{j=0}^{7} j^2 = 0^2 + 1^2 + 2^2 + 3^2 + 4^2 + 5^2$

   $+ 6^2 + 7^2$

   $= 0 + 1 + 4 + 9 + 16 + 25$

   $+ 36 + 49$

5. $\displaystyle\sum_{j=0}^{4} (3 - j)^2 = (3 - 0)^2 + (3 - 1)^2 + (3 - 2)^2$

   $+ (3 - 3)^2 + (3 - 4)^2$

   $= 9 + 4 + 1 + 0 + 1$

7. $\displaystyle\sum_{j=1}^{6} (-1)^{j-1} j = [(-1)^{1-1} \cdot 1] + [(-1)^{2-1} \cdot 2]$

   $+ [(-1)^{3-1} \cdot 3] + [(-1)^{4-1} \cdot 4]$

   $+ [(-1)^{5-1} \cdot 5] + [(-1)^{6-1} \cdot 6]$

   $= 1 - 2 + 3 - 4 + 5 - 6$

9. $\displaystyle\sum_{j=1}^{5} \frac{j-1}{j+1}(-1)^{j+1}$

$= \left[\left(\frac{1-1}{1+1}\right)(-1)^{1+1}\right] + \left[\left(\frac{2-1}{2+1}\right)(-1)^{2+1}\right]$

$+ \left[\left(\frac{3-1}{3+1}\right)(-1)^{3+1}\right] + \left[\left(\frac{4-1}{4+1}\right)(-1)^{4+1}\right]$

$+ \left[\left(\frac{5-1}{5+1}\right)(-1)^{5+1}\right]$

$= 0 - \frac{1}{3} + \frac{1}{2} - \frac{3}{5} + \frac{2}{3}$

11. $1 - 3 + 5 - 7 + 9 - 11 = \displaystyle\sum_{j=i}^{6} (-1)^{j+1}(2j-1)$

13. $1 + \dfrac{x}{2} + \dfrac{x^2}{3} + \dfrac{x^3}{4} + \dfrac{x^4}{5} + \dfrac{x^5}{6}$

$+ \dfrac{x^6}{7} = \displaystyle\sum_{j=1}^{7} \dfrac{x^{j-1}}{j}$

15. $1 - \sqrt{2}x^2 + \sqrt{3}x^4 - 2x^6 + \sqrt{5}x^8 - \sqrt{6}x^{10}$

$+ \sqrt{7}x^{12} = \displaystyle\sum_{j=1}^{7} (-1)^{j-1}\sqrt{j}x^{2j-2}$

17. $\displaystyle\sum_{j=0}^{4} 2(j-5) = [2(0-5)] + [2(1-5)]$

$+ [2(2-5)] + [2(3-5)]$

$+ [2(4-5)]$

$= -10 - 8 - 6 - 4 - 2$

19. $\displaystyle\sum_{j=1}^{7} 2^{j-1} = 2^{1-1} + 2^{2-1} + 2^{3-1}$

$+ 2^{4-1} + 2^{5-1} + 2^{6-1} + 2^{7-1}$

$= 1 + 2 + 4 + 8 + 16 + 32 + 64$

21. $\displaystyle\sum_{j=0}^{5} (j-1)^3 = (0-1)^3 + (1-1)^3 + (2-1)^3$

$+ (3-1)^3 + (4-1)^3 + (5-1)^3$

$= -1 + 0 + 1 + 8 + 27 + 64$

23. $\displaystyle\sum_{j=1}^{5} (-1)^j 2^{j-1}$

$= (-1)^1 2^{1-1} + (-1)^2 2^{2-1}$

$+ (-1)^3 2^{3-1} + (-1)^4 2^{4-1} + (-1)^5 2^{5-1}$

$= -1 + 2 - 4 + 8 - 16$

25. $3 + 4 + 5 + 6 + 7 + 8 + 9 + 10 + 11$

$+ 12 + 13 + 14$

$= \displaystyle\sum_{j=3}^{14} j$

27. $1 + \dfrac{x^2}{3} + \dfrac{x^4}{5} + \dfrac{x^6}{7} + \dfrac{x^8}{9}$

$= \displaystyle\sum_{j=0}^{4} \dfrac{x^{2j}}{2j+1}$

29. $-\pi + \dfrac{1-\pi}{10} + \dfrac{2-\pi}{100} + \dfrac{3-\pi}{1000} + \dfrac{4-\pi}{10000}$

$+ \dfrac{5-\pi}{100000} + \dfrac{6-\pi}{1000000}$

$= \displaystyle\sum_{j=0}^{6} \dfrac{j-\pi}{10^j}$

31. $\displaystyle\sum_{i=1}^{5} i = \sum_{j=1}^{5} j = \sum_{k=1}^{5} k = 1 + 2 + 3 + 4 + 5$

33. $\displaystyle\sum_{j=1}^{4} 5A_j = 5A_1 + 5A_2 + 5A_3 + 5A_4$

$= 5(A_1 + A_2 + A_3 + A_4)$

$5\displaystyle\sum_{j=1}^{4} A_j = 5(A_1 + A_2 + A_3 + A_4)$

$\displaystyle\sum_{j=1}^{n} kA_j = k\sum_{j=1}^{n} A_j$

Problem Set 10.3

1. $6! = 1 \cdot 2 \cdot 3 \cdot 4 \cdot 5 \cdot 6 = 720$

3. $12! = 1 \cdot 2 \cdot 3 \cdot 4 \cdot 5 \cdot 6 \cdot 7 \cdot 8 \cdot 9 \cdot 10 \cdot 11 \cdot 12$

$\qquad = 479{,}001{,}600$

5. $\dfrac{14!}{13!} = \dfrac{14 \cdot 13!}{13!} = 14$

7. $\dfrac{8!}{4! \cdot 4!} = \dfrac{4! \cdot 5 \cdot 6 \cdot 7 \cdot 8}{4! \cdot 1 \cdot 2 \cdot 3 \cdot 4}$

$\qquad = 5 \cdot 7 \cdot 2$

$\qquad = 70$

9. $\dfrac{8!}{2! \cdot 6!} = \dfrac{6! \cdot 7 \cdot 8}{1 \cdot 2 \cdot 6!}$

$\qquad = 7 \cdot 4$

$\qquad = 28$

11. $\dfrac{20!}{1! \cdot 19!} = \dfrac{19! \cdot 20}{1 \cdot 19!}$

$\qquad = 20$

13. $\dbinom{8}{5} = \dfrac{8!}{5!3!}$

$\qquad = \dfrac{5! \cdot 6 \cdot 7 \cdot 8}{5! \cdot 1 \cdot 2 \cdot 3}$

$\qquad = 7 \cdot 8$

$\qquad = 56$

15. $\dbinom{5}{3} = \dfrac{5!}{3!2!}$

$\qquad = \dfrac{3! \cdot 4 \cdot 5}{3! \cdot 1 \cdot 2}$

$\qquad = 2 \cdot 5$

$\qquad = 10$

17. $\dbinom{6}{1} = \dfrac{6!}{1!5!}$

$\qquad = \dfrac{5! \cdot 6}{1 \cdot 5!}$

$\qquad = 6$

19. $\dbinom{6}{3} = \dfrac{6!}{3!3!} = \dfrac{3! \cdot 4 \cdot 5 \cdot 6}{3! \cdot 1 \cdot 2 \cdot 3}$

$\qquad = 4 \cdot 5$

$\qquad = 20$

21. $\dbinom{6}{5} = \dfrac{6!}{5!1!} = \dfrac{5! \cdot 6}{5! \cdot 1}$

$\qquad = 6$

23. $\dbinom{500}{0} = \dfrac{500!}{0!500!} = \dfrac{500!}{1 \cdot 500!}$

$\qquad = 1$

25. $7! = 1 \cdot 2 \cdot 3 \cdot 4 \cdot 5 \cdot 6 \cdot 7 = 5040$

27. $11! = 1 \cdot 2 \cdot 3 \cdot 4 \cdot 5 \cdot 6 \cdot 7 \cdot 8 \cdot 9 \cdot 10 \cdot 11$

$\qquad = 39{,}916{,}800$

29. $\dfrac{17!}{18!} = \dfrac{17!}{17! \cdot 18} = \dfrac{1}{18}$

31. $\dfrac{9!}{4!5!} = \dfrac{5! \cdot 6 \cdot 7 \cdot 8 \cdot 9}{1 \cdot 2 \cdot 3 \cdot 4 \cdot 5!}$

$\qquad = 2 \cdot 7 \cdot 9$

$\qquad = 126$

33. $\dfrac{9!}{2!7!} = \dfrac{7! \cdot 8 \cdot 9}{1 \cdot 2 \cdot 7!}$

$\qquad = 4 \cdot 9$

$\qquad = 36$

35. $\dfrac{30!}{1! \cdot 29!} = \dfrac{29! \cdot 30}{1 \cdot 29!}$

$\qquad = 30$

37. $\dbinom{9}{6} = \dfrac{9!}{6!3!}$

$\qquad = \dfrac{6! \cdot 7 \cdot 8 \cdot 9}{6!1 \cdot 2 \cdot 3}$

$\qquad = 7 \cdot 4 \cdot 3$

$\qquad = 84$

39. $\binom{4}{1} = \dfrac{4!}{1!3!} = \dfrac{3! \cdot 4}{1 \cdot 3!}$

$\qquad = 4$

41. $\binom{7}{1} = \dfrac{7!}{1!6!} = \dfrac{6! \cdot 7}{1 \cdot 6!}$

$\qquad = 7$

43. $\binom{7}{3} = \dfrac{7!}{3!4!}$

$\qquad = \dfrac{4! \cdot 5 \cdot 6 \cdot 7}{1 \cdot 2 \cdot 3 \cdot 4!}$

$\qquad = 5 \cdot 7$

$\qquad = 35$

45. $\binom{7}{5} = \dfrac{7!}{5!2!} = \dfrac{5! \cdot 6 \cdot 7}{5! \cdot 1 \cdot 2}$

$\qquad = 3 \cdot 7$

$\qquad = 21$

47. $\binom{7}{7} = \dfrac{7!}{7!0!} = 1$

49. $\binom{n}{r} = \dfrac{n!}{r!(n-r)!}$

$\binom{n}{n-r} = \dfrac{n!}{(n-r)![n-(n-r)]!}$

$\qquad = \dfrac{n!}{(n-r)!r!}$

$\qquad = \dfrac{n!}{r!(n-r)!}$

$\binom{n}{r} = \binom{n}{n-r}$

Problem Set 10.4

1. $(a+b)^{12}$

1st 3 terms:

$\binom{12}{0}a^{12}b^0 + \binom{12}{1}a^{11}b^1 + \binom{12}{2}a^{10}b^2$

$= \dfrac{12!}{0!12!}a^{12} + \dfrac{12!}{1!11!}a^{11}b + \dfrac{12!}{2!10!}a^{10}b^2$

$= a^{12} + \dfrac{11! \cdot 12}{1 \cdot 11!}a^{11}b + \dfrac{10! \cdot 11 \cdot 12}{1 \cdot 2 \cdot 10!}a^{10}b^2$

$= a^{12} + 12a^{11}b + 66a^{10}b^2$

3. $(s+2)^{15}$

1st 3 terms:

$\binom{15}{0}s^{15} \cdot 2^0 + \binom{15}{1}s^{14} \cdot 2^1 + \binom{15}{2}s^{13} \cdot 2^2$

$= \dfrac{15!}{0!15!}s^{15} + \dfrac{15!}{1!14!}s^{14} \cdot 2 + \dfrac{15!}{2!13!}s^{13} \cdot 4$

$= s^{15} + \dfrac{14! \cdot 15}{1 \cdot 14!}s^{14} \cdot 2 + \dfrac{13! \cdot 14 \cdot 15}{1 \cdot 2 \cdot 13!}s^{13} \cdot 4$

$= s^{15} + 30s^{14} + 420s^{13}$

5. $(2a+3b)^6$

1st 3 terms:

$\binom{6}{0}(2a)^6(3b)^0 + \binom{6}{1}(2a)^5(3b)^1 + \binom{6}{2}(2a)^4(3b)^2$

$= \dfrac{6!}{0!6!}(64a^6) + \dfrac{6!}{1!5!}(32a^5)(3b) + \dfrac{6!}{2!4!}(16a^4)(9b^2)$

$= 64a^6 + \dfrac{5! \cdot 6}{1 \cdot 5!}(96a^5b) + \dfrac{4! \cdot 5 \cdot 6}{1 \cdot 2 \cdot 4!}(144a^4b^2)$

$= 64a^6 + 576a^5b + 2160a^4b^2$

7. $(a+b)^{13}$

fifth term: $\binom{13}{4}a^9b^4$

$= \dfrac{13!}{4!9!}a^9b^4$

$= \dfrac{9! \cdot 10 \cdot 11 \cdot 12 \cdot 13}{1 \cdot 2 \cdot 3 \cdot 4 \cdot 9!}a^9b^4$

$= 715a^9b^4$

9. $\left(3x + \dfrac{1}{2}\right)^{11}$

fourth term: $\dbinom{11}{3}(3x)^8\left(\dfrac{1}{2}\right)^3$

$= \dfrac{11!}{3!8!}(6561x^8)\left(\dfrac{1}{8}\right)$

$= \dfrac{8! \cdot 9 \cdot 10 \cdot 11}{1 \cdot 2 \cdot 3 \cdot 8!}\left(\dfrac{6561x^8}{8}\right)$

$= 165\left(\dfrac{6561x^8}{8}\right)$

$= \dfrac{1082565}{8}x^8$

11. $(a + b)^6$

$= \dbinom{6}{0}a^6b^0 + \dbinom{6}{1}a^5b^1 + \dbinom{6}{2}a^4b^2$

$\quad + \dbinom{6}{3}a^3b^3 + \dbinom{6}{4}a^2b^4 + \dbinom{6}{5}ab^5 + \dbinom{6}{6}a^0b^6$

$= \dfrac{6!}{0!6!}a^6 + \dfrac{6!}{1!5!}a^5b + \dfrac{6!}{2!4!}a^4b^2 + \dfrac{6!}{3!3!}a^3b^3$

$\quad + \dfrac{6!}{4!2!}a^2b^4 + \dfrac{6!}{5!1!}ab^5 + \dfrac{6!}{6!0!}b^6$

$= a^6 + 6a^5b + 15a^4b^2 + 20a^3b^3 + 15a^2b^4$

$\quad + 6ab^5 + b^6$

13. $(x + 1)^5$

$= \dbinom{5}{0}x^5 \cdot 1^0 + \dbinom{5}{1}x^4 \cdot 1^1 + \dbinom{5}{2}x^3 \cdot 1^2 + \dbinom{5}{3}x^2 \cdot 1^3$

$\quad + \dbinom{5}{4}x \cdot 1^4 + \dbinom{5}{5}x^0 \cdot 1^5$

$= \dfrac{5!}{0!5!}x^5 + \dfrac{5!}{1!4!}x^4 + \dfrac{5!}{2!3!}x^3 + \dfrac{5!}{3!2!}x^2 + \dfrac{5!}{4!1!}x$

$\quad + \dfrac{5!}{5!0!}$

$= x^5 + 5x^4 + 10x^3 + 10x^2 + 5x + 1$

15. $(2x + y)^6$

$= \dbinom{6}{0}(2x)^6y^0 + \dbinom{6}{1}(2x)^5y^1 + \dbinom{6}{2}(2x)^4y^2$

$\quad + \dbinom{6}{3}(2x)^3y^3 + \dbinom{6}{4}(2x)^2y^4 + \dbinom{6}{5}(2x)y^5$

$\quad + \dbinom{6}{6}(2x)^0y^6$

$= \dfrac{6!}{0!6!}(64x^6) + \dfrac{6!}{1!5!}(32x^5)y + \dfrac{6!}{2!4!}(16x^4)y^2$

$\quad + \dfrac{6!}{3!3!}(8x^3)y^3 + \dfrac{6!}{4!2!}(4x^2)y^4 + \dfrac{6!}{5!1!}(2x)y^5$

$\quad + \dfrac{6!}{6!0!}y^6$

$= 64x^6 + 192x^5y + 240x^4y^2 + 160x^3y^3$

$\quad + 60x^2y^4 + 12xy^5 + y^6$

17. $(a + b)^9$

$= \dbinom{9}{0}a^9b^0 + \dbinom{9}{1}a^8b^1 + \dbinom{9}{2}a^7b^2$

$\quad + \dbinom{9}{3}a^6b^3 + \dbinom{9}{4}a^5b^4 + \dbinom{9}{5}a^4b^5$

$\quad + \dbinom{9}{6}a^3b^6 + \dbinom{9}{7}a^2b^7 + \dbinom{9}{8}ab^8$

$\quad + \dbinom{9}{9}a^0b^9$

$= \dfrac{9!}{0!9!}a^9 + \dfrac{9!}{1!8!}a^8b + \dfrac{9!}{2!7!}a^7b^2$

$\quad + \dfrac{9!}{3!6!}a^6b^3 + \dfrac{9!}{4!5!}a^5b^4 + \dfrac{9!}{5!4!}a^4b^5$

$\quad + \dfrac{9!}{6!3!}a^3b^6 + \dfrac{9!}{7!2!}a^2b^7 + \dfrac{9!}{8!1!}ab^8$

$\quad + \dfrac{9!}{9!0!}b^9$

$= a^9 + 9a^8b + 36a^7b^2 + 84a^6b^3$

$\quad + 126a^5b^4 + 126a^4b^5 + 84a^3b^6$

$\quad + 36a^2b^7 + 9ab^8 + b^9$

19. $(x + 1)^9$

$$= \binom{9}{0}x^9 \cdot 1^0 + \binom{9}{1}x^8 \cdot 1^1 + \binom{9}{2}x^7 \cdot 1^2$$

$$+ \binom{9}{3}x^6 \cdot 1^3 + \binom{9}{4}x^5 \cdot 1^4 + \binom{9}{5}x^4 \cdot 1^5$$

$$+ \binom{9}{6}x^3 \cdot 1^6 + \binom{9}{7}x^2 \cdot 1^7 + \binom{9}{8}x \cdot 1^8 + \binom{9}{9}x^0 \cdot 1^9$$

$$= \frac{9!}{0!9!}x^9 + \frac{9!}{1!8!}x^8 + \frac{9!}{2!7!}x^7 + \frac{9!}{3!6!}x^6$$

$$+ \frac{9!}{4!5!}x^5 + \frac{9!}{5!4!}x^4 + \frac{9!}{6!3!}x^3 + \frac{9!}{7!2!}x^2$$

$$+ \frac{9!}{8!1!}x + \frac{9!}{9!0!}$$

$$= x^9 + 9x^8 + 36x^7 + 84x^6 + 126x^5 + 126x^4$$

$$+ 84x^3 + 36x^2 + 9x + 1$$

21. $(2x + y)^7$

$$= \binom{7}{0}(2x)^7y^0 + \binom{7}{1}(2x)^6y + \binom{7}{2}(2x)^5y^2$$

$$+ \binom{7}{3}(2x)^4y^3 + \binom{7}{4}(2x)^3y^4 + \binom{7}{5}(2x)^2y^5$$

$$+ \binom{7}{6}(2x)y^6 + \binom{7}{7}(2x)^0y^7$$

$$= \frac{7!}{0!7!}(128x^7) + \frac{7!}{1!6!}(64x^6)y + \frac{7!}{2!5!}(32x^5)y^2$$

$$+ \frac{7!}{3!4!}(16x^4)y^3 + \frac{7!}{4!3!}(8x^3)y^4 + \frac{7!}{5!2!}(4x^2)y^5$$

$$+ \frac{7!}{6!1!}(2x)y^6 + \frac{7!}{7!0!}y^7$$

$$= 128x^7 + 448x^6y + 672x^5y^2 + 560x^4y^3$$

$$+ 280x^3y^4 + 84x^2y^5 + 14xy^6 + y^7$$

23. $(a + b)^{13}$

last 3 terms:

$$\binom{13}{11}a^2b^{11} + \binom{13}{12}ab^{12} + \binom{13}{13}a^0b^{13}$$

$$= \frac{13!}{11!2!}a^2b^{11} + \frac{13!}{12!1!}ab^{12} + \frac{13!}{13!0!}b^{13}$$

$$= 78a^2b^{11} + 13ab^{12} + b^{13}$$

25. $(s + 3)^5$

last 3 terms:

$$\binom{5}{3}s^2 \cdot 3^3 + \binom{5}{4}s \cdot 3^4 + \binom{5}{5}s^03^5$$

$$= \frac{5!}{3!2!}s^2(27) + \frac{5!}{4!1!}s \cdot 81 + \frac{5!}{5!0!} \cdot 243$$

$$= 270s^2 + 405s + 243$$

27. $(3x + 2y)^7$

last 3 terms:

$$\binom{7}{5}(3x)^2(2y)^5 + \binom{7}{6}(3x)(2y)^6 + \binom{7}{7}(3x)^0(2y)^7$$

$$= \frac{7!}{5!2!}(9x^2)(32y^5) + \frac{7!}{6!1!}(3x)(64y^6)$$

$$+ \frac{7!}{7!0!}(128y^7)$$

$$= 6048x^2y^5 + 1344xy^6 + 128y^7$$

29. $(a + b)^{12}$

last 3 terms:

$$\binom{12}{10}a^2b^{10} + \binom{12}{11}ab^{11} + \binom{12}{12}a^0b^{12}$$

$$= \frac{12!}{10!2!}a^2b^{10} + \frac{12!}{11!1!}ab^{11} + \frac{12!}{12!0!}b^{12}$$

$$= 66a^2b^{10} + 12ab^{11} + b^{12}$$

31. $(s + 3)^6$

last 3 terms:

$$\binom{6}{4}s^2(3)^4 + \binom{6}{5}s(3)^5 + \binom{6}{6}s^0(3)^6$$

$$= \frac{6!}{4!2!}s^2(81) + \frac{6!}{5!1!}s(243) + \frac{6!}{6!0!}(729)$$

$$= 1215s^2 + 1458s + 729$$

33. $(2x + 3y)^7$

last 3 terms:

$$\binom{7}{5}(2x)^2(3y)^5 + \binom{7}{6}(2x)(3y)^6 + \binom{7}{7}(2x)^0(3y)^7$$

$$= \frac{7!}{5!2!}(4x^2)(243y^5) + \frac{7!}{6!1!}(2x)(729y^6)$$

$$+ \frac{7!}{7!0!}(2187y^7)$$

$$= 20412x^2y^5 + 10206xy^6 + 2187y^7$$

35. $(a + b)^{50}$

1st 2 terms:

$$\binom{50}{0}a^{50}b^0 + \binom{50}{1}a^{49}b^1$$

$$= \frac{50!}{0!50!}a^{50} + \frac{50!}{1!49!}a^{49}b$$

$$= a^{50} + 50a^{49}b$$

37. $(a + 3)^{500}$

1st 2 terms:

$$\binom{500}{0}s^{500} \cdot 3^0 + \binom{500}{1}s^{499} \cdot 3^1$$

$$= \frac{500!}{0!500!}s^{500} + \frac{500!}{1!499!}s^{499} \cdot 3$$

$$= s^{500} + 1500s^{499}$$

39. $(1 + x)^{100}$

1st 2 terms:

$$\binom{100}{0}1^{100}x^0 + \binom{100}{1}1^{99}x^1$$

$$= \frac{100!}{0!100!} + \frac{100!}{1!99!}x$$

$$= 1 + 100x$$

41. $\left(\dfrac{x}{2} + \dfrac{1}{3}\right)^{12}$

4th term:

$$\binom{12}{3}\left(\frac{x}{2}\right)^9\left(\frac{1}{3}\right)^3$$

$$= \frac{12!}{3!9!} \cdot \frac{x^9}{512} \cdot \frac{1}{27}$$

$$= \frac{220x^9}{512 \cdot 27} = \frac{55x^9}{3456}$$

43. $\left(\dfrac{1}{2} + \dfrac{1}{3}\right)^6$

$$= \binom{6}{0}\left(\frac{1}{2}\right)^6\left(\frac{1}{2}\right)^0 + \binom{6}{1}\left(\frac{1}{2}\right)^5\left(\frac{1}{2}\right)^1 + \binom{6}{2}\left(\frac{1}{2}\right)^4\left(\frac{1}{2}\right)^2$$

$$+ \binom{6}{3}\left(\frac{1}{2}\right)^3\left(\frac{1}{2}\right)^3 + \binom{6}{4}\left(\frac{1}{2}\right)^2\left(\frac{1}{2}\right)^4$$

$$+ \binom{6}{5}\left(\frac{1}{2}\right)^1\left(\frac{1}{2}\right)^5 + \binom{6}{6}\left(\frac{1}{2}\right)^0\left(\frac{1}{2}\right)^6$$

$$= \frac{6!}{0!6!}\left(\frac{1}{2}\right)^6 + \frac{6!}{1!5!}\left(\frac{1}{2}\right)^6 + \frac{6!}{2!4!}\left(\frac{1}{2}\right)^6 + \frac{6!}{3!3!}\left(\frac{1}{2}\right)^6$$

$$+ \frac{6!}{4!2!}\left(\frac{1}{2}\right)^6 + \frac{6!}{5!1!}\left(\frac{1}{2}\right)^6 + \frac{6!}{6!0!}\left(\frac{1}{2}\right)^6$$

$$= 1\left(\frac{1}{2}\right)^6 + 6\left(\frac{1}{2}\right)^6 + 15\left(\frac{1}{2}\right)^6 + 20\left(\frac{1}{2}\right)^6 + 15\left(\frac{1}{2}\right)^6$$

$$+ 6\left(\frac{1}{2}\right)^6 + 1\left(\frac{1}{2}\right)^6$$

$$= \left(\frac{1}{2}\right)^6[1 + 6 + 15 + 20 + 15 + 6 + 1] = \frac{1}{64}(64) = 1$$

45. $(a - b)^n = \displaystyle\sum_{j=0}^{n} (-1)^j\binom{n}{j}a^{n-j}b^j$

47. $(x - y)^6$

$$= (-1)^0\binom{6}{0}x^6y^0 + (-1)^1\binom{6}{1}x^5y^1 + (-1)^2\binom{6}{2}x^4y^2$$

$$+ (-1)^3\binom{6}{3}x^3y^3 + (-1)^4\binom{6}{4}x^2y^4$$

$$+ (-1)^5\binom{6}{5}x^1y^5 + (-1)^6\binom{6}{6}x^0y^6$$

$$= \frac{6!}{0!6!}x^6 - \frac{6!}{1!5!}x^5y + \frac{6!}{2!4!}x^4y^2$$

$$- \frac{6!}{3!3!}x^3y^3 + \frac{6!}{4!2!}x^2y^4 - \frac{6!}{5!1!}xy^5 + \frac{6!}{6!0!}y^6$$

$$= x^6 - 6x^5y + 15x^4y^2 - 20x^3y^3 + 15x^2y^4 - 6xy^5 + y^6$$

49. $(q - 2)^8$

$= (-1)^0\binom{8}{0}q^8 \cdot 2^0 + (-1)^1\binom{8}{1}q^7 \cdot 2^1 + (-1)^2\binom{8}{2}q^6 \cdot 2^2$

$\quad + (-1)^3\binom{8}{3}q^5 \cdot 2^3 + (-1)^4\binom{8}{4}q^4 \cdot 2^4 + (-1)^5\binom{8}{5}q^3 \cdot 2^5$

$\quad + (-1)^6\binom{8}{6}q^2 \cdot 2^6 + (-1)^7\binom{8}{7}q^1 \cdot 2^7 + (-1)^8\binom{8}{8}q^0 \cdot 2^8$

$= \dfrac{8!}{0!8!}q^8 - \dfrac{8!}{1!7!}q^7 \cdot 2 + \dfrac{8!}{2!6!}q^6 \cdot 4 - \dfrac{8!}{3!5!}q^5 \cdot 8$

$\quad + \dfrac{8!}{4!4!}q^4 \cdot 16 - \dfrac{8!}{5!3!}q^3 \cdot 32 + \dfrac{8!}{6!2!}q^2 \cdot 64$

$\quad - \dfrac{8!}{7!1!}q \cdot 128 + \dfrac{8!}{8!0!} \cdot 256$

$= q^8 - 16q^7 + 112q^6 - 448q^5 + 1120q^4 - 1792q^3$

$\quad + 1792q^2 - 1024q + 256$

51. $(3x^2 - 2y^2)^4$

$= (-1)^0\binom{4}{0}(3x^2)^4(2y^2)^0 + (-1)^1\binom{4}{1}(3x^2)^3(2y^2)^1$

$\quad + (-1)^2\binom{4}{2}(3x^2)^2(2y^2)^2 + (-1)^3\binom{4}{3}(3x^2)^1(2y^2)^3$

$\quad + (-1)^4\binom{4}{4}(3x^2)^0(2y^2)^4$

$= \dfrac{4!}{0!4!}(81x^8) - \dfrac{4!}{1!3!}(27x^6)(2y^2) + \dfrac{4!}{2!2!}(9x^4)(4y^4)$

$\quad - \dfrac{4!}{3!1!}(3x^2)(8y^6) + \dfrac{4!}{4!0!}(16y^8)$

$= 81x^8 - 216x^6y^2 + 216x^4y^4 - 96x^2y^6 + 16y^8$

<u>Problem Set 10.5</u>

1. $P(4, 2) = \dfrac{4!}{(4 - 2)!} = \dfrac{4!}{2!} = \dfrac{4 \cdot 3 \cdot 2!}{2!} = 12$

3. $P(5, 1) = \dfrac{5!}{(5 - 1)!} = \dfrac{5!}{4!} = \dfrac{5 \cdot 4!}{4!} = 5$

5. $P(5, 3) = \dfrac{5!}{(5 - 3)!} = \dfrac{5!}{2!} = \dfrac{5 \cdot 4 \cdot 3 \cdot 2!}{2!} = 60$

7. $P(5, 5) = \dfrac{5!}{(5 - 5)!} = \dfrac{5!}{0!} = \dfrac{120}{1} = 120$

9. $C(9, 0) = \dfrac{9!}{(9 - 0)!0!} = \dfrac{9!}{9!0!} = 1$

11. $C(9, 2) = \dfrac{9!}{(9 - 2)!2!}$

$\quad = \dfrac{9!}{7!2!}$

$\quad = \dfrac{9 \cdot 8 \cdot 7!}{7! \cdot 2 \cdot 1}$

$\quad = 36$

13. $C(9, 9) = \dfrac{9!}{(9 - 9)!9!} = \dfrac{9!}{0!9!} = 1$

15. $P(6, 3) = \dfrac{6!}{3!}$

$\quad = \dfrac{6 \cdot 5 \cdot 4 \cdot 3!}{3!}$

$\quad = 120$

120 "words"

17. $C(13, 7)$

$\quad = \dfrac{13!}{(13 - 7)!7!}$

$\quad = \dfrac{13!}{6!7!}$

$\quad = \dfrac{13 \cdot 12 \cdot 11 \cdot 10 \cdot 9 \cdot 8 \cdot 7!}{6 \cdot 5 \cdot 4 \cdot 3 \cdot 2 \cdot 1 \cdot 7!}$

$\quad = 1716$

There are 1716 ways the players can be chosen.

19. $P(8, 4) = \dfrac{8!}{(8 - 4)!}$

$\quad = \dfrac{8!}{4!}$

$\quad = \dfrac{8 \cdot 7 \cdot 6 \cdot 5 \cdot 4!}{4!}$

$\quad = 1680$

21. $P(6, 0) = \dfrac{6!}{(6 - 0)!} = \dfrac{6!}{6!} = 1$

23. $P(6, 2) = \dfrac{6!}{(6 - 2)!}$

$= \dfrac{6!}{4!}$

$= \dfrac{6 \cdot 5 \cdot 4!}{4!}$

$= 30$

25. $P(6, 4) = \dfrac{6!}{(6 - 4)!}$

$= \dfrac{6!}{2!}$

$= \dfrac{6 \cdot 5 \cdot 4 \cdot 3 \cdot 2!}{2!}$

$= 360$

27. $P(6, 6) = \dfrac{6!}{(6 - 6)!} = \dfrac{6!}{0!} = \dfrac{720}{1} = 720$

29. $C(7, 5) = \dfrac{7!}{(7 - 5)!5!}$

$= \dfrac{7!}{2!5!}$

$= \dfrac{7 \cdot 6 \cdot 5!}{2 \cdot 1 \cdot 5!}$

$= 21$

31. $C(8, 0) = \dfrac{8!}{(8 - 0)!0!} = \dfrac{8!}{8!0!} = 1$

33. $C(8, 2) = \dfrac{8!}{(8 - 2)!2!}$

$= \dfrac{8!}{6! \cdot 2!}$

$= \dfrac{8 \cdot 7 \cdot 6!}{6! \cdot 2 \cdot 1}$

$= 28$

35. $C(8, 8) = \dfrac{8!}{(8 - 8)!8!} = \dfrac{8!}{0!8!} = 1$

37. $C(8, 6) = \dfrac{8!}{(8 - 6)! \cdot 6!}$

$= \dfrac{8!}{2! \cdot 6!}$

$= \dfrac{8 \cdot 7 \cdot 6!}{2 \cdot 1 \cdot 6!}$

$= 28$

39. $P(14, 4) = \dfrac{14!}{(14 - 4)!}$

$= \dfrac{14!}{10!}$

$= \dfrac{14 \cdot 13 \cdot 12 \cdot 11 \cdot 10!}{10!}$

$= 24{,}024$

41. $C(14, 4) = \dfrac{14!}{(14 - 4)!4!}$

$= \dfrac{14!}{10!4!}$

$= \dfrac{14 \cdot 13 \cdot 12 \cdot 11 \cdot 10!}{10! \cdot 4 \cdot 3 \cdot 2 \cdot 1}$

$= 1001$

43. $P(8, 4) = \dfrac{8!}{(8 - 4)!}$

$= \dfrac{8!}{4!}$

$= \dfrac{8 \cdot 7 \cdot 6 \cdot 5 \cdot 4!}{4!}$

$= 1680$

1680 "words"

45. $C(16, 9) = \dfrac{16!}{(16 - 9)!9!}$

$= \dfrac{16!}{7!9!}$

$$= \frac{16 \cdot 15 \cdot 14 \cdot 13 \cdot 12 \cdot 11 \cdot 10 \cdot 9!}{7 \cdot 6 \cdot 5 \cdot 4 \cdot 3 \cdot 2 \cdot 1 \cdot 9!}$$

$$= 11440$$

They can be chosen 11,440 ways.

47. $P(N, N) = \dfrac{N!}{(N - N)!} = \dfrac{N!}{0!} = \dfrac{N!}{1} = N!$

49. $P(N, N - 2) = \dfrac{N!}{[N - (N - 2)]!}$

$$= \frac{N!}{2!}$$

$$= \frac{N!}{2}$$

51. $C(N, 0) = \dfrac{N!}{(N - 0)!0!}$

$$= \frac{N!}{N! \cdot 1}$$

$$= 1$$

53. $C(N + 2, N) = \dfrac{(N + 2)!}{(N + 2 - N)!N!}$

$$= \frac{(N + 2)!}{2!N!}$$

$$= \frac{(N + 2)(N + 1)N!}{2 \cdot 1 \cdot N!}$$

$$= \frac{(N + 2)(N + 1)}{2}$$

55. $C(N + 1, N - 1)$

$$= \frac{(N + 1)!}{[(N + 1) - (N - 1)]!(N - 1)!}$$

$$= \frac{(N + 1)!}{2!(N - 1)!}$$

$$= \frac{(N + 1)N(N - 1)!}{2 \cdot 1 \cdot (N - 1)!}$$

$$= \frac{N(N + 1)}{2}$$

Chapter 10 Review Problems

1. $a_n = 3n + 1$

$a_1 = 3(1) + 1 = 4$

$a_2 = 3(2) + 1 = 7$

$a_3 = 3(3) + 1 = 10$

$a_4 = 3(4) + 1 = 13$

$a_5 = 3(5) + 1 = 16$

4, 7, 10, 13, 16

3. $a_n = 3 \cdot 5^{n-1}$

$a_1 = 3 \cdot 5^{1-1} = 3$

$a_2 = 3 \cdot 5^{2-1} = 15$

$a_3 = 3 \cdot 5^{3-1} = 75$

$a_4 = 3 \cdot 5^{4-1} = 375$

$a_5 = 3 \cdot 5^{5-1} = 1875$

3, 15, 75, 375, 1875

5. $c_n = (-1)^n \cdot 3n$

$c_1 = (-1)^1 \cdot 3(1) = -3$

$c_2 = (-1)^2 \cdot 3(2) = 6$

$c_3 = (-1)^3 \cdot 3(3) = -9$

$c_4 = (-1)^4 \cdot 3(4) = 12$

$c_5 = (-1)^5 \cdot 3(5) = -15$

-3, 6, -9, 12, 15

7. $a_n = 10 + 3n$

13, 16, 19, ...

Increasing

9. $c_n = 3 - n^{-1} = 3 - \dfrac{1}{n}$

$2, \dfrac{5}{2}, \dfrac{8}{3}, \ldots$

$2, 2\dfrac{1}{2}, 2\dfrac{2}{3}, \ldots$

Increasing

11. $-2, 0, 2, \ldots$

$d = 0 - (-2) = 2$

next term: $2 + 2 = 4$

13. $1, 4, 16, \ldots$

$r = \dfrac{4}{1} = 4$

next term: $16 \cdot 4 = 64$

15. $40, 60, 90, \ldots$

$r = \dfrac{60}{40} = \dfrac{3}{2}$

next term: $90 \cdot \dfrac{3}{2} = 135$

17. $\displaystyle\sum_{j=1}^{6} (1 - j)^2$

$= (1 - 1)^2 + (1 - 2)^2 + (1 - 3)^2$
$\quad + (1 - 4)^2 + (1 - 5)^2 + (1 - 6)^2$

$= 0^2 + (-1)^2 + (-2)^2 + (-3)^2 + (-4)^2 + (-5)^2$

$= 0 + 1 + 4 + 9 + 16 + 25$

19. $\displaystyle\sum_{j=0}^{5} (2j)^2$

$= [2(0)]^2 + [2(1)]^2 + [2(2)]^2 + [2(3)]^2$
$\quad + [2(4)]^2 + [2(5)]^2$

$= 0^2 + 2^2 + 4^2 + 6^2 + 8^2 + 10^2$

$= 0 + 4 + 16 + 36 + 64 + 100$

21. $\displaystyle\sum_{j=1}^{4} (-1)^{j+1} j^j$

$= (-1)^{1+1} \cdot 1^1 + (-1)^{2+1} \cdot 2^2$
$\quad + (-1)^{3+1} \cdot 3^3 + (-1)^{4+1} \cdot 4^4$

$= 1 - 4 + 27 - 256$

23. $1 + 2k + 3k^2 + 4k^3 + 5k^4 + 6k^5 + 7k^6$

$= \displaystyle\sum_{j=0}^{6} (j + 1)k^j$

25. $9! = 1 \cdot 2 \cdot 3 \cdot 4 \cdot 5 \cdot 6 \cdot 7 \cdot 8 \cdot 9$

$= 362{,}880$

27. $0! \cdot 1! \cdot 2! \cdot 3!$

$= 1 \cdot 1 \cdot 1 \cdot 2 \cdot 1 \cdot 2 \cdot 3$

$= 12$

29. $\dfrac{41!}{39!}$

$= \dfrac{41 \cdot 40 \cdot 39!}{39!}$

$= 1640$

31. $\dbinom{9}{3} = \dfrac{9!}{3!6!}$

$= \dfrac{9 \cdot 8 \cdot 7 \cdot 6!}{3 \cdot 2 \cdot 1 \cdot 6!}$

$= 84$

33. $\dbinom{10}{5} = \dfrac{10!}{5!5!}$

$= \dfrac{10 \cdot 9 \cdot 8 \cdot 7 \cdot 6 \cdot 5!}{5 \cdot 4 \cdot 3 \cdot 2 \cdot 1 \cdot 5!}$

$= 252$

35. $(x + 1)^7$

$$= \binom{7}{0}x^7 \cdot 1^0 + \binom{7}{1}x^6 \cdot 1^1 + \binom{7}{2}x^5 \cdot 1^2$$

$$+ \binom{7}{3}x^4 \cdot 1^3 + \binom{7}{4}x^3 \cdot 1^4 + \binom{7}{5}x^2 \cdot 1^5$$

$$+ \binom{7}{6}x^1 \cdot 1^6 + \binom{7}{7}x^0 \cdot 1^7$$

$$= \frac{7!}{0!7!}x^7 + \frac{7!}{1!6!}x^6 + \frac{7!}{2!5!}x^5$$

$$+ \frac{7!}{3!4!}x^4 + \frac{7!}{4!3!}x^3 + \frac{7!}{5!2!}x^2$$

$$+ \frac{7!}{6!1!}x + \frac{7!}{7!0!}$$

$$= x^7 + 7x^6 + 21x^5 + 35x^4$$

$$+ 35x^3 + 21x^2 + 7x + 1$$

37. $(2 + 3y)^6$

$$= \binom{6}{0}2^6(3y)^0 + \binom{6}{1}2^5(3y)^1 + \binom{6}{2}2^4(3y)^2$$

$$+ \binom{6}{3}2^3(3y)^3 + \binom{6}{4}2^2(3y)^4 + \binom{6}{5}2^1(3y)^5$$

$$+ \binom{6}{6}2^0(3y)^6$$

$$= \frac{6!}{0!6!} \cdot 64 + \frac{6!}{1!5!} \cdot 32(3y) + \frac{6!}{2!4!} \cdot 16(9y^2)$$

$$+ \frac{6!}{3!3!} \cdot 8(27y^3) + \frac{6!}{4!2!} \cdot 4(81y^4)$$

$$+ \frac{6!}{5!1!} \cdot 2(243y^5) + \frac{6!}{6!0!}(729y^6)$$

$$= 64 + 576y + 2160y^2 + 4320y^3$$

$$+ 4860y^4 + 2916y^5 + 729y^6$$

39. $(1 + s)^{10}$

$$= \binom{10}{0}1^{10}s^0 + \binom{10}{1}1^9s^1 + \binom{10}{2}1^8s^2$$

$$+ \binom{10}{3}1^7s^3 + \binom{10}{4}1^6s^4 + \binom{10}{5}1^5s^5$$

$$+ \binom{10}{6}1^4s^6 + \binom{10}{7}1^3s^7 + \binom{10}{8}1^2s^8$$

$$+ \binom{10}{9}1^1s^9 + \binom{10}{10}1^0s^{10}$$

$$= \frac{10!}{0!10!} + \frac{10!}{1!9!}s + \frac{10!}{2!8!}s^2$$

$$+ \frac{10!}{3!7!}s^3 + \frac{10!}{4!6!}s^4 + \frac{10!}{5!5!}s^5$$

$$+ \frac{10!}{6!4!}s^6 + \frac{10!}{7!3!}s^7 + \frac{10!}{8!2!}s^8$$

$$+ \frac{10!}{9!1!}s^9 + \frac{10!}{10!0!}s^{10}$$

$$= 1 + 10s + 45s^2 + 120s^3 + 210s^4$$

$$+ 252s^5 + 210s^6 + 120s^7 + 45s^8$$

$$+ 10s^9 + s^{10}$$

41. $(2 + 5t)^8$

1st 3 terms:

$$\binom{8}{0}2^8(5t)^0 + \binom{8}{1}2^7(5t)^1 + \binom{8}{2}2^6(5t)^2$$

$$= \frac{8!}{0!8!} \cdot 256 + \frac{8!}{1!7!} \cdot 640t + \frac{8!}{2!6!} \cdot 1600t^2$$

$$= 256 + 5120t + 44800t^2$$

43. $(a + b)^{14}$

middle term: 8th term

$$\binom{14}{7}a^7b^7 = \frac{14!}{7!7!}a^7b^7 = 3432a^7b^7$$

45. $C(10, 4) = \dfrac{10!}{(10 - 4)!4!}$

$$= \frac{10!}{6!4!}$$

$$= 210$$

47. $P(11, 6) = \dfrac{11!}{(11 - 6)!}$

$$= \frac{11!}{5!}$$

$$= \frac{11 \cdot 10 \cdot 9 \cdot 8 \cdot 7 \cdot 6 \cdot 5!}{5!}$$

$$= 332,640$$

49. $8x^3 - 27 = (2x)^3 - 3^3$

$= (2x - 3)(4x^2 + 6x + 9)$

51. Factored

53. $x^3 - 3x^2 + 3x - 1 = (x - 1)^3$

55. 3 terms;

$t(7 - y) + y - 7$

$= t(7 - y) - 1(7 - y)$

$= (7 - y)(t - 1)$

$7 - y$ is a factor.

57. $-2^{-4} = -\dfrac{1}{2^4} = -\dfrac{1}{16}$

59. $\dfrac{16x^2 - 25}{125 - 64x^3}$

$= \dfrac{(4x - 5)(4x + 5)}{(5 - 4x)(25 + 20x + 16x^2)}$

$= \dfrac{-(5 - 4x)(4x + 5)}{(5 - 4x)(25 + 20x + 16x^2)}$

$= -\dfrac{4x + 5}{25 + 20x + 16x^2}$

61. $\dfrac{2^{-3} + 8}{x^{-3} + 2^{-2}} = \dfrac{\dfrac{1}{2^3} + 8}{\dfrac{1}{x^3} + \dfrac{1}{2^2}}$

$= \dfrac{\dfrac{1}{8} + 8}{\dfrac{1}{x^3} + \dfrac{1}{4}}$

$= \dfrac{\dfrac{1}{8} + 8}{\dfrac{1}{x^3} + \dfrac{1}{4}} \cdot \dfrac{8x^3}{8x^3}$

$= \dfrac{x^3 + 64x^3}{8 + 2x^3}$

$= \dfrac{65x^3}{8 + 2x^3}$

63. $(3x + 2y)^3$

$= (3x)^3 + 3(3x)^2(2y) + 3(3x)(2y)^2 + (2y)^3$

$= 27x^3 + 54x^2y + 36xy^2 + 8y^3$

65. $\dfrac{-14 + \sqrt{-28}}{14} = \dfrac{-14 + 2\sqrt{7}i}{14}$

$= -1 + \dfrac{\sqrt{7}}{7}i$

67. $\dfrac{1}{x} + \dfrac{1}{x^2} = \dfrac{1 \cdot x}{x \cdot x} + \dfrac{1}{x^2}$

$= \dfrac{x}{x^2} + \dfrac{1}{x^2}$

$= \dfrac{x + 1}{x^2}$

69. $\dfrac{\dfrac{1}{x} - 1}{\dfrac{1}{x^2} - 1} = \dfrac{\dfrac{1}{x} - 1}{\dfrac{1}{x^2} - 1} \cdot \dfrac{x^2}{x^2}$

$= \dfrac{x - x^2}{1 - x^2}$

$= \dfrac{x(1 - x)}{(1 - x)(1 + x)}$

$= \dfrac{x}{1 + x}$

71. Expression;

$\dfrac{1 - \dfrac{x}{x + 1}}{1 + \dfrac{1}{x + 1}} = \dfrac{1 - \dfrac{x}{x + 1}}{1 + \dfrac{1}{x + 1}} \cdot \dfrac{x + 1}{x + 1}$

$= \dfrac{x + 1 - x}{x + 1 + 1}$

$= \dfrac{1}{x + 2}$

73. Equation;

$$\left|\frac{2 - 3x}{4}\right| = -3$$

No solution since absolute value is always nonnegative.

$\emptyset$

75. $\dfrac{i}{2 - 3i} = \dfrac{i}{2 - 3i} \cdot \dfrac{2 + 3i}{2 + 3i}$

$$= \frac{2i + 3i^2}{4 - 9i^2}$$

$$= \frac{2i + 3(-1)}{4 - 9(-1)}$$

$$= \frac{2i - 3}{13}$$

$$= \frac{2}{13}i - \frac{3}{13}$$

$$= -\frac{3}{13} + \frac{2}{13}i$$

Chapter 10 Test

1. $a_n = (-1)^n n^2$

$a_1 = (-1)^1 \cdot 1^2 = -1$

$a_2 = (-1)^2 \cdot 2^2 = 4$

$a_3 = (-1)^3 \cdot 3^2 = -9$

$-1, 4, -9, \ldots$

D

2. $a_n = 3 \cdot 2^{n-1}$

$a_1 = 3 \cdot 2^{1-1} = 3$

$a_2 = 3 \cdot 2^{2-1} = 6$

$a_3 = 3 \cdot 2^{3-1} = 12$

$3, 6, 12, \ldots$

B

3. $\displaystyle\sum_{j=0}^{4} j \cdot (j + 1)$

$= 0(0 + 1) + 1(1 + 1) + 2(2 + 1)$

$\quad + 3(3 + 1) + 4(4 + 1)$

$= 0(1) + 1(2) + 2(3) + 3(4) + 4(5)$

D

4. $\displaystyle\sum_{j=1}^{6} (-1)^{j+1} j$

$= (-1)^{1+1} \cdot 1 + (-1)^{2+1} \cdot 2 + (-1)^{3+1} \cdot 3$

$\quad + (-1)^{4+1} \cdot 4 + (-1)^{5+1} \cdot 5 + (-1)^{6+1} \cdot 6$

$= 1 - 2 + 3 - 4 + 5 - 6$

$= -3$

B

5. $6! = 1 \cdot 2 \cdot 3 \cdot 4 \cdot 5 \cdot 6 = 720$

D

6. $\dfrac{12!}{10!} = \dfrac{12 \cdot 11 \cdot 10!}{10!} = 132$

B

7. $\dbinom{9}{3} = \dfrac{9!}{3!6!}$

$$= \frac{9 \cdot 8 \cdot 7 \cdot 6!}{3 \cdot 2 \cdot 1 \cdot 6!}$$

$$= 84$$

A

8. $(x + 2)^5$

$$= \binom{5}{0}x^5 \cdot 2^0 + \binom{5}{1}x^4 \cdot 2^1 + \binom{5}{2}x^3 \cdot 2^2$$

$$+ \binom{5}{3}x^2 \cdot 2^3 + \binom{5}{4}x \cdot 2^4 + \binom{5}{5}x^0 \cdot 2^5$$

$$= \frac{5!}{0!5!}x^5 + \frac{5!}{1!4!}x^4 \cdot 2 + \frac{5!}{2!3!}x^3 \cdot 4$$

$$+ \frac{5!}{3!2!}x^2 \cdot 8 + \frac{5!}{4!1!}x \cdot 16 + \frac{5!}{5!0!} \cdot 32$$

$$= x^5 + 10x^4 + 40x^3 + 80x^2$$

$$+ 80x + 32$$

C

9. seventh term:

$$\binom{12}{6}a^6b^6 = \frac{12!}{6!6!}a^6b^6$$

$$= \frac{12 \cdot 11 \cdot 10 \cdot 9 \cdot 8 \cdot 7 \cdot 6!}{6 \cdot 5 \cdot 4 \cdot 3 \cdot 2 \cdot 1 \cdot 6!}$$

$$= 924a^6b^6$$

D

10. $4, 7, 10, 13, \ldots, 3n + 1$

11. $a_n = (-1)^{n+1}2^n$

$$a_1 = (-1)^{1+1}2^1 = 2$$

$$a_2 = (-1)^{2+1}2^2 = -4$$

$$a_3 = (-1)^{3+1}2^3 = 8$$

$$a_4 = (-1)^{4+1}2^4 = -16$$

$$a_5 = (-1)^{5+1}2^5 = 32$$

$$2, -4, 8, -16, 32$$

12. $\displaystyle\sum_{j=1}^{5} (2j - 1)$

$$= [2(1) - 1] + [2(2) - 1] + [2(3) - 1]$$

$$+ [2(4) - 1] + [2(5) - 1]$$

$$= 1 + 3 + 5 + 7 + 9$$

$$= 25$$

13. $P(8, 5) = \dfrac{8!}{(8 - 5)!}$

$$= \frac{8!}{3!}$$

$$= \frac{8 \cdot 7 \cdot 6 \cdot 5 \cdot 4 \cdot 3!}{3!}$$

$$= 6720$$

6,720 "words"

14. $(3 + t)^4$

$$= \binom{4}{0}3^4t^0 + \binom{4}{1}3^3t^1 + \binom{4}{2}3^2t^2 + \binom{4}{3}3^1t^3$$

$$+ \binom{4}{4}3^0t^4$$

$$= \frac{4!}{0!4!}(81) + \frac{4!}{1!3!}(27t) + \frac{4!}{2!2!}(9t^2)$$

$$+ \frac{4!}{3!1!}3t^3 + \frac{4!}{4!0!}t^4$$

$$= 81 + 108t + 54t^2 + 12t^3 + t^4$$

15. $(a + b)^{18}$

last 3 terms:

$$\binom{18}{16}a^2b^{16} + \binom{18}{17}a^1b^{17} + \binom{18}{18}a^0b^{18}$$

$$= \frac{18!}{16!2!}a^2b^{16} + \frac{18!}{17!1!}ab^{17} + \frac{18!}{18!0!}b^{18}$$

$$= 153a^2b^{16} + 18ab^{17} + b^{18}$$